T0360538

GENERALIZATIONS OF FINITE METRICS AND CUTS

GENERALIZATIONS OF FINITE METRICS AND CUTS

Elena Deza
Moscow State Pedagogical University, Russia

Michel Deza
Ecole Normale Supérieure, France

Mathieu Dutour Sikirić
Ruđer Bošković Institute, Croatia

NEW JERSEY • LONDON • SINGAPORE • BEIJING • SHANGHAI • HONG KONG • TAIPEI • CHENNAI • TOKYO

Published by

World Scientific Publishing Co. Pte. Ltd.

5 Toh Tuck Link, Singapore 596224

USA office: 27 Warren Street, Suite 401-402, Hackensack, NJ 07601

UK office: 57 Shelton Street, Covent Garden, London WC2H 9HE

Library of Congress Cataloging-in-Publication Data
Names: Deza, M., 1939– | Deza, Elena. | Dutour Sikirić, Mathieu
Title: Generalizations of finite metrics and cuts / by Michel Deza (Ecole Normale Supérieure, France),
Elena Deza (Moscow State Pedagogical University, Russia),
Mathieu Dutour Sikirić (Ruder Bošković Institute, Croatia).
Description: New Jersey : World Scientific, 2016. | Includes bibliographical references.
Identifiers: LCCN 2016013680 | ISBN 9789814740395 (hardcover : alk. paper)
Subjects: LCSH: Graph theory. | Metric spaces.
Classification: LCC QA166 .D485 2016 | DDC 511/.5--dc23
LC record available at http://lccn.loc.gov/2016013680

British Library Cataloguing-in-Publication Data
A catalogue record for this book is available from the British Library.

Printed in Singapore

Contents

Introduction

The concept of *distance* is one of the basic ones in whole human experience. In everyday life it means usually some degree of closeness of two physical objects or ideas, i.e., length, time interval, gap, rank difference, coolness or remoteness, while the term *metric* is often used as a standard for a measurement.

The mathematical notions of *distance metric*, i.e., a function $d(x, y)$ from $X \times X$ to the set of real numbers, satisfying to $d(x, y) \geq 0$ with equality only for $x = y$, $d(x, y) = d(y, x)$, and $d(x, y) \leq d(x, z) + d(z, y)$, and of *metric space* (X, d) were originated a century ago by Fréchet (1906) [Frec06] and Hausdorff (1914) [Haus14] as a special case of an infinite topological space.

Fréchet M.
1878–1973

Hausdorff F.
1868–1942

The *triangle inequality* $d(x, y) \leq d(x, z) + d(z, y)$ appears already in Euclid, but was first formalized as a central property of distances in the classical paper by Fréchet (1906) [Frec06], and later treated by Hausdorff (1914) [Haus14]. The notion of a metric space was formalized also in [Frec06], but the term "metric" was first proposed in [Haus14].

The infinite metric spaces are seen usually as a generalization of the *natural metric* $|x - y|$ on the real numbers. Their main classes are the *measurable spaces* (add measure) and *Banach spaces* (add norm and completeness).

However, starting from Menger (1928) [Meng28] and, especially, Blumenthal (1953) [Blum53], an explosion of interest in finite metric spaces occurred.

Blumenthal
L.M.
1876–1944

Menger K.
1902–1985

Another trend: many mathematical theories, in the process of their generalization, settled on the level of metric space.

Now, finite distance metrics have become an essential tool in many areas of Mathematics and its applications including Geometry, Probability, Statistics, Coding/Graph Theory, Combinatorial Optimization, Clustering, Data Analysis, Pattern Recognition, Networks, Engineering, Computer Graphics/Vision, Astronomy, Cosmology, Molecular Biology, and many other areas of Science. Devising the most suitable distance metrics has become a standard task for many researchers. Especially intense ongoing search for such distances occurs, for example, in Genetics, Image Analysis, Speech Recognition, Information Retrieval. Often the same distance metric appears independently in several different areas; for example, the *edit distance* between words, the *evolutionary distance* in Biology, the *Levenstein distance* in Coding Theory, and the *Hamming+Gap* or *shuffle-Hamming distance.*

For general theory of metrics see Blumenthal (1953) [Blum53] and Deza and Laurent (1997) [DeLa97]. The large list of metrics can be found in Deza and Deza (2014) [DeDe14].

If we drop the strict condition $d(x, y) > 0$ for distinct x, y, we obtain the notion of *semimetric*, a weaker version of the notion of metric, often more convenient for practical applications. The notion of a semimetric was first formalized by Fréchet (1906) [Frec06], too.

If we exclude the symmetry condition $d(x, y) = d(y, x)$, we obtain the definitions of *quasi-metric* and *quasi-semimetric*, respectively. These notions are asymmetric analogues of (symmetric) notions of metrics and semimetrics. Such constructions, failing the symmetry condition, are encountered, for example, as quasi-distances on one-way routes, mountain walks or rivers with quick flow.

Asymmetric definition of distance have been used already in Hausdorff (1914) [Haus14], pages 145–146, but the first detailed topological analysis of quasi-metrics was given by Wilson (1931) [Wils31].

Quasi-metrics are used in Semantics of Computation (see, for example, Seda (1997) [Seda97]) and are of interest in Computational Geometry (see, for example, Aichholzer, Aurenhammer, Chen, Lee, Mukhopadnyay and Papadopoulou (1997) [AACLMP9797]).

For general Theory of quasi-metrics see, for example, Fréchet (1906) [Frec06], Hausdorff (1914) [Haus14], Stoltenberg (1969) [Stol69], Deza and Deza (2014) [DeDe14]; see also Deza and Panteleeva (1999) [DePa99] and Deza, Dutour and Panteleeva (2003) [DDP03].

The notions of *m-metrics* and *m-hemimetrics* are multidimensional (in fact, $(m+1)$-ary) analogues of the (binary) notions of metrics and semimetrics. A basic example of a metric is $(\mathbb{R}^2, d)$, where $d(x, y)$ is the *Euclidean distance* between $x, y \in \mathbb{R}^2$, i.e., the length of the segment, joining x and y. An immediate extension is $(\mathbb{R}^3, d)$, where $d(x, y, z)$ is the area of the triangle with verticies x, y, z; it is a basic example of an 2-*metric.* On the same way, the simplest example of m-metric is the m-dimensional volume in $\mathbb{R}^n$ (with $n \geq m$).

The notion of an m-metric, i.e., a function d from X^{m+1} to the set of real numbers, satisfying to $d(x_1, ..., x_{m+1}) \geq 0$ whenever $x_1, ..., x_{m+1}$ are not pairwise distinct, the existence of x_{m+1} such that $d(x_1, ...x_m, x_{m+1}) > 0$ whenever $x_1, ..., x_m$ are distinct, $d(x_1, ..., x_{m+1}) = d(x_{\pi(1)}, ..., x_{\pi(m+1)})$ for any permutation π of the set $\{1, ..., m+1\}$ (*totally symmetry*), and $d(x_1, ..., x_{m+1}) \leq \sum_{i=1}^{m+1} d(x_1, ..., x_{i-1}, x_{i+1}, ..., x_{m+2})$ (*m-simplex inequality*), which directly extend the notion of an usual metric, have been around for 70 years and their 1990 year's bibliography [Gahl90] lists over 260 items. The motivation for m-metrics came mostly from Geometry and so, the main interest has been in their topological aspects.

Independently of m-metrics, there appeared recently papers on weaker versions, motivated (apart, of course, that it represents an extension of metrics) by various discrete applications, e.g., in Statistics and Data Analysis; see, for example, [Diat96], [JoCa95]. Their common trait seems to be the *total symmetry* and the *simplex inequality*. The strong m-metrics axioms $d(x_1, x_1, x_2, ..., x_m) = 0$ and the existence of x_{m+1} such that $d(x_1, ..., x_{m+1}) > 0$ whenever $x_1, ..., x_{m+1}$ are distinct, which make perfect sense in Geometry, are too restrictive for applications and hence, are substantially weakened in various ways. The notion of m-hemimetric, which only assume that $d(x_1, ..., x_{m+1}) \geq 0$ and $d(x_1, x_1, x_2, ..., d_m) = 0$, is one of such new versions. This notion was introduced in [DeRo00] and studied, using the program *cdd* [Fuku95], for small parameters m, n in [DeRo02].

The definitions of metrics, semimetrics and their asymmetric (quasi-metrics and quasi-semimetrics) and multidimensional (m-metrics, m-hemimetrics) analogues are considered (after short representation, in Part I, main notions on Graphs, Vector Spaces, Matrices, Convex Cones and Polytopes) in Part II of this book together with numerous and various examples and comments. In Part III similar definitions and examples for two important special cases of semimetrics - cuts and hypermetrics (as well as their asymmetric and multidimensional analogues) - are given.

In Part IV and V of the book we consider cones and polytopes, the points of which are various specifications, analogs and generalisations of finite semimetrics, defined on the set $V_n = \{1, 2, ..., n\}$. The polyhedra considered here are important in Analysis, Measure Theory, Geometry of Numbers, Combinatorics and Optimisation.

Given two metric spaces (X, d) and (X', d'), we say that (X, d) is *isometrically embeddable into* (is a *metric subspace of*) (X', d'), if there exists a mapping $\phi : X \to X'$, such that $d(x, y) = d(\phi(x), \phi(y))$ for all $, y \in X$.

Given $1 \leq p \leq \infty$ and integer $m \geq 1$, the l_p^m*-space* is the metric space $(\mathbb{R}^m, d_p(x, y) = ||x - y||_p)$, where the l_p*-norm* is

$$||x||_p = \left(\sum_{1 \leq i \leq m} |x_i|^p\right)^{\frac{1}{p}} \quad \text{for finite } p, \text{ and } \quad ||x||_\infty = \max_{1 \leq i \leq m} |x_i|.$$

The most prominent is l_2-metric, called also *Euclidean* (*Pythagorean, as-the-crow-flies, beeline*) *distance.*

A (semi)metric d on X is called l_p*-metric* if (X, d) is a (semi)metric subspace of the l_p^m-space for some $X \subset \mathbb{R}^m$, i.e., d is the restriction of d_p on X.

Every l_2-semimetric is an l_p-semimetric for any $p \geq 1$ (indeed, any n-points semimetric d is realized in the l_∞^{n-1}-space by n points $(d_{1i}, ..., d_{n-1i})$ for $i = 1, 2, ..., n$). So, the n-points l_∞-semimetrics form a *convex cone*, which coincides with the *metric cone* MET_n of all n-points semimetrics. The cone MET_n is defined by $3\binom{n}{3}$ triangle inequalities.

Every l_p-semimetric with $1 < p \leq 2$ is ([Scho38]) an l_1-semimetric. The n-points l_1-semimetrics form a convex cone, called *cut cone* and denoted CUT_n. It is generated by $2^{n-1} - 1$ *cut semimetrics* $\delta_S = \delta_{\overline{S}} = (((\delta_S)_{ij}))$, defined, for any subset $S \subset \{1, 2, ..., n\}$, by $(\delta_S)_{ij} = 1$, if $|\{i, j\} \cup S| = 1$, and $(\delta_S)_{ij} = 0$, otherwise.

Polyhedral cones CUT_n and MET_n are the central prototypes for polyhedra, considered in this book. We will also consider corresponding polytopes $CUT_n^\square$ and $MET_n^\square$, since they have much more symmetries and so, are easier to describe.

The cut cone and polytope also admit ([AsDe80], [AsDe82]) a characterisation in terms of measure spaces: $d \in CUT_n$ ($d \in CUT_n^{\square}$) if and only if there exist a measure (respectively, a probability) space $(\Omega, \mathcal{A}, \mu)$ and n events $A_1, ..., A_n \in \mathcal{A}$, such that $d(i,j) = \mu(A_i \Delta A_j)$ for all $1 \leq i, j \leq n$.

The connection of dual CUT_n and MET_n to Capacities and Multi-commodity Flows was given in [AvDe91].

Connection of CUT_n to Geometry of Numbers is given by the *hypermetrics*, i.e., distance matrices $d = ((d_{ij}))$ with $\sum_{1 \leq i,j \leq n} b_i b_j d_{ij} \leq 0$ for any $(b_1, ..., b_n) \in \mathbb{Z}^n$, $\sum_{i=1}^{n} b_i = 1$. Clearly, permutations of $b = (1, 1, -1, 0, ..., 0)$ give, that d is an n-points semimetric. Any l_1-metric is ([Deza60]) a hypermetric. The hypermetrics are ([Asso84]) exactly squared l_2-metrics of the *constellations* in $(n-1)$-dimensional lattices, i.e., the sets of lattice points on the spheres, bounding maximal balls without lattice points inside.

So, l_1-metrics are l_2^2-metrics, while generalising l_2-metrics. The study of l_1-metrics meets ongoing (especially, in Riemannian Geometry, Real Analysis, Approximation Theory, Statistics) process: to extend many prominent l_2-defined notions and results on more general l_1-setting.

The main connection of CUT_n to Combinatorics was given in [AsDe80]: the rational-valued l_1-metrics d are exactly those, for which λd isometrically embeds into the path-metric of an N-cube for some integers $N, \lambda \geq 1$.

In Part IV we collect known computations of semimetric (including cuts) polyhedra and their asymmetric and multidimensional relatives.

The most important for the asymmetric case are the cones $QMET_n$ of all quasi-semimetrics on n points (its facets correspond to the non-negativity and oriented triangle inequalities) and the cone $OMCUT_n$, generated by all non-zero *oriented multicuts quasi-semimetrics* $\delta^{O}_{S_1,...,S_q} = ((d_{ij}))$, defined, for any oriented partition $S_1, ..., S_q$ of $\{1, 2, ..., n\}$, by $d_{ij} = 1$ if $i \in S_\alpha, j \in S_\beta$, $\alpha < \beta$, and $d_{ij} = 0$, otherwise.

The most important for the multidimensional case are the cones $HMET_n^m$ of all m-hemimetrics on n points (its facets correspond to the non-negativity and m-simplex inequalities) and the cone $HCUT_n$, generated by all non-zero *partition m-hemimetrics* $\delta^{O}_{S_1,...,S_{m+1}} = ((d_{i_1 i_2 ... i_{m+1}}))$, defined, for any partition $S_1, ..., S_{m+1}$ of $\{1, 2, ..., n\}$, by $d_{i_1 ... i_{m+1}} = 1$, if $i_1 \in S_{\alpha_1}, ..., i_{m+1} \in S_{\alpha_{m+1}}$, $\alpha_i \neq \alpha_j$, and $d_{i_1 ... i_{m+1}} = 0$, otherwise.

In Part V we consider similar data for partial semimetrics and weightable quasi-semimetrics, as well as for hypermetrics and their generalizations. One can find here also the information about generalizations of cut and semimetric polyhedra over general (instead of K_n) connected finite graphs. Such polyhedra over graphs are related to Maximum-Cut Problem, which asks to find a cut of maximum weight in a graph. It is a prominent problem in Combinatorial Optimization with many applications.

The polyhedral cones and polytopes, that show up in the study of finite semimetrics, have a complexity that grows very fast with the number of points. For example, it does not appear hopeful that one could compute the facets of, say, $CUT_{20}^{\square}$ in any reasonable time. On the other hand, for too small number of points, no useful information is available. Therefore, when searching for new kind of objects, it is advantageous to search for what the limit of the computational power offers.

One striking feature of the considered polytopes, such as $CUT_n^{\square}$, is their fairly large symmetry group. The methodology of choice for this computation is the Adjacency Decomposition Method, which proceeds when one has a facet to compute the adjacent ones and add it to the list if not equivalent to a known one. The methodology was first introduced in [ChRe96] and applied in [DDP03], [DeDu03] to quasi-semimetric and m-hemimetric cones. Then it was applied in [DSV07], [DSV09], [DSV10] to several polytopes coming from Geometry of Numbers; see [BDS09] for an overview of this kind of techniques. One should keep in mind that yes, symmetry helps the computational tasks, but it is not a panacea: some cases are simply too hard to treat. Similarly to using a faster programming language (like `C++`) or having a parallel program, it helps, but does not radically change the nature of the problem treated.

However, for other tasks the picture is not so bleak. One is computing automorphism groups of polytopes and cones, while knowing only their vertices (facets). Cones defined by $1,000$ vertices do not pose any problems; see [BDPRS14] for details on the procedure used. Another area which is not so much affected by combinatorial explosion is computing the adjacency between vertices (facets). There are several techniques for doing it, that rely on Linear Programming and we are able to scale up to big dimension.

The software is programmed in the `GAP` programming language [Gap] and is available on [Dutol0].

We are grateful for useful cooperation to several people, including Viacheslv Grishukhin, Ivo Rosenberg, Janoš Vidali.

Part I

Preliminaries

Chapter 1

Short preview of the book

This chapter gives an overview of the topics treated in this book. The central objects in the book are cones and polytopes related to asymmetric and multidimensional analogues of such classical (symmetric and binary) objects, as cuts and metrics. Interesting problems concerning these polyhedra arise in many different areas of Mathematics and its Applications. These polyhedra have been considered independently by a number of authors with various mathematical backgrounds and motivations. One of our objectives is to show, on the one hand, the richness and diversity of the results in connection with these polyhedra, and on the other hand, how they can be treated in a unified way as various aspects of a common set of objects. Research on cuts, metrics and their generalizations profits greatly from the variety of subjects where the problems arise. Observations made in different areas by independent authors turn out to be equivalent, but views from different perspectives provide new interpretations, connections and insights.

This book is subdivided into five parts, each treating seemingly diverse topics. Namely, Parts I to V contain results relevant to the following areas:

- short preview of the book and basic questions from different fields of Mathematics and its Applications (Graph Theory, Theory of Vector Spaces, basic Matrix Theory, main notions of the Theory of Polyhedral Cones and Polytopes), tied together through the notions of metrics, cuts and their generalizations;

- the main facts of the theory of metrics, quasi-metrics and m-metrics, as well as partial semimetrics and their relatives, together with numerous definitions, examples and comments;

- the main facts of the theory of cuts and their generalizations (multicuts, oriented multicuts, partition m-hemimetrics), and hypermetrics and their generalizations (quasi-hypermetrics, weightable hypermetrics, partial hypermetrics, infinite hypermetrics, etc.);

- the structures of cones and polytopes of semimetrics and cuts, their asymmetric (quasi-semimetric, oriented cut and oriented multicut polyhedra) and multidimensional (m-hemimetric and partition m-hemimetric polyhedra) analogues, and some natural generalizations of these polyhedra;

- important cases of polyhedra of generalized finite semimetrics, including cones of partial semimetrics, weightable quasi-semimetrics, hypermetrics and their relatives;

the structures and main properties of metric polyhedra over general (instead of K_n) connected finite graphs, and connections between generalized metric polyhedra.

We have made each of the five parts as self-contained as possible. For this reason, some notions and definitions may be repeated in different parts if they are central there. In principle, a reader who is interested, for instance, only in the aspects of generalized metric polyhedra, may consult Parts IV and V without any prior reading of Part II and III. Chapter 2, however, contains some basic notations on graphs, polyhedra and matrices that will be used throughout the book.

In what follows we give a brief overview of the material covered in Parts I to V. This introductory treatment is meant to provide an orientation map through the book for the reader.

1.1 Outline of Part I. Preliminaries

Part I contains two chapters.

In Chapter 1 we give short preview of the book, which should provide for the reader a tool of the navigation throw the text.

Chapter 2 is a list of the basic definitions and notions from different parts of Mathematics, which will be used in the next parts of the book. We consider main facts from the Graph Theory, Theory of Vector Spaces, and basic definitions concerning to the Matrix Theory. Moreover, we give definitions of polyhedral convex cones and polytopes, provide all main notions concerning to them, explain the structure of the 1-skeleton and the ridge graphs of a given polyhedra. Finally, we list all polyhedral structures, considered in the book, as well as main characteristics of those objects for small values of parameters.

1.2 Outline of Part II. Main notions and examples

In the Part II (Chapters 3, 4, 5, 6) the main facts of the theory of metrics, quasi-metrics and m-metrics with numerous definitions, examples and comments are considered.

In Chapter 3 we consider the classical binary non-oriented case. After the list of definitions of distances, semi-metrics and metrics, we give the basic examples of these objects, including *discrete metric*, *Hausdorff metric*, *symmetric difference metric*, *natural metric*, *Hamming metric*, *path metric of connected graph*, *circular railroad distance*, and consider main transforms of (semi)metrics. After we give definitions and examples of most important metric spaces: metrics on Normed Structures (including l_p-metrics and L_p-metrics), metrics on the classical Number Systems (in fact, on the sets $\mathbb{N}$, $\mathbb{Z}$, $\mathbb{Q}$, $\mathbb{R}$ of positive integer, integer, rational and real numbers), metrics on Real and Digital Planes (including, besides classical *Chebyshev metric* and *city-block metric* - in fact, the l_∞- and the l_1-*metrics* on $\mathbb{R}^2$, relatively, such exotic metrics, as *French metro metric*, *Moscow metric*, *lift metric*, *flower-shop metric*, etc.).

Chapter 4 is organized similarly. Here we consider the binary oriented case. After the list of definitions of quasidistances, quasi-semimetrics and quasimetrics, we give the basic examples of these objects, including oriented analogues of *discrete metric* and *Hausdorff metric*, *asymmetric difference quasi-semimetric*, several variants of *natural quasi-semimetric* and *Hamming quasi-metric*, *path metric of strongly connected digraph*, *one way circular railroad distance*, and consider main transforms of quasi-(semi)metrics. After we give definitions and examples of most important quasi-(semi)metric spaces, including those based on the notion of *oriented l_p-norm* of a vector space.

Chapter 5 is devoted to the multidimensional (in fact, $(m+1)$-ary) generalizations of ordinary metric spaces. At first, we give a large list of definitions of 3-dimensional and $(m+1)$-dimensional analogues of metrics: 2-distances, 2-semimetrics, 2-hemimetrics, 2-metrics (as well as $(m+1)$-distances, $(m+1)$-semimetrics, $(m+1)$-hemimetrics, $(m+1)$-metrics), and (m,s)-supermetrics. The diversity of possible definitions comes from different applications of considered objects: for example, some properties, natural for geometrical structures, can be too strong for polyhedral methods, and we drop them.

After we give the basic examples of these objects, including 3-dimensional and multidimensional analogues of *discrete metric*, *Hausdorff metric*, *symmetric difference metric*, *natural metric*, *Hamming metric*, *path metric* (*path metric of hypergraphs*), and *circular railroad distance*, and consider main transforms of m-(hemi)metrics. In the last part of Chapter 5 we put three useful examples of multidimensional analogs of metrics: *m-metrics between subsets*, closely connected to the symmetric difference metric; *n-way m-metrics*; multidimensional volumes, which provide interesting and natural cases of (m,s)-supermetrics.

Finally, in Chapter 6 one can find information on partial semimetrics, weightable quasi-semimetrics and weighted semimetrics: their definitions, examples, properties, applications, natural ways to transform one from these three objects to another.

For general theory of metrics see Blumenthal (1953) [Blum53] and Deza and Laurent (1997) [DeLa97]. The large list of metrics can be found in Deza and Deza (2014) [DeDe14].

For general theory of quasi-metrics see, for example, Fréchet (1906) [Frec06], Hausdorff (1914) [Haus14], and Stoltenberg (1969) [Stol69]. For theory of finite quasi-metrics, quasi-semimetrics and oriented multicuts and their polyhedral aspects, see Deza and Panteleeva (2000) [DePa99], Deza, Dutour and Panteleeva (2003) [DDP03], Deza, Deza and Grishukhin (2010) [DGD12].

For general theory of m-metrics and their relatives, see Menger (1928) [Meng28], Blumenthal (1953) [Blum53], Froda (1958) [Frod58], Gähler (1963, 1990) [Gahl63], [Gahl90]; possible applications were considered in Diatta (1996) [Diat96], Joly and Le Calvé (1995) [JoCa95]; polyhedral aspects of the theory were discussed in Deza and Rosenberg (2000, 2002) [DeRo00], [DeRo02], and in Deza and Dutour (2003) [DeDu03].

Partial semimetrics were introduced by Matthews (1992) in [Matt92] for treatment of partially defined objects in Computer Science. He also remarked that a quasi-semimetric $q = ((q_{ij}))$ is weightable if and only if the function $((q_{ij} + w_i))$ is a partial semimetric. Weak partial semimetrics were studied by Heckmann(1999) in [Heck99]. For applications of partial semimetrics see, for example, Gierz, Hofmann, Keimel, Lawson, Mislove and Scott (2003) [GHKLMS03], Güldürek and Richmond (2005) [GuRi05]. Polyhedral aspects of the theory are considered in Deza and Dutour (2010) [DeDe10], Deza, Deza and Vidali (2011) [DDV11], Deza, Dutour and Deza (2015) [DDD15].

1.3 Outline of Part III. Cuts, hypermetrics and their generalizations

In Part III (Chapters 7, 8) we collect main facts of the theory of two most important for our consideration classes of semimetrics and their oriented and multidimensional generalizations: cuts and hypermetrics.

Both chapters are organized similarly. After short introduction in the theory of classical binary symmetric case, we give its natural asymmetric and/or multidimensional analogies and discuss common and different properties of considered structures.

In Chapter 7 questions, connected with cuts, multicuts (binary symmetric case), oriented cuts, oriented multicuts (binary asymmetric case) and partition m-semimetrics (multidimensional symmetric case) are considered.

In Chapter 8 the information about hypermetric spaces and negative type spaces is given; several possible asymmetric analogues of hypermetrics are considered; possibilities of construction of infinite hypermetrics are discussed.

One can find a huge list of references on theory of cuts in Deza and Laurent (1997) [DeLa97]. The notions of oriented cuts and oriented multicuts were considered, for example, in Shi and Li (1995) [ShLi95]; see also Deza and Panteleeva (Deza) [DePa99], Deza, Dutour and Panteleeva (Deza) (2003) [DDP03]. The multidimensional analogues of cut and multicuts semimetrics has been studied in Deza and Rosenberg (2000, 2002) [DeRo00], [DeRo02], Deza and Dutour (2003) [DeDu03].

Main facts of the theory of hypermetrics and their connections with Geometry of Numbers and Analysis see, for example, in Tylkin (Deza) (1960, 1962) [Deza60], [Deza62], Deza and Terwilliger (1987) [DeTe87], Deza and Grishukhin (1993) [DeGr93], Deza, Grishukhin and Laurent (1995) [DGL95], Deza, Dutour and Deza (2015) [DDD15], and Chapters 13-17, 28 in [DeLa97]. Some generalizations of hypermetrics were considered in Deza and Deza (2010) [DeDe10] (quasi-hypermetrics), Deza, Deza and Vidali (2011) [DDV11] (partial hypermetric and their relatives), Deza and Dutour (2013) [DeDu13a] (infinite hypermetrics).

1.4 Outline of Part IV. Cones and polytopes of generalized finite metrics

In Part IV (Chapters 9, 10, 11) the structures of cones and polytopes of semimetrics and cuts, their asymmetric and multidimensional analogues and some natural generalizations of these polyhedra are considered.

For each three main cases (classical binary symmetric case of semimetrics and cuts, asymmetric case of quasi-semimetrics and oriented multicuts and multidimensional case of m-hemimetrics and partition m-hemimetrics) we present the same structure of consideration.

After short preliminary information, including the list of main polyhedra under consideration, we collect main facts about corresponding cones and polytopes: definitions of underling metric structures, dimensions of considered polyhedra, different possibilities to represent our objects in a real vector space, simplest examples, natural inclusions and other connections between constructed cones and polytopes, questions of their possible symmetries, Tables with basic data on small polyhedra under consideration.

In the next part the structure of cones and polytopes for small values of parameters is considered: dimension of polyhedra; number of its extreme rays (vertices) and facets and their orbits under chosen symmetry group; Tables of adjacencies and incidences of extreme rays (vertices) and facets; properties of corresponding 1-skeleton and ridge graphs and their local subgraphs. In last part we collect known theorems for general case and make some natural conjectures.

Chapter 9 contains such information about *metric cone* MET_n of all semimetrics on n points and *cut cone* CUT_n, generated by all non-zero cut semimetrics on n points, as well as about corresponding polytopes $MET_n^{\square}$ and $CUT_n^{\square}$. See [DeLa97], [DDF96], [DeDe95], [Duto08] for details on MET_n and CUT_n.

In Chapter 10 we collect known information about *quasi-metric cone* $QMET_n$ of all quasi-semimetrics on n points and *oriented multicut cone* $OMCUT_n$, generated by all non-zero oriented multicut semimetrics on n points, and some questions about structure of corresponding polytopes $QMET_n^{\square}$ and $OMCUT_n^{\square}$. (The cones $QMET_n$ and $OMCUT_n$ were introduced and studied, for small n, in [DePa99] and [DDP03], see also [DDD15].)

Multidimensional case is considered in Chapter 10. Here we study the behavior of cones of m-hemimetrics: the *m-hemimetric cone* $HMET_n^m$ of all m-hemimetrics on n points, the *partition m-hemimetric cone* $HCUT_n^m$, generated by all partition m-hemimetrics on n points, the *(m, s)-supermetric cone* $SMET_n^{m,s}$ of all (m, s)-supermetrics on n points, and the *binary (m, s)-supermetric cone* $SCUT_n^{m,s}$, generated by all $\{0, 1\}$-valued extreme rays of $SMET_n^{m,s}$. (The notion of m-hemimetric was introduced in [DeRo00] and studied for small parameters m, n, using the program *cdd* [Fuku95], in [DeRo02]; see also [DeDe10], [DDD15].)

1.5 Outline of Part V. Important cases of polyhedra of generalized finite semimetrics

In Part V (Chapters 12, 13, 14, 15) several important cases of polyhedra of generalized finite semimetrics are considered, including cones of partial semimetrics, weightable quasi-semimetrics, hypermetrics and their relatives; moreover, the structures and main properties of metric polyhedra over general (instead of K_n) connected finite graphs are given. Finally, deep connections between generalized metric polyhedra under consideration are studied.

Chapter 12 collects various information about cones and polytopes of partial semimetrics, weightable quasi-semimetrics and weighted metrics. The main polyhedral structures under consideration are: the *partial semimetric cone* $PMET_n$ of all partial semimetrics on n points; the *weightable quasi-semimetric cone* $WQMET_n$ of all weightable quasi-semimetrics on n points; the *weighted semimetric cone* $WMET_n$ of all weighted semimetrics on n points; the cones $\{0,1\}$-$PMET_n$, $\{0,1\}$-$WQMET_n$, $\{0,1\}$-$WMET_n$, generated by all $\{0,1\}$-valued extreme rays of the cones $PMET_n$, $WQMET_n$, $WMET_n$, respectively. For an additional information see [DeDe10], [DDV11], [Hitz01], [Seda97].

In Chapter 13 we consider the structure of *hypermetric cone* HYP_n of all hypermetrics on n points and corresponding polytope $HYP_n^{\square}$. Moreover, we construct and study some possible generalizations of these objects, including the *quasi-hypermetric cone* $QHYP_n$ of all quasi-hypermetrics on n points, the *weighted hypermetric cone* $WHYP_n$ of all weighted hypermetrics on n points, the *partial hypermetric cone* $PHYP_n$ of all partial hypermetrics on n points. (See for details [DeTe87], [DeGr93], [DGL95] and Chapters 13–17, 28 in [DeLa97].)

In Chapter 14 one can find new information about metric and cut polyhedra $MET(G)$, $CUT(G)$, $MET^{\square}(G)$, $CUT^{\square}(G)$ over some connected graphs: skeletons of Platonic and semiregular polyhedra, bipartite graphs $K_{m,n}$ and several types of multipartite graphs, Möbius ladders, Peterson graph, etc. Mainly, the information comes from [DDD15]; see also [PiSv01], [Fine82], and [DeLa97], section 5.2.

In Chapter 15 we consider important connections between the classical symmetric case of semimetric polyhedra (in particular, of the cut semimetric polyhedra), and the asymmetric case of quasi-semimetric polyhedra (in particular, of the oriented cut semimetric polyhedra). We show, that the cone $WQMET_n$ of weighted n-point quasi-semimetrics and the cone $OCUT_n$, generated by all non-zero oriented cuts on n points, are projections along an extreme ray of the metric cone MET_{n+1} and of the cut cone CUT_{n+1}, respectively. This projection is such that if one knows all faces of an original cone, then one knows all faces of the projected cone(see [DGD12]).

In Appendixes we present all orbits of extreme rays for quasi-semimetric cone $QMET_5$ and all orbits of facets for oriented multicut cone $OMCUT_5$. For representation matrices detailing orbit-wise adjacencies and additional information on those cones see http://mathieudutour.altervista.org/.

Chapter 2

Main definitions

2.1 Graphs

In Mathematics, and more specifically in Graph Theory, a *graph* is a representation of a set of objects, where some pairs of objects are connected by links. Graphs are the basic subject studied by Graph Theory. The word "graph" was first used in this sense by Sylvester (1878) [Sylv78], p. 65 (see also [Sylv78a], p. 284, [GrYe04], p. 35).

The interconnected objects are represented by mathematical abstractions called *vertices*, and the links that connect some pairs of vertices are called *edges*. Typically, a graph is depicted in diagrammatic form as a set of dots for the vertices, joined by lines or curves for the edges.

Undirected graphs

In the most common sense of the term, a *graph* is an ordered pair $G = \langle V, E \rangle$ comprising a set V of *vertices* (*nodes*) together with a set E of *edges*, which are 2-element subsets of V (i.e., an edge is related with two vertices, and this relation is represented as an unordered pair of the vertices with respect to the particular edge). This type of graph may be described precisely as *undirected* and *simple*.

So, an *undirected graph* is one in which edges have no orientation. The edge $\{u, v\}$ is identical to the edge $\{v, u\}$. For an edge $\{u, v\}$, graph theorists usually use the notation (u, v) or the somewhat shorter notation uv. The vertices belonging to an edge are called the *ends, endpoints* or *end vertices* of the edge. A vertex may exist in a graph and not belong to an edge. In an undirected graph G, two vertices u and v are called *connected* if G contains a *path* $uu_1, u_1u_2, ..., u_{k-1}u_k, u_kv$, $u_i \in V$, from u to v. Otherwise, they are called *disconnected.*

V and E are usually taken to be finite, and many of the well-known results are not true (or rather different) for infinite graphs because many of the arguments fail in the infinite case. Moreover, V is often assumed to be non-empty, but E is allowed to be the empty set. So, a *finite graph* is a graph $G = \langle V, E \rangle$ such that V and E are finite sets. An *infinite graph* is one with an infinite set of vertices or edges or both. Most commonly, in modern Graph Theory, unless stated otherwise, *graph* means *undirected simple finite graph.*

For undirected simple finite graph $G = \langle V, E \rangle$, the *order* of G is $|V|$, i.e., the number of its vertices, and the *size* of G is $|E|$, i.e., the number of its edges. For $V = \{v_1, v_2, ..., v_n\}$

we get $|V| = n$; in this case the maximum number of edges is $\frac{n(n-1)}{2}$: $|E| \leq \frac{n(n-1)}{2}$. We can suppose that the vertex set of any such graph is the set $V_n = \{1, 2, ..., n\}$.

The *degree* $d(v)$ of a vertex $v \in V$ is the number of edges that connect to it. For undirected simple finite graph $G = \langle V, E \rangle$ it holds $\sum_{v \in V} d(v) = 2|E|$. In a undirected simple graph with n vertices, the degree of every vertex is at most $n-1$: $d(v) \leq n-1$.

An undirected graph can be seen as a *simplicial complex* consisting of *1-simplices* (the edges) and *0-simplices* (the vertices). As such, complexes are generalizations of graphs since they allow for higher-dimensional simplices.

A *loop* is an edge which starts and ends on the same vertex; these may be permitted or not permitted according to the application. In this context, an edge with two different ends is called a *link*. In some cases *parallel edges* for a given pair of vertices may be permitted: two edges are said to be *parallel* (*multiple*) if they have the same end verticies. The term *multigraph* is generally understood to mean that multiple edges, but not loops, are allowed. The term *pseudograph* is often defined to mean a "graph" which can have both multiple edges and loops. In this generalized notion, E is a *multiset* of unordered pairs of (not necessarily distinct) vertices. As opposed to a pseudograph, a simple graph is a graph that has no loops and no more than one edge between any two different vertices. In a simple graph the edges of the graph form a set (rather than a multiset) and each edge is a pair of distinct vertices.

Directed graphs

In general, edges of a graph-construction may be directed or undirected. For example, if the vertices represent people at a party, and there is an edge between two people if they shake hands, then this is an undirected graph, because if person A shook hands with person B, then person B also shook hands with person A. In contrast, if there is an edge from person A to person B when person A knows of person B, then this graph is *directed*, because knowledge of someone is not necessarily a symmetric relation (that is, one person knowing another person does not necessarily imply the reverse; for example, many fans may know of a celebrity, but the celebrity is unlikely to know all their fans). This type of graph is called a *directed graph* and the edges are called *directed edges* or *arcs*. In Computer Science, directed graphs are used to represent knowledge (e.g., *conceptual graph*), finite state machines, and many other discrete structures.

Formally, a *directed graph* (*digraph*) is an ordered pair $D = \langle V, A \rangle$ with V a set whose elements are called *vertices*, and A a set of ordered pairs of vertices, called *arcs* (*directed edges*, *arrows*).

An arc $\langle x, y \rangle$, $x, y \in V$, is considered to be directed from x to y; y is called the *head* and x is called the *tail* of the arc; y is said to be a *direct successor* of x, and x is said to be a *direct predecessor* of y. If there exists an *ordered path* $\langle x, x_1 \rangle, \langle x_1, x_2 \rangle, ..., \langle x_{k-1}, x_k \rangle, \langle x_k, y \rangle$, $x_i \in V$, leading from x to y, then y is said to be a *successor* of x and *reachable* from x, and x is said to be a *predecessor* of y. The arc $\langle y, x \rangle$ is called the *arc* $\langle x, y \rangle$ *inverted*.

A directed graph D is called *symmetric* if, for every arc in D, the corresponding inverted arc also belongs to D. A symmetric loopless directed graph $D = \langle V, A \rangle$ is equivalent to a simple undirected graph $G = \langle V, E \rangle$, where the pairs of inverse arcs in A correspond with the edges in E; thus the number of edges in G is half the number of arcs in D: $|E| = \frac{|A|}{2}$.

An *oriented* graph is a directed graph in which at most one of $\langle x, y\rangle$ and $\langle y, x\rangle$ may be arcs. A *mixed graph* G is a graph in which some edges may be directed and some may be undirected. It is written as an ordered triple $G = \langle V, E, A\rangle$ with V, E and A defined as above. Directed and undirected graphs are special cases of mixed graphs.

A *quiver* (*multidigraph*) is a directed graph which may have more than one arrow from a given source to a given target. Some quivers may also have directed loops.

Any binary relation ω on a set V defines a directed graph (in general, with loops) with the vertex set V. An element x of V is a direct predecessor of an element y of X if and only if $x\omega y$. For symmetric antireflexive binary relation ω on V one gets an undirected simple graph with the vertex set V.

Hypergraphs

A *hypergraph* is a generalization of a graph, where edges can connect any number of vertices. Formally, a *hypergraph* is a pair $\langle V, E\rangle$, where V is a set of elements, called *vertices*, and E is a set of non-empty subsets of V, called *hyperedges*. Therefore, E is a subset of the power set $P(V)$ of V.

A hypergraph is called k-*uniform* (an k-*hypergraph*) if every edge has cardinality k. A graph is just an 2-uniform hypergraph.

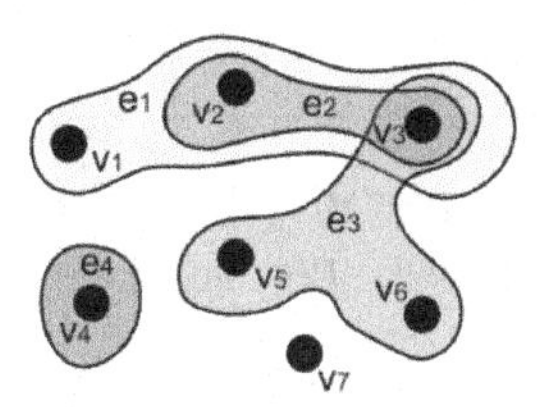

Figure 2.1: Example of an hypergraph

Weighted graphs

A graph is a *weighted graph* if a number (*weight*) is assigned to each its edge. In other words, for a weighted graph (or digraph), each edge is associated with some value, representing its cost, weight, length, capacity or other term depending on the application; such constructions arise in many contexts, for example in Optimal Routing Problems such as the Traveling Salesman Problem. Some authors call such a graph a *network*. Weighted correlation networks can be defined by soft-thresholding the pairwise correlations among variables (e.g., gene measurements). In some cases a *weight* can be assigned also to each vertex of a graph.

Labeled graphs

Normally, the vertices of a graph, by their nature as elements of a set, are distinguishable. This kind of graph may be called *vertex-labeled*. However, for many questions it is better to treat vertices as indistinguishable; then the graph may be called *unlabeled*. Of course, the vertices may be still distinguishable by the properties of the graph itself. The same remarks apply to edges, so graphs with labeled edges are called *edge-labeled graphs*. Note that in the literature the term *labeled* may apply to other kinds of labeling, besides that which serves only to distinguish different vertices or edges.

Adjacency and incidency relations in graphs

Two edges of a graph are called *adjacent* if they share a common vertex. Two arcs of a directed graph are called *consecutive* if the head of the first one is at the tail of the second one.

Similarly, two vertices are called *adjacent* if they share a common edge, and *consecutive*, if they are at the tail and at the head of an arc, in which case the common edge (arc) is said to join the two vertices. An edge (arc) and a vertex on that edge (arc) are called *incident*.

So, the edges E of an undirected graph G induce a symmetric binary relation $\sim$ on V, that is called the *adjacency relation* of G. Specifically, for each edge uv the vertices u and v are adjacent to one another, which is denoted $u \sim v$.

Connectivity of a graph

A graph is called *connected* if every pair u, v of distinct vertices in the graph is connected, i.e., there exists a path $uu_1, u_1u_2, ..., u_{k-1}u_k, u_kv$, $u_i \in V$, between them; otherwise, it is called *disconnected*.

The number of edges in a shortest path connecting two vertices u and v is called *geodesic distance* between u and v. The *eccentricity* $\epsilon(v)$ of a vertex v is the greatest geodesic distance between v and any other vertex. It can be thought of as how far a vertex is from the vertex most distant from it in the graph.

The *radius* $r = r(G)$ of a graph G is the minimum eccentricity of any vertex: $r = \min_{v \in V} \epsilon(v)$.

The *diameter* $d = d(G)$ of a graph G is the maximum eccentricity of any vertex in the graph: $d = \max_{v \in V} \epsilon(v)$. That is, d it is the greatest geodesic distance between any pair of vertices. To find the diameter of a graph, first find the shortest path between each pair of vertices; the greatest length of any of these paths is the diameter of the graph.

A graph is called *k-vertex-connected* (*k-edge-connected*) if no set of $k-1$ vertices (respectively, edges) exists that, when removed, disconnects the graph. A k-vertex-connected graph is often called simply *k-connected*.

A directed graph is called *weakly connected* if replacing all of its directed edges with undirected edges produces a connected undirected graph. It is called *strongly connected* (*strong*) if it contains a directed path from u to v and a directed path from v to u for every pair of vertices u, v.

Classical frequently used graphs

The graph with only one vertex and no edges is called a *trivial graph*. A graph with only vertices and no edges is known as an *empty graph*. In a *complete graph*, each pair of vertices is joined by an edge; that is, the graph contains all possible edges. The complete graph on n vertices is denoted as K_n. It has $\frac{n(n-1)}{2}$ edges.

In a *path graph* P_n, $n \geq 1$, the vertices can be listed in order, $v_1, ..., v_n$, so that the edges are $v_{i-1}v_i$ for each $i = 2, ..., n$. In this case we denote this path by $P_{v_1,...,v_n}$. If a path graph occurs as a subgraph of another graph, it is a *path* in that graph.

In a *circuit* C_n, $n \geq 3$, vertices can be named $v_1, ..., v_n$ so that the edges are $v_{i-1}v_i$ for each $i = 2, ..., n$ in addition to v_nv_1. In this case we denote this circuit by $C_{v_1,...,v_n}$. If a circuit graph occurs as a subgraph of another graph, it is a *circuit* in that graph.

A *cycle* or *Eulerian graph* is a graph which can be decomposed as an edge disjoint union of circuits (equivalently, it is a connected graph in which every node has an even degree). A cycle C is called *s-cycle*, $s \geq 3$, if it has exactly s edges. The cycle C is called *hordless*, if there are no edges, joining two vertices of C and not belonging to C.

A *tree* is a connected graph with no cycles.

A *forest* is a graph with no cycles (i.e., the disjoint union of one or more trees).

In a *bipartite graph*, the vertex set V can be partitioned into two sets, V_1 and V_2, so that no two vertices in V_1 are adjacent and no two vertices in V_2 are adjacent. In a *complete bipartite graph*, the vertex set V is the union of two disjoint sets, V_1 and V_2, so that every vertex in V_1 is adjacent to every vertex in V_2 but there are no edges within V_1 or V_2. If $|V_1| = n_1$ and $|V_2| = n_2$, the corresponding complete bipartite graph is denoted by K_{n_1,n_2}. In a *complete t-partite graph*, the vertex set V is the union of t disjoint sets, V_1, ..., V_t, so that every vertex in V_i is adjacent to every vertex in V_j for any $i \neq j$, but there are no edges within V_i for any i. If $|V_1| = n_1, ..., |V_t| = n_t$, the corresponding complete t-partite graph is denoted by $K_{n_1,...,n_t}$.

Given a vertex subset $S \subseteq V$ in a graph G, let $\delta_S(G)$ denote the set of edges in G having one endpoint in S and the other endpoint in $\overline{S} = V \setminus S$; $\delta_S(G)$ is called the *cut determined by* S. (Thus, the symbol $\delta_S(G)$ denotes here an edge set. In fact, the symbol δ_S will be mostly used in the book for denoting the *cut semimetric*, determined by S, when G is the complete graph K_n. The graph notation $\delta_S(G)$ will be used only locally in the book and the reader will then be reminded that it is an edge set.)

A *regular graph* is a graph where each vertex has the same number of neighbours, i.e., every vertex has the same degree. A regular graph with vertices of degree k is called a *k-regular graph*.

A *planar graph* is a graph whose vertices and edges can be drawn in a plane such that no two of the edges intersect (i.e., a planar graph is a graph, embedded in a plane). The *Euler's formula* states that if a finite connected planar graph is drawn in the plane without any edge intersections, and V is the number of vertices, E is the number of edges and F is the number of *faces* (regions bounded by edges, including the outer, infinitely large region) of this *plane graph*, then $V - E + F = 2$.

The *hypercube graph* $H(n, 2)$ has the vertex set $V = \{0, 1\}^n$, and its edges are the pairs of vectors $x, y \in \{0, 1\}^n$, such that $|\{i \in \{1, ..., n\} \,|\, x_i \neq y_i\}| = 1$.

The *half-cube graph* $\frac{1}{2}H(n, 2)$, has the vertex set $V = \{x \in \{0, 1\}^n \,|\, \sum_{i=1}^n x_i$ is even$\}$, and its edges are the pairs $x, y \in \{0, 1\}^n$ such that $|\{i \in \{1, ..., n\} \,|\, x_i \neq y_i\}| = 2$.

The *cocktail-party graph* $K_{n\times 2}$ has the vertex set $V = \{v_1, \ldots, v_n, v_{n+1}, \ldots, v_{2n}\}$, and its edges are all pairs of nodes in V except the n pairs $v_1v_{n+1}, \ldots, v_nv_{2n}$; in other words, $K_{n\times 2}$ is the complete graph K_{2n} in which a *perfect matching* has been deleted.

Two graphs $G = \langle V, E \rangle$ and $G' = (V', E')$ are said to be *isomorphic* if there exists a bijection $f : V \longrightarrow V'$ such that

$$uv \in E \Longleftrightarrow f(u)f(v) \in E';$$

we write $G \simeq G'$ if G and G' are isomorphic. There are some isomorphisms among the above graphs; for instance,

$$H(2,2) \simeq C_4,\ K_{2\times 2} \simeq C_4,\ \frac{1}{2}H(2,2) \simeq K_2,\ \frac{1}{2}H(3,2) \simeq K_4,\ \frac{1}{2}H(4,2) \simeq K_{3\times 2}.$$

Operations on graphs

There are several operations that produce new graphs from old ones, which might be classified into the following categories:

- elementary unary operations, sometimes called "editing operations" on graphs create a new graph from the original one by a simple, local change, such as *addition* or *deletion* of a vertex or an edge, *merging* and *splitting* of vertices, etc.;
- advanced unary operations, which create a significantly new graph from the old one, such as *line graph*, *dual graph*, *complement graph*, etc.;
- binary operations, which create new graph from two initial graphs, such as *disjoint union of graphs*, *Cartesian product of graphs*, *tensor product of graphs*, *strong product of graphs*, etc.

A *subgraph* $H = \langle V', E' \rangle$ of a graph $G = \langle V, E \rangle$ is a graph whose vertex set V' is a subset of the vertex set V of G, and whose edge set E' is a subset of the edge set E of G. In reverse, a *supergraph* of a graph G is a graph of which G is a subgraph. Any subgraph H of a graph G can be obtained by deletion of some edges and vertices of G.

Given an edge subset $F \subseteq E$ in G, the graph $G\backslash F = \langle V, E \setminus F \rangle$ is called the *graph, obtained from G by deleting F*. When $F = \{e\}$, we also denote $G\backslash\{e\}$ by $G\backslash e$ or $G - e$.

A subgraph H of a graph G is said to be *induced* (*full*) if, for every pair of vertices x and y of H, xy is an edge of H if and only if xy is an edge of G. In other words, H is an induced subgraph of G if it has exactly the edges that appear in G over the same vertex set. If the vertex set of H is the subset S of $V(G)$, then H can be written as $H = G[S]$ and is said to be *induced by S*.

The set W is said to induce a *clique* in G if any two vertices in W are adjacent, i.e., if $G[W]$ is a complete graph.

Contracting an edge $e = uv$ in G means identifying the end vertices u and v of e and deleting the parallel edges that may be created while identifying u and v; G/e denotes the graph obtained from G by contracting the edge e. For an edge set $F \subseteq E$, G/F denotes the graph obtained from G by contracting all edges of F (in any order). A graph H is said to be a *minor* of G if it can be obtained from G by a sequence of deletions and/or contractions of edges, and deletions of nodes.

The *suspension graph* ∇G for a graph G is obtained from G by adding a new vertex (called the *apex* of ∇G), which is adjacent to all the vertices of G.

The graph $K_n - tK_2$ is obtained from the complete graph K_n with t disjoint edges removed.

The *complement* of a graph $G = \langle V, E \rangle$ is a graph $\overline{G} = \langle V, \overline{E} \rangle$ on the same vertices such that two distinct vertices of $\overline{G}$ are adjacent if and only if they are not adjacent in G. That is, to generate the complement of a graph, one fills in all the missing edges required to form a complete graph, and removes all the edges that were previously there.

Given a graph G, its *line graph* $L(G)$ is a graph such that each vertex of $L(G)$ represents an edge of G, and two vertices of $L(G)$ are adjacent if and only if their corresponding edges are adjacent in G.

The *Petersen graph* is the complement of the line graph of K_5. It has 10 vertices and 15 edges and is named after Petersen (1898) [Pete98], who constructed it to be the smallest bridgeless cubic graph with no three-edge-coloring.

Johnson graphs form a special class of undirected graphs defined from systems of sets. The vertices of the Johnson graph $J(n, k)$ are the k-element subsets of an n-element set; two vertices are adjacent when they meet in a $(k-1)$-element set. $J(n, 1)$ is the complete graph K_n. $J(4, 2)$ is the *octahedral graph*. $J(5, 2)$ is the complement graph of the *Petersen graph*, hence, the line graph of K_5.

The *dual graph* of a *plane graph* G is a graph that has a vertex for each face of G, and has an edge whenever two faces of G are separated from each other by an edge. Thus, each edge e of G has a corresponding dual edge: the edge that connects the two faces on either side of e.

The *Cartesian product* $G \times H$ of graphs G and H is a graph such that the vertex set of $G \times H$ is the Cartesian product $V(G) \times V(H)$, and any two vertices (u, u') and $(v, v') \in V(G \times H)$ are adjacent in $G \times H$ if and only if either $u = v$ and u' is adjacent with v' in H, or $u' = v'$ and u is adjacent with v in G.

The Cartesian product of K_2 and a path graph P_n is the *ladder graph* L_n: a planar undirected graph with $2n$ vertices $u_1, ..., u_n, v_1, ..., v_n$ and $n + 2(n-1)$ edges $u_i u_{i+1}$, $v_i v_{i+1}$, $i = 1, 2, ..., n-1$, $u_j v_j$, $j = 1, 2, ..., n$. By construction, the ladder graph L_n is isomorphic to the *grid graph* $G_{2,n}$ and looks like a ladder with n rungs.

The Cartesian product of K_2 and a circuit C_n is the *Prism graph* $Prism_n$; it is the skeleton of corresponding n-prism. The skeleton of an n-antiprism (a polyhedron, composed of two parallel copies of some particular n-gon, connected by an alternating band of triangles) is *Antiprism graph* $APrism_n$.

The *Möbius ladder* M_{2n} is a graph, obtained by introducing a twist in the prism graph $Prism_n$.

Given a set S of q elements and a positive integer d, the *Hamming graph* $H(d, q)$ is the Cartesian product of d complete graphs K_q; its vertex set is the set of sequences of length d from S; two vertices are adjacent if they differ in precisely one coordinate. More generally, $H(q_1, ..., q_d)$ is the Cartesian product of d complete graphs K_{q_i}, $i = 1, ..., d$.

Let $G = \langle V, E \rangle$ be a graph. Let V_1 and V_2 be subsets of V such that $V = V_1 \cup V_2$ and such that the set $W = V_1 \cap V_2$ induces a clique in G. Suppose moreover that there is no edge joining a node in $V_1 \setminus W$ to a node in $V_2 \setminus W$. Then, G is called the *clique k-sum* of the graphs $G_1 = G[V_1]$ and $G_2 = G[V_2]$, where $k = |W|$. One may say simply that G is the *clique sum* of G_1 and G_2 if one does not wish to specify the size of the common clique.

2.2 Vector spaces

Vector space over a field $\mathbb{F}$

A *vector space* (*linear space*) over a *field* $\mathbb{F}$ is a set V equipped with operations of *vector addition* $+ : V \times V \to V$ and *scalar multiplication* $\cdot : F \times V \to V$ such that $\langle V, +, 0\rangle$ forms an *Abelian group* (it means, that $x + (y + z) = (x + y) + z$, $x + y = y + x$, $0 + x = x$ for any $x, y, z \in V$, and, for any $x \in V$, there exists $-x \in V$ with $x + (-x) = 0$, where $0 \in V$ is the *zero vector*), and, for all *vectors* $x, y \in V$ and any *scalars* $a, b \in \mathbb{F}$, we have the following properties: $1 \cdot x = x$ (where 1 is the multiplicative unit of $\mathbb{F}$), $(a \cdot b) \cdot x = a \cdot (b \cdot x)$, $(a + b) \cdot x = a \cdot x + b \cdot x$, and $a \cdot (x + y) = a \cdot x + a \cdot y$.

The most simple example of a vector space over a field $\mathbb{F}$ is the field itself, equipped with its standard addition and multiplication. More generally, a vector space can be composed of *n-tuples* (sequences of length n) of elements of $\mathbb{F}$, such as $(x_1, x_2, ..., x_n)$, where each x_i is an element of $\mathbb{F}$. A vector space composed of all the n-tuples of a field $\mathbb{F}$ is known as a *coordinate space*, usually denoted $\mathbb{F}^n$. For any $x, y \in \mathbb{F}^n$ and any $a \in \mathbb{F}$, it holds

$$x + y = (x_1, x_2, ..., x_n) + (y_1, y_2, ..., y_n) = (x_1 + y_1, x_2 + y_2, ..., x_n + y_n),$$

$$a \cdot x = a \cdot (x_1, x_2, ..., x_n) = (ax_1, ax_2, ..., ax_n).$$

A *basis* of a vector space V is a (finite or infinite) set $B = \{b_i\}_i$ of vectors $b_i \in V$, that *spans the whole space* and is *linearly independent*. *Spanning the whole space* means that any vector $v \in V$ can be expressed as a finite sum (called a linear combination) of the basis elements with coefficients from $\mathbb{F}$:

$$v = a_1 \cdot b_{i_1} + a_2 \cdot b_{i_2} + ... + a_n \cdot b_{i_n},$$

where $a_k \in \mathbb{F}$, $b_{i_k} \in B$, $k = 1, 2, ..., n$. The coefficients $a_k \in \mathbb{F}$ are called the *coordinates* (*components*) of the vector v *with respect to the basis* B. *Linear independence* means that the coordinates a_k are uniquely determined for any vector in the vector space. All bases of a given vector space have the same number of elements (*cardinality*). It is called the *dimension of the vector space* V, denoted $dim\, V$.

The dimension of the coordinate space $\mathbb{F}^n$ is n: the *coordinate vectors* $e_1 = (1, 0, 0, ..., 0)$, $e_2 = (0, 1, 0, ..., 0)$, ..., $e_n = (0, 0, 0, ..., 0, 1)$ form a basis of $\mathbb{F}^n$, called the *standard basis*, since any vector $x = (x_1, x_2, ..., x_n) \in \mathbb{F}^n$ can be uniquely expressed as a linear combination of these vectors:

$$x = (x_1, x_2, ..., x_n) = x_1 \cdot (1, 0, ..., 0) + x_2 \cdot (0, 1, 0, ..., 0) + ... + x_n \cdot (0, ..., 0, 1) =$$

$$= x_1 \cdot e_1 + x_2 \cdot e_2 + ... + x_n \cdot e_n.$$

The corresponding coordinates $x_1, x_2, ..., x_n$ are just the *Cartesian coordinates* of the vector $x = (x_1, x_2, ..., x_n) \in \mathbb{F}^n$.

Subspaces of a vector space

A non-empty subset W of a vector space V that is closed under addition and scalar multiplication (and, therefore, contains the 0-vector of V) is called a *subspace* of V. Subspaces of a vector space V over a field $\mathbb{F}$ are vector spaces over the same field.

The intersection of all subspaces containing a given set S of vectors is called its *span*, and it is the smallest subspace of V containing the set S. Expressed in terms of elements, the span is the subspace consisting of all the linear combinations of elements of S. The dimension of the span of a given set S of vectors from V is called the *rank* of S. It is equal to the number of elements in a maximal linear independent subset of S.

A vector space over the field $\mathbb{C}$ of complex numbers is called *complex vector space.* A vector space over the field $\mathbb{R}$ of real numbers is called *real vector space.* The space $\mathbb{R}^n$ ($\mathbb{C}^n$) is called the *real (complex) coordinate space of n dimensions.*

Normed vector spaces

For a real (complex) vector space V, a function $||.|| : V \to \mathbb{R}$ is called a *norm* on V, if for all $x, y \in V$ and for any scalar $a \in \mathbb{F}$ we have the following properties:

1. $||x|| \geq 0$, with $||x|| = 0$ if and only if $x = 0$ (*positive-definiteness*);
2. $||a \cdot x|| = |a| \cdot ||x||$ (*1-homogenity*);
3. $||x + y|| \leq ||x|| + ||y||$ (*triangle inequality*).

The vector space $\langle V, ||.||\rangle$ is called *normed vector space* or, simply, *normed space.*

An *inner product space* is a real (complex) vector space V with an additional structure, that associates each pair of vectors in the space with a scalar quantity known as the *inner product* of the vectors.

Formally, an *inner product* is a map $\langle \cdot, \cdot \rangle : V \times V \to \mathbb{F}$, $\mathbb{F} \in \{\mathbb{C}, \mathbb{R}\}$, that satisfies for all vectors $x, y, z \in V$ and all scalars $a \in \mathbb{F}$ the following properties:

1. $\langle x, y\rangle = \overline{\langle y, x\rangle}$ (*conjugate symmetry*);
2. $\langle ax, y\rangle = a\langle x, y\rangle$, and $\langle x + y, z\rangle = \langle x, z\rangle + \langle y, z\rangle$ (*linearity in the first argument*);
3. $\langle x, x\rangle \geq 0$, and $\langle x, x\rangle = 0$ i and only if $x = 0$ (*positive-definiteness*).

An inner product naturally induces an associated norm $||x|| = \sqrt{\langle x, x\rangle}$, thus an inner product space is also a normed vector space. A complete space with an inner product is called a *Hilbert space.*

Two vectors, x and y, in an inner product space are called *orthogonal* if their inner product $\langle x, y\rangle$ is zero. Two orthogonal vectors x and y are called *orthonormal,* if, in addition, they are *unit vectors,* i.e., $||x|| = ||y|| = 1$. A set of vectors form an *orthogonal set,* if all vectors in the set are mutually orthogonal.

Two vector subspaces, A and B, of an inner product space V are called *orthogonal subspaces* if each vector in A is orthogonal to each vector in B. The largest subspace of V that is orthogonal to a given subspace is called its *orthogonal complement.*

Inner product spaces over the field of complex numbers are sometimes referred to as *unitary spaces*. In the real coordinate space $\mathbb{R}^n$ one can define for any $x, y \in \mathbb{R}^n$ the *dot product*

$$x \cdot y = (x_1, x_2, ..., x_n) \cdot (y_1, y_2, ..., y_n) = x_1y_1 + x_2y_2 + ... + x_ny_n.$$

It is an inner product. It defines on $\mathbb{R}^n$ the *Euclidean norm*

$$||x||_2 = \sqrt{x \cdot x} = \sqrt{x_1^2 + x_2^2 + ... + x_n^2}.$$

As every vector has its Euclidean norm, then for any pair of points the *Euclidean distance*

$$d(x, y) = \|x - y\|_2 = \sqrt{(x_1 - y_1)^2 + (x_2 - y_2)^2 + ... + (x_n - y_n)^2}$$

is defined, providing a standard metric space structure on $\mathbb{R}^n$.

Linear transformations

The relation of two vector spaces V and W over a field $\mathbb{F}$ can be expressed by *linear map* (*linear transformation*). It is a function $f : V \to W$, that reflects the vector space structure, i.e., preserves sums and scalar multiplication: for all $x, y \in V$ and all $a \in \mathbb{F}$, it holds

$$f(x + y) = f(x) + f(y), \text{ and } f(a \cdot x) = a \cdot f(x).$$

Linear maps $V \to W$ between two vector spaces form a vector space, denoted by $L(V, W)$. Once a basis of V is chosen, linear maps $f : V \to W$ are completely determined by specifying the images of the basis vectors, because any element of V is expressed uniquely as a linear combination of them. In particular, any $m \times n$ *matrix* A *over a field* $\mathbb{F}$, i.e., a table consisting of m rows and n columns with the entries a_{ij} from $\mathbb{F}$, gives rise to a linear map f from F^n to F^m, by the rule $f(x) = A \cdot x$:

$$x = (x_1, x_2, \cdots, x_n) \to \left(\sum_{j=1}^{n} a_{1j}x_j, \sum_{j=1}^{n} a_{2j}x_j, \cdots, \sum_{j=1}^{n} a_{mj}x_j\right).$$

Moreover, after choosing bases of V and W, any linear map $f : V \to W$ is uniquely represented by a matrix via this assignment.

An *isomorphism* is a *bijective* (i.e., *injective* and *surjective*) linear map $f : V \to W$; it means that there exists an inverse map $g : W \to V$, which is a map such that the two possible compositions $f \circ g : W \to W$ and $g \circ f : V \to V$ are identity maps. If there exists an isomorphism between V and W, the two spaces are said to be *isomorphic*; they are then essentially identical as vector spaces, since all identities holding in V are, via f, transported to similar ones in W, and vice versa via g.

On the other hand, if $dim\, V = dim\, W$, a bijective correspondence between fixed bases of V and W gives rise to a linear map that maps any basis element of V to the corresponding basis element of W. It is an isomorphism between spaces V and W. In particular, any n-dimensional vector space V over a field $\mathbb{F}$ is isomorphic to $\mathbb{F}^n$.

An *isometry* of $\mathbb{R}^n$ is a linear mapping $\mathbb{R}^n \to \mathbb{R}^n$, preserving the Euclidean distance.

The space of linear maps from V to $\mathbb{F}$ is called the *dual vector space*, denoted V^*. Via the injective natural map $V \to (V^*)^*$, any vector space can be embedded into its *bidual*; the map is an isomorphism if and only if the space V is finite-dimensional.

2.3 Matrices

Rectangular matrices

An $m \times n$ *matrix* $A = ((a_{ij}))$ *over a field* $\mathbb{F}$ is a table consisting of m rows and n columns with the entries a_{ij} from $\mathbb{F}$.

If $m \neq n$, a matrix is called *rectangular*. The set of all rectangular $m \times n$ matrices with real (complex) entries is denoted by $M_{m,n}$. It forms a *group* $\langle M_{m,n}, +, 0_{m,n}\rangle$, where $((a_{ij})) + ((b_{ij})) = ((a_{ij} + b_{ij}))$, and the matrix $0_{m,n} \equiv 0$, i.e., all its entries are equal to 0. It is also an mn-dimensional vector space over $\mathbb{R}$ (over $\mathbb{C}$).

The *transpose* of a matrix $A = ((a_{ij})) \in M_{m,n}$ is the matrix $A^T = ((a_{ji})) \in M_{n,m}$. The *conjugate transpose* (*adjoint*) of a matrix $A = ((a_{ij})) \in M_{m,n}$ is the matrix $A^* = ((\overline{a}_{ji})) \in M_{n,m}$.

A *product* of a matrix $A = ((a_{ij})) \in M_{m,n}$ and a matrix $B = ((b_{ij})) \in M_{n,k}$ is a matrix $C = ((c_{ij})) \in M_{m,k}$ defined by $((c_{ij})) = ((\sum_{k=1}^{n} a_{ik}b_{kj}))$.

Any vector $x = (x_1, x_2, ..., x_n)$ from the n-dimensional real (complex) coordinate space can be considered as a matrix $((x_{1j})) \in M_{1,n}$ (*row vector*), or, equivalently, as a matrix $((x_{i1})) \in M_{n,1}$ (*column vector*). In this case a *product* of a matrix $A = ((a_{ij})) \in M_{m,n}$ and a matrix $x = ((x_{i1})) \in M_{n,1}$ is a matrix $y = ((y_{i1})) \in M_{m,1}$, defined by $((y_{i1})) = ((\sum_{k=1}^{n} a_{ik}x_{kj}))$. In other words, a *product* of a matrix $A = ((a_{ij})) \in M_{m,n}$ and an n-dimensional vector $x = (x_1, ..., x_n)$ is an m-dimensional vector $y = (y_1, ..., y_m)$, where $y_i = \sum_{k=1}^{n} a_{ik}x_k$. The product of a matrix $x = ((x_{1j})) \in M_{1,n}$ and a matrix $y = ((y_{i1})) \in M_{n,1}$ is a scalar $\sum_{k=1}^{n} x_{1k}y_{k1} \in \mathbb{F}$. In other words, the dot product $x \cdot y$ of two n-dimensional vectors $x = (x_1, ..., x_n)$ and $y = (y_1, ..., y_n)$ can be considered as a special case of matrix product and can be written as

$$x \cdot y^T = x_1y_1 + x_2y_2 + ... + x_ny_n.$$

Any matrix $A \in M_{n,m}$ corresponds to a linear map $x \to Ax$ from the n-dimensional real (complex) coordinate space to the m-dimensional real (complex) coordinate space.

Square Matrices

An $m \times n$ matrix is called *square matrix*, if $m = n$. The set of all square $n \times n$ matrices with real (complex) entries is denoted by M_n. It forms a *ring* $\langle M_n, +, \cdot, 0_n\rangle$, where + and 0_n are defined as above, and $((a_{ij})) \cdot ((b_{ij})) = ((\sum_{k=1}^{n} a_{ik}b_{kj}))$. It is also an n^2-dimensional vector space over $\mathbb{R}$ (over $\mathbb{C}$).

A matrix $A = ((a_{ij})) \in M_n$ is called *symmetric* if $a_{ij} = a_{ji}$ for all $i, j \in \{1, 2, ..., n\}$, i.e., if $A = A^T$.

Special types of square $n \times n$ matrices include the *identity matrix* $1_n = ((c_{ij}))$ with $c_{ii} = 1$, and $c_{ij} = 0$, $i \neq j$.

An *unitary matrix* $U = ((u_{ij}))$ is a square matrix, defined by $U^{-1} = U^*$, where $U^* = ((\overline{u}_{ji})) \in M_n$ is the conjugate transpose of U, and U^{-1} is the *inverse matrix* for U, i.e., $U \cdot U^{-1} = 1_n$.

An *orthogonal matrix* is a matrix $A = ((a_{ij})) \in M_n$, such that $A^*A = 1_n$, where $A^* = ((\overline{a_{ji}})) \in M_n$. So, for a matrix A over $\mathbb{R}$, $A^* = A^T$; in other words, a square matrix with real entries is *orthogonal*, if its columns and rows are orthogonal unit vectors (i.e., orthonormal vectors). The orthogonal matrices are the isometries of the Euclidean space;

that is, the linear transformations of $\mathbb{R}^n$ preserving the Euclidean distance. A well-known basic fact is that any two congruent sets of points can be matched by some orthogonal transformation.

If for a matrix $A \in M_n$ there is a vector x such that $Ax = \lambda x$ for some scalar λ, then λ is called *eigenvalue* of A with corresponding *eigenvector* x. Given a complex matrix $A \in M_{m,n}$, its *singular values* $s_i(A)$ are defined as the square roots of the *eigenvalues* of the matrix A^*A, where A^* is the conjugate transpose of A. They are non-negative real numbers $s_1(A) \geq s_2(A) \geq ....$

The linear transformation of $\mathbb{R}^n$ (of $\mathbb{C}^n$) corresponding to a real (complex) $n \times n$ matrix. The *determinant* $det(A)$ of a square matrix A is a scalar that tells whether the associated map is an isomorphism or not: to be so it is sufficient and necessary that the determinant is non-zero. Moreover, the linear transformation of $\mathbb{R}^n$ is *orientation preserving* if and only if the determinant of corresponding matrix is positive.

An $n \times n$ symmetric matrix $A = ((a_{ij}))$ is said to be *positive semidefinite* if $xAx^T \geq 0$ holds for all $x \in \mathbb{R}^n$ (or, equivalently, for all $x \in \mathbb{Z}^n$); then we write $A \succeq 0$. The matrix A is positive semidefinite if and only if all its eigenvalues are non-negative, or if $det(A_I) \geq 0$ for every *principal submatrix* $A_I = ((a_{ij}))_{i,j \in I}$, $I \subseteq \{1, 2, ..., n\}$.

2.4 Cones and polytopes

Cones

A *convex cone* in $\mathbb{R}^n$ (see [Schr86], [DeLa97]) is defined either by *generators* $v_1, ..., v_N \in \mathbb{R}^n$ as

$$\left\{ \sum_{i=1}^{N} \lambda_i v_i \,\middle|\, \lambda_i \geq 0 \right\},$$

or by *homogeneous linear inequalities* $f_1, \dots, f_M$, as

$$\{x \in \mathbb{R}^n \mid f_i(x) \leq 0, 1 \leq i \leq M\}.$$

We consider only *polyhedral* convex cones, i.e., the cones for which the number of generators and, alternatively, the number of defining inequalities is finite. Moreover, all convex cones under consideration are *pointed*, i.e., $0 \in C$.

The first representation of a cone means, that any convex cone C in $\mathbb{R}^n$ is the *conic hull*

$$\mathbb{R}_{\geq 0}(X) = \left\{ \sum_{x \in X} \lambda_x x \,\middle|\, \lambda_x \geq 0 \right\}$$

of the set $X = \{v_1, ..., v_N\} \subset \mathbb{R}^n$ of its generators, and $\mathbb{R}_{\geq 0}(C) = C$. As any homogeneous linear inequality $f(x) \leq 0, x \in \mathbb{R}^n$, can be written in the form $f_1x_1 + ... + f_nx_n \leq 0$, the set $\{x \in \mathbb{R}^n : f_i(x) \leq 0, 1 \leq i \leq M\}$ can be rewritten as $\{x \in \mathbb{R}^n \mid A \cdot x \leq 0\}$, where $A = ((f_{ij}))$ is an $n \times M$ matrix over $\mathbb{R}$. So, a convex cone C in $\mathbb{R}^n$ can be expressed as a solution set of a system $A \cdot x \leq 0$; such representation is called a *linear description* of C.

We call a convex cone in $\mathbb{R}^n$ *full-dimensional* if the rank of the set of its generators is n. In this case the rank of the system $A \cdot x \leq 0$ of inequalities from the linear description of C (i.e., the rank of the matrix A), is also equals to n.

The space $\mathbb{R}^n$, and any vector subspace of $\mathbb{R}^n$, including the trivial subspace $\{0\}$, are convex cones by this definition. Other examples are the set of all positive multiples of an arbitrary vector v of $\mathbb{R}^n$, or the positive (non-negative) *orthant* of $\mathbb{R}^n$ - the set of all vectors whose coordinates are all positive (non-negative). A *simplex cone* is a cone of the form $\mathbb{R}_{\geq 0}(X)$, where the set X is linearly independent.

The intersection of two convex cones in the same vector space is again a convex cone, but their union may fail to be one. The class of convex cones is also closed under arbitrary linear maps. In particular, if C is a convex cone, so is its opposite $-C$, and $C \cap -C$ is the largest linear subspace contained in C.

Let $C \subset V$ be a convex cone in $\mathbb{R}^n$. The *dual cone* C^* to C is the set

$$C^* = \{v \in \mathbb{R}^n \mid \forall w \in C, \langle w, v \rangle \geq 0\},$$

where $\langle w, v \rangle = w \cdot v^T = \sum_{i=1}^n w_i v_i$ is the dot product of w and v. This is also a convex cone in $\mathbb{R}^n$. If C is equal to its dual cone, C is called *self-dual.*

Let C be a full-dimensional polyhedral cone in $\mathbb{R}^n$. Given $v \in \mathbb{R}^n$, the inequality $\sum_{i=1}^n v_i x_i \leq 0$ is said to be *valid* for C, if it holds for all $x \in C$. Then the set

$$f = \left\{ x \in C \;\middle|\; \sum_{i=1}^n v_i x_i = 0 \right\}$$

is called the *face of C, induced by the valid inequality* $\sum_{i=1}^n v_i x_i \leq 0$. In other words, a face of C is any subset f of C such that, for every $x \in f$, any decomposition $x = y+z, y, z \in C$, implies that $y, z \in f$. A face of dimension $\dim(C) - 1 = n - 1$ is called a *facet* of C. A face of dimension $\dim(C) - 2 = n - 2$ is called a *ridge* of C. A face r of dimension 1 is called an *extreme ray* of C; it is a set $\mathbb{R}_{\geq 0}(x)$, where x is a vector, belonging to the cone C: an *representative* of r, denoted by $v(r)$; so, by abuse of language, we can identify r with its representative $v(r)$.

An extreme ray is called $\{0, 1\}$*-valued,* if it contains a vector with only values $0, 1$. For any cone C denote by $\{0, 1\}$-C the cone, generated by all extreme rays of C, containing a non-zero $\{0, 1\}$-valued point.

Denote by $F(C)$ the set of facets of a cone C and by $R(C)$ the set of its extreme rays. It is easy to see, that for any polyhedral cone the numbers $|R(C)|$ and, alternatively, $|F(C)|$, are finite.

So, any extreme ray r of a full-dimensional polyhedral cone C in $\mathbb{R}^n$ can be defined using its representative $v(r)$ - some vector from $\mathbb{R}^n$. Any facet f of the cone C also can be represented by some vector from $\mathbb{R}^n$; in fact, the facet $f : \sum_{i=1}^n v_i x_i = 0$, induced by the valid inequality $\sum_{i=1}^n v_i x_i \leq 0$, can be identified using the vector $v = (v_i) \in \mathbb{R}^n$, which is orthogonal to this facet.

Two extreme rays of C are said to be *adjacent* on C, if they generate a two-dimensional face of C (or, respectively, their intersection has dimension $n - 2$). Two facets of C are said to be *adjacent,* if their intersection has dimension $dim(C) - 2 = n - 2$ (*codimension* 2). The *adjacency number* of a facet (of an extreme ray) is the number of facets adjacent to this facet (or, respectively, the number of extreme rays, adjacent to this extreme ray).

A facet and an extreme ray are *incident,* if this extreme ray lying on this facet (or, respectively, the facet contains this extreme ray). For an extreme ray $r \subset C$ denote by $F(r)$ the set $\{f \in F(C) \mid r \subset f\}$. For a facet $f \subset C$ denote by $R(f)$ the set $\{r \in R(C) \mid r \subset f\}$. The *incidence number* $Inc(f)$ of a facet f is the number

$$|R(f)| = |\{r \in R(C) : r \subset f\}|;$$

the *incidence number* $Inc(r)$ of an extreme ray r is the number

$$|F(r)| = |\{f \in F(C) \,|\, r \subset f\}|.$$

The *rank* rank(f) of a facet f (the *rank* rank(r) of an extreme ray r) is the dimension of $\{r \in R(C) \,|\, r \subset f\}$ (of $\{f \in F(C) \,|\, r \subset f\}$, respectively).

The 1-*skeleton graph* G_C of C is a graph, whose vertices are the extreme rays of C and whose edges are the pairs of adjacent vertices. The *ridge graph* G^*_C of C is a graph, whose vertices are the facets of C and with an edge between two facets if they are adjacent on C. So, the ridge graph C^*_C of a cone C is the 1-skeleton graph G_{C^*} of its dual cone C^*.

For any cone C we will call *diameter of* C and denote $d(C)$ the diameter $d(G_C)$ of its 1-skeleton graph G_C, and will call *diameter of dual* C and denote $d(C^*)$ the diameter of its dual, i.e., the diameter $d(G^*_C)$ of its ridge graph G^*_C.

An isometry of $\mathbb{R}^n$ is a linear mapping preserving the Euclidean distance. An isometry $f : \mathbb{R}^n \longrightarrow \mathbb{R}^n$ is called a *symmetry* of the cone C, if it is an isometry, satisfying the condition $f(C) = C$. The set of all symmetries of C forms an group, called the *full symmetry group* of the cone C; it is denoted by $Is(C)$.

Given a face f (an extreme ray r) of C, the *orbit* $\Omega(f)$ of f (the *orbit* $\Omega(r)$ of r) consists of all facets (all extreme rays) of C, that can be obtained from f (from r) by a symmetry.

All cones C considered in this book will be symmetric under permutations of the set $V_n = \{1, 2, ..., n\}$, and in the symmetric case on usually has $Is(C) = Sym(n)$. In asymmetric case appears also a *reversal* symmetry (see [DDP03]), corresponding to transposition of matrix $((q_{ij}))$, and we mostly get that in this situation $Is(C) = Z_2 \times Sym(n)$.

The *representation matrix* of the 1-skeleton graph (of the ridge graph) of a cone C is the square matrix, where on the place i, j we put the number of members of orbit O_j of extreme rays (of orbit F_j of facets, respectively), which are adjacent to a fixed representative of orbit O_i (of orbit F_i).

For any cone C, let I_{O_i,F_j} and I_{F_j,O_i} denote the number of facets from the orbit F_j, incident to an extreme ray of the orbit O_i, and, respectively, the number of extreme rays from the orbit O_i, incident to a facet from the orbit F_j. Clearly,

$$|O_i| I_{O_i,F_j} = |F_j| I_{F_j,O_i},$$

where $|O_j|$ and $|F_i|$ are the orbit sizes. So, in general, Table of incidences of extreme rays and facets of the cone C can be obtained from Table of incidences of its facets and extreme rays by the above formula.

Polytopes

A *polytope* P in $\mathbb{R}^n$ is defined either by *generators* $v_1, \ldots, v_N$, as

$$\left\{ \sum \lambda_i v_i \,\middle|\, \lambda_i \geq 0, \sum_{i=1}^{n} \lambda_i = 1 \right\},$$

or by (in general, *non-homogeneous*) linear inequalities $f_1, \ldots, f_M$ as

$$\{x \in \mathbb{R}^n \,|\, f_i(x) \leq b_i, b_i \in \mathbb{R}, 1 \leq i \leq M\}.$$

We consider only *polyhedral* polytopes, i.e., the polytopes for which the number of generators and, alternatively, the number of defining inequalities is finite.

The first representation means, that any polytope P in $\mathbb{R}^n$ is the *convex hull*

$$Conv(X) = \left\{ \sum_{x \in X} \lambda_x x \,\middle|\, \lambda_x \geq 0, \sum_{x \in X}^{n} \lambda_x = 1 \right\}$$

of the set $X = \{v_1, ..., v_N\} \subset \mathbb{R}^n$ of its generators, and $Conv(P) = P$. As any (in general, non-homogeneous) linear inequality $f(x) \leq b, x \in \mathbb{R}^n, b \in \mathbb{R}$, can be written in the form $f_1x_1 + ... + f_nx_n \leq b$, the set $\{x \in \mathbb{R}^n | f_i(x) \leq b_i, b_i \in \mathbb{R}, 1 \leq i \leq M\}$ can be rewritten as $\{x \in \mathbb{R}^n : A \cdot x \leq b\}$, where $A = ((f_{ij}))$ is an $n \times M$ matrix over $\mathbb{R}$, and $b \in \mathbb{R}^M$ is an M-dimensional real vector.

So, a polytope P in $\mathbb{R}^n$ can be expressed as a solution set of a system $A \cdot x \leq b$; such representation is called a *linear description* of P.

The classical examples of polytopes are:

- the n-dimensional *simplex* α_n; it is the polytope

$$Conv(0, e_1, \ldots, e_n) = \left\{ x \in \mathbb{R}^n \,\middle|\, 0 \leq x_i \leq 1, \sum_{i=1}^{n} x_i \leq 1, 1 \leq i \leq n \right\};$$

- the n-dimensional *cross-polytope* β_n; it is the polytope

$$Conv(\pm e_1, \ldots, \pm e_n) = \left\{ x \in \mathbb{R}^n \,\middle|\, \sum_{i=1}^{n} a_i x_i \leq 1 \text{ for all } a \in \{\pm 1\}^n \right\};$$

- the n-dimensional *hypercube* γ_n; it is the polytope

$$Conv(\{0, 1\}^n) = [0, 1]^n.$$

Here, $e_1, \ldots, e_n$ denote the coordinate vectors in $\mathbb{R}^n$. In general, a *simplex* is any polytope of the form $Conv(X)$, where the set X is *affinely independent*. In fact, we use the symbol α_n for denoting any n-dimensional simplex. Similarly, β_n and γ_n denote the above cross-polytope and hypercube, up to affine bijection. Note that α_1, β_1 and γ_1 coincide; $\beta_2 = \gamma_2$ (up to affine bijection); α_3 is the tetrahedron, β_3 is the octahedron, and γ_3 is the usual cube.

Let P be a full-dimensional polytope in $\mathbb{R}^n$. Given $v \in \mathbb{R}^n$ and $v_0 \in \mathbb{R}$, the inequality $v^T x = \sum_{i=1}^{n} v_i x_i \leq v_0$ is said to be *valid* for P, if it holds for all $x \in P$. Then the set

$$f = \left\{ x \in P \,\middle|\, v^T x = \sum_{i=1}^{n} v_i x_i = v_0 \right\}$$

is called the *face of P, induced by the valid inequality* $v^T x = \sum_{i=1}^{n} v_i x_i \leq v_0$. In other words, a face of P is any subset f of P such that, for every $x \in F$, any decomposition $x = \alpha y + (1 - \alpha)z$, where $0 \leq \alpha \leq 1$, and $y, z \in P$, implies that $y, z \in F$.

The only face of dimension $dim(P)$ is P itself. A face of dimension $\dim(P) - 1 = n - 1$ is called a *facet* of P. A face of dimension $\dim(P) - 2 = n - 2$ is called a *ridge* of P. A face of dimension 1 is called an *edge* of P; it has the form $\{\alpha x + (1-\alpha)y \,|\, x, y \in P\}$. A face of dimension 0 is called an *vertex* of P; it is of the form $\{x \,|\, x \in P\}$.

Two vertices of $x, y \in P$ are said to be *adjacent* on P, if the set

$$\{\alpha x + (1-\alpha)y \,|\, 0 \leq \alpha \leq 1\}$$

is an edge of P. Two facets of P are said to be *adjacent*, if their intersection has dimension $dim(P) - 2 = n-2$ (*codimension* 2). The *adjacence number* of a facet (of an vertex) is the number of facets adjacent to this facet (or, respectively, the number of vertices, adjacent to this vertex).

The *incidence number* of a facet (of an vertex) is the number of vertices lying on this facet (or, respectively, of facets containing this vertex).

The 1-*skeleton graph* G_P of P is a graph, whose vertices are the vertices of P and whose edges are the pairs of adjacent vertices. The *ridge graph* G^*_P of P is a graph, whose vertices are the facets of P and with an edge between two facets if they are adjacent on P. So, the ridge graph G^*_P of a polytope P is the 1-skeleton graph G_{P^*} of its dual P^*.

For any polytope P we will call *diameter of* P and denote $d(P)$ the diameter $d(G_P)$ of its 1-skeleton graph G_P. We will call *diameter of dual* P and denote $d(P^*)$ the diameter of its dual, i.e., the diameter $d(G^*_P)$ of its ridge graph G^*_P.

An isometry $f : \mathbb{R}^n \longrightarrow \mathbb{R}^n$ is called a *symmetry* of P, if it is an isometry, satisfying the condition $f(P) = P$. The set of all symmetries of P forms an group, called the *full symmetry group* of the polytope P; it is denoted by $Is(P)$.

Given a face f (an vertex v) of P, the *orbit* $\Omega(f)$ of f (the *orbit* $\Omega(v)$ of v) consists of all facets (all vertices) of P, that can be obtained from f (from v) by a symmetry.

All polytopes P, considered in this book, will be symmetric under permutations of the set $V_n = \{1, 2, ..., n\}$. In the classical non-oriented case there are some additional symmetries, defined by so-called *switching operation* (see [DeLa97] and the Chapter 9 of this book). In oriented case we have a *reversal* symmetry (see [DDP03]), corresponding to transposition of matrix $((q_{ij}))$.

The *representation matrix* of the 1-skeleton graph (of the ridge graph) of a polytope P is the square matrix, where on the place i, j we put the number of members of orbit O_j of vertices (orbit F_j of facets, respectively), which are adjacent to a fixed representative of orbit O_i (of orbit F_i, respectively). Table of incidences of vertices and facets of the polytope P (as well as Table of incidences of its facets and vertices) can be obtained similarly to the case of a cone C.

Polyhedra under consideration

In this book we will consider the following cones and polytopes.

- *Metric cone* MET_n: the set of all *semimetrics* on n points, i.e., the symmetric functions $d : \{1, 2, ..., n\}^2 \rightarrow \mathbb{R}_{\geq 0}$, satisfying all $d_{ii} = 0$ and all *triangle inequalities*

$$d_{ik} \leq d_{ij} + d_{jk}.$$

- *Metric polytope* $MET^{\square}_n$: the set of all $d \in MET_n$, satisfying, in addition, all *perimeter inequalities*

$$d_{ik} + d_{ij} + d_{jk} \leq 2.$$

- *Cut cone* CUT_n: the conic hull of all $2^{n-1}-1$ non-zero *cut semimetrics* on n points, where the *cut semimetric* δ_S for a set $S \subseteq \{1,2,...,n\}$ is defined by

$$(\delta_S)_{ij} = \begin{cases} 1, & \text{if} \quad |S \cap \{i,j\}| = 1, \\ 0, & \quad \text{otherwise.} \end{cases}$$

- *Cut polytope* $CUT_n^{\square}$: the convex hull of all 2^{n-1} cut semimetrics on n points.

- *Multicut cone* $MCUT_n$: the conic hull of all non-zero *multicut semimetrics* on n points, where the *multicut semimetric* (more exactly, the *m-multicut semimetric*) $\delta_{S_1,...,S_m}$ for a partition $\cup_{i=1}^m S_i = \{1,2,...,n\}$ is defined by

$$(\delta_{S_1,...,S_m})_{ij} = \begin{cases} 1, & \text{if } i \in S_a, j \in S_b, a \neq b, \\ 0, & \text{otherwise.} \end{cases}$$

- *Multicut polytope* $MCUT_n^{\square}$: the convex hull of all multucut semimetrics on n points.

- *Metric cone of a graph* $MET(G)$: given a graph $G = \langle V, E\rangle$, the projection of $MET_{|V|}$ on the subspace $\mathbb{R}^{|E|}$, indexed by the edge set E of G.

- *Metric polytope of a graph* $MET^{\square}(G)$: given a graph $G = \langle V, E\rangle$, the projection of $MET_{|V|}^{\square}$ on $\mathbb{R}^{|E|}$, indexed by the edge set of G.

- *Cut cone of a graph* $CUT(G)$: the positive span of all cut vectors of a given graph $G = \langle V, E\rangle$, or, equivalently, the projection of $CUT_{|V|}^{\square}$ on the subspace $\mathbb{R}^{|E|}$, indexed by the edge set E of G.

- *Cut polytope of the graph* $CUT^{\square}(G)$: given a graph $G = \langle V, E\rangle$, the convex hull of all cut vectors of G, or, equivalently, the projection of $CUT_{|V|}^{\square}$ on the subspace $\mathbb{R}^{|E|}$, indexed by the edge set E of G.

- *Quasi-semimetric cone* $QMET_n$: the set of all *quasi-semimetrics* on n points, i.e., the functions $q : \{1,2,...,n\}^2 \to \mathbb{R}_{\geq 0}$, satisfying all $q_{ii} = 0$ and all *oriented triangle inequalities*

$$q_{ik} \leq q_{ij} + q_{jk}.$$

- *Quasi-semimetric polytope* $QMET_n^{\square}$: the set of all $q \in QMET_n$, satisfying, in addition, all *perimeter inequalities*

$$q_{ij} + q_{ji} \leq 2.$$

- *Oriented multicut cone* $OMCUT_n$: the conic hull of all non-zero *oriented multicut quasi-semimetrics* on n points, where the *oriented multicut quasi-semimetric* (more exactly, the *oriented m-multicut quasi-semimetric*) $\delta^O_{S_1,...,S_m}$ for an ordered partition $\{1,2,...,n\}$ into m parts $S_1,...,S_m$ is defined by

$$(\delta^O_{S_1,...,S_m})_{ij} = \begin{cases} 1, & \text{if } i \in S_a, j \in S_b, \text{ and } a < b, \\ 0, & \text{otherwise.} \end{cases}$$

- *Oriented multicut polytope* $OMCUT_n^{\square}$: the convex hull of all oriented multicut quasi-semimetrics on n points.

- *Oriented cut cone* $OCUT_n$: the conic hull of all $2^n - 2$ non-zero *oriented cut quasi-semimetrics*, i.e., 2-multicut quasi-semimetrics $\delta^O_{S,\overline{S}}$, on n points.

- *Oriented cut polytope* $OCUT_n^{\square}$: the convex hull of all $2^n - 1$ oriented cut quasi-semimetrics on n points.

- *m-hemimetric cone* $HMET_n^m$: the set of all *m-hemimetrics* on n points, i.e., the totally symmetric functions $d : \{1, 2, ..., n\}^{m+1} \longrightarrow \mathbb{R}_{\geq 0}$, satisfying the condition $d_{x_1...x_{m+1}} = 0$ whenever $x_1, ..., x_{m+1}$ are not pairwise distinct, and all *m-simplex inequalities*

$$d_{x_1...x_{m+1}} \leq \sum_{i=1}^{m+1} d_{x_1...x_{i-1}x_{i+1}...x_{m+2}}.$$

- *Partition m-hemimetric cone* $HCUT_n^m$: the conic hull of all *partition m-hemimetrics* on n points, where, for an ordered partition of $\{1, 2, ..., n\}$ into $m+1$ parts $S_1, ..., S_{m+1}$, the *partition m-hemimetric* $\alpha_{S_1,...,S_{m+1}}$ is defined as

$$(\alpha_{S_1,...,S_{m+1}})_{x_1...x_{m+1}} = \begin{cases} 1, & \text{if no two } x_i \text{ belong to the same } S_k, \\ 0, & \text{otherwise.} \end{cases}$$

- *(m, s)-supermetric cone* $SMET_n^{m,s}$: the set of all *(m, s)-supermetrics* on n points, i.e., for a given number $s > 0$ and a given integer $m \geq 1$, the totally symmetric functions $h : \{1, 2, ..., n\}^{m+1} \to \mathbb{R}_{\geq 0}$, satisfying all the conditions $h_{i_1...i_{m+1}} = 0$ if $i_a = i_b$ for $a \neq b$, and, for all distinct $i_1, ..., i_{m+2}$, the *$(s; m)$-simplex inequalities*

$$s \times h_{i_2...i_{m+2}} \leq h_{i_1 i_3...i_{m+2}} + ... + h_{i_1...i_{m+1}}.$$

- *Binary (m, s)-supermetric cone* $SCUT_n^{m,s}$: the conic hull of all $\{0, 1\}$-valued extreme rays of $SMET_n^{m,s}$.

- *Weak weightable quasi-semimetric cone* $wWQMET_n$: the set of all *weak weightable quasi-semimetrics* on n points, i.e., the functions $q : \{1, 2, ..., n\}^2 \to \mathbb{R}$, satisfying all $q_{ij} + q_{ji} \geq 0$ (so, non-positive values of q_{ij} are allowed), all $q_{ii} = 0$, and all *oriented triangle inequalities*

$$q_{ik} \leq q_{ij} + q_{jk}.$$

- *Weightable quasi-semimetric cone* $WQMET_n$: the set of all *weightable quasi-semimetrics* on n points, i.e., quasi-semimetrics $q \in QMET_n$, for which exists a (weight) function $w = (w_i) : \{1, 2, ..., n\} \to \mathbb{R}_{\geq 0}$, satisfying all

$$q_{ij} + w_i = q_{ji} + w_j.$$

- *Strong weightable quasi-semimetric cone* $sWQMET_n$: the set of all *strong weightable quasi-semimetrics* on n points, i.e., all weightable quasi-semimetrics on V_n, satisfying, in addition, all $q_{ij} \leq w_j$.

- *Weightable quasi-semimetric polytope* $WQMET_n^{\square}$: the set $WQMET_n \cap QMET_n^{\square}$.

- *Weak partial semimetric cone* $wPMET_n$: the set of all *weak partial semimetrics* on n points, i.e., the symmetric functions $p : \{1, 2, ..., n\}^2 \to \mathbb{R}_{\geq 0}$, satisfying all *sharp triangle inequalities*
$$p_{ik} \leq p_{ij} + p_{jk} - p_{jj}.$$
- *Partial semimetric cone* $PMET_n$: the set of all *partial semimetrics* on n points, i.e., weak partial semimetrics on n points, satisfying, in addition, all $p_{ii} \leq p_{ij}$.
- *Strong partial semimetric cone* $sPMET_n$: the set of all *strong partial semimetrics* on n points, i.e., partial semimetrics on n points, satisfying, in addition, all
$$p_{ii} + p_{jj} - p_{ij} \geq 0.$$
- *Partial semimetric convex body* $PMET_n^{\square}$: the set of all $p \in PMET_n$, satisfying, in addition, all $p_{ij} \leq 1 + p_{ii}$, and all *perimeter inequalities*
$$p_{ij} + p_{jk} + p_{ki} \leq 2 + p_{ii} + p_{jj} + p_{kk}.$$
- *Weighted semimetric cone* $WMET_n$: the set of all *weighted semimetrics* on n points, i.e, semimetrics d on n points with weight functions $w : V_n \to \mathbb{R}_{\geq 0}$ on their points.
- *Down-weighted semimetric cone* $dWMET_n$: the set of all *down-weighted semimetrics* on n points, i.e., weighted semimetrics d on n points, satisfying, in addition, all $d_{ij} \geq w_i - w_j$.
- *Strongly weighted semimetric cone* $dWMET_n$: the set of all *strongly weighted semimetrics* on n point, i.e., down-weighted semimetrics d on n points, satisfying, in addition, all $d_{ij} \leq w_i + w_j$.
- Cones $\{0,1\}$-$wPMET_n$, $\{0,1\}$-$PMET_n$, $\{0,1\}$-$sPMET_n$, $\{0,1\}$-$wWQMET_n$, $\{0,1\}$-$WQMET_n$, $\{0,1\}$-$sWQMET_n$, $\{0,1\}$-$WMET_n$, $\{0,1\}$-$dWMET_n$, $\{0,1\}$-$sWMET_n$, generated by all $\{0,1\}$-valued extreme rays of the cones $wPMET_n$, $PMET_n$, $sPMET_n$, $wWQMET_n$, $WQMET_n$, $sWQMET_n$, $WMET_n$, $dWMET_n$, $sWMET_n$, respectively.

In Table 2.1 we present some data on most important listed cones. For each cone C we collect its dimension, the number of its extreme rays and their orbits, the number of its facets and their orbits, and diameters of its 1-skeleton graph and its ridge graph.

Cone	Dimension	# of ext. rays (orbits)	# of facets (orbits)	Diameters
CUT_3=MET_3	3	3(1)	3(1)	1; 1
CUT_4=MET_4	6	7(2)	12(1)	1; 2
CUT_5	10	15(2)	40(2)	1; 2
MET_5	10	25(3)	30(1)	2; 2
CUT_6	15	31(3)	210(4)	1; 3
MET_6	15	296(7)	60(1)	2; 2
CUT_7	21	63(3)	38780(36)	1; 3
MET_7	21	55226(46)	105(1)	3; 2
CUT_8	28	127(4)	49604520(2169)	1; ?
MET_8	28	119269588(3918)	168(1)	?; 2
$OMCUT_3$=$QMET_3$	6	12(2)	12(2)	2; 2
$OMCUT_4$	12	74(5)	72(4)	2; 2
$QMET_4$	12	164(10)	36(2)	3; 2
$OMCUT_5$	20	540(9)	35320(194)	2; 3
$QMET_5$	20	43590(229)	80(2)	3; 2
$OMCUT_6$	30	4682(19)	$\geq$ 217847040($\geq$ 163822)	2; ?
$QMET_6$	30	$\geq$ 182403032($\geq$ 127779)	150(2)	?; 2
$HCUT_5^2$	10	25(2)	120(4)	2; 3
$HMET_5^2$	10	37(3)	30(2)	2; 2
$HCUT_6^3$	15	65(2)	4065(16)	2; 3
$HMET_6^3$	15	287(5)	45(2)	3; 2
$HCUT_7^4$	21	140(2)	474390(153)	2; 3
$HMET_7^4$	21	3692(8)	63(2)	3; 2
$HCUT_8^5$	28	266(2)	$\geq$ 409893148($\geq$ 11274)	2; ?
$HMET_8^5$	28	55898(13)	84(2)	3; 2
$HMET_9^6$	36	864174(20)	108(2)	?; 2
$HCUT_6^2$	20	90(3)	2095154(3086)	2; ?
$HMET_6^2$	20	12492(41)	80(2)	3; 2
$HMET_7^2$	35	$\geq$ 454191608($\geq$ 91836)	175(2)	?; 2
$HMET_7^3$	35	$\geq$ 551467967($\geq$ 110782)	140(2)	?; 2
$SMET_{m+2}^{m,s}$, $1 \leq s \leq m-1$	m+2	$\binom{m+2}{s+1}$(1)	2m+4(2)	min(s+1,m-s+1); 2 but 2; 3 if m=2,s=1
$SMET_{m+2}^{m,m}$	m+2	m+2(1)	m+2(1)	1; 1
$SMET_5^{2,2}$	10	132(6)	20(1)	2; 1
$SCUT_5^{2,2}$	10	20(2)	220(6)	1; 3
$SMET_6^{3,3/2}$	15	331989(596)	45(2)	6; 2
$SMET_6^{3,2}$	15	12670(40)	45(2)	4; 2
$SCUT_6^{3,2}$	15	247(5)	866745(1345)	2; ?
$SMET_6^{3,5/2}$	15	85504(201)	45(2)	6; 2
$SMET_6^{3,3}$	15	1138(12)	30(1)	3; 1
$SCUT_6^{3,3}$	15	21(2)	150(3)	1; 3
$SMET_7^{4,2}$	21	2561166(661)	63(2)	?; 2
$SMET_7^{4,3}$	21	838729(274)	63(2)	?; 2
$SMET_7^{4,4}$	21	39406(37)	42(1)	3; 1
$SCUT_7^{4,4}$	21	112(2)	148554(114)	1; 4
$SMET_8^{5,2}$	28	$\geq$ 222891598($\geq$ 6228)	84(2)	?; 2
$SMET_8^{5,3}$	28	$\geq$ 881351739($\geq$ 23722)	84(2)	?; 2
$SMET_8^{5,4}$	28	$\geq$ 136793411($\geq$ 4562)	84(2)	?; 2
$SMET_8^{5,5}$	28	775807(92)	56(1)	?; 1
$SMET_9^{6,6}$	36	30058078(335)	72(1)	?;1
$SMET_{10}^{7,7}$	45	923072558(1067)	90(1)	?;1
$SMET_6^{2,2}$	20	21775425(30827)	60(1)	?; 1
$SCUT_6^{2,2}$	20	96(3)	$\geq$ 243692840($\geq$ 341551)	1; ?
$SMET_7^{3,3}$	35	$\geq$ 594481939($\geq$ 119732)	105(1)	?; 1
$SMET_7^{2,2}$	35	$\geq$ 465468248($\geq$ 93128)	140(1)	?; 1

Table 2.1: Some parameters of cones on n points for small n

Part II

Main notions and examples

Chapter 3

Non-oriented case: metrics

3.1 Preliminaries

The mathematical notions of *distance metric*, i.e., a function d from $X \times X$ to the set of real numbers satisfying to $d(x, y) \geq 0$ with equality only for $x = y$, $d(x, y) = d(y, x)$, and $d(x, y) \leq d(x, z) + d(z, y)$, and of *metric space* (X, d), were originated a century ago by Fréchet (1906) [Frec06] and Hausdorff (1914) [Haus14] as a special case of an infinite topological space.

The *triangle inequality* $d(x, y) \leq d(x, z) + d(z, y)$ appears already in Euclid, but was first formalized as a central property of distances in the classical paper by Fréchet (1906) [Frec06] and later treated by Hausdorff (1914) [Haus14]. The notion of a metric space was formalized also in [Frec06], but the term "metric" was first proposed in [Haus14], page 211.

The infinite metric spaces are seen usually as a generalization of the *natural metric* $|x - y|$ on the real numbers. Their main classes are the measurable spaces (add measure) and Banach spaces (add norm and completeness).

However, starting from Menger (1928) [Meng28] and, especially, Blumenthal (1953) [Blum53], an explosion of interest in finite metric spaces occurred. Another trend: many mathematical theories, in the process of their generalization, settled on the level of metric space.

Now, finite distance metrics have become an essential tool in many areas of Mathematics and its applications including Geometry, Probability, Statistics, Coding/Graph Theory, Combinatorial Optimization, Clustering, Data Analysis, Pattern Recognition, Networks, Engineering, Computer Graphics/Vision, Astronomy, Cosmology, Molecular Biology, and many other areas of Science. Devising the most suitable distance metrics has become a standard task for many researchers.

For general theory of metrics see Blumenthal (1953) [Blum53] and Deza and Laurent (1997) [DeLa97]. The large list of metrics can be found in Deza and Deza (2014) [DeDe14].

If we drop the strict condition $d(x, y) > 0$ for distinct x, y, we obtain the notion of *semimetric*, a weaker version of the notion of metric, often more convenient for practical applications, in particular for polygonal goals. The notion of a semimetric was first formalized by Fréchet (1906) [Frec06], too. The first detailed topological analysis of semimetrics was given by Wilson (1931) [Wils31].

3.2 Definitions

Definition 3.1 *Let X be a set. A function $d : X \times X \to \mathbb{R}$ is called distance (dissimilarity) on X if for all $x, y \in X$ it holds:*

1. *$d(x, y) \geq 0$ (non-negativity);*
2. *$d(x, x) = 0$;*
3. *$d(x, y) = d(y, x)$ (symmetry).*

A *distance space* (X, d) is a set X equipped with a distance d.

Because of the symmetry and since $d(i, i) = 0$ for $i \in V_n$, we can view a distance d as a vector $(d_{ij})_{1 \leq i < j \leq n} \in \mathbb{R}^{E_n}$, where $E_n = \binom{n}{2} = \frac{n(n-1)}{2}$: $d(i, j) = d_{ij}$. Alternatively, the distance d can be viewed as an $n \times n$ symmetric matrix with $d_{ii} = 0$ on the main diagonal. On the other hand, any $n \times n$ symmetric matrix $((d_{ij}))$ with non-negative entries and with zeros on the main diagonal (any vector $d = (d_{ij}) \in \mathbb{R}^{E_n}$, $1 \leq i < j \leq n$, with non-negative coordinates) can be considered as a distance on V_n: $d(i, j) = d_{ij}$.

If we add the *triangle axiom* to the properties of a distance, we obtain the notion of *semimetric.*

Definition 3.2 *Let X be a set. A function $d : X \times X \to \mathbb{R}$ is called semimetric (écart, pseudo-metric) on X if for all $x, y, z \in X$ it holds:*

1. *$d(x, y) \geq 0$ (non-negativity);*
2. *$d(x, x) = 0$;*
3. *$d(x, y) = d(y, x)$ (symmetry);*
4. *$d(x, y) \leq d(x, z) + d(z, y)$ (triangle inequality).*

A *semimetric space* (X, d) is a set X equipped with a semimetric d.

One can define a semimetric on X as a function $d : X \times X \to \mathbb{R}$, such that it holds $d(x, x) = 0$; $d(x, y) = d(y, x)$; $d(x, y) \leq d(x, z) + d(z, y)$. It means, that the non-negativity condition follows from the triangle inequality and symmetry.

It is easy to see also that symmetry and non-negativity follow from the condition $d(x, x) = 0$ and $(4')$, where the last one is

$$(4') \qquad d(x, y) \leq d(z, x) + d(z, y) \text{ for all } x, y, z \in X.$$

For a semimetric d, the triangle inequality is equivalent, for each fixed $n \geq 4$, to the following *n-gon inequality*

$$d(x, y) \leq d(x, z_1) + d(z_1, z_2) + \dots + d(z_{n-2}, y).$$

A semimetric d on V_n can be viewed alternatively as a vector $(d_{ij})_{1 \leq i < j \leq n} \in \mathbb{R}^{E_n}$, or as an $n \times n$ symmetric matrix with $d_{ii} = 0$ on the main diagonal: $d(i, j) = d_{ij}$.

For any distance d, the function $D(x, y) = \inf \sum_i d(z_i, z_{i+1})$, where the infimum is taken over all sequences $x = z_0, \dots, z_{n+1} = y$, is a *semimetric.*

If we add *separation axiom* to the properties of a semimetric, we obtain the notion of *metric.*

Definition 3.3 *Let X be a set. A function $d : X \times X \to \mathbb{R}$ is called metric on X if for all $x, y, z \in X$ it holds:*

1. $d(x, y) \geq 0$ *(non-negativity);*

2. $d(x, y) = 0$ *if and only if* $x = y$ *(separation axiom);*

3. $d(x, y) = d(y, x)$ *(symmetry);*

4. $d(x, y) \leq d(x, z) + d(z, y)$ *(triangle inequality).*

A *metric space* (X, d) is a set X equipped with a metric d.

A metric d on V_n can be viewed alternatively as a vector $(d_{ij})_{1 \leq i < j \leq n} \in \mathbb{R}^{E_n}$ with positive coordinates, or as $n \times n$ symmetric matrix with $d_{ii} = 0$ on the main diagonal, and with $d_{ij} > 0$ for $i \neq j$: $d(i, j) = d_{ij}$.

For any distance d, the function D, defined for $x \neq y$ by $D(x, y) = d(x, y) + c$, where $c = \max_{x,y,z \in X}\{d(x, y) - d(x, z) - d(y, z)\}$, and $D(x, x) = 0$, otherwise, is a metric. Also, $D_1(x, y) = d(x, y)^c$ is a metric for sufficiently small $c \geq 0$.

For a semimetric d on X, define an equivalence relation by $x \sim y$ if $d(x, y) = 0$; equivalent points are equidistant from all other points. Let $[x]$ denote the equivalence class containing x; then $D_3([x], [y]) = d(x, y)$ is a metric on the set $\{[x] \,|\, x \in X\}$ of classes.

3.3 Examples

Basic examples

Example 1. Discrete metric. Given a set X, the *discrete metric* (*trivial metric, sorting distance*) is a metric on X, defined by

$$d(x, y) = \begin{cases} 1, & \text{if } x \neq y, \\ 0, & \text{otherwise.} \end{cases}$$

The function $d(x, y) = 0$ for all $x, y \in X$ is a semimetric on X, called *non-discrete semimetric*.

Example 2. Hausdorff metric. Let (X, D) be a finite metric space and let $X = X_1 \cup \ldots \cup X_n$ be a *partition* of X, i.e., a decomposition of X into the union of pairwise disjoint sets. The *Hausdorff metric* is a metric d on the set $Y = \{X_1, \ldots, X_n\}$, defined by

$$d(X_i, X_j) = \max_{x \in X_i, y \in X_j} D(x, y).$$

Example 3. Symmetric difference metric. Given an *anti-chain* of sets $Z = \{x, y, z, \ldots \,|\, x \not\subset y \text{ for all } x \neq y\}$, the *symmetric difference metric* (*measure metric*) is a metric $d_\triangle$ on Z, defined by

$$d_\triangle(x, y) = |x \triangle y|.$$

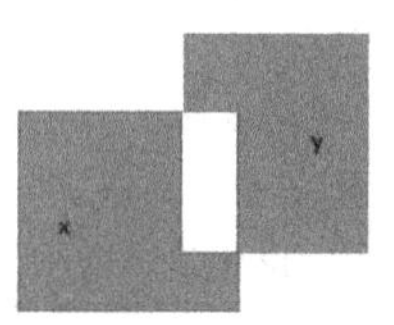

Figure 3.1: Symmetric difference: $x\triangle y$

It is the cardinality of the *symmetric difference*

$$x\triangle y = (x\backslash y) \cup (y\backslash x)$$

of sets x and y from Z.

In fact, $|x\triangle y| = |y\triangle x|$, $|x\triangle y| \geq 0$, $|x\triangle x| = 0$, $|x\triangle z| - |x\triangle y| - |y\triangle z| \leq -2(y\backslash(x \cup z)) - 2((x \cap z)\backslash y) \leq 0$.

Similarly, if $P(V_n)$ is the set of all subsets of $V_n = \{1, 2, ..., n\}$, then the function $|x\triangle y|$ - the symmetric difference of sets x and y from $P(V_n)$ - is a metric on $P(V_n)$.

If we represent each $x \subseteq V_n$ by its characteristic zero-one n-vector $\chi_x = (a_1, ..., a_n)$ (with $a_i = 1$ if $i \in x$ and $a_i = 0$ if $i \notin x$), then

$$|x\triangle y| = |a_1 - b_1| + \cdots + |a_n - b_n|$$

is the *Hamming distance* between χ_x and χ_y.

In general, for a given *measure space* $(\Omega, \mathcal{A}, \mu)$, the *symmetric difference semimetric* (*measure semimetric*) $d_\triangle$ is a semimetric on the set $\mathcal{A}_\mu = \{A \in \mathcal{A} : \mu(A) < \infty\}$, defined by

$$d_\triangle(A, B) = \mu(A\triangle B),$$

where $A\triangle B = (A \cup B)\backslash(A \cap B)$ is the *symmetric difference* of the sets $A, B \in \mathcal{A}_\mu$.

The value $d_\triangle(A, B) = 0$ if and only if $\mu(A\triangle B) = 0$, i.e., if A and B are equal *almost everywhere*. Identifying two sets $A, B \in \mathcal{A}_\mu$ if $\mu(A\triangle B) = 0$, we obtain the *symmetric difference metric* (*Fréchet-Nikodym-Aronszyan distance, measure metric*).

If μ is the *cardinality measure*, i.e., $\mu(A) = |A|$ is the number of elements in A, then $d_\triangle(A, B) = |A\triangle B|$. In this case $|A\triangle B| = 0$ if and only if $A = B$.

Similarly, the *Johnson distance* between k-sets A and B is $\frac{|A\triangle B|}{2} = k - |A \cap B|$.

Here Ω is a set, $\mathcal{A}$ is an σ-*algebra* of subsets of Ω, i.e., a collection of subsets of Ω, satisfying the following properties: $\Omega \in \mathcal{A}$; if $A \in \mathcal{A}$, then $\Omega\backslash A \in \mathcal{A}$; if $A = \cup_{i=1}^{\infty} A_i$ with $A_i \in \mathcal{A}$, then $A \in \mathcal{A}$. The function $\mu : \mathcal{A} \to \mathbb{R}_{\geq 0}$ is a *measure* on $\mathcal{A}$: it is *additive*, i.e., $\mu(\cup_{i\geq 1} A_i) = \sum_{i\geq 1} \mu(A_i)$ for all pairwise disjoint sets $A_i \in \mathcal{A}$, and satisfies $\mu(\emptyset) = 0$.

Example 4. Natural metric. For the set $\mathbb{R}$ of real numbers, the ordinary metric on $\mathbb{R}$ is the *natural metric*, defined by

$$d(x, y) = |x - y|.$$

It is the *Euclidean metric* on $\mathbb{R}$. In fact, for $\mathbb{R}$ all l_p-*metrics* coincide with the natural metric $|x - y|$.

Example 5. Hamming metric. The *Hamming metric* is a metric d_H on $\mathbb{R}^n$ (in general, on any set of sequences of length n), defined by

$$d_H(x,y) = |\{i \,|\, 1 \leq i \leq n, x_i \neq y_i\}|.$$

On binary vectors $x, y \in \{0,1\}^n$ the Hamming metric is equal to

$$|I(x)\Delta I(y)| = |I(x)\backslash I(y)| + |I(y)\backslash I(x)|,$$

where $I(z) = \{i \in \{1,2,...,n\} \,|\, z_i = 1\}$. In fact, $\max\{|I(x)\backslash I(y)|, |I(y)\backslash I(x)|\}$ is also a metric on $\{0,1\}^n$.

Example 5. Path metric in graphs. Given a *connected graph* $G = \langle V, E\rangle$, the *path metric* (*graphic metric, shortest path metric*) d_{path} is a metric on the vertex set V of G, defined, for any $u, v \in V$, as the length (i.e., the number of edges) of a shortest $(u - v)$ path in G. The corresponding metric space, associated with the graph G, is called *graphic metric space.*

The path metric of the *Cayley graph* Γ of a finitely-generated group $\langle G, \cdot, e\rangle$ is called a *word metric.*

The path metric of a graph $G = \langle V, E\rangle$, such that V can be cyclically ordered in a *Hamiltonian cycle* (a circuit containing each vertex exactly once), is called a *Hamiltonian metric.*

The *hypercube metric* is the path metric of a *hypercube graph* $H(m,2)$ with the vertex set $V = \{0,1\}^m$, and whose edges are the pairs of vectors $x, y \in \{0,1\}^m$ such that $|\{i \in \{1,2,...,n\} \,|\, x_i \neq y_i\}| = 1$; it is equal to

$$|\{i \in \{1,2,...,n\} \,|\, x_i = 1\}\Delta\{i \in \{1,...,n\} \,|\, y_i = 1\}|.$$

The graphic metric space associated with a hypercube graph is called a *hypercube metric space.*

Given an integer $n \geq 1$, the *line metric* on $\{1,...,n\}$ is the path metric of the path P_n with vertices $v_1 = 1, ..., v_n = n$.

The *weighted path metric* d_{wpath} is a metric on the vertex set V of a connected weighted graph $G = \langle V, E\rangle$ with positive edge-weights $(w(e))_{e\in E}$, defined by

$$d_{wpath}(v,v) = \min_P \sum_{e\in P} w(e),$$

where the minimum is taken over all $(u - v)$ paths P in G.

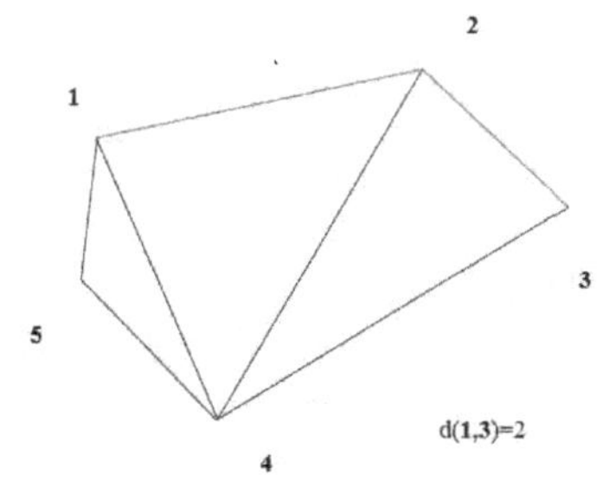

Figure 3.2: Path metric of a graph

Example 6. Circular railroad distance. Consider a circular railroad line, which moves around a circular track, represented by the unit circle $C_1 = \{x = (x_1, x_2) \in \mathbb{R}^2 \mid x_1^2 + x_2^2 = 1\}$. The *circular railroad distance* d_c is a metric on C_1, defined as the length of the shortest circular arc from x to y on C_1.

Transforms of metrics

In this section we consider some classical examples of *metric transforms*: the ways to obtain a new metric (semimetric) from other metrics (semimetrics).

- **Transform metric**

  A *transform metric* is a metric on a set X which is a *metric transform*, i.e., is obtained as a function of a given metric (given metrics) on X. In particular, transform metrics can be obtained from a given metric d (given metrics d_1 and d_2) on X by any of the following operations (here $t > 0$):

  1. $td(x, y)$ (*t-scaled metric, dilated metric, similar metric*);
  2. $\min\{t, d(x, y)\}$ (*t-truncated metric*);
  3. $\max\{t, d(x, y)\}$ for $x \neq y$ (*t-uniformly discrete metric*);
  4. $d(x, y) + t$ for $x \neq y$ (*t-translated metric*);
  5. $\frac{d(x,y)}{1+d(x,y)}$;
  6. $d^p(x, y) = \frac{2d(x,y)}{d(x,p)+d(y,p)+d(x,y)}$, where p is an fixed element of X (*biotope transform metric*, or *Steinhaus transform metric*);
  7. $\max\{d_1(x, y), d_2(x, y)\}$;
  8. $\alpha d_1(x, y) + \beta d_2(x, y)$, where $\alpha, \beta > 0$.

  Remark 1. Given an integer $2 \leq s \leq D$, call a finite metric space (X, d) of diameter D *s-truncated embeddable*, if $(X, \min(d, s))$ is isometrically embeddable in some m-half-cube $\frac{1}{2}H(m, 2)$. In [AAADDS15], eventual s-truncated embeddability of many network topologies was investigated.

  Remark 2. Easy to see, that if d and d' are metrics on X and $\alpha, \beta \in \mathbb{R}_{\geq 0}$, then $\alpha d + \beta d'$ is an semimetric on X (it is a metric, if $\alpha \neq 0$ or $\beta \neq 0$). Here, as usual, for all $x, y \in X$ it holds

  $$(\alpha d + \beta d')(x, y) = \alpha d(x, y) + \beta d'(x, y).$$

- **Power transform metric**

  Let $0 < \alpha \leq 1$. Given a metric space (X, d), the *power transform metric* (*snowflake transform metric*) is a *functional transform metric* on X, defined by

  $$(d(x, y))^\alpha.$$

  The distance $d(x, y) = (\sum_1^n |x_i - y_i|^p)^{\frac{1}{p}}$ with $0 < p = \alpha < 1$ is not a metric on $\mathbb{R}^n$, but its power transform $d(x, y)^\alpha$ is a metric.

For a given metric d on X and any $\alpha > 1$, the function d^α is, in general, only a distance on X. It is a metric, for any positive α, if and only if d is an *ultrametric*, i.e., it holds the following strengthened version of the triangle inequality(Hausdorff, 1934), called *ultrametric inequality*:

$$d(x, y) \leq \max\{d(x, z), d(z, y)\}.$$

- **Schoenberg transform metric**

Let $\lambda > 0$. Given a metric space (X, d), the *Schoenberg transform metric* is a *functional transform metric* on X, defined by

$$1 - e^{-\lambda d(x,y)}.$$

The Schoenberg transform metrics are exactly P*-metrics*: a metric d on X is a P*-metric*, if it has values in $[0, 1]$ and satisfies the *correlation triangle inequality*

$$d(x, y) \leq d(x, z) + d(y, z) - d(x, z)d(z, y).$$

The equivalent inequality $(1 - d(x, y)) \geq (1 - d(x, z))(1 - d(z, y))$ express that the probability, say, to reach x from y via z is either equal to $(1 - d(x, z))(1 - d(z, y))$ (independence of reaching z from x and y from z), or greater than it (positive correlation).

- **Product metric**

Given finite number n of metric spaces (X_1, d_1), (X_2, d_2), ..., (X_n, d_n), the *product metric* is a metric on the *Cartesian product* $X_1 \times X_2 \times ... \times X_n = \{x = (x_1, x_2, ..., x_n) \mid x_1 \in X_1, ..., x_n \in X_n\}$, defined as a function of $d_1, ..., d_n$. The simplest finite product metrics are defined by:

1. $\sum_{i=1}^{n} d_i(x_i, y_i)$;
2. $(\sum_{i=1}^{n} d_i^p(x_i, y_i))^{\frac{1}{p}}$, $1 < p < \infty$;
3. $\max_{1 \leq i \leq n} d_i(x_i, y_i)$;
4. $\sum_{i=1}^{n} \frac{1}{2^i} \frac{d_i(x_i, y_i)}{1 + d_i(x_i, y_i)}$.

The last metric is *bounded* and can be extended to the product of countably many metric spaces.

If $X_1 = ... = X_n = \mathbb{R}$, and $d_1 = ... = d_n = d$, where $d(x, y) = |x - y|$ is the *natural metric* on $\mathbb{R}$, all product metrics above induce the Euclidean topology on the n-dimensional space $\mathbb{R}^n$. They do not coincide with the Euclidean metric on $\mathbb{R}^n$, but they are equivalent to it. In particular, the set $\mathbb{R}^n$ with the Euclidean metric can be considered as the Cartesian product $\mathbb{R} \times ... \times \mathbb{R}$ of n copies of the *real line* $(\mathbb{R}, d)$ with the product metric, defined by $\sqrt{\sum_{i=1}^{n} d^2(x_i, y_i)}$.

- **Box metric**

  Let (X, d) be a metric space and I the unit interval of $\mathbb{R}$. The *box metric* d is the *product metric* on the Cartesian product $X \times I$, defined by

$$d((x_1, t_1), (x_2, t_2)) = \max\{d(x_1, x_2), |t_1 - t_2|\}.$$

- **Fréchet product metric**

  Let (X, d) be a metric space with a *bounded* metric d. Let $X^\infty = X \times ... \times X ... = \{x = (x_1, ..., x_n, ...), |, x_1 \in X_1, ..., x_n \in X_n, ...\}$ be the *countable Cartesian product space* of X.

  The *Fréchet product metric* is a *product metric* on X^∞, defined by

$$\sum_{n=1}^{\infty} A_n d(x_n, y_n),$$

  where $\sum_{n=1}^{\infty} A_n$ is any convergent series of positive terms. Usually, $A_n = \frac{1}{2^n}$ is used.

  A metric (sometimes called the *Fréchet metric*) on the set of all sequences $\{x_n\}_n$ of real (complex) numbers, defined by

$$\sum_{n=1}^{\infty} A_n \frac{|x_n - y_n|}{1 + |x_n - y_n|},$$

  where $\sum_{n=1}^{\infty} A_n$ is any convergent series of positive terms, is a Fréchet product metric of countably many copies of $\mathbb{R}$ ($\mathbb{C}$). Usually, $A_n = \frac{1}{n!}$ or $A_n = \frac{1}{2^n}$ are used.

Metrics on normed spaces

In this section we consider a special class of metrics, defined on some *normed structures*, as the norm of difference between two given elements. This structure can be a group (with a *group norm*), a vector lattice (with a *Riesz norm*), a field (with a *valuation*), etc. We list here only the most important case *norm metrics*, defined on vector spaces with a *vector norm* (or, simply, a *norm*).

- **Norm metric**

  A *norm metric* is a metric on a real (complex) vector space V, defined by

$$||x - y||,$$

  where $||.||$ is a *norm* on V, i.e., a function $||.|| : V \to \mathbb{R}$ such that, for all $x, y \in V$ and for any scalar a, we have the following properties:

  1. $||x|| \geq 0$, with $||x|| = 0$ if and only if $x = 0$;
  2. $||ax|| = |a|||x||$;
  3. $||x + y|| \leq ||x|| + ||y||$ (*triangle inequality*).

The vector space $\langle V, ||.||\rangle$ is called a *normed vector space* or, simply, *normed space*.

On any given finite-dimensional vector space all norms are equivalent. Every finite-dimensional normed space is *complete*, i.e., every its *Cauchy sequence* converges. Any metric space can be embedded isometrically in some normed vector space as a closed linearly independent subset.

A *Banach space* is a *complete* metric space $\langle V, ||x-y||\rangle$ on a vector space V with a norm metric $||x-y||$. In this case, the norm $||.||$ on V is called the *Banach norm*. Some examples of Banach spaces are:

1. l_p^n*-spaces*, l_p^∞*-spaces*, $1 \leq p \leq \infty$, $n \in \mathbb{N}$;
2. The space C of convergent numerical sequences with the norm $||x|| = \sup_n |x_n|$;
3. The space C_0 of numerical sequences which converge to zero with the norm $||x|| = \max_n |x_n|$;
4. The space $C^p_{[a,b]}$, $1 \leq p \leq \infty$, of continuous functions on $[a,b]$ with the L_p*-norm* $||f||_p = (\int_a^b |f(t)|^p dt)^{\frac{1}{p}}$;
5. The space C_K of continuous functions on a compactum K with the norm $||f|| = \max_{t\in K} |f(t)|$;
6. The space $(C_{[a,b]})^n$ of functions on $[a,b]$ with continuous derivatives up to and including the order n with the norm $||f||_n = \sum_{k=0}^n \max_{a\leq t\leq b} |f^{(k)}(t)|$;
7. The space $C^n[I^m]$ of all functions defined in an m-dimensional cube that are continuously differentiable up to and including the order n with the norm of uniform boundedness in all derivatives of order at most n;
8. The space $M_{[a,b]}$ of bounded measurable functions on $[a,b]$ with the norm
$$||f|| = ess \sup_{a\leq t\leq b} |f(t)| = \inf_{e, \mu(e)=0} \sup_{t\in[a,b]\setminus e} |f(t)|;$$
9. The space $A(\Delta)$ of functions analytic in the open *unit disk* $\Delta = \{z \in \mathbb{C} \,|\, |z| < 1\}$ and continuous in the closed disk $\overline{\Delta}$ with the norm $||f|| = \max_{z\in\overline{\Delta}} |f(z)|$;
10. The *Lebesgue spaces* $L_p(\Omega)$, $1 \leq p \leq \infty$;
11. The *Sobolev spaces* $W^{k,p}(\Omega)$, $\Omega \subset \mathbb{R}^n$, $1 \leq p \leq \infty$, of functions f on Ω such that f and its derivatives, up to some order k, have a finite L_p*-norm*, with the norm $||f||_{k,p} = \sum_{i=0}^k ||f^{(i)}||_p$;
12. The *Bohr space* AP of almost periodic functions with the norm
$$||f|| = \sup_{-\infty<t<+\infty} |f(t)|.$$

A finite-dimensional real Banach space is called a *Minkowskian space*. A norm metric of a Minkowskian space is called a *Minkowskian metric*. In particular, any l_p*-metric* is a Minkowskian metric.

All n-dimensional Banach spaces are pairwise isomorphic; the set of such spaces becomes compact if one introduces the *Banach-Mazur distance* by
$$d_{BM}(V,W) = \ln \inf_T ||T|| \cdot ||T^{-1}||,$$
where the infimum is taken over all operators which realize an isomorphism $T : V \to W$.

- **l_p-metric**

The *l_p-metric* d_{l_p}, $1 \le p \le \infty$, is a norm metric on $\mathbb{R}^n$ (on $\mathbb{C}^n$), defined by

$$||x - y||_p,$$

where the *l_p-norm* $||.||_p$ is defined by

$$||x||_p = \left(\sum_{i=1}^{n} |x_i|^p \right)^{\frac{1}{p}}.$$

For $p = \infty$, we obtain $||x||_\infty = \lim_{p \to \infty} \sqrt[p]{\sum_{i=1}^{n} |x_i|^p} = \max_{1 \le i \le n} |x_i|$. The metric space $(\mathbb{R}^n, d_{l_p})$ is abbreviated as l_p^n and is called *l_p^n-space.*

The *l_p-metric,* $1 \le p \le \infty$, on the set of all sequences $x = \{x_n\}_{n=1}^{\infty}$ of real (complex) numbers, for which the sum $\sum_{i=1}^{\infty} |x_i|^p$ (for $p = \infty$, the sum $\sum_{i=1}^{\infty} |x_i|$) is finite, is defined by

$$\left(\sum_{i=1}^{\infty} |x_i - y_i|^p \right)^{\frac{1}{p}}.$$

For $p = \infty$, we obtain $\max_{i \ge 1} |x_i - y_i|$. This metric space is abbreviated as l_p^∞ and is called *l_p^∞-space.*

Most important are l_1-, l_2- and l_∞-metrics; the l_2-metric on $\mathbb{R}^n$ is also called the *Euclidean metric.* The l_2-metric on the set of all sequences $\{x_n\}_n$ of real (complex) numbers, for which $\sum_{i=1}^{\infty} |x_i|^2 < \infty$, is also known as the *Hilbert metric.* On $\mathbb{R}$ all l_p-metrics coincide with the *natural metric* $|x - y|$.

- **Euclidean metric**

The *Euclidean metric* (*Pythagorean distance, as-the-crow-flies distance, beeline distance*) d_E is the metric on $\mathbb{R}^n$, defined by

$$d_E(x, y) = ||x - y||_2 = \sqrt{(x_1 - y_1)^2 + ... + (x_n - y_n)^2}.$$

It is the ordinary *l_2-metric* on $\mathbb{R}^n$. The metric space $(\mathbb{R}^n, d_E)$ is abbreviated as $\mathbb{E}^n$ and is called *Euclidean space* (*real Euclidean space*). Sometimes, the expression "Euclidean space" stands for the case $n = 3$, as opposed to the *Euclidean plane* for the case $n = 2$. The *Euclidean line* (*real line*) is obtained for $n = 1$, i.e., it is the metric space $(\mathbb{R}, |x - y|)$ with the *natural metric* $|x - y|$.

In fact, $\mathbb{E}^n$ is an *inner product space* (and even a *Hilbert space*), i.e.,

$$d_E(x, y) = ||x - y||_2 = \sqrt{\langle x - y, x - y \rangle},$$

where $\langle x, y \rangle$ is the *inner product* on $\mathbb{R}^n$ which is given in a suitably chosen (Cartesian) coordinate system by the formula $\langle x, y \rangle = \sum_{i=1}^{n} x_i y_i$. In a standard coordinate system one has $\langle x, y \rangle = \sum_{i,j} g_{ij} x_i y_j$, where $g_{ij} = \langle e_i, e_j \rangle$, $e_1, \dots, e_n$ constitute the *standard basis* of $\mathbb{R}^n$ and the *metric tensor* $((g_{ij}))$ is a positive-definite symmetric $n \times n$ matrix.

In general, a Euclidean space is defined as a space, the properties of which are described by the axioms of *Euclidean Geometry.*

- **Unitary metric**

The *unitary metric* (*complex Euclidean metric*) is the *l_2-metric* on $\mathbb{C}^n$, defined by

$$||x-y||_2 = \sqrt{|x_1-y_1|^2 + \ldots + |x_n-y_n|^2}.$$

The metric space $(\mathbb{C}^n, ||x-y||_2)$ is called the *unitary space* (*complex Euclidean space*). For $n=1$, we obtain the *complex plane* (*Argand plane*), i.e., the metric space $(\mathbb{C}, |z-u|)$ with the *complex modulus metric* $|z-u|$; here $|z| = |z_1+z_2i| = \sqrt{z_1^2+z_2^2}$ is the *complex modulus*.

- **Norm-related metrics on $\mathbb{R}^n$**

On the vector space $\mathbb{R}^n$, there are many well-known metrics related to a given norm $||.||$ on $\mathbb{R}^n$, especially, to the Euclidean norm $||.||_2$. Some examples are given below.

1. The *British Rail metric*, defined by

$$||x|| + ||y||$$

for $x \neq y$ (and is equal to 0, otherwise).

2. The *radar screen metric*, defined by

$$\min\{1, ||x-y||\}.$$

3. The (p,q)*-relative metric*, defined by

$$\frac{||x-y||_2}{(\frac{1}{2}(||x||_2^p + ||y||_2^p))^{\frac{q}{p}}}$$

for x or $y \neq 0$ (and equal to 0, otherwise), where $0 < q \leq 1$, and $p \geq \max\{1-q, \frac{2-q}{3}\}$. For $q=1$ and any $1 \leq p < \infty$, one obtains the *p-relative metric*; for $q=1$ and $p=\infty$, one obtains the *relative metric*.

4. The *M-relative metric*, defined by

$$\frac{||x-y||_2}{f(||x||_2)\cdot f(||y||_2)}$$

for x or $y \neq 0$, where $f:[0,\infty) \to (0,\infty)$ is a convex increasing function, such that $\frac{f(x)}{x}$ is decreasing for $x>0$. In particular, the distance $\frac{||x-y||_2}{\sqrt[p]{1+||x||_2^p}\sqrt[p]{1+||y||_2^p}}$ is a metric on $\mathbb{R}^n$ if and only if $p \geq 1$. A similar metric on $\mathbb{R}^n \backslash \{0\}$ can be defined by $\frac{||x-y||_2}{||x||_2 \cdot ||y||_2}$.

- **L_p-metric**

An *L_p-metric* d_{L_p}, $1 \leq p \leq \infty$, is a norm metric on $L_p(\Omega, \mathcal{A}, \mu)$, defined by

$$||f-g||_p$$

for any $f,g \in L_p(\Omega,\mathcal{A},\mu)$. The metric space $(L_p(\Omega,\mathcal{A},\mu), d_{L_p})$ is called the *L_p-space* (*Lebesgue space*).

Here, as before, Ω is a set, $\mathcal{A}$ is an *σ-algebra* of subsets of Ω (a collection of subsets of Ω, such that it holds $\Omega \in \mathcal{A}$; $A \in \mathcal{A} \Rightarrow \Omega \backslash A \in \mathcal{A}$; $A = \cup_{i=1}^{\infty} A_i$ with $A_i \in \mathcal{A} \Rightarrow A \in \mathcal{A}$), and μ is a *measure* on $\mathcal{A}$: a function $\mu : \mathcal{A} \to \mathbb{R}_{\geq 0}$ which is *additive*, i.e., $\mu(\cup_{i \geq 1} A_i) = \sum_{i \geq 1} \mu(A_i)$ for all pairwise disjoint sets $A_i \in \mathcal{A}$, and satisfies $\mu(\emptyset) = 0$.

Given a function $f : \Omega \to \mathbb{R}(\mathbb{C})$, its *$L_p$-norm* is defined by

$$||f||_p = \left(\int_{\Omega} |f(\omega)|^p \mu(d\omega) \right)^{\frac{1}{p}}.$$

Let $L_p(\Omega, \mathcal{A}, \mu) = L_p(\Omega)$ denote the set of all functions $f : \Omega \to \mathbb{R}$ ($\mathbb{C}$) such that $||f||_p < \infty$. Strictly speaking, $L_p(\Omega, \mathcal{A}, \mu)$ consists of equivalence classes of functions, where two functions are *equivalent* if they are equal *almost everywhere*, i.e., if the set on which they differ has measure zero. The set $L_{\infty}(\Omega, \mathcal{A}, \mu)$ is the set of equivalence classes of measurable functions $f : \Omega \to \mathbb{R}$ ($\mathbb{C}$) whose absolute values are bounded almost everywhere.

The most classical example of an L_p-metric is d_{L_p} on the set $L_p(\Omega, \mathcal{A}, \mu)$, where Ω is the open interval $(0, 1)$, $\mathcal{A}$ is the *Borel σ-algebra* on $(0, 1)$, and μ is the *Lebesgue measure*. This metric space is abbreviated by $L_p(0, 1)$ and is called *$L_p(0, 1)$-space*.

In the same way, one can define the L_p-metric on the set $C_{[a,b]}$ of all real (complex) continuous functions on $[a, b]$:

$$d_{L_p}(f, g) = \left(\int_a^b |f(x) - g(x)|^p dx \right)^{\frac{1}{p}}.$$

For $p = \infty$,

$$d_{L_\infty}(f, g) = \max_{a \leq x \leq b} |f(x) - g(x)|.$$

This metric space is abbreviated by $C_{[a,b]}^p$ and is called *$C_{[a,b]}^p$-space*.

If $\Omega = \mathbb{N}$, $\mathcal{A} = 2^{\Omega}$ is the collection of all subsets of Ω, and μ is the *cardinality measure* (i.e., $\mu(A) = |A|$ if A is a finite subset of Ω, and $\mu(A) = \infty$, otherwise), then the metric space $(L_p(\Omega, 2^{\Omega}, |.|), d_{L_p})$ coincides with the space l_p^{∞}.

If $\Omega = V_n$ is a set of cardinality n, $\mathcal{A} = 2^{V_n}$, and μ is the cardinality measure, then the metric space $(L_p(V_n, 2^{V_n}, |.|), d_{L_p})$ coincides with the space l_p^n.

Metrics on numbers

Here we consider some of the most important metrics on the classical Number Systems: the semiring $\mathbb{N}$ of positive integers, the ring $\mathbb{Z}$ of integers, and the fields $\mathbb{Q}$, $\mathbb{R}$, and $\mathbb{C}$ of rational, real and complex numbers, respectively.

- **Metrics on $\mathbb{N}$**

  There are several well-known metrics on the set $\mathbb{N}$ of positive integers:

  1. $|n - m|$; the restriction of the *natural metric* (from $\mathbb{R}$) on $\mathbb{N}$;
  2. $p^{-\alpha}$, where α is the highest power of a given prime number p dividing $m - n$, for $m \neq n$ (and equal to 0 for $m = n$); the restriction of the *p-adic metric* (from $\mathbb{Q}$) on $\mathbb{N}$;

3. $\ln \frac{l.c.m.(m,n)}{g.c.d.(m,n)}$; an example of the *lattice valuation metric*;
4. $w_r(n-m)$, where $w_r(n)$ is the *arithmetic r-weight* of n; the restriction of the *arithmetic r-norm metric* (from $\mathbb{Z}$) on $\mathbb{N}$;
5. $\frac{|n-m|}{mn}$; this metric space is not complete (for example, $\{n\}_n$ is a non-convergent Cauchy sequence) and has discrete topology;
6. $1+\frac{1}{m+n}$ for $m \neq n$ (and equal to 0 for $m=n$); the *Sierpinski metric*.

Most of these metrics on $\mathbb{N}$ can be extended on $\mathbb{Z}$. Moreover, any one of the above metrics can be used in the case of an arbitrary countable set X. For example, the *Sierpinski metric* is defined, in general, on a countable set $X = \{x_n \,|\, n \in \mathbb{N}\}$ by $1+\frac{1}{m+n}$ for all $x_m, x_n \in X$ with $m \neq n$ (and is equal to 0, otherwise).

- **Metrics on $\mathbb{Z}$**

Besides of the metrics on $\mathbb{Z}$, which can be obtained as an extension of some metric on $\mathbb{N}$, listed above, and, for example, the natural metric $|x-y|$ on $\mathbb{Z}$, which is the restriction on $\mathbb{Z}$ of the natural metric from $\mathbb{R}$, the *arithmetic r-norm metric* (see, for example, [Ernv85]) is a metric on $\mathbb{Z}$, defined by

$$w_r(x-y),$$

where the *arithmetic r-weight* $w_r(x)$ of an integer x is the number of non-zero coefficients in a *minimal r-ary form* of x, in particular, in the generalized non-adjacent form.

Let $r \in \mathbb{N}, r \geq 2$. The *modified r-ary form* of an integer x is a representation

$$x = e_n r^n + ... + e_1 r + e_0,$$

where $e_i \in \mathbb{Z}$, and $|e_i| < r$ for all $i = 0, ..., n$. An r-ary form is called *minimal* if the number of non-zero coefficients is minimal. The minimal form is not unique, in general. But if the coefficients e_i, $0 \leq i \leq n-1$, satisfy the conditions $|e_i + e_{i+1}| < r$, and $|e_i| < |e_{i+1}|$ if $e_i e_{i+1} < 0$, then the above form is unique and minimal; it is called *generalized non-adjacent form.*

- **Metrics on $\mathbb{Q}$**

Besides of the metrics on $\mathbb{Q}$, which can be obtained as an extension of some metric on $\mathbb{Z}$, listed above, and, for example, the natural metric $|x-y|$ on $\mathbb{Q}$, which is the restriction on $\mathbb{Q}$ of the natural metric from $\mathbb{R}$, the most important metric on $\mathbb{Q}$ is the *p-adic metric.*

Let p be a prime number. Any non-zero rational number x can be represented as $x = p^\alpha \frac{c}{d}$, where c and d are integers not divisible by p, and α is an unique integer. The *p-adic norm* of x is defined by $|x|_p = p^{-\alpha}$. Moreover, $|0|_p = 0$ is defined.

The *p-adic metric* is a *norm metric* on the set $\mathbb{Q}$ of rational numbers, defined by

$$|x-y|_p.$$

This metric forms the basis for the algebra of p-adic numbers. In fact, the *Cauchy completion* of the metric space $(\mathbb{Q}, |x-y|_p)$ gives the field $\mathbb{Q}_p$ of *p-adic numbers*;

also the Cauchy completion of the metric space $(\mathbb{Q}, |x-y|)$ with the *natural metric* $|x-y|$ gives the field $\mathbb{R}$ of real numbers. It is an *ultrametric*, i.e., for any rational numbers x, y, z, it holds:

$$|x-y|_p \leq \max\{|x-z|_p, |z-y|_p\}.$$

The *Gajić metric* is an *ultrametric* on the set $\mathbb{Q}$ of rational numbers defined, for $x \neq y$ (via the integer part $\lfloor z \rfloor$ of a real number z), by

$$\inf\{2^{-n} \,|\, n \in \mathbb{Z},\ \lfloor 2^n(x-e) \rfloor = \lfloor 2^n(y-e) \rfloor\},$$

where e is any fixed irrational number. This metric is *equivalent* to the natural metric $|x-y|$ on $\mathbb{Q}$.

- **Metrics on $\mathbb{R}$**

  The most used metric on $\mathbb{R}$ is the *natural metric* (*absolute value metric, the distance between numbers*): a metric d on $\mathbb{R}$, defined by

$$d(x,y) = |x-y| = \begin{cases} y-x, & \text{if } x-y<0, \\ x-y, & \text{if } x-y \geq 0. \end{cases}$$

  On $\mathbb{R}$ all l_p*-metrics*, including the *Euclidean metric*, coincide with the natural metric. The metric space $(\mathbb{R}, |x-y|)$ is called the *real line* (*Euclidean line*).

  There exist many other metrics on $\mathbb{R}$, coming from $|x-y|$ by some *metric transform*. For example: $\min\{1, |x-y|\}$, $\frac{|x-y|}{1+|x-y|}$, $|x|+|x-y|+|y|$ (for $x \neq y$), and, for a given $0 < \alpha < 1$, the *generalized absolute value metric* $|x-y|^\alpha$.

Metrics on real and digital planes

In the plane $\mathbb{R}^2$ we can use many various metrics. In particular, any l_p*-metric* (as well as any *norm metric* for a given norm $||.||$ on $\mathbb{R}^n$) can be used on the plane, and the most natural is the l_2-metric, i.e., the *Euclidean metric* $d_E(x,y) = \sqrt{(x_1-y_1)^2 + (x_2-y_2)^2}$, which gives the length of the straight line segment $[x,y]$, and is the *intrinsic metric* of the plane.

However, there are other, often "exotic", metrics on $\mathbb{R}^2$. Many of them are used for the construction of *generalized Voronoi diagrams* on $\mathbb{R}^2$ (see, for example, *Moscow metric, network metric, nice metric*). Some of them are used in Digital Geometry.

- **City-block metric**

  The *city-block metric* is the l_1*-metric* on $\mathbb{R}^2$, defined by

$$||x-y||_1 = |x_1-y_1| + |x_2-y_2|.$$

  This metric has many different names, for example, it is called *taxicab metric, Manhattan metric, rectilinear metric, right-angle metric*; on $\mathbb{Z}^2$ it is called *greed metric*, and 4*-metric*.

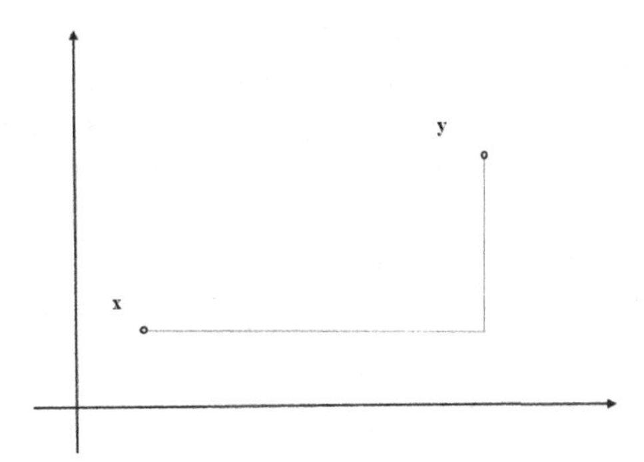

Figure 3.3: City-block metric

- **Chebyshev metric**

  The *Chebyshev metric* is the l_∞-*metric* on $\mathbb{R}^2$, defined by

$$||x-y||_\infty = \max\{|x_1-y_1|, |x_2-y_2|\}.$$

  This metric is called also *uniform metric*, *sup metric*, and *box metric*; on $\mathbb{Z}^2$ it is called *lattice metric*, *chessboard metric*, *king-move metric*, and 8-*metric*.

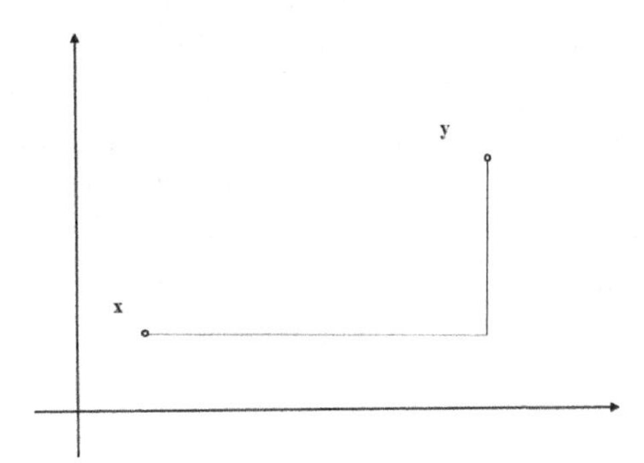

Figure 3.4: Chebyshev metric

- **French Metro metric**

  The *French metro metric* is a metric on $\mathbb{R}^2$, defined by

$$||x-y||$$

  if $x = cy$ for some $c \in \mathbb{R}$, and by

$$||x|| + ||y||,$$

  otherwise, where $||.||$ is a *norm* on $\mathbb{R}^2$.

For the Euclidean norm $||.||_2$, it is called *hedgehog metric*, *Paris metric*, or *radial metric*. In this case it can be defined as the minimum Euclidean length of all *admissible* connecting curves between two given points x and y, where a curve is called *admissible* if it consists of only segments of straight lines passing through the origin.

In graph terms, this metric is similar to the *path metric* of the tree consisting of a point from which radiate several disjoint paths.

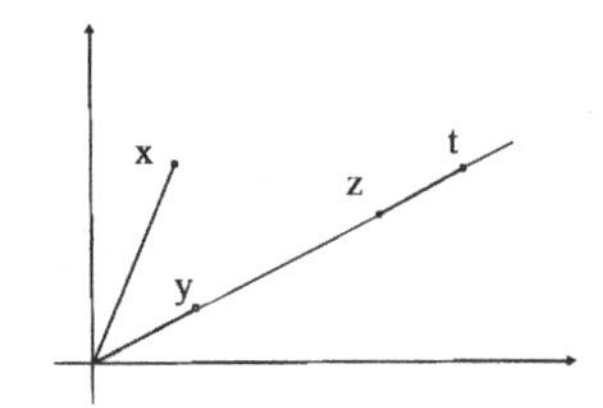

Figure 3.5: French Metro metric

- **Moscow metric**

  The *Moscow metric* (*Karlsruhe metric*) is a metric on $\mathbb{R}^2$, defined as the minimum Euclidean length of all *admissible* connecting curves between x and $y \in \mathbb{R}^2$, where a curve is called *admissible* if it consists of only segments of straight lines passing through the origin, and of segments of circles centered at the origin (see, for example, [Klei88]).

  If the polar coordinates for points $x, y \in \mathbb{R}^2$ are (r_x, θ_x), (r_y, θ_y), respectively, then the distance between them is equal to $\min\{r_x, r_y\}\Delta(\theta_x, \theta_y) + |r_x - r_y|$, if $0 \leq \Delta(\theta_x, \theta_y) < 2$, and is equal to $r_x + r_y$, if $2 \leq \Delta(\theta_x, \theta_y) < \pi$, where

$$\Delta(\theta_x, \theta_y) = \min\{|\theta_x - \theta_y|, 2\pi - |\theta_x - \theta_y|\}, \theta_x, \theta_y \in [0, 2\pi),$$

  is the *metric between angles*.

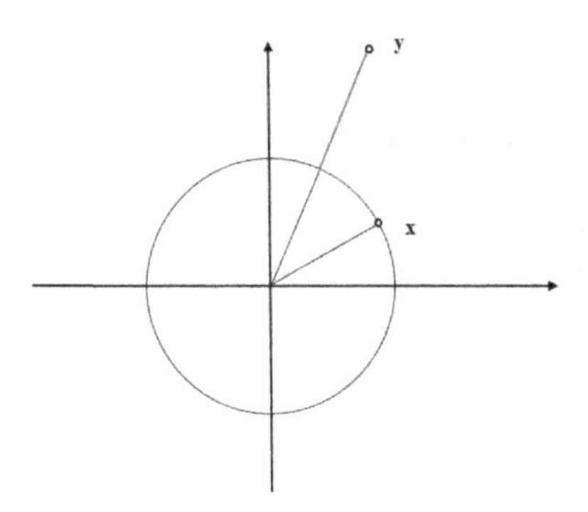

Figure 3.6: Moscow metric

- **Lift metric**

The *lift metric* (*raspberry picker metric*) is a metric on $\mathbb{R}^2$, defined by

$$|x_1 - y_1|,$$

if $x_2 = y_2$, and by

$$|x_1| + |x_2 - y_2| + |y_1|,$$

if $x_2 \neq y_2$ (see, for example, [Brya85]).

It can be defined as the minimum Euclidean length of all *admissible* connecting curves between two given points x and y, where a curve is called *admissible* if it consists of only segments of straight lines parallel to x_1-axis, and of segments of x_2-axis.

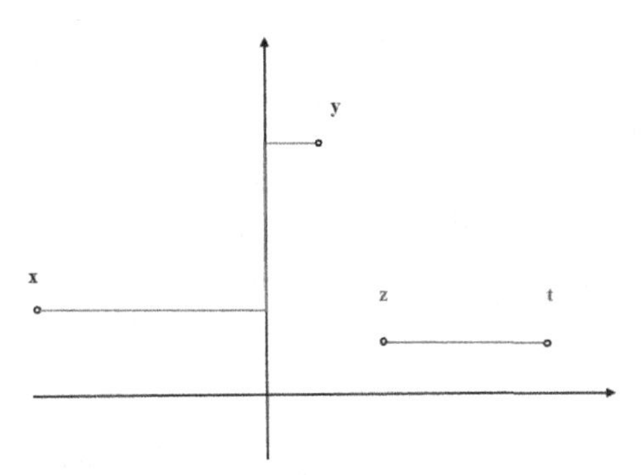

Figure 3.7: Lift metric

- **Flower-shop metric**

The *flower-shop metric* is a metric on $\mathbb{R}^2$, defined by

$$d(x, f) + d(f, y)$$

for $x \neq y$ (and is equal to 0, otherwise). Here d is a metric on $\mathbb{R}^2$, and f is a fixed point (a *flower-shop*) in the plane.

So, a person living at point x, who wants to visit someone else living at point y, first goes to f, to buy some flowers. In the case $d(x, y) = ||x - y||$ and $f = (0, 0)$, it is the *British rail metric*. In this case it is also called *caterpillar metric*, and *shuttle metric*. For the Euclidean norm $||.||_2$ it is called *post-office metric*.

If $k > 1$ flower-shops $f_1, ..., f_k$ are available, one buys the flowers, where the detour is a minimum, i.e., the distance between distinct points x, y is equal to

$$\min_{1 \leq i \leq k} (d(x, f_i) + d(f_i, y)).$$

A *digital metric* is any metric on a digital nD space. Usually, it should take integer values. The metrics on $\mathbb{Z}^n$ that are mainly used are the l_1-*metric* (called here *taxicab metric*, *Manhattan metric*, *rectilinear metric*, *right-angle metric*, *greed metric*, *4-metric*) and l_∞-*metric* (which is called also *uniform metric*, *sup metric*, *box metric*, *lattice metric*, *chessboard metric*, *king-move metric*, *8-metric*), as well as the l_2-*metric* after rounding to the nearest upper (lower) integer. In general, given a list of *neighbors* of a pixel, it can be seen as a list of permitted *one-step moves* on $\mathbb{Z}^2$. Let associate a *prime distance*, i.e., a positive weight, to each type of such move. Many digital metrics can be obtained now as the minimum, over all admissible paths (i.e., sequences of permitted moves), of the sum of corresponding prime distances.

In practice, the subset $(\mathbb{Z}_m)^n = \{0, 1, ..., m-1\}^n$ is considered instead of the full space $\mathbb{Z}^n$. $(\mathbb{Z}_m)^2$ and $(\mathbb{Z}_m)^3$ are called *m-grill* and *m-framework*, respectively.

The most used metrics on $(\mathbb{Z}_m)^n$ are the *Hamming metric*

$$|\{i \mid 1 \le i \le n, x_i \ne y_i\}|,$$

and the *Lee metric*

$$\sum_{1 \le i \le n} \min\{|x_i - y_i|, m - |x_i - y_i|\}.$$

The metric space (Z_m^n, d_{Lee}) is a discrete analog of the *elliptic space*.

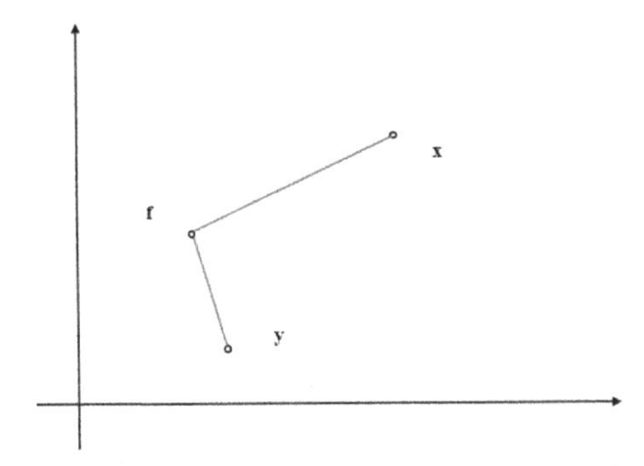

Figure 3.8: Flower-shop metric

- **Knight metric**

  The *knight metric* is a metric on $\mathbb{Z}^2$, defined as the minimum number of moves a chess knight would take to travel from x to $y \in \mathbb{Z}^2$. Its *unit sphere* S^1_{knight}, centered at the origin, contains exactly 8 integral points $\{(\pm 2, \pm 1), (\pm 1, \pm 2)\}$, and can be written as $S^1_{knight} = S^3_{l_1} \cap S^2_{l_\infty}$, where $S^3_{l_1}$ denotes the l_1-sphere of radius 3, and $S^2_{l_\infty}$ denotes the l_∞-sphere of radius 2, centered at the origin (see [DaCh88]).

  The distance between x and y is equal to 3 if $(M, m) = (1, 0)$, is equal to 4 if $(M, m) = (2, 2)$, and is equal to $\max\{\lceil \frac{M}{2} \rceil, \lceil \frac{M+m}{3} \rceil\} + (M+m) - \max\{\lceil \frac{M}{2} \rceil, \lceil \frac{M+m}{3} \rceil\}$ $(mod\ 2)$, otherwise, where $M = \max\{|u_1|, |u_2|\}$, $m = \min\{|u_1|, |u_2|\}$, $u_1 = x_1 - y_1$, $u_2 = x_2 - y_2$.

- **Super-knight metric**

  An (p, q)*-super-knight* ((p, q)*-leaper*) is a (variant) chess piece a move of which consists of a leap p squares in one orthogonal direction followed by a 90 degree direction change, and q squares leap to the destination square. (Here $p, q \in \mathbb{N}$ such that $p + q$ is odd, and $(p, q) = 1$.) Chess-variant terms exist for an $(p, 1)$-leaper with $p = 0, 1, 2, 3, 4$ (*Wazir*, *Ferz*, usual *Knight*, *Camel*, *Giraffe*), and for an $(p, 2)$-leaper with $p = 0, 1, 2, 3$ (*Dabbaba*, usual *Knight*, *Alfil*, *Zebra*).

  An (p, q)*-super-knight metric* ((p, q)*-leaper metric*) is a metric on $\mathbb{Z}^2$, defined as the minimum number of moves a chess (p, q)-super-knight would take to travel from x to $y \in \mathbb{Z}^2$. Thus, its *unit sphere* $S^1_{p,q}$, centered at the origin, contains exactly 8 integral points $\{(\pm p, \pm q), (\pm q, \pm p)\}$. (See [DaMu90].)

  The *knight metric* is the $(1, 2)$-super-knight metric. The *city-block metric* can be considered as the *Wazir metric*, i.e., $(0, 1)$-super-knight metric.

- **Rook metric**

  The *rook metric* is a metric on $\mathbb{Z}^2$, defined as the minimum number of moves a chess rook would take to travel from x to $y \in \mathbb{Z}^2$. This metric can take only the values $\{0, 1, 2\}$, and coincides with *Hamming metric* on $\mathbb{Z}^2$. (Remind, that the *Hamming metric* d_H is a metric on $\mathbb{R}^n$, defined by $|\{i \mid 1 \leq i \leq n, x_i \neq y_i\}|$ for all $x, y \in \mathbb{R}^n$. On binary vectors $x, y \in \{0, 1\}^n$ the Hamming metric and the l_1*-metric* coincide.)

Many applications consider *chamfer metric* on $\mathbb{Z}^2$ (in general, on $\mathbb{Z}^n$), which is defined as follow.

- **Chamfer metric**

  Given two positive numbers α, β with $\alpha \leq \beta < 2\alpha$, consider (α, β)*-weighted* l_∞*-grid*, i.e., the infinite graph with the vertex set $\mathbb{Z}^2$, two vertices being adjacent if their l_∞-distance is one, while horizontal/vertical and diagonal edges having *weights* α and β, respectively.

  A *chamfer metric* ((α, β)*-chamfer metric*, [Borg86]) is the *weighted path metric* in this graph. For any $x, y \in \mathbb{Z}^2$ it can be written as

  $$\beta m + \alpha(M - m),$$

  where $M = \max\{|u_1|, |u_2|\}$, $m = \min\{|u_1|, |u_2|\}$, $u_1 = x_1 - y_1$, $u_2 = x_2 - y_2$.

  If the weights α and β are equal to the Euclidean lengths $1, \sqrt{2}$ of horizontal/vertical and diagonal edges, respectively, then one obtains the Euclidean length of the shortest chessboard path between x and y. If $\alpha = \beta = 1$, one obtains the *chessboard metric*. The $(3, 4)$-chamfer metric is the most used one for digital images; it is called simply $(3, 4)$*-metric*.

  An $3D$*-chamfer metric* is the *weighted path metric* of the graph with the vertex set $\mathbb{Z}^3$ of *voxels*, two voxels being adjacent if their l_∞-distance is one, while weights α, β and γ are associated, respectively, to the distance from 6 face neighbors, 12 edge neighbors, and 8 corner neighbors.

Chapter 4

Oriented case: quasi-metrics

4.1 Preliminaries

The notions of directed distances, quasi-metrics and oriented cuts are generalizations of the notions of distances, metrics and cuts, respectively (see, for example, [DeLa97]), which are central objects in Graph Theory, Combinatorial Optimization and, more generally, in Discrete Mathematics.

Oriented (directed, asymmetric) distances are encountered very often, for example, these are one-way transport routes, rivers with quick flow and so on.

Quasi-metrics are used in Semantics of Computation (see, for example, Seda (1997) [Seda97]) and are of interest in Computational Geometry (see, for example, Aichholzer, Aurenhammer, Chen, Lee, Mukhopadnyay and Papadopoulou (1997) [AACLMP9797]).

Asymmetric definition of distance have been used already in Hausdorff (1914) [Haus14], pages 145–146, but the first detailed topological analysis of quasi-metrics was given by Wilson (1931) [Wils31].

For general theory of quasi-metrics, see, for example, Fréchet (1906) [Frec06], Hausdorff (1914) [Haus14], and Stoltenberg (1969) [Stol69]. For theory of finite quasi-metrics, quasi-semimetrics, oriented multicuts and their polyhedral aspects, see Deza and Panteleeva (2000) [DePa99], Deza, Dutour and Panteleeva (2003) [DDP03], Deza, Deza and Grishukhin (2010), [DGD12].

4.2 Definitions

Definition 4.1 *Let X be a set. A function $q : X \times X \to \mathbb{R}$ is called quasi-distance on X if for all $x, y \in X$ it holds:*

1. *$q(x, y) \geq 0$ (non-negativity);*
2. *$q(x, x) = 0$.*

In Topology, it is also called a *premetric.*

A *quasi-distance space* (X, q) is a set X equipped with a quasi-distance q.

Since $q(x, x) = 0$, we can view a quasi-distance q on the set $V_n = \{1, 2, ..., n\}$ as a vector $(q_{ij})_{1 \leq i \neq j \leq n} \in \mathbb{R}^{I_n}$, where $I_n = n(n-1)$; alternatively, q can be represented as an (in general, non-symmetric) $n \times n$ matrix with $q_{ii} = 0$ on the main diagonal: $q(x, y) = q_{ij}$.

On the other hand, any $n \times n$ matrix $((q_{ij}))$ with non-negative entries and with zeros on the main diagonal (any vector $(q_{ij})_{1 \le i \neq j \le n} \in \mathbb{R}^{I_n}$ with non-negative coordinates) can be considered as a quasi-distance on the set V_n: $q(i,j) = q_{ij}$.

For any quasi-distance q, the function $D(x,y) = q(x,y) + q(y,x)$ is a distance; it called the *symmetrization distance* of q.

If we add the *oriented triangle axiom* to the properties of a distance, we obtain the notion of *quasi-semimetric.*

Definition 4.2 *Let X be a set. A function $q : X \times X \to \mathbb{R}$ is called quasi-semimetric (weak metric) on X if for all $x, y, z \in X$ it holds:*

1. *$q(x,y) \ge 0$ (non-negativity);*
2. *$q(x,x) = 0$;*
3. *$q(x,y) \le q(x,z) + q(z,y)$ (oriented triangle inequality).*

A *quasi-semimetric space* (X, q) is a set X equipped with a quasi-semimetric q.

If in the symmetric case the non-negativity property of a semimetric follows from the triangle inequality and the symmetry, for a quasi-semimetric it is not a case. However, for a quasi-distance q, the *strong triangle inequality*

$$q(x,y) \le q(x,z) + q(y,z)$$

implies that q is symmetric and so, is a semimetric.

A quasi-semimetric q on V_n can be viewed alternatively as a vector $(q_{ij})_{1 \le i \neq j \le n} \in \mathbb{R}^{I_n}$, or as an $n \times n$ (in general, non-symmetric) matrix with $q_{ii} = 0$ on the main diagonal: $q(i,j) = q_{ij}$.

A *weak quasi-metric* is a quasi-semimetric q on X with *weak symmetry*, i.e., for all $x, y \in X$ the equality $q(x,y) = 0$ implies $q(y,x) = 0$.

An *Albert quasi-metric* is a quasi-semimetric q on X with *weak definiteness*, i.e., for all $x, y \in X$, the equality $q(x,y) = q(y,x) = 0$ implies $x = y$.

For a quasi-semimetric space (X, q), the set X can be partially ordered by the *specialization order*: $x \preceq y$ if and only if $q(x,y) = 0$.

For any quasi-semimetric q, the function $D(x,y) = q(x,y) + q(y,x)$ is a semimetric: the *symmetrization semimetric* of d.

If we add the *separation axiom* to the properties of a quasi-semimetric, we obtain the notion of *quasi-metric.*

Definition 4.3 *Let X be a set. A function $q : X \times X \to \mathbb{R}$ is called quasi-metric on X if for all $x, y, z \in X$ it holds:*

1. *$q(x,y) \ge 0$ (non-negativity);*
2. *$q(x,y) = 0$ if and only if $x = y$ (definiteness, or separation axiom);*
3. *$q(x,y) \le q(x,z) + q(z,y)$ (oriented triangle inequality).*

A *quasi-metric space* (X, q) is a set X equipped with a quasi-metric q.

A *non-Archimedean quasi-metric* q is a quasi-distance on X which satisfies the following strengthened version of the oriented triangle inequality:

$$q(x,y) \leq \max\{q(x,z), q(z,y)\}.$$

A quasi-metric q on V_n can be viewed alternatively as a vector $(q_{ij})_{1\leq i\neq j\leq n} \in \mathbb{R}^{I_n}$ with positive coordinates, or as $n \times n$ matrix with $q_{ii} = 0$ on the main diagonal, and with $q_{ij} > 0$ for $i \neq j$: $q(i,j) = q_{ij}$.

For any quasi-metric q, the function $D(x,y) = q(x,y) + q(y,x)$ is a metric; it called the *symmetrization metric* of q.

For any quasi-metric q, the functions $\max\{q(x,y), q(y,x)\}$, $\min\{q(x,y), q(y,x)\}$ and $\frac{1}{2}(q^p(x,y) + q^p(y,x))^{\frac{1}{p}}$ with $p \geq 1$ (usually, $p = 1$ is taken) are metrics.

It is known (see Proposition 8 in [LLR94]), that any quasi-metric on n points embeds isometrically into $\mathbb{R}^n$ equipped with some directed norm.

4.3 Examples

Basic examples

Example 1. Discrete quasi-semimetric. Let $X = \{1, 2, ..., n\}$. The *discrete quasi-semimetric* (not a quasi-metric) q on X is defined as

$$q(x,y) = \begin{cases} 1, & \text{if } x > y, \\ 0, & \text{otherwise.} \end{cases}$$

Example 2. Hausdorff quasi-metric. Let (X, D) be a finite metric space and let $X = X_1 \cup \ldots \cup X_n$ be a *partition* of X, i.e., a decomposition of X into the union of pairwise disjoint sets. The *Hausdorff quasi-metric* q on the set $Y = \{X_1, \ldots, X_n\}$ is defined by

$$q(X_i, X_j) := \min_{x\in X_i} \max_{y\in X_j} D(x,y).$$

Example 3. Asymmetric difference quasi-metric. Given any anti-chain of sets $Z = \{x, y, z, ... \mid x \not\subset y \text{ for all } x \neq y\}$, the *asymmetric difference quasi-metric* (*measure quasi-metric*) q is a quasi-metric on Z, defined by

$$q(x,y) = |x \backslash y|.$$

It is the cardinality of the asymmetric difference $x\backslash y$ of sets x and y from Z.

In fact, $|x\backslash y| \geq 0$, $|x\backslash x| = 0$, $|x\backslash z| - |x\backslash y| - |y\backslash z| = -|(x\cap z)\backslash y| - |y\backslash(x\cup z)| \leq 0$.

If $P(V_n)$ is the set of all subsets of $V_n = \{1, 2, ..., n\}$, then the function $|x\backslash y|$ - the cardinality of the asymmetric difference of sets x and y from $P(V_n)$ - is only a quasi-semimetric (not a quasi-metric) on $P(V_n)$.

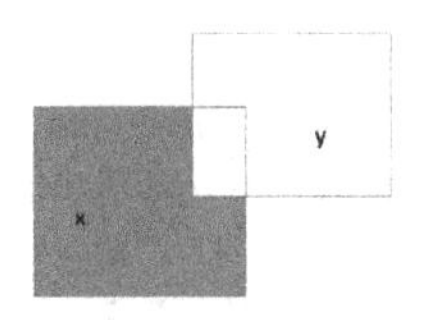

Figure 4.1: Asymmetric difference: $x \backslash y$

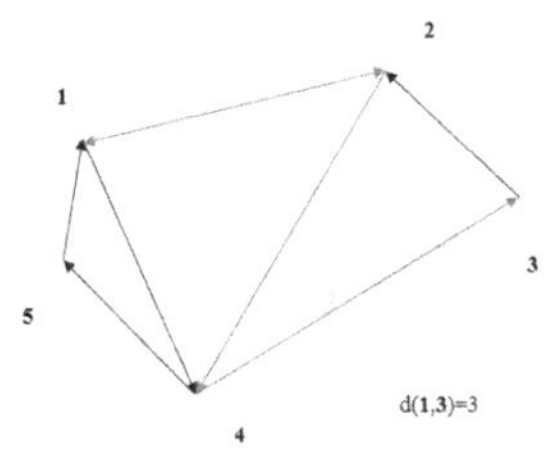

Figure 4.2: Path quasi-metric of a digraph

Example 4. Natural quasi-semimetric. The *natural quasi-semimetric* is a quasi-semimetric (not a quasi-metric) q on $\mathbb{R}$, defined by

$$q(x,y) = \begin{cases} \min\{1, y-x\}, & \text{if } x \leq y, \\ 1, & \text{otherwise.} \end{cases}$$

It is an example of not metrizable quasi-metric space, such that $q(x,y)$, for any fixed x, is a continuous function of y (see [Stol69]).

Example 5. Hamming quasi-semimetric. The *Hamming quasi-metric* is a quasi-metric q_H on $\mathbb{R}^n$, defined as

$$q_H(x,y) = |\{i \,|\, 1 \leq i \leq n, x_i > y_i\}|.$$

On binary vectors $x, y \in \{0,1\}^n$ the Hamming quasi-metric is equal to $|I(x) \backslash I(y)|$, where $I(z) = \{i \in \{1,2,...,n\} \,|\, z_i = 1\}$.

Example 5. Path quasi-metric in digraphs. A *directed graph* $D = \langle V', E' \rangle$ is called *strongly connected*, if there exist both directed paths $(u - v)$ and $(v - u)$ between any two vertices $u, v \in V'$. Given a strongly connected directed graph $D = \langle V', E' \rangle$, the *path quasi-metric* (*directed distance*) q_{dpath} is a quasi-metric on V', defined, for any $u, v \in V'$, as the length of the shortest directed path from u to v in D; see, for example, [CJTW93]. The presence of one-way streets in city street systems produces exactly the same type of distances, which are triangular but fail to be symmetric.

Figure 4.2 gives an example of quasi-distance between two points in a digraph, while Figure 4.3 contains the full information on path quasi-metric of *oriented circuit* $C_4^O =$

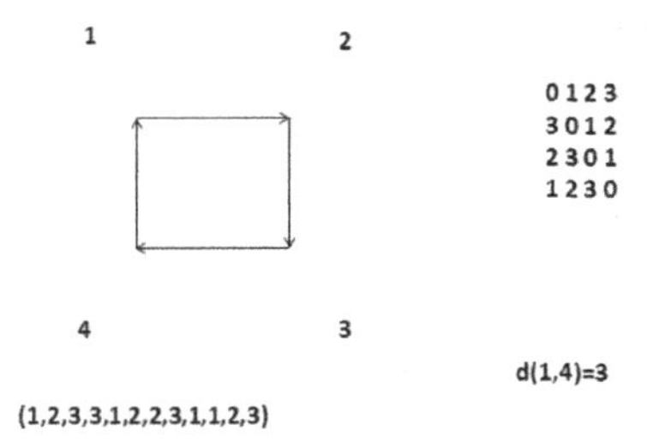

Figure 4.3: Description of path quasi-metric of C_4^O

$\langle V, A\rangle$ with vertex set $V' = \{1, 2, 3, 4\}$ and arc set $A = \{\langle 1, 2\rangle, \langle 2, 3\rangle, \langle 3, 4\rangle, \langle 4, 1\rangle\}$, including its matrix and vector representation.

The *circular metric in digraphs* is the symmetrization of the path quasi-metric in digraphs; it is a metric on the vertex set V of a strongly connected directed graph $D = \langle V, A\rangle$, defined by

$$q_{dpath}(u, v) + q_{dpath}(v, u),$$

where q_{dpath} is the path quasi-metric in D.

Example 6. One way circular railroad distance. Consider a circular railroad line, which moves only in a counter-clockwise direction around a circular track, represented by the unit circle $C_1 = \{x = (x_1, x_2) \in \mathbb{R}^2 \,|\, x_1^2 + x_2^2 = 1\}$. The *one way circular railroad distance* (*circular railroad quasi-distance*) is a quasi-metric q_c on C_1, defined for any $x, y \in C_1$ as the length of the counter-clockwise circular arc from x to y on C_1.

It is easy to see, that q_c is not symmetric ($q_c(x, y) + q_c(y, x) = 2\pi$), but it always satisfies triangle inequality.

Note, that Examples 5 and 6 represent the much wider class of "one-way path" distances, which commonly occur in practice. For example, the presence of one-way streets in a city produces exactly the same type of distances (such as shortest travel-time distance), which satisfy the triangle inequality, but fail to be symmetric.

Transforms of quasi-metrics

In this section we consider some classical examples of *quasi-metric transforms*: the ways to obtain a new quasi-metric (quasi-semimetric) from other quasi-metrics (quasi-semimetrics).

- **Transform quasi-metric**

  A *transform quasi-metric* is a quasi-metric on a set X which is a *quasi-metric transform*, i.e., is obtained as a function of a given quasi-metric (given quasi-metrics) on X. In particular, transform quasi-metrics can be obtained from a given quasi-metric q (given quasi-metrics q_1 and q_2) on X by any of the following operations (here $t > 0$):

1. $tq(x, y)$ (*t-scaled quasi-metric*);
2. $\min\{t, q(x, y)\}$ (*t-truncated quasi-metric*);
3. $\max\{t, q(x, y)\}$ for $x \neq y$ (*t-uniformly discrete quasi-metric*);
4. $q(x, y) + t$ for $x \neq y$ (*t-translated quasi-metric*);
5. $\max\{q_1(x, y), q_2(x, y)\}$;
6. $\alpha q_1(x, y) + \beta q_2(x, y)$, where $\alpha, \beta > 0$.

Moreover, for any quasi-metric q, the functions $\max\{q(x, y), q(y, x)\}$, $\min\{q(x, y), q(y, x)\}$ and $(q^p(x, y) + q^p(y, x))^{\frac{1}{p}}$ with $p \geq 1$ (usually, $p = 1$ is taken) are metrics.

Remark. Easy to see, that if q and q' are quasi-metrics on X and $\alpha, \beta \in \mathbb{R}_{\geq 0}$, then $\alpha q + \beta q'$ is a quasi-semimetric on X (it is a quasi-metric, if $\alpha \neq 0$ or $\beta \neq 0$). Here, as usual, for all $x, y \in X$ it holds

$$(\alpha q + \beta q')(x, y) = \alpha q(x, y) + \beta q'(x, y).$$

- **Product quasi-metric**

  Given finite number n of quasi-metric spaces (X_1, q_1), (X_2, q_2), ..., (X_n, q_n), the *product quasi-metric* is a quasi-metric on the *Cartesian product* $X_1 \times X_2 \times ... \times X_n = \{x = (x_1, x_2, ..., x_n) \mid x_1 \in X_1, ..., x_n \in X_n\}$, defined as a function of $q_1, ..., q_n$. The simplest finite product quasi-metrics are defined by:

  1. $\sum_{i=1}^{n} q_i(x_i, y_i)$;
  2. $(\sum_{i=1}^{n} q_i^p(x_i, y_i))^{\frac{1}{p}}$, $1 < p < \infty$;
  3. $\max_{1 \leq i \leq n} q_i(x_i, y_i)$;
  4. $\sum_{i=1}^{n} \frac{1}{2^i} \frac{q_i(x_i, y_i)}{1 + q_i(x_i, y_i)}$.

Quasi-metrics and normed spaces

In this section we consider a special class of quasi-metrics (quasi-semimetrics), defined on *oriented normed spaces* as the oriented norm of difference between two given elements.

- **Oriented norm quasi-metric**

  A *oriented norm quasi-metric* is a quasi-metric on a real (complex) vector space V, defined by

  $$||x - y||_{or},$$

  where $||.||_{or}$ is an *oriented norm* on V, i.e., a function $||.||_{or} : V \to \mathbb{R}$ such that, for all $x, y \in V$ and for any scalar a, we have the following properties:

  1. $||x||_{or} \geq 0$;
  2. $||x||_{or} = 0$ if and only if $x = 0$;
  3. $||x + y||_{or} \leq ||x||_{or} + ||y||_{or}$ (*oriented triangle inequality*).

  If we will use instead of the condition $||x||_{or} = 0 \Leftrightarrow x = 0$ a weaker version $x = 0 \Rightarrow ||x||_{or} = 0$, we obtain the notion of *oriented norm quasi-semimetric*.

- **l_p-quasi-semimetric**

  Define for $p \geq 1$ *oriented l_p-norm* on $\mathbb{R}^n$ as

$$||x||_{p,or} = \left(\sum_{k=1}^{n}(\max\{x_k, 0\})^p\right)^{\frac{1}{p}},$$

  and *oriented l_∞-norm* as

$$||x||_{\infty,or} = \max_{1\leq k\leq n} \max\{x_k, 0\},$$

  where x_k is the k-th coordinate of x. The *l_p-quasi-semimetric* is a quasi-semimetric $l_{p,or}$, defined by

$$l_{p,or}(x,y) = ||x-y||_{p,or}.$$

  The quasi-semimetric space $(\mathbb{R}^m, l_{p,or})$ is abbreviated as $l^m_{p,or}$.

  It is known (see Chapter 10), that any quasi-semimetric q on V_n is embeddable in $l^n_{\infty,or}$, and any quasi-semimetric q on V_n is embeddable in $l^m_{1,or}$ for some m if and only if $q \in OCUT_n$.

  In [CMM06], the *oriented l_p-semimetric* on $\mathbb{R}^n$ is given by

$$||x-y||^{CMM}_{p,or} = \left(\sum_{i=1}^{n}|x_i - y_i|^p\right)^{\frac{1}{p}} + \left(\sum_{i=1}^{n}|y_i|^p\right)^{\frac{1}{p}} - \left(\sum_{i=1}^{n}|x_i|^p\right)^{\frac{1}{p}}.$$

  Those two definitions are very similar on $\mathbb{R}^n_{\geq 0}$ for $p=1$:

$$||x-y||^{CMM}_{1,or} = 2||y-x||_{1,or}.$$

Other examples of quasi-metrics

- **Hitting time quasi-metric**

  Let $G = \langle V, E\rangle$ be a connected graph with m edges. Consider random walks on G, where at each step the walk moves to a vertex randomly with uniform probability from the neighbors of the current vertex. The *hitting (first-passage) time quasi-metric* $H(u,v)$ from $u \in V$ to $v \in V$ is the expected number of steps (edges) for a random walk on G beginning at u to reach v for the first time; it is 0 for $u = v$.

  This quasi-metric is a *weightable quasi-semimetric* (cf. Chapter 6).

  The *commuting time metric* is $C(u,v) = H(u,v) + H(v,u)$.

  Then $C((u,v) = 2m\Omega(u,v)$, where $\Omega(u,v)$ is the *resistance metric* (*effective resistance*), i.e., 0 if $u = v$ and, otherwise, $\frac{1}{\Omega(u,v)}$ is the current flowing into v, when grounding v and applying a 1 volt potential to u (each edge is seen as a resistor of 1 ohm). Then

$$\Omega(u,v) = \sup_{f:V\to\mathbb{R},\, D(f)>0} \frac{(f(u)-f(v))^2}{D(f)},$$

  where $D(f)$ is the *Dirichlet energy* of f, i. e., $\sum_{st\in E}(f(s)-f(t))^2$.

- **Sorgenfrey quasi-metric**

  The *Sorgenfrey quasi-metric* is a quasi-metric d on $\mathbb{R}$, defined by

$$y - x$$

  if $y \geq x$, and equal to 1, otherwise.

  Some examples of similar quasi-metrics on $\mathbb{R}$ are:

  1. $q_1(x, y) = \max\{y - x, 0\}$;
  2. $q_2(x, y) = \min\{y - x, 1\}$ if $y \geq x$, and equal to 1, otherwise;
  3. $q_3(x, y) = y - x$ if $y \geq x$, and equal to $a(x - y)$ (for fixed $a > 0$), otherwise;
  4. $q_4(x, y) = e^y - e^x$ if $y \geq x$, and equal to $e^{-y} - e^{-x}$, otherwise.

- **Real half-line quasi-semimetric**

  The *real half-line quasi-semimetric* is defined on the half-line $\mathbb{R}_{>0}$ by

$$\max\left\{0, \ln \frac{y}{x}\right\}.$$

- **Convex distance function**

  Given a compact convex region $B \subset \mathbb{R}^n$ which contains the origin in its interior, the *convex distance function* (*gauge*, *Minkowski distance function*) q_B is the quasi-metric on $\mathbb{R}^n$ defined, for $x \neq y$, by

$$q_B(x, y) = \inf\{\alpha > 0 : y - x \in \alpha B\}.$$

  It is also defined, equivalently, as $\frac{||y-x||_2}{||z-x||_2}$, where z is the unique point of the boundary $\partial(x + B)$ hit by the ray from x through y. Then $B = \{x \in \mathbb{R}^n : q_B(0, x) \leq 1\}$ with equality only for $x \in \partial B$. A convex distance function is called *polyhedral*, if B is a polytope, *tetrahedral*, if it is a tetrahedron and so on.

  If B is centrally-symmetric with respect to the origin, then q_B is a *Minkowskian metric*, whose unit ball is B.

- **Contact quasi-distances**

  The *contact quasi-distances* are the following variations of the *distance convex function*, defined on $\mathbb{R}^2$ (in general, on $\mathbb{R}^n$).

  Given a set $B \subset \mathbb{R}^2$, the *first contact quasi-distance* q_B is defined by

$$\inf\{\alpha > 0 : y - x \in \alpha B\}.$$

  Given, moreover, a point $b \in B$ and a set $A \subset \mathbb{R}^2$, the *linear contact quasi-distance* is a *point-set distance* defined by $q_b(x, A) = \inf\{\alpha \geq 0 : \alpha b + x \in A\}$.

  The *intercept quasi-distance* is, for a finite set B, defined by $\frac{\sum_{b \in B} q_b(x,y)}{|B|}$.

- **Quasi-metrics between fuzzy sets**

A *fuzzy subset* of a set S is a mapping $\mu : S \to [0,1]$, where $\mu(x)$ represents the "degree of membership" of $x \in S$. It is an ordinary (*crisp*) subset if all $\mu(x)$ are 0 or 1. Fuzzy sets are good models for *gray-scale images*, random objects and objects with non-sharp boundaries.

If q is a quasi-metric on $[0,1]$, and S is a finite set, then

$$Q(\mu, \nu) = \sup_{x \in S} q(\mu(x), \nu(x))$$

is a quasi-metric on fuzzy subsets of S.

Bhutani-Rosenfeld (2003) [BhRo03] introduced the following two metrics between two fuzzy subsets μ and ν of a finite set S. The *diff-dissimilarity* is a metric (a fuzzy generalization of *Hamming metric*), defined by

$$d(\mu, \nu) = \sum_{x \in S} |\mu(x) - \nu(x)|.$$

The *perm-dissimilarity* is a semimetric, defined by

$$\min\{d(\mu, p(\nu))\},$$

where the minimum is taken over all permutations p of S.

The *Klement-Puri-Ralesku metric* (1988) [KPR86] between two fuzzy numbers μ, ν is

$$\int_0^1 d_{Haus}(A_\mu(t), A_\nu(t))dt,$$

where $d_{Haus}(A_\mu(t), A_\nu(t))$ is the *Hausdorff metric*

$$\max\left\{\sup_{x \in A_\mu(t)} \inf_{y \in A_\nu(t)} |x - y|, \sup_{x \in A_\nu(t)} \inf_{y \in A_\mu(t)} |x - y|\right\}.$$

Several other Hausdorff-like metrics on some families of fuzzy sets were proposed by Boxer (1997) [Boxe97], Fan (1998) [Fan98] and Brass (2002) [Bras02]; Brass also argued the non-existence of a "good" such metric.

Chapter 5

Multidimensional case: m-metrics

5.1 Preliminaries

In this chapter we study m-hemimetrics and their discrete aspects. The notions of m-hemimetric and m-partition hemimetric generalize the notion of m-metric (see [Meng28], [Blum53], [Frod58], [Gahl63]), which have been around for 70 years and whose 1990 bibliography lists over 260 items [Gahl90]. These objects are the direct $(m+1)$-ary extensions of the binary notions of metrics and cuts, which are well-known and central objects in Graph Theory, Combinatorial Optimization and, more generally, in Discrete Mathematics. The motivation for m-metrics came (apart, of course, that it represents an extension of metrics) mostly from Geometry and so, the main interest has been in their topological aspects.

Independently of m-metrics, there appeared recently papers on weaker versions of 2-metrics, motivated by various discrete applications, e.g., in Statistics and Data Analysis (see [Diat96], [JoCa95]). The strong m-metric axioms $d(x_1, x_1, x_2, ..., x_{m+1}) = 0$ and the existence of x_{m+1} such that $d(x_1, ..., x_{m+1}) > 0$ whenever $x_1, ..., x_m$ are distinct, which make perfect sense in Geometry, are too restrictive for applications and, hence, are substantially weakened in various ways, which common trait seems to be the total symmetry and the m-simplex inequality.

The m-hemimetrics on n points provide an example of such weaker version of m-metric's notion, useful for polyhedral methods. They were introduced in [DeRo00] and studied using the program *cdd* [Fuku95] for small parameters m, n in [DeRo02]. The notion of (m, s)-supermetric was introduced in [DeDu03].

In Section 5.2 we define m-metrics and m-hemimetrics. To illustrate them, in Section 5.3 we study in detail basic examples of m-hemimetrics, some m-metrics on the set $\{0,1\}^n$, which can be naturally expressed in terms of the subsets of $V_n = \{1, 2, ..., n\}$, and an extension of the two-way distances from [JoCa95] to n-way distances. We give also a few other examples and constructions of m-hemimetrics.

5.2 Definitions

Definition 5.1 *Let X be a set. A function $d : X^3 \longrightarrow \mathbb{R}$ is called 2-distance on X if for all $x, y, z, t \in X$ it holds:*

1. *$d(x, y, z) \geq 0$ (non-negativity);*

2. $d(x,x,x) = 0$ *(positive definiteness);*

3. $d(x,y,z) = d(\pi(x),\pi(y),\pi(z))$ *for any permutation* π *of the set* $\{x,y,z\}$ *(total symmetry).*

Definition 5.2 *Let* X *be a set. A function* $d : X^3 \longrightarrow \mathbb{R}$ *is called 2-semimetric on* X *if for all* $x,y,z,t \in X$ *it holds:*

1. $d(x,y,z) \geq 0$ *(non-negativity);*

2. $d(x,x,x) = 0$*(positive definiteness);*

3. $d(x,y,z) = d(\pi(x),\pi(y),\pi(z))$ *for any permutation* π *of the set* $\{x,y,z\}$ *(total symmetry);*

4. $d(x,y,z) \leq d(t,y,z) + d(x,t,z) + d(x,y,t)$ *(tetrahedron inequality).*

Definition 5.3 *Let* X *be a set. A function* $d : X^3 \longrightarrow \mathbb{R}$ *is called 2-hemimetric on* X *if for all* $x,y,z,t \in X$ *it holds:*

1. $d(x,y,z) \geq 0$ *(non-negativity);*

2. $d(x,x,y) = 0$ *(weak positive-definiteness);*

3. $d(x,y,z) = d(\pi(x),\pi(y),\pi(z))$ *for any permutation* π *of the set* $\{x,y,z\}$ *(total symmetry);*

4. $d(x,y,z) \leq d(t,y,z) + d(x,t,z) + d(x,y,t)$ *(tetrahedron inequality).*

Definition 5.4 *Let* X *be a set. A function* $d : X^3 \longrightarrow \mathbb{R}$ *is called 2-metric on* X *if for all* $x,y,z,t \in X$ *it holds:*

1. $d(x,y,z) \geq 0$ *(non-negativity);*

2. $d(x,x,y) = 0$ *(weak positive-definiteness);*

2′. $x \neq y \Longrightarrow d(x,y,u) > 0$ *for some* $u \in X$*;*

3. $d(x,y,z) = d(\pi(x),\pi(y),\pi(z))$ *for any permutation* π *of the set* $\{x,y,z\}$ *(total symmetry);*

4. $d(x,y,z) \leq d(t,y,z) + d(x,t,z) + d(x,y,t)$ *(tetrahedron inequality).*

Because of the total symmetry and since $d(i,i,k) = 0$ for $i,k \in V_n$, we can view an 2-metric (as well as 2-hemimetric) as a vector $(d_{ijk})_{1\leq i<j<k\leq n} \in \mathbb{R}^{E_n^3}$, where $E_n^3 = \binom{n}{3} = \frac{n(n-1)(n-2)}{6}$: $d(i,j,k) = d_{ijk}$. On the other hand, any 2-distance(and 2-semimetric), satisfying only the conditions $d(i,i,i) = 0$ for $i \in V_n$, can be represented as a vector $d = (d_{ijk}) \in \mathbb{R}^{n^3-n}$, $1 \leq i < j \leq k \leq n$, with non-negative coordinates. But a matrix representation could not be used in the multidimensional case.

In the last definition, the *total symmetry* means that the value of $d(x,y,z)$ is independent of the order of x, y and z. The *tetrahedron inequality* captures that fact that in $\mathbb{R}^3$ the area of a triangle face of a tetrahedron does not exceed the sum of the areas of the remaining three faces. The axiom $d(x, x, y) = 0$ states that certain degenerate triangles have area 0 while the condition $x \neq y \Longrightarrow d(x, y, u) > 0$ for some $u \in X$ stipulates that each pair of distinct points is on at least one non-degenerate triangle.

An 2-metric allows the introduction of several geometrical and topological concepts – e.g., the betweenness, convexity, line and neighborhood – which lead to interesting results.

For finite 2-metrics, for their polyhedral aspects and for applications, the axiom $x \neq y \Longrightarrow d(x, y, u) > 0$ seems to be too restrictive and so we drop it (see [DeRo00], [DeRo02]). The condition $d(x, y, z) \geq 0$ is convenient for polyhedral methods.

The three definitions given below are generalizations of the three-dimensional notions of 2-distance, 2-semimetric, 2-hemimetric and 2-metric for an arbitrary positive integer m.

Definition 5.5 *Let X be a set. Let $m \geq 1$. A function $d : X^{m+1} \longrightarrow \mathbb{R}$ is called m-distance on X if for all $x_i \in X$ it holds:*

1. $d(x_1, ..., x_{m+1}) \geq 0$ *(non-negativity);*
2. $d(x_1, ..., x_1) = 0$ *(positive definiteness);*
3. $d(x_1, ..., x_{m+1}) = d(\pi(x_1), ..., \pi(x_{m+1}))$ *for any permutation π of the set $\{x_1, ..., x_{m+1}\}$ (total symmetry);*

Definition 5.6 *Let X be a set. Let $m \geq 1$. A function $d : X^{m+1} \longrightarrow \mathbb{R}$ is called m-semimetric on X if for all $x_i \in X$ it holds:*

1. $d(x_1, ..., x_{m+1}) \geq 0$ *(non-negativity);*
2. $d(x_1, ..., x_1) = 0$ *(positive definiteness);*
3. $d(x_1, ..., x_{m+1}) = d(\pi(x_1), ..., \pi(x_{m+1}))$ *for any permutation π of the set $\{x_1, ..., x_{m+1}\}$ (total symmetry);*
4. $d(x_1, ..., x_{m+1}) \leq \sum_{i=1}^{m+1} d(x_1, ..., x_{i-1}, x_{i+1}, ..., x_{m+2})$ *(m-simplex inequality).*

Definition 5.7 *Let X be a set. Let $m \geq 1$. A function $d : X^{m+1} \longrightarrow \mathbb{R}$ is called m-hemimetric on X if for all $x_i \in X$ it holds:*

1. $d(x_1, ..., x_{m+1}) \geq 0$ *(non-negativity);*
2. $d(x_1, ..., x_{m+1}) = 0$*, whenever $x_1, ..., x_{m+1}$ are not pairwise distinct (weak positive definiteness);*
3. $d(x_1, ..., x_{m+1}) = d(\pi(x_1), ..., \pi(x_{m+1}))$ *for any permutation π of the set $\{x_1, ..., x_{m+1}\}$ (total symmetry);*
4. $d(x_1, ..., x_{m+1}) \leq \sum_{i=1}^{m+1} d(x_1, ..., x_{i-1}, x_{i+1}, ..., x_{m+2})$ *(m-simplex inequality).*

Definition 5.8 *Let X be a set. Let $m \geq 1$. A function $d : X^{m+1} \longrightarrow \mathbb{R}$ is called m-metric on X if for all $x_i \in X$ it holds:*

1. $d(x_1, ..., x_{m+1}) \geq 0$ *(non-negativity);*

2. $d(x_1, ..., x_{m+1}) = 0$, *whenever* $x_1, ..., x_{m+1}$ *are not pairwise distinct (weak positive definiteness);*

2′ *if* $x_1, ..., x_m$ *are pairwise distinct, then* $d(x_1, ..., x_m, x_{m+1}) > 0$ *for some* $x_{m+1} \in X$;

3. $d(x_1, ..., x_{m+1}) = d(\pi(x_1), ..., \pi(x_{m+1}))$ *for any permutation* π *of the set* $\{x_1, ..., x_{m+1}\}$ *(total symmetry);*

4. $d(x_1, ..., x_{m+1}) \leq \sum_{i=1}^{m+1} d(x_1, ..., x_{i-1}, x_{i+1}, ..., x_{m+2})$ *(m-simplex inequality).*

An example of an m-hemimetric is the m-dimensional volume in $\mathbb{R}^n$ (with $n \geq m$).

Remark. The definition of m-hemimetric is defective in the following sense: in the case of semimetric, if we have a path $x_1, ..., x_n$, then we have the inequality

$$d(x_1, x_n) \leq \sum_{i=1}^{n-1} d(x_i, x_{i+1})$$

and this path inequality is implied by the triangle inequalities. The equivalent case would be the following in the case of m-hemimetric. If we have an m-dimensional simplicial complex with simplices Δ_1, ..., Δ_n without boundary, then we have the inequality for the m-dimensional volume

$$d_m(\Delta_i) \leq \sum_{i \neq j} d_m(\Delta_j),$$

for $1 \leq i \leq n$.

But for $m > 1$ those inequalities are, in general, not implied by the m-simplex inequality. Therefore, they should be added to the system of inequalities, in order to get a more coherent inequality system.

The notion of (m, s)-*supermetric* is a generalization of the notion of m-hemimetric.

Definition 5.9 *Let* X *be a set. Let* $m \geq 1$, *and let* $s > 0$ *be any positive number. A function* $d : X^{m+1} \longrightarrow \mathbb{R}$ *is called* (m, s)*-supermetric on* X *if for all* $x_i \in X$ *it holds:*

1. $d(x_1, ..., x_{m+1}) \geq 0$ *(non-negativity);*

2. $d(x_1, ..., x_{m+1}) = 0$, *whenever* $x_1, ..., x_{m+1}$ *are not pairwise distinct (zero-condition property);*

3. $d(x_1, ..., x_{m+1}) = d(\pi(x_1), ..., \pi(x_{m+1}))$ *for any permutation* π *of the set* $\{x_1, ..., x_{m+1}\}$ *(total symmetry);*

4. $s \times d(x_1, ..., x_{m+1}) \leq \sum_{i=1}^{m+1} d(x_1, ..., x_{i-1}, x_{i+1}, ..., x_{m+2})$ *for distinct* $x_1, ..., x_{m+1} \in X$ *((m, s)-m-simplex inequality).*

So, an m-hemimetric is just an $(m, 1)$-supermetric and a semimetric is an $(1, 1)$-supermetric. An (m, s)-supermetric is an *m-hemimetric* if $s \geq 1$.

If $T = \{x_1, ..., x_{m+2}\}$, then we will set $d_{x_i} = d(x_1, ..., x_{i-1}, x_{i+1}, ..., x_{m+2})$ and $\Sigma_T = \sum_{i=1}^{m+2} d_{x_i}$. In this case, the (m, s)-simplex inequality can be rewritten as

$$(s + 1)d_{x_{m+2}} \leq \Sigma_T.$$

Because of the total symmetry and since $d(i_1, i_1, i_3, ..., i_{m+1}) = 0$ for $i_1, i_3, ..., i_{m+1} \in V_n$, we can view an m-hemimetric (as well as an m-metric and an (m, s)-supermetric) as a vector

$$(d_{i_1 i_2 \dots i_{m+1}})_{1 \le i_1 < i_2 < \dots < i_{m+1} \le n} \in \mathbb{R}^{E_n^{m+1}},$$

where $E_n^{m+1} = \binom{n}{m+1} = \frac{n(n-1)...(n-m+1)}{(m+1)!}$: $d(i_1, i_2, ..., i_{m+1}) = d_{i_1 i_2 \dots i_{m+1}}$. On the other hand, any m-distance(and m-semimetric), satisfying only the conditions $d(i, i, ..., i) = 0$ for $i \in V_n$, can be represented as a vector

$$d = (d_{i_1 i_2 \dots i_{m+1}}) \in \mathbb{R}^{n^{m+1}-n}, \ 1 \le i_1 < i_2 \le i_3 \le \dots \le i_{m+1} \le n,$$

with non-negative coordinates. A matrix representation could not be used.

The *Vitanyi multiset-metric* (see Chapter 14) represents an other generalization of the notion of ordinary metric; it was proposed by Vitanyi, 2011, and is very close to the multidimensional constrictions of m-metrics.

5.3 Examples

Basic examples

Example 1. Discrete 2-metric. Given a set X, $|X| \ge 4$, the *discrete 2-metric* d on X is defined by

$$d(x, y, z) = \begin{cases} 0, & \text{if } x = y, \text{ or } y = z, \text{ or } x = z, \\ 1, & \text{otherwise.} \end{cases}$$

In general, an *m-metric* d on X, $|X| \ge m + 1$, is defined by

$$d(x_1, ..., x_{m+1}) = \begin{cases} 0, & \text{if } x_i = x_j \text{ for some } 1 \le i < j \le m+1, \\ 1, & \text{otherwise.} \end{cases}$$

In fact, by definition, the function d is non-negative, totally symmetric and zero-conditioned; if $|X| \ge m+1$, there exist $x_{m+1} \in X$, such that $d(x_1, ..., x_{m+1}) > 0$ whenever x_1, ..., x_m are distinct. (if $|X| \le m$, the function d is only an m-hemimetric).

Suppose to the contrary that the m-simplex inequality does not hold. Then there exist $x_1, ..., x_{m+2} \in X$ such that $d(x_1, ..., x_{m+1}) = 1$, while for all $j = 1, ..., m+1$ it holds

$$d(x_1, ..., x_{j-1}, x_{j+1}, ..., x_{m+2}) = 0.$$

These equalities imply that x_{m+1} is equal to at least two elements from $\{x_1, ..., x_{m+1}\}$, and, hence, these elements can not to be pairwise distinct, i.e., $d(x_1, ..., x_{m+1})$ should be equal to zero.

The following example ([DeRo00]) is similar; it extends an example from [CPK96]. Let $X = \{1, ..., m+2\}$ and let $d : X^{m+1} \longrightarrow \{0, 1\}$ be defined by setting

$$d(x_1, ..., x_{m+1}) = \begin{cases} 0, & \text{if } x_i = x_j \text{ for some } 1 \le i < j \le m+1 \\ & \text{or } \{x_1, ..., x_{m+1}\} = \{1, ..., m+1\}, \\ 1, & \text{otherwise.} \end{cases}$$

Then d is an m-metric. Obviously, it satisfies for all $x_1, ..., x_m \in X$ the axioms

$$d(x_1, x_1, x_2, ..., x_m) = 0$$

and $d(x_1, ..., x_{n-1}, n+1) > 0$ whenever $x_1, ..., x_m$ are distinct. Suppose to the contrary that the m-simplex inequality does not hold. Then there exist $x_1, ..., x_{m+2} \in X$, such that $d(x_1, ..., x_{m+1}) = 1$, while for all $j = 1, ..., m+1$ one has

$$d(x_1, ..., x_{j-1}, x_{j+1}, ..., x_{m+2}) = 0.$$

These equalities imply that $Y = \{x_1, ..., x_{m+1}\}$ satisfies $|Y| = m+1$ and $m+2 \in Y$. Without loss of generality we can assume that $Y = \{2, ..., m+2\}$, and $(x_1, ..., x_{m+1}) = (2, 3, ..., m+2)$. The definition and the equation $d(x_1, ..., x_{j-1}, x_{j+1}, ..., x_{m+2}) = 0$ show that $x_{m+2} = 1$. So, we obtain that $d(1, 3, ..., m+2) = 0$ contrary to the definition of d. Thus, d is an m-metric. The last statement follows from the definition.

Example 2. Hausdorff 2-metric. Let (X, D) be a finite metric space and let $X = X_1 \cup \ldots \cup X_n$ be a decomposition of X into the union of pairwise disjoint sets.

The *Hausdorff 2-metric* d on the set $Y = \{X_1, \ldots, X_n\}$ is defined by

$$d(X_i, X_j, X_k) = \frac{1}{3}\left(\max_{x \in X_i, y \in X_j} D(x, y) + \max_{x \in X_i, y \in X_k} D(x, y) + \max_{x \in X_k, y \in X_j} D(x, y)\right).$$

In general, the *Hausdorff m-metric* d on the set $Y = \{X_1, \ldots, X_n\}$ is defined as

$$d(X_{i_1}, X_{i_2}, ..., X_{i_{m+1}}) =$$

$$= \frac{1}{m+1}\left(\max_{x \in X_{i_1}, y \in X_{i_2}} D(x, y) + \max_{x \in X_{i_1}, y \in X_{i_3}} D(x, y) + ... + \max_{x \in X_{i_m}, y \in X_{i_{m+1}}} D(x, y)\right).$$

Example 3. Symmetric difference 2-semimetric. For any anti-chain of sets $Z = \{x, y, z, ... \mid x \not\subset y \text{ for all } x \neq y\}$, the *symmetric difference 2-semimetric* ρ is a 2-semimetric on Z, defined by

$$\rho(x, y, z) = \frac{1}{2}(|x\Delta y| + |y\Delta z| + |x\Delta z|);$$

it is the number of elements, that belong to x, y or z but not to all of x, y, z.

Similarly, if $P(V_n)$ is the set of all subsets of $V_n = \{1, 2, ..., n\}$, then the function

$$\rho(a, b, c) = \frac{1}{2}(|a\Delta b| + |b\Delta c| + |a\Delta c|)$$

is an 2-semimetric on $P(V_n)$. Clearly, ρ is non-negative and totally symmetric. To check the tetrahedron inequality, let $a, b, c, d \subseteq V_n$. From the definition,

$$\rho(b, c, d) + \rho(a, c, d) + \rho(a, b, d) = |a\Delta d| + |b\Delta d| + |c\Delta d| + \rho(a, b, c) \geq \rho(a, b, c).$$

Clearly, it holds with equality if and only if $a = b = c = d$. Moreover, $\rho(a, a, b) = 0 \Longleftrightarrow a = b$ and, therefore, δ does not satisfies the condition $\rho(x, x, y) = 0$. If $m > 2$, the axiom of existence of some $u \in V_n$ such that $\rho(a, b, u) > 0$ for $a \neq b$, holds for the function ρ.

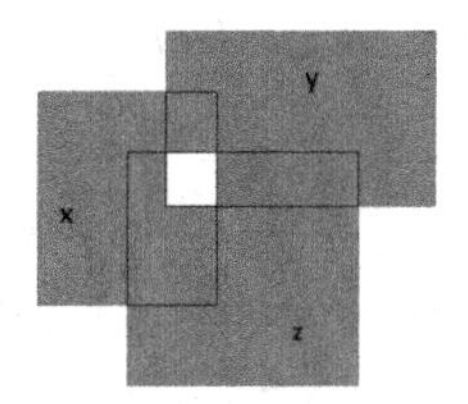

Figure 5.1: Symmetric difference: $(x\Delta y) \cup (y\Delta z) \cup (x\Delta z)$

If we represent each $a \subseteq V_n$ by its characteristic zero-one n-vector $\chi_a = (a_1, ..., a_n)$ with $a_i = 1$, if $i \in a$, and $a_i = 0$, if $i \notin a$, then

$$|a\Delta b| = |a_1 - b_1| + \cdots + |a_m - b_m|$$

is the *Hamming distance* between χ_a and χ_b. For this reason we could refer to ρ as the *half-perimeter* 2*-semimetric*.

This construction has the following immediate extension: if we consider $\rho(x_1, ..., x_{m+1})$ as the number of elements, that belong to some x_i but not to all x_i, we obtain an example of m-semimetric. Formally, the *symmetric difference* m*-semimetric* ρ is an m-semimetric on $P(V_n)$, defined by

$$\rho(x_1, x_2, ..., x_{m+1}) = \frac{1}{m+1}(|x_1\Delta x_2| + |x_1\Delta x_3| + ... + |x_m\Delta x_{m+1}|).$$

Example 4. Natural 2-metric. The ordinary metric on $\mathbb{R}$ - the *natural metric* - is defined as $|x - y|$. In this situation, the *natural 2-metric* d on $\mathbb{R}$ can be defined by

$$d(x, y) = |x - y| + |x - z| + |z - y|.$$

An other 2-metric d on $\mathbb{R}$, relating to natural metric, can be used:

$$d(x, y) = \min\{1, |x - y|, |x - z|, |z - y|\}.$$

So, the *natural* m*-metric* d on $\mathbb{R}$ can be defined by

$$d(x_1, x_2, ..., x_{m+1}) = \sum_{x_i \neq x_j} |x_i - x_j|.$$

Similarly, an other m-metric d on $\mathbb{R}$, relating to natural metric, can be used:

$$d(x_1, x_2, ..., x_{m+1}) = \min\{1, |x_1 - x_2|, |x_1 - x_3|, ..., |x_m - x_{m+1}|\}.$$

Other trend: the *natural 2-metric* d on $\mathbb{R}^2$ can be defined as square of the triangle with the vertices x, y, z.

On the same way, the *natural* m*-metric* on $\mathbb{R}^n$, $n > m$, can be defined as m-dimensional volume, defined by given points $x_1, ..., x_{m+1}$.

Example 5. Hamming 2-metric. The *Hamming 2-metric* d_H is an 2-metric on $\mathbb{R}^n$, defined by

$$d_H(x,y,z) = |\{i \,|\, 1 \le i \le n, x_i \ne y_i\}| + |\{i \,|\, 1 \le i \le n, y_i \ne z_i\}| + |\{i|1 \le i \le n, x_i \ne z_i\}|.$$

The *Hamming m-metric* d_H is an m-metric on $\mathbb{R}^n$, defined by

$$d_H(x_1, x_2, ..., x_{m+1}) = \sum_{x_i \ne x_j} |\{k \,|\, 1 \le k \le n, x_i^k \ne x_j^k\}|,$$

where $x_i = (x_i^1, ..., x_i^n) \in \mathbb{R}^n$.

Example 6. Path 2-metric in hypergraphs. Given a *connected graph* $G = \langle V, E\rangle$, the *path 2-metric* (*graphic 2-metric, shortest path 2-metric*) D_{path} is an 2-metric on the vertex set V of G, defined, for any $u, v, t \in V$, as

$$D_{path}(u,v,t) = d_{path}(u,v) + d_{path}(u,t) + d_{path}(v,t),$$

where d_{path} is the classical *path metric* in G, i.e., for vertices $u, v \in V$, the length (the number of edges) of a shortest $(u - v)$ path in G.

On the other hand, we can obtain an example of 2-metric from an *3-uniform hypergraph*, i.e., a generalization of a graph, where and *hyperedge* connects exactly three vertices. Formally, an *3-uniform hypergraph* H is a pair $\langle V, E\rangle$, where V is a set of elements, called *vertices*, and $E = \{(u,v,z) \,|\, u,v,z \in V\}$ is a set of non-ordered triples of distinct elements from V, called *hyperedges*.

We say, that there is a *hyperpath* between vertices u and v in an hypergraph H, if there exists an alternating sequence of distinct vertices and hyperedges $u = v_1$; e_1; v_2; e_2; ... ;e_{k-1}; $v_k = v$, such that $v_i, v_{i+1} \in e_i$ for all $1 \le i \le k-1$. An hypergraph is called *connected*, if there exists a hyperpath for every pair of vertices.

Given an 3-uniform (in general, k-uniform, $k \ge 3$) connected hypergraph $H = \langle V, E\rangle$, the *hyperpath 2-metric* in H is an 2-metric on the set V, defined, for any distinct $u, v, t \in V$, as the length (number of hyperedges) of the shortest hyperpath in H, including all three vertices u, v, t (and is equal to zero, otherwise).

Other approach: we can say, that there is an *2-hyperpath* between vertices u, v and t of an hypergraph H, if there exists an alternating sequence of vertices and hyperedges $u = v_1$, $v = v_1^{'}$, e_1; v_2, $v_2^{'}$, e_2; ... ; v_{k-1}, $v_{k-1}^{'}$, e_{k-1}; $v_k = v$, $v_k^{'} = t$, such that $v_i, v_i^{'}, v_{i+1}, v_{i+1}^{'} \in e_i$ for all $1 \le i \le k-1$. A hypergraph is called *2-connected*, if there is an 2-hyperpath for every triple of its distinct vertices.

Given an 3-uniform (in general, k-uniform, $k \ge 3$) 2-connected hypergraph $H = \langle V, E\rangle$, the *2-hyperpath 2-metric* in H is an 2-metric on the set V, defined, for any distinct vertices $u, v, t \in V$, as the length (number of hyperedges) of the shortest 2-hyperpath between vertices u, v and t (and is equal to zero, otherwise).

Similarly, given a connected graph $G = \langle V, E\rangle$, the *path m-metric* (*graphic m-metric, shortest path m-metric*) D_{path} is an m-metric on the vertex set V of G, defined, for any $u_1, ..., u_{m+1} \in V$, as

$$D_{path}(u_1, ..., u_{m+1}) = \sum_{1 \le i < j \le m+1} d_{path}(u_i, u_j),$$

where d_{path} is the path metric in G.

On the other hand, given $(m+1)$-uniform (in general, k-uniform, $k \geq m+1$) connected hypergraph $H = \langle V, E \rangle$, the *hyperpath m-metric* is an m-metric on the set V, defined, for any $u_1, u_2, ..., u_{m+1} \in V$, as the length of the shortest hyperpath in H, including all the vertices $u_1, u_2, ..., u_{m+1}$.

Example 7. Circular railroad 2-distance. Consider a circular railroad line, which moves around a circular track, represented by the unit circle $C_1 = \{x = (x_1, x_2) \in \mathbb{R}^2 \mid x_1^2 + x_2^2 = 1\}$. If we define $d(x, y, z), x, y, z \in C_1$, as the area of the triangle with the vertices x, y, z, we obtain an 2-metric on C_1.

Similarly, if we define $d(a_1, a_2, ..., a_{m+1}), a_i \in C_1$, as the area of the polygon with the vertices a_i, we obtain an m-metric on C_1.

Another trend: we can define an 2-metric on the unit sphere $C_2 = \{x = (x_1, x_2, x_3) \in \mathbb{R}^3 \mid x_1^2 + x_2^2 + x_3^2 = 1\}$ as the area of the triangle with the vertices $x, y, z \in C_2$. Then a m-metric on the unit sphere $C_m = \{x = (x_1, x_2, ..., x_{m+1}) \in \mathbb{R}^{m+1} \mid x_1^2 + x_2^2 + ... + x_{m+1}^2 = 1\}$ can be defined as the volume of the polytope with the vertices $a_1, a_2, ..., a_{m+1} \in C_{m+1}$.

Transforms of m-metrics

- **Transform m-metric**

  A *transform m-metric* is an m-metric on a set X which is a *metric transform*, i.e., is obtained as a function of a given m-metric (given m-metrics) on X. In particular, transform m-metrics can be obtained from a given m-metric d (given m-metrics d_1 and d_2) on X by any of the following operations (here $t > 0$):

  1. $td(x_1, ..., x_{m+1})$ (*t-scaled m-metric*);
  2. $\min\{t, d(x_1, ..., x_{m+1})\}$ (*t-truncated m-metric*);
  3. $\max\{t, d(x_1, ..., x_{m+1})\}$ for pairwise distinct x_i (*t-uniformly discrete m-metric*);
  4. $d(x_1, ..., x_{m+1}) + t$ for pairwise distinct x_i (*t-translated m-metric*);
  5. $\max\{d_1(x_1, ..., x_{m+1}), d_2(x_1, ..., x_{m+1})\}$;
  6. $\alpha d_1(x_1, ..., x_{m+1}) + \beta d_2(x_1, ..., x_{m+1})$, where $\alpha, \beta > 0$.

  Remark. Easy to see, that if d and d' are m-hemimetrics on X and $\alpha, \beta \in \mathbb{R}_{\geq 0}$, then $\alpha d + \beta d'$ is an m-hemimetric on X. Here, as usual, for all $x_i \in X$ it holds

  $$(\alpha d + \beta d')(x_1, ..., x_{m+1}) = \alpha d(x_1, ..., x_{m+1}) + \beta d'(x_1, ..., x_{m+1}).$$

- **Product m-metric**

  Given finite number n of m-metric spaces (X_1, d_1), (X_2, d_2), ..., (X_n, d_n), the *product m-metric* is an m-metric on the *Cartesian product* $X_1 \times X_2 \times ... \times X_n = \{x = (x_1, x_2, ..., x_n) \mid x_1 \in X_1, ..., x_n \in X_n\}$, defined as a function of $d_1, ..., d_n$. The simplest finite product m-metrics are defined by:

  1. $\sum_{i=1}^{n} d_i(x_1, ..., x_{m+1})$;
  2. $(\sum_{i=1}^{n} d_i^p(x_1, ..., x_{m+1}))^{\frac{1}{p}}$, $1 < p < \infty$;

3. $\max_{1 \leq i \leq n} d_i(x_1, ..., x_{m+1})$.

- **Perimeter m-hemimetric**

Let $m > 1$ and let $d : X^m \longrightarrow \mathbb{R}_{\geq 0}$ be an $(m-1)$-distance, i.e., totally symmetric function with all $d(x, ..., x) = 0$. The *perimeter m-semimetric* d^* is an m-semimetric on X, defined for all $x_1, ..., x_{m+1} \in X$ by

$$d^*(x_1, ..., x_{m+1}) = \frac{1}{m} \sum_{i=1}^{m+1} d(x_1, ..., x_{i-1}, x_{i+1}, ..., x_{m+1}).$$

Clearly, d^* is totally symmetric. To prove the m-simplex inequality, let $x_1, ..., x_{m+2} \in X$. For any $1 \leq j \leq m+1$ it holds

$$\frac{1}{m} d(x_1, ..., x_{j-1}, x_{j+1}, ..., x_{m+1}) \leq d^*(x_1, ..., x_{j-1}, x_{j+1}, ..., x_{m+2}),$$

and so by the construction it holds

$$d^*(x_1, ..., x_{m+1}) \leq \sum_{j=1}^{m+1} d^*(x_1, ..., x_{j-1}, x_{j+1}, ..., x_{m+2}).$$

Moreover, $d^*(x, x, ..., x) = 0$ as the sum of $m+1$ zero items $d(x, ..., x)$.

Note, that for an $(m-1)$-hemimetric d the function d^* can fail the zero-condition property; so, in general, it is not an m-hemimetric.

- **Maximum m-hemimetric**

Let d be an $(m-1)$-hemimetric on X. Define $d^* : X^{m+1} \to \mathbb{R}_{\geq 0}$ by setting

$$d^*(x_1, ..., x_{m+1}) = \begin{cases} \max\{d(x_1, ..., x_{i-1}, x_{i+1}, ..., x_{m+1}) \mid 1 \leq i \leq m+1\}, \\ \qquad \text{if } x_1, ..., x_{m+1} \text{ are pairwise distinct}, \\ 0, \qquad \text{otherwise}. \end{cases}$$

Then d^* is an m-hemimetric on X.

To prove the m-simplex inequality, let $x_1, ..., x_{m+2} \in X$ be such that

$$\delta = d^*(x_1, ..., x_{m+1}) > 0.$$

Then $\delta = d(x_1, ..., x_{i-1}, x_{i+1}, ..., x_{m+1})$ for some $1 \leq i \leq m+1$, and so, using the $(m-1)$-simplex inequality for $d(x_1, ..., x_{i-1}, x_{i+1}, ..., x_{m+1})$ and the properties of the construction, we obtain, that

$$d^*(x_1, ..., x_{i-1}, x_{i+1}, ..., x_{m+2}) = d(x_1, ..., x_{i-1}, x_{i+1}, ..., x_{m+1})$$

$$\leq \sum_{j=1}^{m+2} d^*(x_1, ..., x_{j-1}, x_{j+1}, ..., x_{m+2}).$$

m-metrics between subsets

To illustrate the definition of m-metric, we study in detail some examples of m-semimetrics on $\{0,1\}^n$. They can be naturally expressed in terms of the subsets of $V_n = \{1, 2, ..., n\}$. In particular, we completely characterize the m-semimetrics ∇^K_{mn} on $P(V_n)$, which for any $a_1, ..., a_{m+1} \in P(V_n)$ count the number of elements of V_n that belong to exactly i sets $a_1, ..., a_{m+1}$ for some $i \in K$, where K is a given family of pairwise disjoint intervals in $[1, m+1]$.

Proposition 5.1 *Let* $n > 1, V_n = \{1, 2, ..., n\}$, *and*

$$1 \leq r \leq s \leq m+1, \quad r \leq m, \quad \tau = \min\{m+1-r, s\}. \qquad (I)$$

For all (not necessarily pairwise distinct) subsets $a_1, ..., a_{m+1}$ *of* V_n *denote by* $\nabla^{rs}_{mn}(a_1, ..., a_{m+1})$ *(shortly by* $\nabla(a_1, ..., a_{m+1})$*) the number of elements of* V_n *that belong to exactly* i *sets amongst* $a_1, ..., a_{m+1}$ *for some* $r \leq i \leq s$. *Then*
(i) for all $a_1, ..., a_{m+1} \subseteq V_n$ *it holds*

$$\tau \times \nabla(a_1, ..., a_{m+1}) \leq \sum_{i=1}^{m+1} \nabla(a_1, ..., a_{i-1}, a_{i+1}, ..., a_{m+2}); \qquad (*)$$

(ii) the coefficient τ *in (*) is the largest possible;*
(iii) ∇ *is an* m*-semimetric on* $P(V_n)$.

Proof. (i) Let $a_1, ..., a_{m-2} \subseteq V_n$. Let $x \in V_n$ belonging to exactly t sets amongst $a_1, ..., a_{m+1}$. If $t < r$ or $t > s$, then x is not counted in $\nabla(a_1, ..., a_{m+1})$. Thus, let $r \leq t \leq s$. We can arrange the notation so that

$$x \in (a_1 \cap ... \cap a_t) \setminus (a_{t+1} \cup ... \cup a_{m+1}). \qquad (**)$$

For $i = 1, ..., m+1$ set

$$A_i = \{a_1, ..., a_{i-1}, a_{i+1}, ..., a_{m+2}\}, \; d_i = \nabla(a_1, ..., a_{i-1}, a_{i+1}, ..., a_{m+2}). \qquad (***)$$

We distinguish the following cases.

1) Let $x \in a_{m+2}$. If $1 \leq i \leq t$, then in view of (**) and (***) the element x belongs to exactly t sets from A_i (namely, $a_1, ..., a_{i-1}, a_{i+1}, ..., a_t, a_{m+2}$), while for $t < i \leq m+1$ the element x belongs to exactly $t+1$ sets from A_i (namely, $a_1, ..., a_t, a_{m+2}$).

a). Let $t < s$. Then by the definition of ∇ the element x contributes 1 to d_i for all $i = 1, ..., m+2$, and, hence, it contributes $m+1$ to the right-hand side of (*). It contributes τ to its left-hand side whereby by hypothesis $\tau \leq s \leq m+1$.

b). Thus, let $t = s$. Then x contributes 1 to d_i exactly if $1 \leq i \leq t$; this is due to the fact that $t+1 > t = s$ and, therefore, x contributes 0 to d_i for $i > t$. In this case the contribution of x to the left-hand side of (*) is τ and to its right-hand side is t whereby $\tau \leq s = t$.

2) Thus, let $x \notin a_{m+2}$. Notice that for every $i = t+1, ..., m$, the element x belongs to exactly t sets from A_i (namely, $a_1, ..., a_t$), while for $i = 1, ..., t$, it belongs to exactly $t-1$ sets from A_i (namely, $a_1, ..., a_{i-1}, a_{i+1}, ..., a_t$).

a). Let $r < t$. Again x contributes τ to the left-hand side of (*) and $m+1$ to the right-hand side whereby $\tau \leq s \leq m+1$.

b). Let $t = r$. Then x contributes τ and $m+1-t$ to the left and right-hand sides of (*) whereby $\tau \leq m+1-r \leq m+1-t$. This proves (i).

(ii) Choose

$$a_1 = ... = a_r = \{1\}, a_{r-1} = ... = a_{m+2} = \emptyset.$$

As $\bigtriangledown(a_1, ..., a_{m+1}) = \bigtriangledown(\{1\}, ..., \{1\}, \emptyset, ..., \emptyset) = 1$, while $d_i = 1$ for $r < i \leq m$ and $d_i = 0$ for $1 \leq i \leq r$, the value of τ in (*) satisfies $\tau \leq m+1-r$. Finally choose $a_1 = ... = a_s = a_{m+2} = \{1\}$, and $a_{s+1} = ... = a_{m-1} = \emptyset$. Proceeding as above we obtain $\tau \leq s$. This shows that $\tau = \min\{m+1-r, s\}$ is the best possible and proves (ii).

(iii) From (I) clearly $s \geq 1$ and $n+1-r \geq 1$ whence $\tau \geq 1$ and the m-simplex inequality follows from (*).

□

Example 1. Consider $\bigtriangledown = \bigtriangledown^{1m}_{mn}$. Clearly, for all subsets $a_1, ..., a_{m+1}$ of V_n it holds

$$\bigtriangledown(a_1, ..., a_{m+1}) = |(a_1 \cup ... \cup a_{m+1}) \setminus (a_1 \cap ... \cap a_{m+1})|, \qquad (II)$$

i.e., it counts the number of elements that belong to some a_i but not to all a_i. Notice that $\tau = n$. From (II) it follows that

$$\bigtriangledown(a_1, ..., a_{m+1}) = 0 \iff a_1 = ... = a_{m+1};$$

whence, the m-semimetric $\bigtriangledown$ satisfies only the weakest variant $d(x_1, ..., x_1) = 0$ of the axiom *"$d(x_1, ..., x_{m+1}) = 0$ whenever $x_1, ..., x_{m+1}$ are not pairwise distinct"*. However, for all $a_1, ..., a_m \subseteq V_n$, there exists $u \subseteq V_n$ such that $\bigtriangledown(a_1, ..., a_m, u) > 0$ and so $\bigtriangledown$ satisfies the strongest variant of the axiom *"if $x_1, ..., x_m$ are pairwise distinct, then $d(x_1, ..., x_m, x_{m+1}) > 0$ for some $x_{m+1} \in X$"*.

For a set A and a non-negative integer k denote by $C(A, k)$ the family of all k-element subsets of A. Set $K = \{1, ..., m+1\}$. For $a_1, ..., a_{m+1} \subseteq V_n$ we have the formula

$$\bigtriangledown(a_1, ..., a_{m+1}) = \sum_{i=1}^{m+1} (-1)^{i-1} \sum_{G \in C(K,i)} |\bigcap_{g \in G} a_g|,$$

obtained by an inclusion-exclusion, based on (II). For $n = 1$ the metric $\bigtriangledown$ is the Hamming distance $|a_1 \triangle a_2|$, mentioned before. The case $m = 2$ was considered in [Dill77], Example 3.

Let $\mathcal{C} \subseteq P(V_n)$. In [Bass95] Bassalygo introduced the following "function of supports" of $\mathcal{C}$. For $0 < m \leq |\mathcal{C}|$ denote by s_m the smallest value of $\bigtriangledown^{1,m-1}_{m-1,n}(a_1, ..., a_m)$ for pairwise distinct $a_1, ..., a_m \in \mathcal{C}$. We have $s_1 \leq s_2 \leq ...$. Let $d_1, ..., d_k$ denote the longest strictly increasing subsequence of $s_1, s_2, ...$. Bassalygo calls d_j the j-th *generalized Hamming distance* of $\mathcal{C}$.

Example 2. Consider $\bigtriangledown = \bigtriangledown^{11}_{mn}$ (i.e., $K = \{1\}$). Clearly, for all subsets $a_1, ..., a_{m+1}$ of V_n the number $\bigtriangledown(a_1, ..., a_{m+1})$ counts the number of elements of V_n that belong to exactly one of the sets $a_1, ..., a_{m+1}$. Notice that $\tau = 1$. For pairwise disjoint $a_1, ..., a_{m+1}$ and $a_{m+2} = a_1 \cup ... \cup a_{m+1}$ the m-simplex inequality is sharp. There is again an inclusion-exclusion formula

$$\bigtriangledown(a_1, ..., a_{m+1}) = \sum_{i=1}^{m+1} (-1)^{i-1} \sum_{G \in C(K,i)} |\bigcap_{g \in G} a_g|$$

(see [Rior68], §1.3). Again $\bigtriangledown$ was considered in [Dill77], Example 2.

Example 3. The next example settles the case $r = s = n + 1$ not covered by proved proposition: *the m-distance $\bigtriangledown = \bigtriangledown^{m+1,m+1}_{mn}$ is not an m-semimetric.*

Indeed, $\bigtriangledown(a_1, ..., a_{m+1}) = |a_1 \cap ... \cap a_{m+1}|$. For $a_1 = ... = a_{m+1} = \{1\}$ and $a_{m+2} = \emptyset$, the m-simplex inequality becomes $1 \leq 0$.

We can extend the above proposition to a family of disjoint intervals in $\{1, ..., m + 1\}$.

Corollary 5.1 *Let $n > 1, V_n = \{1, 2, ..., n\}$, and let*

$$K = \bigcup_{i=1}^{k} \{r_i, r_i + 1, ..., s_i\},$$

where

$$1 \leq r_1 \leq s_1 < ... < r_k \leq s_k \leq m + 1, \quad r_k \leq m.$$

For all (not necessarily distinct) subsets $a_1, ..., a_{m+1}$ of V_n denote by $\bigtriangledown^K_{mn}(a_1, ..., a_{m+1})$ (shortly by $\bigtriangledown(a_1, ..., a_{m+1})$) the number of elements of V_n that belong to exactly i sets amongst $a_1, ..., a_{m+1}$ for some $i \in K$. Set $\tau = \min\{m + 1 - r_k, s_1\}$. Then
(i) for all $a_1, ..., a_{m+2} \subseteq V_n$ it holds

$$\tau \times \bigtriangledown(a_1, ..., a_{m+1}) \leq \sum_{i=1}^{m+1} \bigtriangledown(a_1, ..., a_{i-1}, a_{i+1}, ..., a_{m+2});$$

(ii) the coefficient τ above is the largest possible;
(iii) $\bigtriangledown$ is an m-semimetric on $P(V_n)$.

Proof. Clearly, $\bigtriangledown = \bigtriangledown^{r_1 s_1}_{mn} + ... + \bigtriangledown^{r_k s_k}_{mn}$ and the statement follows from the proposition. □

Remark 1. For $\alpha_1, ..., \alpha_k \in \mathbb{R}_{\geq 0}$ the same result holds for $\bigtriangledown = \alpha_1 \bigtriangledown^{r_1 s_1}_{mn} + \cdots + \alpha_k \bigtriangledown^{r_k s_k}_{mn}$.

Remark 2. If in the corollary we set $r_1 = s_1 = 1, r_2 = s_2 = 3, ..., r_k = s_k = 2k + 1$, where $k = \lfloor \frac{1}{2} n \rfloor$, then we obtain that $\bigtriangledown^K_{mn}(a_1, ..., a_{m+1})$ is the number of elements of V_n, contained in precisely an odd number of sets amongst $a_1, ..., a_{m+1}$. The m-semimetric $\bigtriangledown^K_{mn}$ has been introduced in [PoTo96]; the paper gives an inclusion-exclusion type formula for $\bigtriangledown^K_{mn}$ and Bonferroni-type inequalities.

m-way m-metrics

In this section we extend the three-way distance from [JoCa95] to m-metrics.

Definition 5.10 *Let $m > 0$. A totally symmetric map $d : X^m \longrightarrow \mathbb{R}_{\geq 0}$ is called a weak m-way distance, if for all $x_1, ..., x_{m+1} \in X$ it holds:*

- $d(x_1, ..., x_1) = 0$;
- $d(x_1, ..., x_m) \leq \sum\limits_{i=2}^{m+1} d(x_1, ..., x_{i-1}, x_{i+1}, ..., x_{m+1})$.

Notice that the condition $d(x_1, ..., x_m) \leq \sum\limits_{i=2}^{m+1} d(x_1, ..., x_{i-1}, x_{i+1}, ..., x_{m+1})$ is stronger than the m-simplex inequality, because the summation only starts at $i = 2$.

Clearly, every weak m-way distance determines an $(m-1)$-semimetric.

Definition 5.11 *A weak m-way distance d is called an m-way distance if for all $x_1, ..., x_m \in X$ it holds*

- $d(x_1, x_1, x_3, ..., x_m) = d(x_1, x_3, x_3, ..., x_m) \leq d(x_1, x_2, ..., x_m)$.

In view of the total symmetry, the last condition implies that $d(x_1, ..., x_m)$ only depends on the k-element set $\{x_{i_1}, ..., x_{i_k}\}$, where $1 \leq i_1 < ... < i_k \leq m$, such that $\{x_1, ..., x_m\} = \{x_{i_1}, ..., x_{i_k}\}$.

The following construction extends a concept from [JoCa95].

Definition 5.12 *Given a set X, let $\alpha : X \longrightarrow \mathbb{R}_{\geq 0}$, and $m > 2$. Let $x_1, ..., x_m \in X$, and $0 \leq i_1 < ... < i_k \leq m$ be such that $|\{x_1, ..., x_m\}| = |\{x_{i_1}, ..., x_{i_k}\}| = k$. The star m-distance is a function $d_\alpha : X^m \longrightarrow \mathbb{R}_{\geq 0}$, defined by*

$$d_\alpha(x_1, ..., x_m) = \begin{cases} \alpha(x_{i_1}) + \cdots + \alpha(x_{i_k}), & \text{if } k > 1, \\ 0, & \text{if } k = 1. \end{cases}$$

Proposition 5.2 *The star m-distance d_α is an m-way distance. Moreover, d_α satisfies for all $x_1, ..., x_{m+1} \in X$ the condition*

$$(m-2)d_\alpha(x_1, ..., x_m) \leq \sum_{i=2}^{m} d_\alpha(x_1, ..., x_{i-1}, x_{i+1}, ..., x_{m+1})$$

with equality if and only if $\alpha(x_i) > 0$ implies, that $\alpha(x_{m+1}) = 0$ and that x_i appears only once amongst $x_1, ..., x_m$ $(i = 1, ..., m+1)$.

Proof. There is nothing to prove in

$$(m-2)d_\alpha(x_1, ..., x_m) \leq \sum_{i=2}^{m} d_\alpha(x_1, ..., x_{i-1}, x_{i+1}, ..., x_{m+1})$$

if $x_1 = ... = x_m$. Thus, we may assume that the sequence $\langle x_1, ..., x_m \rangle$ has the form $\langle x_1, ..., x_1, ..., x_k, ..., x_k \rangle$, where $x_1, ..., x_k$ are distinct and x_i appears with the frequency φ_i

$(i = 1, ..., k)$. Suppose $\alpha(x_i) > 0$ for some $1 \leq i \leq k$. If $\varphi_i > 1$ or $\alpha(x_{m+1}) = \alpha(x_i)$, then $\alpha(x_i)$ appears in each $d_\alpha(x_1, ..., x_{j-1}, x_{j+1}, ..., x_{m+1})$ with $2 \leq j \leq m$, and, hence, $\alpha(x_i)$ appears $m-1$ times on the right-hand side of the equality. If $\varphi_i = 1$ and $\alpha(x_{m+1}) \neq \alpha(x_i)$, then $\alpha(x_i)$ appears only $m-2$ times on the right hand side of the inequality. For the equality we need the latter case and $\alpha(x_{m+1}) = 0$.

The conditions $d_\alpha(x_1, ..., x_1) = 0$ and

$$d(x_1, x_1, x_3, ..., x_m) = d(x_1, x_3, x_3, ..., x_m) \leq d(x_1, x_2, ..., x_m)$$

follow from the definition of d_α, and the inequality

$$d(x_1, ..., x_m) \leq \sum_{i=2}^{m+1} d(x_1, ..., x_{i-1}, x_{i+1}, ..., x_{m+1})$$

follows from the inequality

$$(m-2)d_\alpha(x_1, ..., x_m) \leq \sum_{i=2}^{m} d_\alpha(x_1, ..., x_{i-1}, x_{i+1}, ..., x_{m+1}).$$

□

Proposition 5.3 *Let d be a weak m-way distance on X. Define $d' : X^{m+1} \to \mathbb{R}_{\geq 0}$ by setting for all $x_1, ..., x_{m+1} \in X$*

$$d'(x_1, ..., x_{m+1}) = \sum_{i=1}^{m+1} d(x_1, ..., x_{i-1}, x_{i+1}, ..., x_{m+1}).$$

Then d' is a weak $(m+1)$-way distance on X.

Proof. Clearly, d' is totally symmetric and satisfies the condition $d'(x_1, ..., x_1) = 0$. To prove the second condition from the definition of a weak $(m+1)$-way distance, let $x_1, ..., x_{m+2} \in X$. Then from $d'(x_1, ..., x_{m+1}) = \sum_{i=1}^{m+1} d(x_1, ..., x_{i-1}, x_{i+1}, ..., x_{m+1})$ and

$d(x_1,...,x_m) \leq \sum_{i=2}^{m+1} d(x_1,...,x_{i-1},x_{i+1},...,x_{m+1})$ it holds:

$$\begin{aligned}
d'(x_1,...,x_{m+1}) &= d(x_2,...,x_{m+1}) + \sum_{i=2}^{m+1} d(x_1,...,x_{i-1},x_{i+1},...,x_{m+1}) \\
&\leq \sum_{j=3}^{m+1} d(x_2,...,x_{j-1},x_{j+2},...,x_{m+2}) \\
&\quad + \sum_{i=2}^{n+1} d(x_1,...,x_{i-1},x_{i+1},...,x_{m+1}) \\
&\leq \sum_{j=3}^{m+1} d(x_2,...,x_{j-1},x_{j+2},...,x_{m+2}) \\
&\quad +2 \sum_{2\leq k<l\leq m+1} d(x_1,...,x_{k-1},x_{k+1},...,x_{l-1},x_{l+1},...,x_{m+2}) \\
&\leq \sum_{p=2}^{m+1} (\sum_{q=1}^{p-1} d(x_1,...,x_{q-1},x_{q+1},...,x_{p-1},x_{p+1},...,x_{m+2}) \\
&\qquad + \sum_{q=p+1}^{m+2} d(x_1,...,x_{p-1},x_{p+1},...,x_{q-1},x_{q+1},...,x_{m+2})) \\
&= \sum_{p=2}^{m+1} d'(x_1,...,x_{p-1},x_{p+1},...,x_{m+2}).
\end{aligned}$$

□

We can also construct an $(m-1)$-way distance from an m-way distance.

Proposition 5.4 *Let $m > 2$ and let d be an m-way distance on X. For all $x_1,...,x_{m-1} \in X$ set*

$$d'(x_1,...,x_{m-1}) = d(x_1,x_1,x_2,...,x_{m-1}).$$

Then d' is an $(m-1)$-way distance on X.

Proof. The condition $d'(x_1,...,x_1) = 0$ is obvious. To prove the second condition from the definition of an $(m-1)$-way distance, let $x_1,...,x_m \in X$. Then, using that d is an m-way distance, we get, that

$$\begin{aligned}
d'(x_1,...,x_m) &= d(x_1,x_1,x_2,...,x_m) \leq d(x_1,...,x_{m+1}) \\
&\leq \sum_{j=2}^{m+1} d(x_1,...,x_{j-1},x_{j+1},...,x_{m+1},x_{m+1}) \\
&= \sum_{j=2}^{m+1} d(x_1,x_1,x_2,...,x_{j-1},x_{j+1},...,x_{m+1}) \\
&= \sum_{j=2}^{m+1} d(x_1,...,x_{j-1},x_{j+1},...,x_{m+1}).
\end{aligned}$$

To show third condition from the definition of an $(m-1)$-way distance for d', let $x_1,...,x_m \in X$. Then

$$\begin{aligned}
d'(x_1,x_1,x_3,...,x_m) &= d(x_1,x_1,x_1,x_3,...,x_m) = d(x_1,x_1,x_3,x_3,...,x_m) \\
&\leq d(x_1,x_1,x_2,...,x_m)
\end{aligned}$$

and so above condition holds for d'.

□

The next result holds only for X finite.

Proposition 5.5 *Let $m > 1$, and let d be an m-way distance on a finite set X. For all $x_1, ..., x_m \in X$ set*

$$d'(x_1, ..., x_m) = \sum_{x \in X} d(x, x_1, ..., x_m).$$

Then d' is an $(m-1)$-semimetric on X.

Proof. To prove the $(m-1)$-simplex inequality for d', let $x_1, ..., x_{m+1} \in X$. Applying the definition of d' and the fact, that d is an m-way distance, we obtain

$$\begin{aligned} d'(x_1, ..., x_m) &= \sum_{x \in X} d(x, x_1, ..., x_m) \\ &\leq \sum_{x \in X} \sum_{j=1}^{m} d(x, x_1, ..., x_{j-1}, x_{j+1}, ..., x_{m+1}) \\ &= \sum_{j=1}^{m} \sum_{x \in X} d(x, x_1, ..., x_{j-1}, x_{j+1}, ..., x_{m+1}) \\ &= \sum_{j=1}^{m} d'(x_1, ..., x_{j-1}, x_{j+1}, ..., x_{m+1}). \end{aligned}$$

□

Volumes and (m, s)-supermetrics

An natural example of an m-hemimetric is the m-dimensional volume in $\mathbb{R}^n$ with $n \geq m$.

The same notion of m-volume, restricted to some subset X of $\mathbb{R}^n$, gives examples of (m, s)-supermetrics with $s \geq 1$ (see [DDM04)]). One can check, for example, that the m-dimensional volume is, moreover, an (m, s)-supermetric:

(i) if X is $m+2$ vertices of regular $(m+1)$-simplex, and $s = m+1$;

(ii) if X is six vertices of octahedron, $m = 2$, and $s = 1 + \sqrt{3}$.

Let m be a positive integer, and X be a finite point-set in a Euclidean space with size $|X| \geq m+2$. Let $\rho : X^{m+1} \to [0, \infty)$ be an m-hemimetric on X. If ρ is not identically zero, the maximum value of s such that ρ is an (m, s)-supermetric on X is called the *super-bound* of ρ and denoted by $s(\rho, X)$. If ρ is identically zero, then we put $s(\rho, X) = \infty$.

For $B = \{x_1, ..., x_{m+1}\}$, we may write $\rho(B)$ instead $\rho(x_1, ..., x_{m+1})$. Notice that if $s(\rho, X) < \infty$, then $s(\rho, X)$ is the minimum value of $s(\rho, A)$ for $(m+2)$-set $A \subset X$, where

$$s(\rho, A) = \min_{y \in A} \left\{ \frac{\sum_{x \in A \backslash \{y\}} \rho(A \backslash \{x\})}{\rho(A \backslash \{y\})} \right\}.$$

Theorem 5.1 *If ρ is an m-hemimetric on X that is not identically zero, then*

$$s(\rho, X) \leq m+1.$$

Proof. Since ρ is not identically zero, there is an $(m+2)$-set $A \subset X$, such that ρ is not identically zero on A. Let $B_1, B_2, ..., B_{m+2}$ be the $(m+1)$-subsets of A. We may suppose that

$$\rho(B_1) \leq \rho(B_2) \leq ... \leq \rho(B_{m+2}).$$

Then $\rho(B_{m+2}) > 0$, and

$$s(\rho, A) \leq \frac{\rho(B_1) + ... + \rho(B_{m+1})}{\rho(B_{m+2})} \leq m+1,$$

which proves the Theorem.

□

An example of an m-hemimetric is the m-dimensional Euclidean volume μ_m, that is, $\mu_m(x_1, ..., x_{m+1})$ is the m-dimensional volume of the convex hull of the point-set $\{x_1, ..., x_{m+1}\}$. Thus, μ_1 is the usual Euclidean distance.

Let us denote the super-bound $s(\mu_m, X)$ simply by $s_m(X)$. The following two Theorems show, that, in some cases, the value of $s_m(X)$ determines the "configuration" X to some extent.

Theorem 5.2 *Let A be a set of $m+2$ points. Then $s_m(A) = 1$ if and only if the convex hull of A is an m-dimensional simplex.*

Proof. Denote the convex hull of X by $Conv(X)$.

Suppose $s_m(A) = 1$. Then $\dim(Conv(A)) \geq m$, for otherwise, we should have $s_m(A) = \infty$ by definition. If $\dim(Conv(A)) \geq m+1$, then $Conv(A)$ is an $(m+1)$-dimensional simplex. In this case, it is easy to see that $s_m(A) > 1$, a contradiction. Therefore, we can deduce that $\dim(Conv(A)) = m$. If $Conv(A)$ is not an m-dimensional simplex, $m+2$ points are all vertices of the polytope $Conv(A)$. Then, it is not difficult to see that for any $y \in A$, $Conv(A \backslash \{y\})$ is a proper subset of $\bigcup_{x \in A \backslash \{y\}} Conv(A \backslash \{x\})$. Hence, we also have $s_m(A) > 1$, a contradiction. Thus, $Conv(A)$ must be an m-dimensional simplex.

Conversely, if $Conv(A)$ is an m-dimensional simplex, then there is $y \in A$, such that $y \in Conv(A \backslash \{y\})$. Then $Conv(A \backslash \{y\}) = \bigcup_{x \in A \backslash \{y\}} Conv(A \backslash \{x\})$. Hence, we have $s_m(A) = 1$.

□

Theorem 5.3 *For a point set X containing at least 5 points, the following three conditions are equivalent:*

(0) *X is the vertex set of a regular simplex;*
(1) *$s_1(X) = 2$;*
(2) *$s_2(X) = 3$.*

Remark.

(i) The assumption $|X| \geq 5$ is necessary, because the condition (2) also holds for the 4 vertices of a parallelogram.

(ii) For any $k \geq 5$, there is an X of size k that is *not* the vertex set of a regular simplex, but $s_3(X) = 4$.

Proof. Since (0) $\Leftrightarrow$ (1) and (0) $\Rightarrow$ (2) are obvious, we show (2) $\Rightarrow$ (0). Let x, y, z, w be four distinct points in X, and suppose that among the triples in $\{x, y, z, w\}$, μ_2 becomes maximum at $\{x, y, z\}$. Then the inequality

$$3 \times \mu_2(x, y, z) \leq \mu_2(x, y, w) + \mu_2(x, z, w) + \mu_2(y, z, w)$$

implies that μ_2 is constant on the triples in $\{x, y, z, w\}$. Thus, for any four distinct points x, y, z, w, we have $\mu_2(x, y, z) = \mu_2(x, y, w)$. Therefore, for any two triples $\{x, y, z\}$ and $\{u, v, w\}$ in X, we have $\mu_2(x, y, z) = \mu_2(u, v, w)$. Since $s_2(X) < \infty$ implies that μ_2 is not identically zero, the value of μ_2 is a positive constant.

Now the Theorem follows from the result obtained in [FrMa90]: *If $|X| \geq 5$ and all triangles, each spanned by a triple of X, have the same non-zero area, then X forms the vertex set of a regular simplex.*

□

As given in an example in [FrMa90], the set

$$X = \{(\pm 1, 0, ..., 0), (0, 1, 0, ..., 0), (0, 0, 1, 0, ..., 0), ..., (0, ..., 0, 1)\}$$

satisfies $s_3(X) = 4$. Since we cannot think of any other example, it seems true that if $s_3(X) = 4$, then $X \backslash \{\text{one point}\}$ is the vertex set of a regular simplex.

Problem. For $m > 3$, does the condition

$(m) \quad s_m(X) = m + 1$

imply the condition (0) of Theorem 5.3?

It will be of some interest to find $s_m(X)$ for several typical point-sets X. Using computer, we found the values of s_m for the vertex set of five Platonic solids.

vertex set of	s_1	s_2	s_3
tetrahedron	2	3	-
octahedron	$\sqrt{2}$	$1 + \sqrt{3}$	3
cube	$\frac{1+\sqrt{2}}{\sqrt{3}}$	$\sqrt{3}$	2
dodecahedron	$\sqrt{2}\frac{\sqrt{5}+2}{\sqrt{5}+3}$	$\sqrt{3}\frac{\sqrt{22\sqrt{5}+50}}{3\sqrt{5}+7}$	$\frac{27\sqrt{5}+61}{21\sqrt{5}+47}$
icosahedron	$\frac{4}{1+\sqrt{5}}$	$\frac{\sqrt{6\sqrt{5}+30+2\sqrt{24\sqrt{5}+60}}}{3\sqrt{5}+9}$	$\frac{2\sqrt{5}+3}{\sqrt{5}+2}$

For the vertex set V of the 4-dimensional polytope 24-*cell*, we have

$$s_1(V) = \frac{2}{\sqrt{3}}, \; s_2(V) = \frac{1 + 2\sqrt{2}}{\sqrt{5}}, \; s_3(V) = \frac{1 + 3\sqrt{2}}{\sqrt{7}}, \; s_4(V) = \frac{5}{3}.$$

Let O_n be the vertex set of the n-dimensional octahedron and let Q_n denote the vertex set of the n-dimensional cube in $\mathbb{R}^n$.

Theorem 5.4 *For $n > m$,*

$$s_m(O_n) = \begin{cases} \sqrt{2}, & m = 1, \\ 1 + \sqrt{3}, & m = 2, \\ 3, & m \geq 3; \end{cases}$$

$$s_m(Q_n) \leq \begin{cases} \sqrt{3}, & m = 2, \\ 2, & m \geq 3. \end{cases}$$

Remark. For the cube-case, we have the following exact values:
$s_2(Q_4) = \frac{1+2\sqrt{2}}{\sqrt{5}}$, $s_2(Q_5) = \frac{1+2\sqrt{3}}{\sqrt{7}}$, $s_2(Q_6) = \frac{1+2\sqrt{4}}{\sqrt{9}}$,
$s_2(Q_7) = \frac{1+2\sqrt{5}}{\sqrt{11}}$, $s_2(Q_8) = \frac{1+2\sqrt{6}}{\sqrt{13}}$,
$s_3(Q_4) = \frac{1+3\sqrt{2}}{\sqrt{7}}$, $s_3(Q_5) = \frac{1+3\sqrt{3}}{\sqrt{10}}$, $s_3(Q_6) = \frac{1+3\sqrt{4}}{\sqrt{13}}$, $s_3(Q_7) = \frac{1+3\sqrt{5}}{\sqrt{16}}$,
$s_4(Q_5) = \frac{1+4\sqrt{2}}{\sqrt{13}}$, $s_4(Q_6) = \frac{\sqrt{2}+\sqrt{3}+3\sqrt{4}}{\sqrt{25}}$.

The exact values for s_2, s_3 seem to suggest the formulas

$$\frac{1+2\sqrt{n-2}}{\sqrt{2n-3}}, \quad \frac{1+3\sqrt{n-2}}{\sqrt{3n-5}}$$

for $s_2(Q_n), s_3(Q_n)$, respectively. (In fact, by following the patterns of the vertices giving these values, we can prove that they are upper bounds of $s_2(Q_n)$ and $s_3(Q_n)$.) But they do not give the correct values for large n, since we have the following asymptotic result: for any fixed $m > 0$, it holds $\lim_{n\to\infty} s_m(Q_n) = 1$ (see the corresponding theorem below).

Proof. 1. At first, we show the O_n-case of Theorem. We may put

$$O_n = \{(\pm 1, 0, ..., 0), (0, \pm 1, 0, ..., 0), ..., (0, ..., 0, \pm 1)\}.$$

Let us find the minimum value of

$$\frac{\sum_{x\in A\backslash\{y\}} \mu_m(A\backslash\{x\})}{\mu_m(A\backslash\{y\})}$$

for $(m+2)$-set $A \subset O_n$ and $y \in A$. Since $s_1(O_n) = \sqrt{2}$ is obvious, we assume that $m \geq 2$. Denote the vertices of O_n by p_i, q_i, $i = 1, 2, ..., n$, where p_i has i-th coordinate $+1$ and other coordinates 0, and q_i has i-th coordinate -1. Notice that we need not consider the case that the dimension of the convex hull of A is less than m. If the dimension of the convex hull of A is $\geq m$, then A cannot contain three pairs of vertices q_i, p_i; q_j, p_j; q_k, p_k. So, we may consider only such an A that is *congruent* to one of the following three sets A_1, A_2, A_3:

$$\begin{aligned} A_1 &= \{p_1, p_2, p_3, p_4, ..., p_{m+2}\} \text{ (all suffixes are different);} \\ A_2 &= \{q_1, p_1, p_2, p_3, ..., p_{m+1}\} \text{ (only one suffix appears twice);} \\ A_3 &= \{q_1, p_1, q_2, p_2, p_3, ..., p_m\} \text{ (only two suffixes appear twice).} \end{aligned}$$

Since A_1 spans an $(m+1)$-dimensional regular simplex,

$$s_m(A_1) = m + 1.$$

The third set A_3 contains 4 vertices q_1, p_1, q_2, p_2 forming a square, and A_3 spans an m-dimensional polytope. If an $(m+1)$-subset B of A_3 contains the 4 vertices q_1, p_1, q_2, p_2, then B spans $(m-1)$-dimensional polytope, and $\mu_m(B) = 0$. So, in order to get a B with non-zero μ_m value, we have to remove one of q_1, p_1, q_2, p_2 from A_3. Thus, among the $(m+1)$-subsets of A_3, exactly four of them have positive μ_m values, and their values are all equal. Hence,

$$s_m(A_3) = 3.$$

Now we consider the set A_2. Since $A_2 \backslash \{q_1\}$ spans a regular m-simplex of edge-length $\sqrt{2}$, we have

$$\mu_m(A_2 \backslash \{q_1\}) = \mu_m(A_2 \backslash \{p_1\}) = \frac{\sqrt{m+1}}{m!},$$

and

$$\mu_m(A_2 \backslash \{p_2\}) = \mu_m(A_2 \backslash \{p_3\}) = ... = \mu_m(A_2 \backslash \{p_{m+1}\}) = \frac{2}{m!}.$$

Hence, for $m = 2$,

$$s_2(A_2) = \frac{2/2! + 2\sqrt{3}/2!}{2/2!} = 1 + \sqrt{3},$$

and, for $m \geq 3$,

$$s_m(A_2) = \frac{\sqrt{m+1}/m! + m(2/m!)}{\sqrt{m+1}/m!} = 1 + \frac{2m}{\sqrt{m+1}} \geq 4.$$

Therefore, $s_2(O_n) = 1 + \sqrt{3}$, and $s_m(O_n) = 3$ for $m \geq 3$.

2. Let us consider now the Q_n-case of Theorem. We may suppose that Q_n is the vertex set of the n dimensional cube, spanned by the n vectors

$$(1, 0, ..., 0), (0, 1, 0, ..., 0), ..., (0, ..., 0, 1)$$

in $\mathbb{R}^n$. To get the first inequality, consider the set of 4 points

$$\begin{gathered}(0,0,0,0,...,0),\\(1,0,0,0,...,0),\\(0,1,0,0,...,0),\\(0,0,1,0,...,0).\end{gathered}$$

The value of s_2 on this set is

$$\frac{1/2 + 1/2 + 1/2}{\sqrt{3}/2} = \sqrt{3}.$$

Hence, $s_2(Q_n) \leq \sqrt{3}$.

Now suppose $m \geq 3$. To see clearly, let us consider the case $m = 5, n = 8$. Let A be the set of the following $m + 2 = 7$ vertices of Q_8:

$$\begin{aligned}
p_0 &= (0,0,0,0,0,0,0,0),\\
p_1 &= (1,0,0,0,0,0,0,0),\\
p_2 &= (0,1,0,0,0,0,0,0),\\
p_3 &= (0,0,1,0,0,0,0,0),\\
p_4 &= (1,1,1,0,0,0,0,0),\\
p_5 &= (0,0,0,1,0,0,0,0),\\
p_6 &= (0,0,0,0,1,0,0,0).
\end{aligned}$$

To compute $s_5(A)$, we need the values $\sigma_i = \mu_5(A \backslash \{p_i\}), i = 0, 1, ..., 6$. It is obvious that $\sigma_5 = \sigma_6 = 0$ from their dimensions. It is also easy to see that $\sigma_4 = 1/5!$. Now, since the 3-dimensional volume of the simplex spanned by p_1, p_2, p_3, p_4 is $1/3 = 2/3!$, we have $\sigma_0 = \frac{2}{3!} \times \frac{1}{4} \times \frac{1}{5} = \frac{2}{5!}$. Similarly, since the 3-dimensional volume of the simplex spanned by p_0, p_2, p_3, p_4 is $1/3!$, we have $\sigma_1 = \sigma_2 = \sigma_3 = 1/5!$. Therefore,

$$s_5(A) = \frac{1/5! + 1/5! + 1/5! + 1/5!}{2/5!} = 2,$$

which proves $s_5(Q_8) \le 2$.

Similar argument clearly works in any $n > m \ge 3$.

□

Theorem 5.5 *For any fixed $m > 0$, it holds $\lim_{n \to \infty} s_m(Q_n) = 1$.*

Proof. We are going to show that for any fixed $m > 0$, $s_m(Q_n) \to 1$ as $n \to \infty$. To be clear, we put $m = 3, n = 3k + 1$. Let

$$\begin{aligned} p_0 &= (0, \overbrace{0, ..., 0}^{k}, \overbrace{0, ..., 0}^{k}, \overbrace{0, ..., 0}^{k}), \\ p_1 &= (0, 1, ..., 1, 0, ..., 0, 0, ..., 0), \\ p_2 &= (0, 0, ..., 0, 1, ..., 1, 0, ..., 0), \\ p_3 &= (0, 0, ..., 0, 0, ..., 0, 1, ..., 1), \\ p_4 &= (1, 0, ..., 0, 0, ..., 0, 0, ..., 0). \end{aligned}$$

Then, since $p_0p_i = \sqrt{k}$, $p_4p_i = \sqrt{k+1}$ for $i = 1, 2, 3$, and $p_0p_4 = 1$, it is easy to see that

$$\mu_m(p_0, p_1, p_2, p_3) = \frac{k^{m/2}}{m!}, \quad \mu_m(p_1, p_2, p_3, p_4) \le \frac{(k+1)^{m/2}}{m!},$$

$$\begin{aligned} \mu_m(p_0, p_2, p_3, p_4) &= \mu_m(p_0, p_1, p_3, p_4) = \\ &= \mu_m(p_0, p_1, p_2, p_4) = \frac{(k+1)^{(m-1)/2}}{m!}. \end{aligned}$$

Hence,

$$1 \le s_m(Q_{3k+1}) \le \frac{(k+1)^{m/2} + m(k+1)^{(m-1)/2}}{k^{m/2}}.$$

Since the right-most-side tends to 1 as $k \to \infty$, $s_m(Q_n)$ tends to 1 as $n \to \infty$. The same argument works for any $m > 0$.

□

So, it is proven, that if α_n, β_n, γ_n denote the vertex set of regular n-simplex, n-hyperoctahedron and n-cube, respectively, it holds:

- $s_m(\alpha_n) = m + 1$ for all $n \ge 2$, $m \ge 1$;
- $s_2(\beta_n) = 1 + \sqrt{3}$ for all n, and $s_m(\beta_n) = 3$ for all n and for all $m \ge 3$;
- $s_2(\gamma_n) = \frac{1+2\sqrt{n-2}}{\sqrt{2n-3}}$ for $n \le 8$, while for any fixed m, $\lim_{n \to \infty} s_m(\gamma_n) = 1$.

Chapter 6

Important special case: partial metrics and weightable quasi-metrics

6.1 Preliminaries

Partial semimetrics are a generalization of semimetrics, having important applications in Computer Science: Domain Theory, Analysis of Data Flow Deadlock, Complexity Analysis of Programs, etc.

For example, Scott's Domain Theory (see [GHKLMS03]) gives partial order and non-Hausdorff topology on partial objects in Computation. In Quantitative Domain Theory, a "distance" between programs (points of a semantic domain) is used to quantify speed (of processing or convergence) or complexity of programs and algorithms. For instance, $x \preceq y$ (program y contains all information from program x) is the *specialization preorder*, and $x \preceq y$ if and only if $p(x,y)=p(x,x)$ for a partial semimetric p on X. In computation over a metric space of totally defined objects, partial semimetric models partially defined information: $p(x,x) > 0$ or $p(x,x) = 0$ mean that object x is partially or totally defined, respectively. For example, for *vague real numbers* x (i.e., non-empty segments of $\mathbb{R}$ as, say, decimals approximating π), the self-distance $p(x,x)$ can be the length of the segment measuring the extent of ambiguity at point x.

Any topology on a finite set X is defined by $cl\{x\} = \{y \in X : y \preceq x\}$ for $x \in X$, where $x \preceq y$ is the specialization preorder, meaning $p(x,y)=p(x,x)$, for some partial semimetric p on X ([GuRi05]). (Not every finite topology is defined from a semimetric on X by this way.)

Partial semimetrics were introduced by Matthews (1992) [Matt92] for treatment of partially defined objects in Computer Science. He also remarked that a quasi-semimetric $q = ((q_{ij}))$ is weightable if and only if the function $((q_{ij}+w_i))$ is a partial semimetric. Weak partial semimetrics were studied in [Heck99] as a generalization of partial semimetrics introduced in [Matt92].

6.2 Definitions

In this chapter, let us use the symbols p, q, or d for a partial semimetric, quasi-semimetric, or semimetric, respectively.

Definition 6.1 *Let X be a set. A function $p : X^2 \longrightarrow \mathbb{R}$ is called weak partial semimetric* *on X, if for all $x, y, z, t \in X$ it holds:*

1. $p(x, y) \geq 0$ *(non-negativity);*
2. $p(x, y) = p(y, x)$ *(symmetry);*
3. $p(x, y) \leq p(x, z) + p(z, y) - p(z, z)$ *(sharp triangle inequality).*

Because of the symmetry and since, in general, $p(i, i) \neq 0$ for $i \in V_n$, we can view a weak partial semimetric p on $V_n = \{1, 2, ..., n\}$ as a vector

$$(p_{ij})_{1 \leq i \leq j \leq n} \in \mathbb{R}^{E_n'},$$

where $E_n' = \binom{n}{2} + n = \binom{n+1}{2} = \frac{n(n+1)}{2}$: $p(i, j) = p_{ij}$. Alternatively, a weak partial semimetric p can be viewed as an $n \times n$ symmetric matrix, which, in general, has non-zero entries on the main diagonal.

A weak partial semimetric becomes a partial semimetric, if we add the *small self-distance conditions.*

Definition 6.2 *Let X be a set. A function $p : X^2 \longrightarrow \mathbb{R}$ is called partial semimetric on X, if for all $x, y, z, t \in X$ it holds:*

1. $d(x, y) \geq 0$ *(non-negativity);*
2. $p(x, y) = p(y, x)$ *(symmetry);*
3. $p(x, y) \geq p(x, x)$ *(small self-distances);*
4. $p(x, y) \leq p(x, z) + p(z, y) - p(z, z)$ *(sharp triangle inequality).*

A partial semimetric becomes a strong partial semimetric, if we add the *large self-distance conditions.*

Definition 6.3 *Let X be a set. A function $d : X^2 \longrightarrow \mathbb{R}$ is called strong partial semimetric on X if for all $x, y, z, t \in X$ it holds:*

1. $p(x, y) \geq 0$ *(non-negativity);*
2. $p(x, y) = p(y, x)$ *(symmetry);*
3. $p(x, y) \geq p(x, x)$ *(small self-distances);*
4. $p(x, x) + p(y, y) \geq p(x, y)$ *(large self-distances);*
5. $p(x, y) \leq p(x, z) + p(z, y) - p(z, z)$ *(sharp triangle inequality).*

Easy to see, that

$$p(x, x) = (p(x, x) + p(y, y) - p(x, y)) + p(y, x) - p(y, y) \geq 0,$$

i.e., for any strong partial semimetric, the symmetry and the conditions of small and large self-distances imply non-negativity.

Definition 6.4 *A weak partial metric, partial metric or strong partial metric p on X is weak partial, partial or strong partial semimetric, respectively, if, in addition, it holds:*

- $p(x,x) = p(x,y) = p(y,y)$ *imply* $x = y$ *(separation axiom).*

A quasi-semimetric q on X is called *weightable* if there exist a *weight function* $w = (w_x) : X \longrightarrow \mathbb{R}_{\geq 0}$, such that for all different $x, y \in X$ it holds:

$$q(x,y) + w_x = q(y,x) + w_y.$$

It means, that the function $d(x,y) = 2q(x,y) + w_x - w_y$ is the *symmetrization semimetric* of q. It leads to the following formal definition.

Definition 6.5 *Let X be a set. A function $q : X \times X \to \mathbb{R}$ is called weightable quasi-semimetric on X if there exists a weight function $w = (w_x) : X \longrightarrow \mathbb{R}_{\geq 0}$, such that for all $x, y, z \in X$ it holds:*

1. $q(x,y) \geq 0$ *(non-negativity);*
2. $q(x,x) = 0$;
3. $q(x,y) \leq q(x,z) + q(z,y)$ *(oriented triangle inequality);*
4. $q(x,y) + w_x = q(y,x) + w_y$ *for* $x \neq y$.

A weightable quasi-semimetric (q, w) with all $q(x,y) \leq w_y$ is called a *weightable strong quasi-semimetric*. But if, on the contrary, the non-negativity condition is weakened to the condition $q(x,y) + q(y,x) \geq 0$ (so, in general, $q(x,y) < 0$ is allowed), the pair (q, w) is called a *weightable weak quasi-semimetric*. It leads to the following formal definitions.

Definition 6.6 *Let X be a set. A function $q : X \times X \to \mathbb{R}$ is called weak weightable quasi-semimetric on X, if there exists a weight function $w = (w_x) : X \longrightarrow \mathbb{R}_{\geq 0}$, such that for all $x, y, z \in X$ it holds:*

1. $q(x,y) + q(y,x) \geq 0$ *(weak non-negativity);*
2. $q(x,x) = 0$;
3. $q(x,y) \leq q(x,z) + q(z,y)$ *(oriented triangle inequality);*
4. $q(x,y) + w_x = q(y,x) + w_y$ *for* $x \neq y$.

Definition 6.7 *Let X be a set. A function $q : X \times X \to \mathbb{R}$ is called strong weightable quasi-semimetric on X, if there exists a weight function $w = (w_x) : X \longrightarrow \mathbb{R}_{\geq 0}$, such that for all $x, y, z \in X$ it holds:*

1. $q(x,y) \geq 0$ *(non-negativity);*
2. $q(x,x) = 0$;
3. $q(x,y) \leq q(x,z) + q(z,y)$ *(oriented triangle inequality);*

4. $q(x, y) + w_x = q(y, x) + w_y$ *for* $x \neq y$;

5. $q(x, y) \leq w_y$.

A weightable quasi-semimetric q on V_n can be viewed alternatively as a vector

$$(q_{ij})_{1 \leq i \leq j \leq n} \in \mathbb{R}^{E_n'},$$

or as an $n \times n$ (in general, non-symmetric) matrix, which, in general, has non-zero entries on the main diagonal: $q(i, j) = q_{ij}, i \neq j$, and $w_i = q_{ii}$.

A *weighted semimetric* $(d; w)$ is a semimetric d with a weight function $w : V_n \to \mathbb{R}_{\geq 0}$ on its points. Call a weighted semimetric $(d; w)$ *down-weighted* or *up-weighted*, if for all distinct $i, j \in V_n$ it holds $d(x, y) \geq w_x - w_y$, or $d(x, y) \leq w_x + w_y$, respectively. Call a weighted semimetric $(d; w)$ *strongly weighted* if it is both, down-weighted and up-weighted. Formally, we get the following definitions.

Definition 6.8 *Let* X *be a set. A function* $d : X \times X \to \mathbb{R}$ *is called weighted semimetric on* X*, if there exists a weight function* $w = (w_x) : X \longrightarrow \mathbb{R}_{\geq 0}$ *on* X*, and for all* $x, y, z \in X$ *it holds:*

1. $d(x, y) \geq 0$ *(non-negativity);*

2. $d(x, x) = 0$;

3. $d(x, y) = d(y, x)$ *(symmetry);*

4. $d(x, y) \leq d(x, z) + d(z, y)$ *(triangle inequality).*

Definition 6.9 *Let* X *be a set. A function* $d : X \times X \to \mathbb{R}$ *is called down-weighted semimetric on* X*, if there exists a weight function* $w = (w_x) : X \longrightarrow \mathbb{R}_{\geq 0}$ *on* X*, such that for all* $x, y, z \in X$ *it holds:*

1. $d(x, y) \geq 0$ *(non-negativity);*

2. $d(x, x) = 0$;

3. $d(x, y) = d(y, x)$ *(symmetry);*

4. $d(x, y) \leq d(x, z) + d(z, y)$ *(triangle inequality);*

5. $d(x, y) \geq w_x - w_y$ *for* $x \neq y$.

Definition 6.10 *Let* X *be a set. A function* $d : X \times X \to \mathbb{R}$ *is called up-weighted semimetric on* X*, if there exists a weight function* $w = (w_x) : X \longrightarrow \mathbb{R}_{\geq 0}$ *on* X*, such that for all* $x, y, z \in X$ *it holds:*

1. $d(x, y) \geq 0$ *(non-negativity);*

2. $d(x, x) = 0$;

3. $d(x, y) = d(y, x)$ *(symmetry);*

4. $d(x, y) \leq d(x, z) + d(z, y)$ *(triangle inequality);*

5. $d(x, y) \leq w_x + w_y$.

Definition 6.11 *Let X be a set. A function $d : X \times X \to \mathbb{R}$ is called strong weighted semimetric on X, if there exists a weight function $w = (w_x) : X \longrightarrow \mathbb{R}_{\geq 0}$ on X, such that for all $x, y, z \in X$ it holds:*

1. $d(x, y) \geq 0$ *(non-negativity);*

2. $d(x, x) = 0$;

3. $d(x, y) = d(y, x)$ *(symmetry);*

4. $d(x, y) \leq d(x, z) + d(z, y)$ *(triangle inequality);*

5. $d(x, y) \geq w_x - w_y$ *for* $x \neq y$;

6. $d(x, y) \leq w_x + w_y$ *for* $x \neq y$.

A weighted semimetric d on V_n can be viewed alternatively as a vector

$$(d_{ij})_{1 \leq i \leq j \leq n} \in \mathbb{R}^{E_n'},$$

or as an $n \times n$ symmetric matrix, which has non-zero entries on the main diagonal: $d(i, j) = d_{ij}, i \neq j$, and $w_i = d_{ii}$.

Consider a weighted semimetric $(d; w)$ on $V_n = \{1, 2, ..., n\}$ as the $(n+1) \times (n+1)$ matrix $((d'_{ij}))$, $0 \leq i, j \leq n$, with $d'_{00} = 0$, $d'_{0i} = d'_{i0} = w_i$ for $i \in V_n$, and $d'_{ij} = d_{ij}$ for $i, j \in V_n$. In other words, the weight w_i can be considered as distance from the point $i \in V_n$ to an additional point 0: $w_i = d(i, 0) = d(0, i)$. In the case of strong weighted semimetric we get for these additional distances corresponding triangle inequalities: $d(i, j) \leq d(i, 0) + d(0, j)$, and $d(i, 0) \leq d(i, j) + d(j, 0)$. So, any strong weighted semimetric on n points corresponds to an ordinary semimetric on $n + 1$ points.

6.3 Examples

Basic examples

- **Natural partial semimetric**

  The simplest example of a *weak partial semimetric* is the *natural partial semimetric* p on $\mathbb{R}_{\geq 0}$, defined, for any $x, y \in \mathbb{R}_{\geq 0}$, as

$$p(x, y) = x + y.$$

  In fact, $p(x, y)$ is a weak partial semimetric if and only if

$$d(x, y) = 2p(x, y) - p(x, x) - p(y, y)$$

  is a weighted semimetric with the weight function $w = (w_i), w_i = p_{ij}$. In our example, the corresponding weighted semimetric is the *non-discrete semimetric* $d(x, y) = 0$ for all $x, y \in \mathbb{R}_{\geq 0}$; its weight function $\{w_x\}_{x \in \mathbb{R}_{\geq 0}}$ is $w_x = 2x$.

The reader can construct many similar examples of partial semimetrics and weightable quasi-semimetrics, using the ideas of previous section and classical examples of semimetrics.

- **Hitting time quasi-metric**

  Consider random walks on a connected graph $G = \langle V, E \rangle$, where at each step walk moves with uniform probability from current vertex to a neighboring one. The *hitting time quasi-metric* $H(u,v)$ on V is the expected number of steps (edges) for a random walk on beginning at vertex u to reach v for the first time; put $H(u,u) = 0$. The *cyclic tour* property of reversible Markov chains implies that $H(u,v)$ is weightable.

  The *commuting time metric* $C(u,v) = H(u,v) + H(v,u)$ is ([Teta91]) $2|E|R(u,v)$, where $R(u,v)$ is the *effective resistance metric*, i.e., 0 if $u = v$, and, otherwise, $\frac{1}{R(u,v)}$ is the current flowing into v, when grounding v and applying 1 volt potential to u (each edge is seen as a resistor of 1 ohm). It holds

$$R(u,v) = \sup_{f:V \to \mathbb{R},\, D(f)>0} \frac{(f(u) - f(v))^2}{D(f)},$$

  where $D(f) = \sum_{st \in E} (f(s) - f(t))^2$.

Connections between weightable quasi-metrics and partial metrics

Given a weighted semimetric $(d;w)$ on V_n, define the map P by the function $p = P(d;w)$ with

$$p_{ij} = \frac{d_{ij} + w_i + w_j}{2}.$$

Clearly, P is an *automorphism* (invertible linear operator) of the vector space $\mathbb{R}^{\binom{n+1}{2}}$, and $(d;w) = P^{-1}(p)$, where the inverse map P^{-1} is defined by

$$d_{ij} = 2p_{ij} - p_{ii} - p_{jj},\ w_i = p_{ii}.$$

Given a weighted semimetric $(d;w)$ on V_n, define the map Q by the function $(q,w) = Q(d;w)$ with

$$q_{ij} = \frac{d_{ij} - w_i + w_j}{2}.$$

Clearly, Q is also an *automorphism* of the vector space $\mathbb{R}^{\binom{n+1}{2}}$, and $(d;w) = Q^{-1}(q)$, where the inverse map Q^{-1} is defined by

$$d_{ij} = q_{ij} + q_{ji},$$

and the weight functions for d and q coincide. In other words, d is the *symmetrization semimetric* of q.

Moreover, it holds

$$q_{ij} = p_{ij} - p_{ii},$$

so, on can write, that

$$Q(d;w) = P(d;w) - ((1)) \cdot w,$$

where $((1))$ is the $n \times n$ matrix, consisting only of unities.

Clearly,

$$d_{ij} + d_{ik} - d_{jk} = p_{ij} + p_{ik} - p_{jk} - p_{ii} = q_{ji} + q_{ik} - q_{jk},$$

i.e., the triangle inequalities are equivalent on all three levels: d - of semimetrics, p - of would-be partial semimetrics and q - of would-be quasi-semimetrics.

Moreover, it holds:

- $2p_{ij} \geq p_{ii} + p_{jj}$ if and only if $d_{ij} \geq 0$ if and only if $q_{ij} + q_{ji} \geq 0$; so, the *non-negativity* condition for weak partial semimetrics is equivalent to the non-negativity condition for weighted semimetrics and to its weaker version for weak weightable quasi-semimetrics;
- $p_{ij} \geq p_{ii}$ if and only if $d_{ij} \geq w_i - w_j$ if and only if $q_{ij} \geq 0$; so, the condition of *small self-distances* for partial semimetrics is equivalent to the *down-weighted* condition for weighted semimetrics and *non-negativity* condition for (weightable) quasi-semimetrics;
- $p_{ij} \leq p_{ii} + p_{jj}$ if and only if $d_{ij} \leq w_i + w_j$ if and only if $q_{ij} \leq w_j$; so, the condition of *large self-distances* for strong partial semimetrics is equivalent to the *up-weighted* condition for strongly weighted semimetrics and corresponding condition for strong weightable quasi-semimetrics.

So, given a weighted semimetric d, we obtain from it by the map P a weak partial semimetric and, by the map Q, a weak weightable quasi-semimetric. Given a down-weighted semimetric d, we obtain from it by the map P a partial semimetric and, by the map Q, a weightable quasi-semimetric. Given a strongly weighted semimetric d, we obtain from it by the map P a strong partial semimetric and, by the map Q, a strong weightable quasi-semimetric.

Example. Three 6×6 matrices below represent the weighted semimetric d on V_6 with the weight function $w = (0,0,0,0,1,1)$, the partial semimetric $p = P(d; w)$, $p_{ij} = \frac{d_{ij}+w_i+w_j}{2}$ (in particular, $p_{ii} = w_i$), and the weightable quasi-semimetric $q = Q(d; w)$, $q_{ij} = \frac{d_{ij}-w_i+w_j}{2} = p_{ij} - p_{ii}$ with the weight function $w = (0,0,0,0,1,1)$.

$$\begin{matrix} \mathbf{0}&2&2&2&1&1\\ 2&\mathbf{0}&0&2&1&1\\ 2&0&\mathbf{0}&2&1&1\\ 2&2&2&\mathbf{0}&1&1\\ 1&1&1&1&\mathbf{1}&0\\ 1&1&1&1&0&\mathbf{1} \end{matrix} \qquad \begin{matrix} \mathbf{0}&1&1&1&1&1\\ 1&\mathbf{0}&0&1&1&1\\ 1&0&\mathbf{0}&1&1&1\\ 1&1&1&\mathbf{0}&1&1\\ 1&1&1&1&\mathbf{1}&1\\ 1&1&1&1&1&\mathbf{1} \end{matrix} \qquad \begin{matrix} \mathbf{0}&1&1&1&1&1\\ 1&\mathbf{0}&0&1&1&1\\ 1&0&\mathbf{0}&1&1&1\\ 1&1&1&\mathbf{0}&1&1\\ 0&0&0&0&\mathbf{1}&0\\ 0&0&0&0&0&\mathbf{1} \end{matrix}$$

Weightable, relaxed symmetric and digraphic quasi-metrics

We say, that a quasi-semimetric q on X has *relaxed symmetry* if for different $x, y, z \in X$ it holds

$$q(x,y) + q(y,z) + q(z,x) = q(x,z) + q(z,y) + q(y,x).$$

Obviously, if a quasi-semimetric q on X has relaxed symmetry, then

$$q(x,y) + q(y,z) - q(x,z) = q(z,y) + q(y,x) - q(z,x),$$

i.e., the triangle inequalities $q(x,y)+q(y,z)-q(x,z) \geq 0$ and $q(z,y)+q(y,x)-q(z,x) \geq 0$ are equivalent.

Theorem 6.1 *([Vito99])*
A quasi-semimetric q on X has relaxed symmetry if and only if it is weightable.

Proof. Relaxed symmetry means

$$q(x,y) - q(y,x) = (q(z,y) - q(y,z)) - (q(z,x) - q(x,z)).$$

Equivalently, q is weightable: fix point $z_0 \in X$ and define

$$w(x) = q(z_0,x) - q(x,z_0) + \max_z (q(z,z_0) - q(z_0,z)) \geq 0$$

for all $x \in X$. On the other hand, it is easy to see that the above equality holds if q is weightable.

□

We introduce the following short notation for the cyclic sum:

$$\sum_{1 \leq i \leq k-1} q(x_i, x_{i+1}) + q(x_k, x_1) = x_1 x_2 ... x_k x_1.$$

Then the relaxing symmetry property has the form $xyzx = xzyx$.
Given $k \geq 3$, a quasi-semimetric q is called *k-cyclically symmetric* if it holds

$$x_1 x_2 x_3 ... x_k x_1 = x_1 x_k x_{k-1} ... x_2 x_1$$

for any different $x_1 x_2 ... x_k \in X$. So, a quasi-semimetric has relaxed symmetry if and only if it 3-cyclically symmetric.

Theorem 6.2 *A quasi-semimetric q on X has relaxed symmetry if and only if it is k-cyclically symmetric for any $k \geq 3$.*

Proof. In fact, it holds

$$(x_1x_2x_3x_1 - x_1x_3x_2x_1) + (x_1x_3x_4x_1 - x_1x_4x_3x_1) + ...$$

$$+ (x_1x_{k-1}x_kx_1 - x_1x_kx_{k-1}x_1) = (x_1x_2...x_kx_1 - x_1x_k...x_2x_1)$$

for any $k \geq 4$. For example, for $k = 4$ it holds

$$(x_1x_2x_3x_1 - x_1x_3x_2x_1) + (x_1x_3x_4x_1 - x_1x_4x_3x_1) = x_1x_2x_3x_4x_1 - x_1x_4x_3x_2x_1.$$

In other direction, we have:

$$(k-2) \cdot (x_1x_2...x_{k-1}x_1 - x_1x_{k-1}...x_2x_1) = (x_1x_2...x_{k-1}x_kx_1 - x_1x_k...x_2x_1)$$

$$+(x_1x_2...x_kx_{k-1}x_1 - x_1x_{k-1}x_k...x_2x_1) + ... + (x_1x_kx_2...x_{k-1}x_1 - x_1x_{k-1}...x_2x_kx_1).$$

For example,

$$2(x_1x_2x_3x_1 - x_1x_3x_2x_1) = (x_1x_2x_3x_4x_1 - x_1x_4x_3x_2x_1) +$$

$$(x_1x_2x_4x_3x_1 - x_1x_3x_4x_2x_1) + (x_1x_4x_2x_3x_1 - x_1x_3x_2x_4x_1).$$

□

Clearly, any finite quasi-semimetric q can be realized as the (shortest directed) path quasi-semimetric of an $\mathbb{R}_{\geq 0}$-weighted digraph: take the complete digraph and put on each arc $\langle i, j \rangle$ the weight q_{ij}. The earliest known to us occurrence of the notion, but not the term, of relaxed symmetry was in Patrinos and Hakimi (1972) [PaHa72].

Theorem 6.3 *(Theorem 5 in [PaHa72])*
A finite quasi-metric q can be realized as the path quasi-metric of an $\mathbb{R}_{\geq 0}$-weighted bidirectional tree (a tree with all edges replaced by 2 oppositely directed arcs) if and only if q has relaxed symmetry and its symmetrization $((q_{ij} + q_{ji}))$ can be realized as the path metric of an $\mathbb{R}_{\geq 0}$-weighted tree.

Part III

Cuts, hypermetrics and their generalizations

Chapter 7

Cuts and their generalizations

7.1 Preliminaries

In this chapter we consider the notions of cut and multicut semimetrics and their oriented and multidimensional generalizations: oriented cuts and multicuts, and partition m-hemimetrics.

The cuts semimetrics play a central role in the study of (semi)metric spaces that can be isometrically embedded into an ℓ_1-space $(\mathbb{R}^m, d_{\ell_1})$ for some integer $m \geq 1$. Here, d_{ℓ_1} denotes the ℓ_1-distance, defined by

$$d_{\ell_1}(x, y) = \sum_{1 \leq i \leq m} |x_i - y_i| \ \text{ for } x, y \in \mathbb{R}^m.$$

In fact, a semimetric d is isometrically ℓ_1-embeddable if and only if it can be decomposed as a non-negative linear combination of cut semimetrics. The *hypercube embeddable* semimetrics on V_n (i.e., isometrically embeddable into the space $(\{0,1\}^m, d_{\ell_1})$ for some $m \geq 1$) can be written as a non-negative integer combination of cut semimetrics.

Moreover, a semimetric on V_n can be decomposed as a non-negative linear combination (respectively, a convex combination) of cut semimetrics on V_n if and only if there exists a measure space (respectively, a probability space) $(\Omega, \mathcal{A}, \mu)$ and n events $A_1, \dots, A_n \in \mathcal{A}$, such that

$$d(i, j) = \mu(A_i \triangle A_j) \ \text{ for all } i, j \in V_n.$$

There is a vast literature on the cut polyhedra (see [DeLa97] for main results and large list of detailed references).

The notion of cut semimetric can be extended to the oriented case, and one obtains *oriented cuts* and *oriented multicuts*, using oriented partitions of the set V_n. These functions are quasi-semimetrics on V_n. This notion was considered, for example, in [ShLi95]; see also [DePa99], [DDP03].

The multidimensional analogues of cut and multicuts semimetrics has been studied in [DeRo00], [DeRo02], [DeDu03], [DeDu03].

7.2 Classical non-oriented case

We start with the formal definition of cut semimetric.

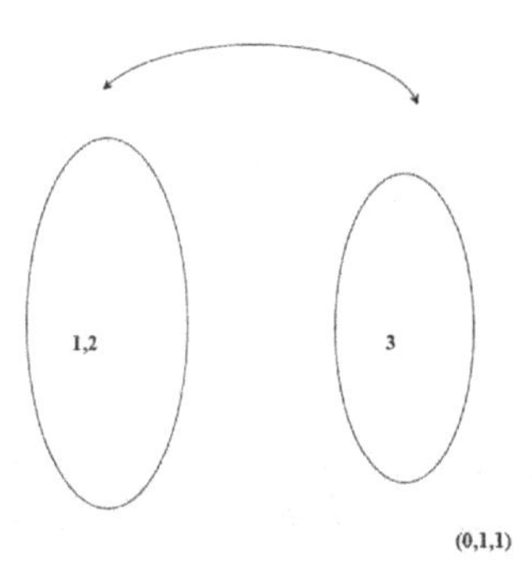

Figure 7.1: Example of a cut semimetric on 3 points

Definition 7.1 *Given a subset S of $V_n = \{1, 2, ..., n\}$, the cut semimetric (split semimetric) δ_S is a semimetric on V_n, defined by*

$$\delta_S(i,j) = \begin{cases} 1, & \text{if } i \neq j, |S \cap \{i,j\}| = 1, \\ 0, & \text{otherwise.} \end{cases}$$

The cut semimetric δ_S can be see as an $\{0,1\}$-vector in $\mathbb{R}^{E_n}$, where $E_n = |\{\{i,j\} \mid 1 \leq i < j \leq n\}|$ (i.e., $E_n = \binom{n}{2} = \frac{n(n-1)}{2}$), defined by $(\delta_S)_{ij} = 1$, if $|S \cap \{i,j\}| = 1$, and $(\delta_S)_{ij} = 0$, otherwise, for $1 \leq i < j \leq n$: $\delta_S(i,j) = (\delta_S)_{ij}$. This vector is called a *cut vector*, or simply a *cut*.

Similarly, the cut semimetric δ_S can be considered as a symmetric $n \times n$ $\{0,1\}$-matrix $(((\delta_S)_{ij}))$ with zeros on the main diagonal, where $(\delta_S)_{ij} = \delta_S(i,j)$.

In fact, any (non-zero) cut can be considered as a *partition* of a set into two (non-empty) parts: given a subset S of $V_n = \{1, 2, ..., n\}$, we obtain the partition $\{S, \overline{S} = V_n \backslash S\}$ of V_n. The cut-semimetric on V_n, defined by this partition, can be seen as a special semimetric on the vertex set of the *complete bipartite graph* $K_{|S|,|\overline{S}|}$, where the distance between vertices is equal to 1 if they belong to different parts of this graph, and is equal to 0, otherwise. Clearly, $\delta_{\overline{S}} = \delta_S$, and the matrix $(((\delta_S)_{ij}))$ is the adjacency matrix of the *cut* (into S and $\overline{S}$) *subgraph* $K_{|S|,|\overline{S}|}$ of K_n.

For example, there are four cuts on 3 points:

$$\begin{array}{ll} S = \emptyset, \overline{S} = V_3 \backslash S = \{1,2,3\}, & \delta_\emptyset = (0,0,0); \\ S = \{1\}, \overline{S} = V_3 \backslash S = \{2,3\}, & \delta_{\{1\}} = (1,1,0); \\ S = \{2\}, \overline{S} = V_3 \backslash S = \{1,3\}, & \delta_{\{2\}} = (1,0,1); \\ S = \{3\}, \overline{S} = V_3 \backslash S = \{1,2\}, & \delta_{\{3\}} = (0,1,1). \end{array}$$

Consider now the definition of multicut semimetric.

Definition 7.2 *Let $\{S_1, ..., S_q\}$, $q \geq 2$, be a partition of the set $V_n = \{1, 2, ..., n\}$, i.e., a collection $S_1, ..., S_q$ of pairwise disjoint non-empty subsets of V_n, such that $S_1 \cup ... \cup S_q = V_n$.*

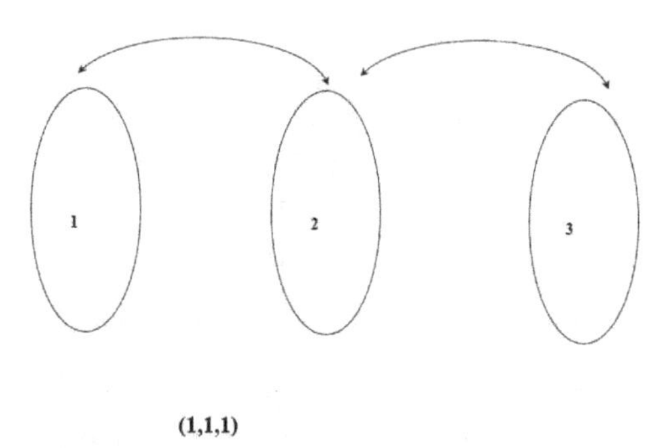

Figure 7.2: Example of multicut semimetric on 3 points

The multicut semimetric $\delta_{S_1,...,S_q}$ *is a semimetric on* V_n*, defined by*

$$\delta_{S_1,...,S_q}(i,j) = \begin{cases} 0, & \text{if } i,j \in S_\alpha \text{ for some } \alpha, 1 \le \alpha \le q, \\ 1, & \text{otherwise.} \end{cases}$$

The multicut semimetric $\delta_{S_1,...,S_q}$ can be seen as the vector in $\mathbb{R}^{E_n}$, defined by

$$(\delta_{S_1,...,S_q})_{ij} = 0,$$

if $i, j \in S_\alpha$ for some α, $1 \le \alpha \le q$, and $(\delta_{S_1,...,S_q})_{ij} = 1$, otherwise, for $1 \le i < j \le n$: $\delta_{S_1,...,S_q}(i,j) = (\delta_{S_1,...,S_q})_{ij}$. This vector is called a *multicut vector*, or simply a *multicut*.

Similarly, the multicut semimetric $\delta_{S_1,...,S_q}$ can be considered as symmetric $n \times n$ $\{0,1\}$-matrix $((\delta_{S_1,...,S_q})_{ij})$ with zeros on the main diagonal, where $(\delta_{S_1,...,S_q})_{ij} = \delta_{S_1,...,S_q}(i,j)$.

In fact, a *multicut* is a *partition* of a set into several parts:

$$V_n = \{1, 2, ..., n\} = S_1 \cup S_2 \cup ... \cup S_q.$$

The multicut semimetric on V_n, defined by this partition, can be seen as a special semimetric on the vertex set of the *complete q-partite graph* $K_{|S_1|,|S_2|,...,|S_q|}$, where the distance between vertices is equal to 1 if they belong to different parts of this graph, and is equal to 0, otherwise.

For example, there are five multicuts on 3 points: four cuts

$$S_1 = \emptyset,\ S_2 = \{1,2,3\}, \qquad \delta_\emptyset = (0,0,0),$$

$$S_1 = \{1\},\ S_2 = \{2,3\}, \qquad \delta_{\{1\}} = (1,1,0),$$

$$S_1 = \{2\},\ S_2 = \{1,3\}, \qquad \delta_{\{2\}} = (1,0,1),$$

$$S_1 = \{3\},\ S_2 = \{1,2\}, \qquad \delta_{\{3\}} = (0,1,1),$$

and one "proper" multicut

$$S_1 = \{1\},\ S_2 = \{2\},\ S_3 = \{3\}, \qquad \delta_{\{1\},\{2\},\{3\}} = (1,1,1).$$

Easy to see, that every cut δ_S can be considered as a multicut $\delta_{S,\overline{S}}$ for the partition $S_1 = S, S_2 = V_n \backslash S$ of V_n. Moreover,

$$\delta_{S_1,\dots,S_q} = \frac{1}{2}\sum_{i=1}^{q} \delta_{S_i},$$

i.e., every multicut is the linear combination of cuts with non-negative coefficients.

The number of all cuts on n points is 2^{n-1} (half of the number 2^n of all subsets of V_n, since $\delta_S = \delta_{\overline{S}}$).

The number $p(n)$ of all multicuts on n points is the number of all partitions of n. In fact, $p(n)$ is equal to n-th *Bell number* $B(n)$: the sequence

$$1, 1, 2, 5, 15, 52, 203, 877, 4140, 21147, 115975, \dots \text{ (A000110)}$$

in [Sloa15]. So, $p(3) = 5$, $p(4) = 15$, $p(5) = 52$, $p(6) = 203$, $p(7) = 877$, $p(8) = 4140$. Table 7.1 gives the values of $p(n)$ for small n.

n	3	4	5	6	7	8	9	10
$p(n)$	5	15	52	203	877	4140	21147	115975

Table 7.1: The number of multicuts for small values of n

Let *q-cut* be any multicut $\delta_{S_1,\dots,S_q}$, defined by a partition $S_1, \dots, S_q$ of V_n into q non-empty sets. Clearly, the number of q-cuts on V_n is the number of ways to partite a set of n objects into q non-empty groups, i.e., it is the *Stirling number of the second kind* $S(n, q)$. In particular, any non-zero cut is 2-cut; so, the number of non-zero cuts on V_n is $S(n, 2)$.

It is known ([GKP94]), that $S(n, 2) = 2^{n-1} - 1$, i.e., the number of all non-zero cuts on n points is equal to $2^{n-1} - 1$; moreover, $S(n, n-1) = \binom{n}{2}$, i.e., the number of all $(n-1)$-cuts on n points is $\frac{n(n-1)}{2}$. In general, the number of all q-cuts on n points can be find by the recurrence

$$S(n, q) = kS(n-1, q) + S(n-1, q-1),\ 0 < q \le n,$$

or by the formula

$$S(n, q) = \frac{1}{q!}\sum_{j=0}^{q}(-1)^j \binom{q}{j}(q-j)^n.$$

The following triangle is constructed using the above recurrence and consists of the numbers $S(n, q)$, $n \ge 0$, $0 \le k \le n$.

$n\backslash k$	0	1	2	3	4	5	6	7	8	9
0	1									
1	0	1								
2	0	1	1							
3	0	1	3	1						
4	0	1	7	6	1					
5	0	1	15	25	10	1				
6	0	1	31	90	65	15	1			
7	0	1	63	301	350	140	21	1		
8	0	1	127	966	1701	1050	266	28	1	
9	0	1	255	3025	7770	6951	2646	462	36	1

The number $p(n)$ of all multicuts of V_n can be calculated as the sum of all elements in n-th row of this triangle. Alternatively, there exist the following recurrence for Bell numbers:

$$B(n) = \sum_{t=0}^{n-1} \binom{n-1}{t} B(t), \; B(0) = 1.$$

The notion of a cut semimetric is connected with the notion of symmetric difference of sets. For example, the cut vector $\delta_{\{1\}} = (1,1,0)$ on V_3 can be defined by the symmetric differences of sets $\{1\}$, $\emptyset$, $\emptyset$:

$$\delta_{\{1\}} = (|\{1\}\triangle\emptyset|, |\{1\}\triangle\emptyset|, |\emptyset\triangle\emptyset|).$$

Moreover, it is easy to see, that the symmetric difference of two cuts is again a cut.

7.3 Oriented case

Definition 7.3 *Given a subset S of $V_n = \{1, 2, ..., n\}$, the oriented cut quasi-semimetric (oriented cut, o-cut) δ_S^O is a quasi-semimetric on V_n, defined by*

$$\delta_S^O(i,j) = \begin{cases} 1, & \text{if } i \in S, j \notin S, \\ 0, & \text{otherwise.} \end{cases}$$

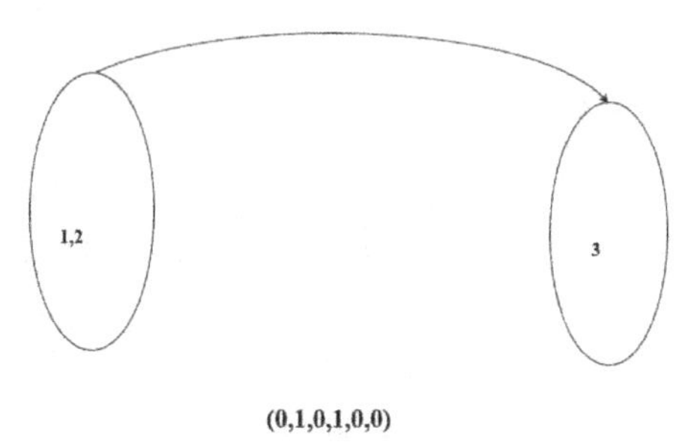

Figure 7.3: Example of an oriented cut on 3 points

Usually, it is considered as an $\{0,1\}$-vector in $\mathbb{R}^{I_n}$, where $I_n = \{\langle i,j\rangle \,|\, 1 \le i \ne j \le n\}$ (i.e., $I_n = 2\binom{2}{n} = n(n-1)$), defined by $(\delta_S^O)_{ij} = 1$, if $i \in S, j \notin S$, and $(\delta_S^O)_{ij} = 0$, otherwise, for $1 \le i \ne j \le n$: $\delta_S^O(i,j) = (\delta_S^O)_{ij}$. It is called an *oriented cut vector*, or an *o-cut vector*, or simply *o-cut.*

Alternatively, any oriented cut on V_n can be view as an $n \times n$ non-symmetric $\{0,1\}$-matrix $(((\delta_S^O)_{ij}))$ with zeros on the main diagonal.

Any non-zero oriented cut can be considered as an *ordered partition* of the set V_n into two non-empty parts. Given a subset S of $V_n = \{1, 2, ..., n\}$, we obtain the ordered partition $\langle S, \overline{S} = V_n \backslash S\rangle$ (with the first part S and the second part $\overline{S}$) of V_n. The o-cut semimetric on V_n, defined by this ordered partition, can be seen as a special quasi-semimetric on the vertex set of the *oriented complete bipartite graph* $K^O_{|S|,|\overline{S}|}$ with all the arcs going from S into $\overline{S}$, where the distance between vertices u and v is equal to 1, if there exist the arc uv, (i.e., $u \in S$, and $v \in V_n \backslash S$), and is equal to 0, otherwise.

For example, there are seven o-cuts on 3 points:

$$\begin{array}{rl}
S = \emptyset, \overline{S} = V_3 \backslash S = \{1,2,3\}, & \delta^O_{\emptyset} = (0,0,0,0,0,0);\\
S = \{1\}, \overline{S} = V_3 \backslash S = \{2,3\}, & \delta^O_{\{1\}} = (1,1,0,0,0,0);\\
S = \{2,3\}, \overline{S} = V_3 \backslash S = \{1\}, & \delta^O_{\{2,3\}} = (0,0,1,0,1,0);\\
S = \{2\}, \overline{S} = V_3 \backslash S = \{1,3\}, & \delta^O_{\{2\}} = (0,0,1,1,0,0);\\
S = \{1,3\}, \overline{S} = V_3 \backslash S = \{2\}, & \delta^O_{\{1,3\}} = (1,0,0,0,0,1);\\
S = \{3\}, \overline{S} = V_3 \backslash S = \{1,2\}, & \delta^O_{\{3\}} = (0,0,0,0,1,1);\\
S = \{1,2\}, \overline{S} = V_3 \backslash S = \{3\}, & \delta^O_{\{1,2\}} = (0,1,0,1,0,0).
\end{array}$$

Obviously, the cut semimetric δ_S can be represented as a sum of two oriented cuts:

$$\delta_S = \delta^O_S + \delta^O_{V_n \backslash \{S\}}.$$

Definition 7.4 *Given an oriented partition $\langle S_1, ..., S_q\rangle$, $q \geq 2$, of V_n, the oriented multicut quasi-semimetric (oriented multicut, o-multicut) $\delta^O_{S_1,...,S_q}$ is a quasi-semimetric on V_n, defined by*

$$\delta^O_{S_1,...,S_q}(i,j) = \begin{cases} 1, & \text{if } i \in S_\alpha, j \in S_\beta, \alpha < \beta,\\ 0, & \text{otherwise.}\end{cases}$$

Usually, o-multicut $\delta^O_{S_1,...,S_q}$ is considered as an $\{0,1\}$-vector in $\mathbb{R}^{I_n}$, defined by

$$(\delta^O_{S_1,...,S_q})_{ij} = 1,$$

if $i \in S_\alpha$, $j \in S_\beta$, where $\alpha < \beta$, and $(\delta^O_{S_1,...,S_q})_{ij} = 0$, otherwise: $\delta^O_{S_1,...,S_q}(i,j) = (\delta^O_{S_1,...,S_q})_{ij}$ It is called an *oriented multicut vector*, or an *o-multicut vector*, or simply an *o-multicut*.

Alternatively, $\delta^O_{S_1,...,S_q}$ can be viewed as an $n \times n$ $\{0,1\}$-valued matrix $(((\delta^O_{S_1,...,S_q})_{ij}))$ with zeros on the mail diagonal.

In fact, an *oriented multicut* is an ordered partition of a set into several parts:

$$V_n = \{1,2,...,n\} = \langle S_1, S_2, ..., S_q\rangle.$$

The o-multicut semimetric on V_n, defined by this partition, can be seen as a special quasi-semimetric on the vertex set of the *oriented complete q-partite graph* $K^O_{|S_1|,|S_2|,...,|S_q|}$ with all the arcs going from the set S_i to the set S_j, $i < j$, where the distance between vertices u and v is equal to 1, if there exists the arc uv (i.e., $u \in S_i, v \in S_j, i < j$), and is equal to 0, otherwise.

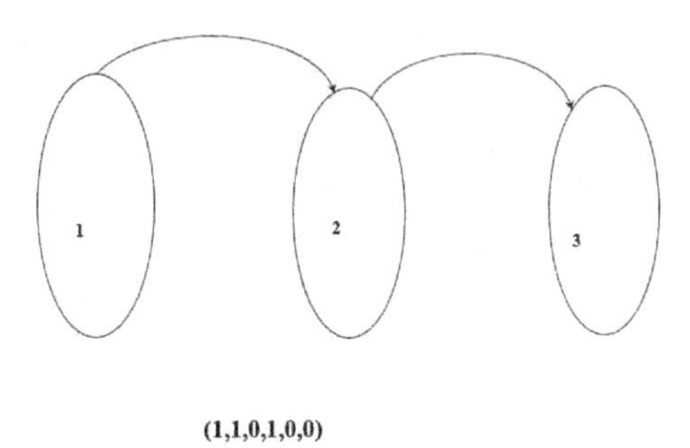

Figure 7.4: Example of an oriented multicut on 3 points

Every non-zero o-cut δ_S^O can be considered as an o-multicut $\delta_{S,\overline{S}}^O$ for the ordered partition $\langle S_1, S_2\rangle$ of V_n with $S_1 = S, S_2 = \overline{S}$.

For example, there are 13 o-multicuts on 3 points: together with 7 o-cuts, considered above, we have 6 "proper" o-multicuts, corresponding to 6 permutations of the non-ordered partition $\{1\}$, $\{2\}, \{3\}$ of V_3:

$$\begin{array}{ll}
S_1 = \{1\},\ S_2 = \{2\},\ S_3 = \{3\}, & \delta^O_{\{1\},\{2\},\{3\}} = (1,1,0,1,0,0);\\
S_1 = \{1\},\ S_2 = \{3\},\ S_3 = \{2\}, & \delta^O_{\{1\},\{3\},\{2\}} = (1,1,0,0,0,1);\\
S_1 = \{2\},\ S_2 = \{1\},\ S_3 = \{3\}, & \delta^O_{\{2\},\{1\},\{3\}} = (0,1,1,1,0,0);\\
S_1 = \{2\},\ S_2 = \{3\},\ S_3 = \{1\}, & \delta^O_{\{2\},\{3\},\{1\}} = (0,0,1,1,1,0);\\
S_1 = \{3\},\ S_2 = \{1\},\ S_3 = \{2\}, & \delta^O_{\{3\},\{1\},\{2\}} = (1,0,0,0,1,1);\\
S_1 = \{3\},\ S_2 = \{2\},\ S_3 = \{1\}, & \delta^O_{\{3\},\{2\},\{1\}} = (0,0,1,0,1,1).
\end{array}$$

We proved, that in non-oriented case every multicut is the linear combination of cuts with non-negative coefficients. Similar property fails for oriented multicuts. For example, in the case of 3 points the o-multicut $\delta^O_{\{1\},\{2\},\{3\}}$ is not a linear combination of oriented cuts on V_3.

The number of all oriented cuts on n points is $2^n - 1$ (the number of all subsets of V_n minus 1, since the partitions $\emptyset, V_n$ and $V_n, \emptyset$ define the zero o-cut), and the number $p^O(n)$ of all oriented multicuts on V_n is the number of all ordered partitions of n. So, $p^O(n)$ coincides with the n-th *Fubini number* (*ordered Bell number*). Starting from $n = 0$, these numbers are 1, 1, 3, 13, 75, 541, 4683, 47293, 545835, 7087261, 102247563, ... (sequence A000670 in [Sloa15]). The ordered Bell numbers can be obtained by the following recurrence:

$$p^O(n) = \binom{n}{1}p^O(n-1) + \binom{n}{2}p^O(n-2) + \binom{n}{3}p^O(n-3) + ... + \binom{n}{n}p^O(0),\ p^O(0) = 1.$$

Moreover, it holds

$$p^O(n) = S(n,0) + 1!S(n,1) + 2!S(n,2) + ... + k!S(n,k) + ... + n!S(n,n),$$

where $S(n,k)$, $k = 0,1,2,...,n$, are the Stirling numbers of the second kind. As

$$S(n,k) = \frac{k^n}{k!} - \frac{(k-1)^n}{1!(k-1)!} + \frac{(k-2)^n}{2!(k-2)!} - ... + (-1)^{k-1}\frac{1^n}{(k-1)!1!},$$

we obtain for $p^O(n)$ the following closed formula:

$$p^O(n) = \sum_{k=0}^{n}(C_k^0 k^n - C_k^1(k-1)^n + C_k^2(k-2)^n - ... + (-1)^{k-1}C_k^k \cdot 1^n).$$

In fact,

$$p^O(n) = \sum_{\pi \in Sym(n)} 2^{d(\pi)}$$

with $d(\pi) = |\{i \leq n \mid a_i > a_{i+1}\}|$ for the permutation $\pi = (a_1, ..., a_n) \in Sym(n)$; in other words, $p^O(n) = \frac{1}{2}A_n(2)$, where $A_n(x)$ is the *Euler's polynomial*

$$A_n(x) = \sum_{\pi \in S_n} x^{1+d(\pi)}.$$

So, $p^O(3) = 13$, $p^O(4) = 75$, $p^O(5) = 541$, $p^O(6) = 4683$, $p^O(7) = 47293$, $p^O(8) = 545835$.
Table 7.2 gives the values of $p^O(n)$ for small n.

n	3	4	5	6	7	8	9	10
$p^O(n)$	13	75	541	4683	47293	545835	7087261	102247563

Table 7.2: The number of o-multicuts for small values of n

The notion of o-cut quasi-semimetric is connected with the notion of asymmetric difference of sets. For example, the o-cut $\delta^O_{\{1\}} = (1,1,0,0,0,0)$ can be defined by the asymmetric difference of sets $\{1\}$, $\emptyset$, $\emptyset$:

$$\delta^O_{\{1\}} = (|\{1\}\backslash\emptyset|, |\{1\}\backslash\emptyset|, |\emptyset\backslash\{1\}|, |\emptyset\backslash\emptyset|, |\emptyset\backslash\{1\}|, |\emptyset\backslash\emptyset|).$$

But, the set of o-multicuts on n points is not closed under the symmetric difference.

7.4 Multidimensional case

Partition m-hemimetric is a multidimensional analogue of the binary notion of multicut semimetric.

Definition 7.5 *Given an $(m+1)$-partition $S_1, \ldots, S_{m+1}$ of the set $V_n = \{1,2,...,n\}$, the partition m-hemimetric $\alpha_{S_1,...,S_{m+1}}$ is defined by*

$$\alpha_{S_1,...,S_{m+1}}(i_1,...,i_{m+1}) = \begin{cases} 1, & \text{if } \; i_1 \in S_{\alpha_1}, ..., i_{m+1} \in S_{\alpha_{m+1}}, \alpha_k \neq \alpha_t \text{ for } k \neq t, \\ 0, & \text{if } \; i_k, i_t \in S_\alpha \text{ for some } k \neq t, 1 \leq \alpha \leq m+1. \end{cases}$$

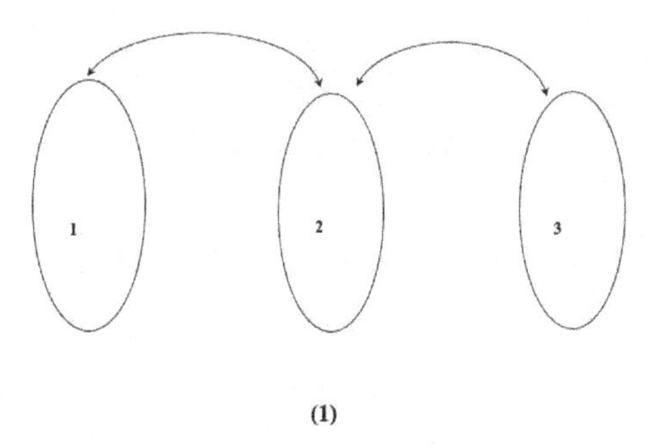

Figure 7.5: Example of a partition 2-hemimetric on 3 points

So, the function $\alpha_{S_1,\dots,S_{m+1}}$ is defined by setting $\alpha_{S_1,\dots,S_{m+1}}(i_1,\dots,i_{m+1})$ is equal to 1, if for no $1 \le j < l \le m+1$ both i_j and i_l belong to the same S_k, and is equal to 0, otherwise. It is easy to see, that $\alpha_{S_1,\dots,S_{m+1}}$ is an m-hemimetric, and that for $m = 1$ it is the usual cut semimetric (see, for example, [DeLa97]).

One can check also that for $n = m + 2$ any partition $(m - s + 1)$-hemimetric, such that all, but one, sets S_i are singletons, is also (m, s)-supermetric.

Any partition m-hemimetric on V_n cam be seen as an $\{0,1\}$-vector in $\mathbb{R}^{E_n^{m+1}}$, where $E_n^{m+1} = \binom{n}{m+1} = \frac{n!}{(m+1)!(n-m-1)!}$; its coordinates $(\alpha_{S_1,\dots,S_q})_{i_1\dots i_{m+1}}$, where $1 \le i_1 < i_2 < \dots < i_{m+1} \le n$, correspond to the values $\alpha_{S_1,\dots,S_q}(i_1,\dots,i_{m+1})$. It is called *partition m-hemimetric vector*. On the other hand, the matrix representation can not be used in multidimensional case.

But for any $\{0,1\}$-vector $v = (v_{i_1\dots i_{m+1}}) \in \mathbb{R}^{E_n^{m+1}}$, indexed by $(m+1)$-subsets of the set V_n, we can use its *representation graph* G_v: the restriction of the *Jonson graph* $J(n, m+1)$ on the *support* of v (i.e., the set of all indices of the vector, on which its components are non-zero): the vertices of this graph are all unordered $(m+1)$-tuples $(i_1,\dots,i_{m+1})$ such that $v_{i_1\dots i_{m+1}} = 1$; two vertices of the Johnson graph (and also of its induced subgraph) are adjacent if the corresponding $(m+1)$-tuples have m common elements.

The representation graph G_v of partition m-hemimetric vector $v = ((\alpha_{S_1,\dots,S_{m+1}})_{i_1\dots i_{m+1}})$ is the graph $H(|S_1|,\dots,|S_{m+1}|)$, i.e., the direct (Cartesian) product

$$K_{|S_1|} \times K_{|S_2|} \times \dots \times K_{|S_{m+1}|}$$

of the cliques $K_{|S_i|}$, $1 \le i \le m+1$.

It is easy to see, that for $m = 1$ any m-hemimetric α_{S_1,S_2} is the usual cut semimetric $\delta_{S_1,S_2} = \delta_{S_1,V_n\setminus S_1} = \delta_{S_1}$.

The connection between multicut $\delta_{S_1,\dots,S_{m+1}}$ and m-hemimetric $\alpha_{S_1,\dots,S_{m+1}}$ is given by the formula

$$\alpha_{S_1,\dots,S_{m+1}}(i_1,\dots,i_{m+1}) = \prod_{1 \le s < t \le m+1} \delta_{S_1,\dots,S_{m+1}}(i_s, i_t)$$

$$= \left\lfloor \frac{\sum_{1 \le s < t \le m+1} \delta_{S_1,\dots,S_{m+1}}(i_s, i_t)}{\binom{q}{2}} \right\rfloor;$$

compare it with the *half-perimeter m-semimmetric* ρ from Chapter 5.

7.5 Important special cases of cut-related constructions

1. A *decomposable semimetric* is a semimetric on $V_n = \{1, 2, ..., n\}$, which can be represented as a non-negative linear combination of cut semimetrics. A semimetric on V_n is decomposable if and only if it is a *finite* l_1*-semimetric*, i.e., a semimetric d such that the metric space (V_n, d) is a semimetric subspace of the l_1^m*-space* $(\mathbb{R}^m, d_{l_1})$ for some $m \in \mathbb{N}$. It means, that we can find n vectors $v_1, \dots, v_n \in \mathbb{R}^m$, such that

$$d(i,j) = d_{\ell_1}(v_i, v_j) \text{ for all } i,j \in V_n.$$

2. A *circular cut* of V_n is defined by a subset $S_{[k+1,l]} = \{k+1, ..., l\} (\text{mod } n) \subset V_n$: if we consider the points $\{1, 2, ..., n\}$ as being ordered along a circle in that circular order, then $S_{[k+1,l]}$ is the set of its consecutive vertices from $k+1$ to l. For a circular cut, the corresponding cut semimetric is called *circular cut semimetric.*

A *circular decomposable semimetric* is a semimetric on $V_n = \{1, 2, ..., n\}$, which can be represented as a non-negative linear combination of circular cut semimetrics. In fact, a semimetric on V_n is circular decomposable if and only if it is a *Kalmanson semimetric* with respect to the same ordering (see [ChFi98]): a semimetric d on V_n, which satisfies the condition

$$\max\{d(i,j) + d(r,s), d(i,s) + d(j,r)\} \le d(i,r) + d(j,s)$$

for all $1 \le i \le j \le r \le s \le n$. (In this definition the ordering of the elements is important; so, d is a Kalmanson semimetric *with respect to the ordering* $1, 2, ..., n$.)

Equivalently, if we consider the points $\{1, 2, ..., n\}$ as being ordered along a circle C_n in that circular order, then the distance d on V_n is a Kalmanson semimetric if the inequality

$$d(i,r) + d(j,s) \le d(i,j) + d(r,s)$$

holds for all $i, j, r, s \in V_n$ such that the segments $[i, j]$ and $[r, s]$ are crossing chords of C_n.

The Euclidean metric, restricted to the points that form a convex polygon in the plane, is a Kalmanson metric.

3. An *even cut semimetric* is a cut semimetric δ_S on V_n with even $|S|$. An *odd cut semimetric* is a cut semimetric δ_S on V_n with odd $|S|$. An *k-uniform cut semimetric* is a cut semimetric δ_S on V_n with $|S| \in \{k, n-k\}$. An *equicut semimetric* is a cut semimetric δ_S on V_n with $|S| \in \{\lfloor \frac{n}{2} \rfloor, \lceil \frac{n}{2} \rceil\}$. An *inequicut semimetric* is a cut semimetric δ_S on V_n with $|S| \notin \{\lfloor \frac{n}{2} \rfloor, \lceil \frac{n}{2} \rceil\}$ (see, for example, [DeLa97]).

4. Given an ordered partition $\langle S_1, ..., S_q \rangle$, the *oriented anti-multicut quasi-semimetric* (*o-anti-multicut*) $\beta^O_{S_1,...,S_q}$ is a quasi-semimetric on V_n, defined by

$$\beta^O_{S_1,...,S_q}(i,j) = \begin{cases} 1 - \delta^O_{S_1,...,S_q}(i,j), & \text{if } 1 \le i \ne j \le n, \\ 0, & \text{if } 1 \le i = j \le n. \end{cases}$$

where $\delta^O_{S_1,...,S_q}$ is the corresponding o-multicut.

The o-anti-multicut $\beta^O_{S_1,S_2} = \beta^O_{S,\overline{S}}$ is called *o-anti-cut* and denoted by β^O_S.

Because $\delta^O_\emptyset = (0,0,...,0)$, it holds that $\beta^O_\emptyset = (1,1,...,1)$, i.e., $\beta^O_\emptyset = d_{path}(K_n)$, where $d_{path}(K_n)$ is the path metric of the complete graph K_n.

Given an ordered partition $\langle S_1,...,S_q\rangle$, the *anti-multicut semimetric* $\beta_{S_1,...,S_q}$ is the *symmetrization* of the quasi-semimetric $\beta^O_{S_1,...,S_t}$:

$$\beta_{S_1,...,S_q} = \beta^O_{S_1,...,S_q} + \beta^O_{S_q,...,S_1}.$$

In fact, it is the path metric $d_{path}(K_{|S_1|,...,|S_q|})$ of the complete q-partite graph $K_{|S_1|,...,|S_q|}$.

In the case $q = 2$, the anti-multicut semimetric β_{S_1,S_2} is called *anti-cut semimetric* and denoted by β_S. Set $\beta_\emptyset = d_{path}(K_n)$.

Theorem 7.1 *O-multicuts $\delta^O_{S_1,...,S_q}$ and o-anti-multicuts $\beta^O_{S_1,...,S_q}$ are $\{0,1\}$-valued quasi-semimetrics, which are weightable if and only if $q \leq 2$. The weight functions of o-cut δ^O_S and o-anti-cut β^O_S are $w_{\delta^O} = (w_i = 1_{i \notin S})$, and $w_{\beta^O} = (w_i = 1_{i\in S})$, respectively.*

Proof. In fact, let $i \in S_1$, $j \in S_2$, $k \in S_3$ in the quasi-semimetric $q{=}\delta^O_{S_1,...,S_q}$. If q is weightable, then $q_{ij} = (q_{ji} + w_j) - w_i = w_j - w_i$. Impossible, since also $q_{ik} = w_k - w_i = 1, q_{jk} = w_k - w_j = 1$. The proof for o-anti-multicuts is similar.

□

Theorem 7.2 *There are the following equalities:*

1. $\delta_{S_1,...,S_q} = \sum_{i=1}^q \delta^O_{S_i} = \sum_{i=1}^q \delta^O_{\overline{S}_i} = \frac{1}{2}\sum_{i=1}^q \delta_{S_i}$;
2. $\beta_\emptyset = d_{path}(K_n)$, *and* $\beta_{S_1,...,S_q} = d_{path}(K_{|S_1|,...,|S_q|})$;
3. $\beta^O_{\{i\}} = \sum_{j\in\{\overline{i}\}} \delta^O_{\{j\}}$;
4. *if $q = n$, i.e., all $|S_i| = 1$, then $\beta^O_{S_1,...,S_n} = \delta^O_{S_n,...,S_1}$ (the reversal of the ordered partition).*

Proof. These equalities are easy to check; some of them we considered above.

□

5. For any $S \subset V_n$, denote by $J(S)$ the $\{0,1\}$-valued function with $J(S)_{ij} = 1$ exactly when $i,j \in S$. So, $J(V_n)$ and $J(\emptyset)$ are all-ones and all-zeros partial semimetrics, respectively.

For any $S_0 \subset V_n$ and a given partition $\{S_1,\ldots,S_t\}$ of $\overline{S}_0$, the *partial t-cut* (or, specifically, the *partial multicut*) $\gamma_{S_0;S_1,...,S_t}$ is a partial semimetric on V_n, defined by

$$\gamma_{S_0;S_1,...,S_t} = J(S_0) + \delta_{S_0,S_1,...,S_t}.$$

Clearly, $\gamma_{S_0;S_1,...,S_t}$ is a partial semimetric: it can be obtained from the weighted semimetric $(d;w)$, where $d = 2\delta_{S_0,S_1,...,S_t} - \delta_{S_0}$, and $w = (w_i = 1_{i\in S_0})$, by the map

$$p = P(d;w): \ p_{ij} = \frac{d_{ij} + w_i + w_j}{2}.$$

Chapter 8

Hypermetrics and their generalizations

8.1 Preliminaries

The hypermetrics form an important special class of finite semimetrics. They have deep connections with Geometry of Numbers and Analysis; see, for example, [DeTe87], [DeGr93], [DGL95] and Chapters 13-17, 28 in [DeLa97]. So, generalizations of notion of hypermetric can put those connections in a more general setting.

In particular, the hypermetrics and the negative type distances closely connect with the problems of L_p-embeddability, $p = 1, 2$. There are many known necessary conditions of L_1- and L_2-embeddability, arising, in particular, from the known valid inequalities for the cut cone CUT_n; they will be described in Chapters 9 and 13. Here we focus on the hypermetric and negative type conditions. These conditions seem indeed to be among the most essential ones. For instance, there are several classes of distance spaces for which these conditions are also sufficient for ensuring L_1-embeddability (see [DeLa97]).

In this chapter the hypermetric inequalities and hypermetrics will be treated in detail for their own sake. The hypermetric and negative type inequalities are introduced in Section 8.2. The definition of a hypermetric, of a negative type distances and connections between them are considered in Section 8.3. Finally, we will introduce in Section 8.4 several generalizations of hypermetrics, including quasi-hypermetrics (see [DDV11], [DeDe10]), partial hypermetrics (see [DDV11], [DDD15]), and infinite hypermetrics (see [DeDu13a]).

8.2 Hypermetric and negative type inequalities

Let $n \geq 2$, and let $b_1, \ldots, b_n$ be integers. Given $b_1, \ldots, b_n \in \mathbb{Z}$, and a given semimetric $d : V_n \times V_n \to \mathbb{R}$, $d(i, j) = d_{ij}$, consider the inequality

$$\sum_{1 \leq i < j \leq n} b_i b_j d_{ij} \leq 0.$$

For convenience, introduce the following notation: for a vector $b \in \mathbb{R}^n$ with $b_i \in \mathbb{Z}$, let $Q_n(b)$ denote the vector from $\mathbb{R}^{E_n}$, defined by

$$Q_n(b)_{ij} = b_i b_j \ \text{ for } 1 \leq i < j \leq n.$$

When the parameter n is clear from the context, we also denote $Q_n(b)$ by $Q(b)$. Hence, the inequality $\sum_{1\leq i<j\leq n} b_ib_jd_{ij} \leq 0$ (in the variable $d = (d_{ij}) \in \mathbb{R}^{E_n}$) can be rewritten as $Q_n(b)^T d \leq 0$. We can suppose that at least two of the b_i's are non-zero; else, $Q_n(b) = 0$ and the inequality $\sum_{1\leq i<j\leq n} b_ib_jd_{ij} \leq 0$ is void.

When $\sum_{i=1}^n b_i = 1$, the inequality

$$Q_n(b)^T d = \sum_{1\leq i<j\leq n} b_ib_jd_{ij} \leq 0$$

is called an *hypermetric inequality* and is denoted by Hyp_b:

$$Hyp_b : Hyp_b(d) = \sum_{1\leq i<j\leq n} b_ib_jd_{ij} \leq 0, \quad \text{with} \quad b \in \mathbb{Z}^n, \sum_{i=1}^n b_i = 1.$$

When $\sum_{i=1}^n b_i = 0$, the inequality

$$Q_n(b)^T d = \sum_{1\leq i<j\leq n} b_ib_jd_{ij} \leq 0$$

is called a *negative type inequality* and is denoted by Neg_b:

$$Neg_b : Neg_b(d) = \sum_{1\leq i<j\leq n} b_ib_jd_{ij} \leq 0, \quad \text{with} \quad b \in \mathbb{Z}^n, \sum_{i=1}^n b_i = 0.$$

The negative type inequalities are classical inequalities in Analysis; they were used, in particular, by Schoenberg [Scho38]. The hypermetric inequalities were considered by several authors, including Deza [Deza60], [Deza62], Kelly [Kell70], Baranovskii [Bara71] and, in the context of correlations or Boolean quadratic programming, by Kelly [Kell70], Erdahl [Erda87] and many others.

The inequality $Q_n(b)^T d \leq 0$ is said to be *pure* if $|b_i| = 0, 1$ for all $i \in V_n$. The inequality $Q_n(b)^T d \leq 0$ is said to be an k*-gonal inequality* if $\sum_{i=1}^n |b_i| = k$ holds. (Note, that k and $\sum_{i=1}^n b_i$ have the same parity.)

In particular, the *2-gonal inequality* is the inequality of negative type, where $b_i = 1$, $b_j = -1$, and $b_h = 0$ for $h \in V_n \setminus \{i, j\}$, for some distinct $i, j \in V_n$; it is nothing but the non-negativity constraint $d_{ij} \geq 0$.

The *pure 3-gonal inequality* is the hypermetric inequality with $b_i = b_j = 1$, $b_k = -1$, and $b_h = 0$ for $h \in V_n \setminus \{i, j, k\}$, for some distinct $i, j, k \in V_n$; it coincides with the triangle inequality $d_{ij} - d_{ik} - d_{kj} \leq 0$.

The *5-gonal inequalities* are the inequalities $Q_n(b)^T d \leq 0$, where b is (up to permutation of its components) one of the following vectors:

$$b = (1, 1, 1, -1, -1, 0, \ldots, 0),\ b = (1, 1, 1, -2, 0, \ldots, 0),$$
$$b = (2, 1, -1, -1, 0, \ldots, 0), b = (3, -1, -1, 0, \ldots, 0),$$
$$b = (2, 1, -2, 0, \ldots, 0),\ b = (3, -2, 0, \ldots, 0).$$

The pure $(2k)$-gonal inequality has the form

$$\sum_{1\leq r<s\leq k} d_{i_r i_s} + \sum_{1\leq r<s\leq k} d_{j_r j_s} - \sum_{\substack{1\leq r\leq k \\ 1\leq s\leq k}} d_{i_r j_s} \leq 0,$$

where $i_1, \ldots, i_k, j_1, \ldots, j_k$ are distinct indices of V_n.

The pure $(2k+1)$-gonal inequality has the form

$$\sum_{1 \le r < s \le k+1} d_{i_r i_s} + \sum_{1 \le r < s \le k} d_{j_r j_s} - \sum_{\substack{1 \le r \le k+1 \\ 1 \le s \le k}} d_{i_r j_s} \le 0,$$

where $i_1, \ldots, i_k, i_{k+1}, j_1, \ldots, j_k$ are distinct indices of V_n.

For example, the pure 5-gonal inequality is called *pentagonal inequality* and has the form

$$\sum_{1 \le r < s \le 3} d_{i_r j_s} + d_{k_1 k_2} - \sum_{\substack{r=1,2,3 \\ s=1,2}} d_{i_r k_s} \le 0.$$

For a pure $(2k+1)$-gonal inequality there exists very simple graph realization. In fact, consider a partition of the vertex set of complete graph K_{2k+1} into two parts S and $\overline{S} = V_n \backslash S$ of cardinalities k and $k+1$; find the sum $\sum_S$ of all distances (each equal to 1) between all points from S, as well as the sum $\sum_{\overline{S}}$ of the distances between all points from $\overline{S}$; find the sum $\sum_{S,\overline{S}}$ between any pair of points, one belonging to S, and another to $\overline{S}$. Then a pure $(2k+1)$-gonal inequality obtains the form $\sum_S + \sum_{\overline{S}} \le \sum_{S,\overline{S}}$. See Figure 6.1.2 in [DeLa97].

In fact, *the negative type inequalities are implied by the hypermetric inequalities* (see, for example, [Deza62]).

Proposition 8.1 *Let $k \ge 1$ be an integer. The $(2k+2)$-gonal inequalities are implied by the $(2k+1)$-gonal inequalities.*

Proof.

Let $b \in \mathbb{Z}^n$ with $\sum_{i=1}^n b_i = 0$ and $\sum_{i=1}^n |b_i| = 2k+2$. We show that the inequality $Q_n(b)^T d \le 0$ can be expressed as a non-negative linear combination of $(2k+1)$-gonal inequalities. We can suppose without loss of generality that $b_1, \ldots, b_p > 0 > b_{p+1}, \ldots, b_n$ for some p, $1 \le p \le n-1$. For $1 \le i \le p$, set

$$c^{(i)} = (-b_1, \ldots, -b_{i-1}, 1 - b_i, -b_{i+1}, \ldots, -b_p, -b_{p+1}, \ldots, -b_n),$$

and, for $p+1 \le i \le n$, set

$$c^{(i)} = (b_1, \ldots, b_p, b_{p+1}, \ldots, b_{i-1}, b_i + 1, b_{i+1}, \ldots, b_n).$$

Then each vector $c^{(i)}$ belongs to $\mathbb{Z}^n$, has sum of entries 1, and sum of absolute values of its entries $2k+1$. Therefore, each inequality $Q_n(c^{(i)})^T d \le 0$ is an $(2k+1)$-gonal inequality. Observe now that

$$\sum_{1 \le i \le n} |b_i| Q_n(c^{(i)}) = 2k Q_n(b).$$

This shows that the $(2k+2)$-gonal inequality $Q_n(b)^T d \le 0$ is implied by the $(2k+1)$-gonal inequalities $Q_n(c^{(i)})^T d \le 0$ $(1 \le i \le n)$.

□

Example. The 4-gonal inequality $Q_4(1,1,-1,-1)^T d \leq 0$ follows by summation of the following 3-gonal inequalities:

$$\begin{array}{c} Q_4(1,1,-1,0)^T d \leq 0; \\ Q_4(1,1,0,-1)^T d \leq 0; \\ Q_4(-1,0,1,1)^T d \leq 0; \\ Q_4(0,-1,1,1)^T d \leq 0. \end{array}$$

Corollary 8.1 *The negative type inequalities are implied by the hypermetric inequalities.*

8.3 Hypermetrics and distances of negative type

Definition 8.1 *Given a distance space (X,d), the distance d is called hypermetric, if it satisfies all the hypermetric inequalities, i.e., if d satisfies*

$$\sum_{1 \leq i < j \leq n} b_i b_j d(x_i, x_j) \leq 0$$

for all $b \in \mathbb{Z}^n$ with $\sum_{i=1}^n b_i = 1$ and for all distinct elements $x_1, \ldots, x_n \in X$, $n \geq 2$.

In this case the distance space (X,d) is called *hypermetric space.*

Since any triangle inequality is an pure 3-gonal inequality, it holds, that any hypermetric is a semimetric.

Definition 8.2 *Given distance space (X,d), the distance d is called a distance of negative type, if it satisfies all the negative type inequalities, i.e., if d satisfies*

$$\sum_{1 \leq i < j \leq n} b_i b_j d(x_i, x_j) \leq 0$$

for all $b \in \mathbb{Z}^n$ with $\sum_{i=1}^n b_i = 0$ and for all distinct elements $x_1, \ldots, x_n \in X$, $n \geq 2$.

In this case the distance space (X,d) is called *negative type space.*

Observe that in the above definitions we can drop the condition that the elements $x_1, \ldots, x_n$ are distinct. Indeed, suppose for instance that $x_1 = x_2$. Then, $d(x_1,x_2) = 0$ and $d(x_1,x_i) = d(x_2,x_i)$ for all i. Therefore, the quantity $\sum_{1 \leq i < j \leq n} b_i b_j d(x_i,x_j)$ can be rewritten as $\sum_{2 \leq i < j \leq n} b'_i b'_j d(x_i,x_j)$, after setting $b'_2 = b_1 + b_2, b'_3 = b_3, \ldots, b'_n = b_n$.

In other words, (X,d) is hypermetric (respectively, of negative type) if and only if d satisfies the inequalities $\sum_{1 \leq i < j \leq n} b_i b_j d(x_i,x_j) \leq 0$ for all $b \in \{0,-1,1\}^n$ with $\sum_{i=1}^n b_i = 1$ (respectively, with $\sum_{i=1}^n b_i = 0$) and all (not necessarily distinct) elements $x_1, \ldots, x_n \in X$, $n \geq 2$.

Remark. Since the negative type inequalities are implied by the hypermetric inequalities, it holds that any hypermetric space is a negative type space. But the negative type inequalities do *not* imply the triangle inequalities. In other words, a distance may be of negative type without being a semimetric.

To see it, consider for instance the distance d on V_n defined by $d_{1i} = 1$ for $i = 2, \ldots, n$, and $d_{ij} = \frac{2n}{n-1}$ for $2 \leq i < j \leq n$. Then, d violates some triangle inequalities, as

$d_{ij} - d_{1j} - d_{1i} = \frac{2}{n-1} > 0$ for any $i \neq j \in \{2, \ldots, n\}$. On the other hand, it is easy to verify that d is adistance of negative type (e.g., because its image $\xi_1(d)$ - under the covariance mapping pointed at position 1 - defines a positive semidefinite matrix, see [DeLa97]).

However, for a negative type distance d on V_n, the condition $d_{1n} = 0$ implies that $d_{1i} = d_{in}$ for all $i = 2, \ldots, n-1$. Hence, the metric condition is partially satisfied. Moreover, letting d' denote the distance on the set $V_{n-1} = V_n \setminus \{n\}$, defined as the projection of d (i.e., $d'_{ij} = d_{ij}$ for all $i, j \in V_{n-1}$), we get, that d is a negative type distance on V_n if and only if d' is a negative type distance on V_{n-1}. In other words, for testing the negative type conditions, we can restrict ourselves to distances taking only positive values.

One of the main motivations for introducing hypermetric inequalities lies in the fact that *they are valid for the cut semimetrics*. In other words, *every distance space that is isometrically ℓ_1-embeddable satisfies all the hypermetric inequalities.*

Proposition 8.2 *Any cut semimetric δ_S on V_n is an hypermetric on V_n.*

Proof. In order to verify that every cut semimetric satisfies all the hypermetric inequalities, let $S \subseteq V_n$ and $b \in \mathbb{Z}^n$ with $\sum_{i=1}^n b_i = 1$. Then the value

$$\sum_{1 \leq i < j \leq n} b_i b_j \delta_S(i,j) = \sum_{i \in S, j \notin S} b_i b_j = \left(\sum_{i \in S} b_i\right)\left(\sum_{j \notin S} b_j\right) = \left(\sum_{i \in S} b_i\right)\left(1 - \sum_{i \in S} b_i\right)$$

is non-positive, since $\sum_{i \in S} b_i$ is an integer.

□

Definition 8.3 *Given an integer $k \geq 1$ and a distance space (X, d), the distance d is called $(2k+1)$-gonal, if it satisfies the inequalities $\sum_{1 \leq i < j \leq n} b_i b_j d(x_i, x_j) \leq 0$ for all $b \in \mathbb{Z}^n$ with $\sum_{i=1}^n b_i = 1$ and $\sum_{i=1}^n |b_i| = 2k+1$, and for all $x_1, \ldots, x_n \in X$, $n \geq 2$.*

In this case the distance space (X, d) is called $(2k+1)$-*gonal space.*

Definition 8.4 *Given an integer $k \geq 1$ and a distance space (X, d), the distance d is called $(2k)$-gonal, if it satisfies the inequalities $\sum_{1 \leq i < j \leq n} b_i b_j d(x_i, x_j) \leq 0$ for all $b \in \mathbb{Z}^n$ with $\sum_{i=1}^n b_i = 0$ and $\sum_{i=1}^n |b_i| = 2k$, and for all $x_1, \ldots, x_n \in X$, $n \geq 2$.*

In this case the distance space (X, d) is called $(2k)$-*gonal space.*

Again we obtain the same definition if we require that d satisfies all these inequalities only for b pure, i.e., with entries in $\{0, 1, -1\}$. For instance, (X, d) is 5-gonal if and only if, for all $x_1, x_2, x_3, y_1, y_2 \in X$, the following *pentagonal inequality* holds:

$$\sum_{1 \leq i < j \leq 3} d(x_i, x_j) + d(y_1, y_2) - \sum_{\substack{i=1,2,3 \\ j=1,2}} d(x_i, y_j) \leq 0.$$

In fact, the notion of k-gonal distance spaces is monotone in k, in the sense that $(k+2)$-gonality implies k-gonality.

Proposition 8.3 *Let (X, d) be a distance space.*

(i) For any integer $k \geq 2$, if (X, d) is $(k+2)$-gonal, then (X, d) is k-gonal.

(ii) For any integer $k \geq 1$, if (X, d) is $(2k+1)$-gonal, then (X, d) is $(2k+2)$-gonal.

Proof. (i) Suppose that (X, d) is $(k+2)$-gonal. Let $b \in \mathbb{Z}^n$ with $\sum_{i=1}^n |b_i| = k$ and $\sum_{i=1}^n b_i = \epsilon$, where $\epsilon = 1$, if k is odd, and $\epsilon = 0$, if k is even. Let $x_1, \ldots, x_n \in X$. We show that $\sum_{1 \leq i < j \leq n} b_i b_j d(x_i, x_j) \leq 0$. For this, set $b' = (b, 1, -1) \in \mathbb{Z}^{n+2}$, and $x_{n+1} = x_{n+2} = x$, where $x \in X$. Then, $\sum_{1 \leq i < j \leq n} b_i b_j d(x_i, x_j) = \sum_{1 \leq i < j \leq n+2} b'_i b'_j d(x_i, x_j)$, which is nonpositive by the assumption that (X, d) is $(k+2)$-gonal.

(ii) The assertion (ii) follows from Proposition 8.1.

□

Remark. Note that the k-gonal inequalities do *not* follow from the $(k+2)$-gonal inequalities $(k \geq 2)$. (The proof of (i) in the Proposition above works indeed at the level of distance spaces since we make the assumption that the two points x_{n+1} and x_{n+2} of X coincide.) For instance, the 5-gonal inequalities do *not* imply the triangle inequalities. To see it, consider the distance d on V_5, defined by $d_{ij} = 1$ for all pairs except $d_{12} = \frac{9}{4}$, and $d_{34} = \frac{3}{2}$. Then, d violates some triangle inequality, as $d_{12} - d_{13} - d_{23} = \frac{1}{4} > 0$; on the other hand, one can verify that d satisfies all 5-gonal inequalities.

8.4 Some generalizations of hypermetrics

Quasi-hypermetrics and quasi-distances of negative type

Let $n \geq 2$, and let $b_1, \ldots, b_n$ be integers. Given $b_1, \ldots, b_n \in \mathbb{Z}$, and a given quasi-semimetric $q : V_n \times V_n \to \mathbb{R}$, $q(i,j) = q_{ij}$, consider the inequality

$$\sum_{1 \leq i \neq j \leq n} b_i b_j q_{ij} \leq 0.$$

For convenience, introduce the following notation: for a vector $b \in \mathbb{R}^n$ with $b_i \in \mathbb{Z}$, let $Q_n^O(b)$ denote the vector from $\mathbb{R}^{I_n}$, defined by

$$Q_n^O(b)_{ij} = b_i b_j \ \text{ for } 1 \leq i \neq j \leq n.$$

When the parameter n is clear from the context, we also denote $Q_n^O(b)$ by $Q^O(b)$. Hence, the inequality

$$\sum_{1 \leq i < j \leq n} b_i b_j q_{ij} \leq 0$$

(in the variable $q = (q_{ij}) \in \mathbb{R}^{I_n}$) can be rewritten as

$$Q_n^O(b)^T q \leq 0.$$

When $\sum_{i=1}^n b_i = 1$, the inequality

$$Q_n^O(b)^T q = \sum_{1 \leq i \neq j \leq n} b_i b_j q_{ij} \leq 0$$

is called an *oriented hypermetric inequality* and is denoted by Hyp_b^O:

$$Hyp_b^O : Hyp_b^O(q) = \sum_{1 \le i \ne j \le n} b_i b_j q_{ij} \le 0, \text{ with } b \in \mathbb{Z}^n, \sum_{i=1}^{n} b_i = 1.$$

When $\sum_{i=1}^{n} b_i = 0$, the inequality

$$Q_n^O(b)^T q = \sum_{1 \le i \ne j \le n} b_i b_j q_{ij} \le 0$$

is called an *oriented negative type inequality* and is denoted by Neg_b^O:

$$Neg_b^O : Neg_b^O(q) = \sum_{1 \le i \ne j \le n} b_i b_j q_{ij} \le 0, \text{ with } b \in \mathbb{Z}^n, \sum_{i=1}^{n} b_i = 0.$$

The inequality $Q_n^O(b)^T q \le 0$ is said to be *pure* if $|b_i| = 0, 1$ for all $i \in V_n$. The inequality $Q_n(b)^T q \le 0$ is said to be an *oriented k-gonal inequality*, if $\sum_{i=1}^{n} |b_i| = k$ holds.

In particular, the *oriented 2-gonal inequality* is the oriented negative type inequality, where $b_i = 1$, $b_j = -1$, and $b_h = 0$ for $h \in V_n \setminus \{i, j\}$, for some distinct $i, j \in V_n$; it is nothing but the weak non-negativity constraint $q_{ij} + q_{ij} \ge 0$.

The *pure oriented 3-gonal inequality* is the oriented hypermetric inequality with $b_i = b_j = 1$, $b_k = -1$, and $b_h = 0$ for $h \in V_n \setminus \{i, j, k\}$, for some distinct $i, j, k \in V_n$; it gives the following version of triangle inequality:

$$(q_{ij} - q_{ik} - q_{kj}) + (q_{ji} - q_{jk} - q_{ki}) \le 0.$$

Easy to see, that for a function $q : V_n \times V_n \to \mathbb{R}$, satisfying to all the oriented hypermetric inequalities, its symmetrization $d : d_{ij} = q_{ij} + q_{ji}$ is a hypermetric. But, however, q itself is, in general, even not a quasi-semimetric. So, in oriented case we define an quasi-hypermetric as a quasi-semimetric, satisfying to all the oriented hypermetric inequalities.

Definition 8.5 *Given a quasi-semimetric space (X, q), the quasi-semimetric q is called quasi-hypermetric, if it satisfies all the oriented hypermetric inequalities, i.e., if it holds*

$$\sum_{1 \le i \ne j \le n} b_i b_j q(x_i, x_j) \le 0$$

for all $b \in \mathbb{Z}^n$ with $\sum_{i=1}^{n} b_i = 1$ and for all distinct elements $x_1, \dots, x_n \in X$, $n \ge 2$.

In this case the quasi-semimetric space (X, q) is called *quasi-hypermetric space.*

One can define similarly an *quasi-distance space of negative type*, using for the construction the oriented negative type inequalities.

One of the main motivations for introducing oriented hypermetric inequalities lies in the fact that *they are valid for the oriented cut semimetrics.*

Proposition 8.4 *Any oriented cut semimetric δ_S^O on V_n is an quasi-hypermetric on V_n.*

Proof. In order to verify that every o-cut semimetric satisfies all the oriented hypermetric inequalities, let $S \subseteq V_n$ and $b \in \mathbb{Z}^n$ with $\sum_{i=1}^n b_i = 1$. Then, the value

$$\sum_{1 \le i \ne j \le n} b_i b_j \delta_S(i,j) = \sum_{i \in S, j \notin S} b_i b_j = \left(\sum_{i \in S} b_i\right)\left(\sum_{j \notin S} b_j\right) = \left(\sum_{i \in S} b_i\right)\left(1 - \sum_{i \in S} b_i\right)$$

is non-positive since $\sum_{i \in S} b_i$ is an integer.

□

Partial hypermetrics and weighted hypermetrics

For a given sequence $b = (b_1, ..., b_n)$ of integers and a symmetric $n \times n$ matrix $A = ((a_{ij}))$, denote by $H_b(a)$ the sum $-\sum_{1 \le i,j \le n} b_i b_j a_{ij}$ of the entries of the matrix $-b \cdot A \cdot b^T$.

In this notation, the *hypermetric* on n points can be defined as a semimetric d on V_n with $H_b(d) \ge 0$, whenever $\sum_{i=1}^n b_i = 1$.

Consider a *weighted semimetric* on V_n, i.e., a symmetric function $d : V_n \times V_n \to \mathbb{R}_{\ge 0}$, such that there exists a *weight function* $w = (w_i) : X \longrightarrow \mathbb{R}_{\ge 0}$ on V_n, and for all $i, j, k \in V_n$ it holds $d_{ii} = 0$, and $d_{ij} \le d_{ik} + d_{kj}$.

For a weighted semimetric $(d; w)$ on n points, we will use the following notations:

- $Hyp_b^w(d; w) = \frac{1}{2} H_b(d) + (1 - \sum_{i=1}^n b_i) \sum_{i=1}^n b_i w_i$;

- $Hyp_b^{w'}(d; w) = \frac{1}{2} H_b(d) + (1 + \sum_{i=1}^n b_i) \sum_{i=}^n b_i w_i$.

Definition 8.6 *A weighted semimetric d on V_n is called weighted hypermetric, if it satisfies the inequalities $Hyp_b^w : Hyp_b^w(d; w) \ge 0$ and $Hyp_b^{w'} : Hyp_b^{w'}(d; w) \ge 0$ for all b with $\sum_{i=1}^n b_i = 1$, or $\sum_{i=1}^n b_i = 0$.*

Consider now a *weak partial semimetric* on the set V_n, i.e., a symmetric function $p : V_n \times V_n \to \mathbb{R}_{\ge 0}$, such that for any $i, j, k \in V_n$ it holds $p_{ik} + p_{kj} - p_{ij} - p_{kk} \ge 0$.

It is known, that weighted semimetrics and weak partial semimetrics are connected by the map $P : (d, w) \to p$, defined by

$$p_{ij} = \frac{d_{ij} + w_i + w_j}{2}.$$

The map P is an automorphism of the vector space $\mathbb{R}^{\binom{n+1}{2}}$, and $(d; w) = P^{-1}(p)$, where the inverse map P^{-1} is defined by

$$d_{ij} = 2p_{ij} - p_{ii} - p_{jj}, \quad w_i = p_{ii}.$$

Definition 8.7 *A weak partial semimetric p on V_n is called partial hypermetric, if it satisfies the inequalities $Hyp_b^w(P^{-1}(p)) \ge 0$ for all b with $\sum_{i=1}^n b_i = 1$, or $\sum_{i=1}^n b_i = 0$.*

It is easy to see, that for $p = P(d; w)$ we have

$$Hyp_b^w(d; w) = H_b(p) + \sum_{i=1}^{n} b_i p_{ii}.$$

So, a *partial hypermetric* on n points as a weak partial semimetric p on V_n, satisfying all inequalities

$$Hyp_b^p : H_b(p) + \sum_{i=1}^{n} b_i p_{ii} \geq 0$$

with $\sum_{i=1}^{n} b_i = 1$, or $\sum_{i=1}^{n} b_i = 0$.

Consider now a *weak weightable quasi-semimetric* q on the set V_n, i.e., a function $q : V_n^2 \to \mathbb{R}$ with a *weight function* $w = (w_x) : V_n \longrightarrow \mathbb{R}_{\geq 0}$, such that for all $i, j, k \in V_n$ it holds $q_{ij} + q_{ji} \geq 0$, $q_{ii} = 0$, $q_{ij} \leq q_{ik} + q_{kj}$, and $q_{ij} + w_i = q_{ji} + w_j$ for $i \neq j$.

It is known, that weighted semimetrics and weak weightable quasi-semimetrics are connected by the map $Q : (d, w) \to q$, defined by

$$q_{ij} = \frac{d_{ij} - w_i + w_j}{2}.$$

The map Q is an automorphism of the vector space $\mathbb{R}^{\binom{n+1}{2}}$, and $(d; w) = Q^{-1}(q)$, where the inverse map Q^{-1} is defined by

$$d_{ij} = q_{ij} + q_{ji}.$$

Similarly, we can define a *weightable quasi-hypermetric* q on V_n as a *weak weightable quasi-semimetric* q on V_n, satisfying the inequalities $Hyp_b^w(Q^{-1}(q)) \geq 0$ for all b with $\sum_{i=1}^{n} b_i = 1$, or $\sum_{i=1}^{n} b_i = 0$.

It is easy to see, that for $(q, w) = Q(d; w)$ we have

$$Hyp_b^w(d; w) = H_b(q) + \left(1 - \sum_{i=1}^{n} b_i\right) \sum_{i=1}^{n} b_i w_i.$$

In other words, a *weightable quasi-hypermetric* q on V_n is a *weak weightable quasi-semimetric* q on V_n, satisfying, in addition, all inequalities

$$Hyp_b^o : H_b(q) + \left(1 - \sum_{i=1}^{n} b_i\right) \sum_{i=1}^{n} b_i w_i \geq 0$$

with $\sum_{i=1}^{n} b_i = 1$, or $\sum_{i=1}^{n} b_i = 0$.

However, for *weightable quasi-semimetrics* on V_n (weak weightable quasi-semimetrics q, which are quasi-semimetrics, i.e., satisfy, in addition, all $q_{ij} \geq 0$) there exists an other approach to the construction of *weightable quasi-hypermetrics.*

In order to introduce it, consider a *tournament* K_n^O, i.e., an oriented version of the complete graph K_n with an unique arc between any distinct $i, j \in V_n$. Call a tournament *admissible*, if its arcs can be partitioned into arc-disjoint directed cycles.

It does not exists for even n, because then the number of arcs involving each vertex is odd, while each cycle provides 0 or 2 such arcs. But for odd n, there are at least $2^{\frac{n-3}{2}}$

admissible tournaments: take the decomposition of K_n into $\frac{n-1}{2}$ disjoint Hamiltonian cycles and, fixing the order on one them, all possible orders on remaining cycles. The *Kelly conjecture* state that the arcs of every *regular* (i.e., the vertices have the same outdegree) tournament can be partitioned into arc-disjoint directed Hamiltonian cycles.

Definition 8.8 *A weightable quasi-semimetric q on n points is called weightable quasi-hypermetric, if it satisfies all o-hypermetric inequalities*

$$Hyp_{b,O} : Hyp_{b,O}(q) = -\sum_{1\leq i<j\leq n} b_i b_j q_{a(ij)} \geq 0$$

with $b = (b_1, ..., b_n) \in \mathbb{Z}^n$, $\sum_{i=1}^n b_i = 1$, *and all o-negative type inequalities*

$$Neg_{b,O} : Neg_{b,O}(q) = -\sum_{1\leq i<j\leq n} b_i b_j q_{a(ij)} \geq 0$$

with $b = (b_1, ..., b_n) \in \mathbb{Z}^n$, $\sum_{i=1}^n b_i = 0$, *where* $O = K_n^O$ *is an admissible tournament, and the arc* $a(ij)$ *on the edge* (i, j) *is the same as in* O *if* $b_i b_j \geq 0$, *or the opposite one, otherwise.*

Infinite hypermetrics

Another interesting question is to define infinite hypermetric spaces. One way to do that is to define an *infinite hypermetric* $d : \mathbb{N} \times \mathbb{N} \to \mathbb{R}$ by imposing that for all $b \in \mathbb{Z}^\infty$ with finite *support* (i.e., the set $\{i \mid b_i \neq 0\}$) and $\sum_{i=1}^n b_i = 1$ it holds

$$\sum_{i\leq i<j} b_i b_j d_{ij} \leq 0.$$

For example, the path metric of the skeleton of the *infinite hyperoctahedron* $K_{2,...,2,...}$ is an infinite hypermetric, which does not embed isometrically into l_1. In general, it is easy to build infinite hypermetrics; it basically suffices to use Delaunay polytopes of infinite lattices. For example, above ∞-dimensional hyperoctahedron can be considered as a Delaunay polytope in the ∞-dimensional root lattice D_∞.

However, as far as we know, there is no general Theory of Delaunay polytopes in infinite dimensional lattices.

Part IV

Cones and polytopes of generalized finite semimetrics

Chapter 9

Non-oriented case: semimetrics and cuts

9.1 Preliminaries

For given two semimetrics d_1 and d_2 on a set X their non-negative linear combination $d = \alpha d_1 + \beta d_2, \alpha, \beta \geq 0$, is a semimetric on X. Here, as usual, for all $x, y \in X$

$$(\alpha d_1 + \beta d_2)(x, y) = \alpha d(x, y) + \beta d'(x, y).$$

Then we can speak about the cone of all semimetrics on n points, in fact, on the set $V_n = \{1, 2, ..., n\}$. We can consider already the cones, generated by all cuts and multicuts on n points, as well as polytopes of semimetrics, cuts and multicuts on n points.

In this chapter we consider, for small values of n, the cone of all semimetrics, the cone, generated by all cut semimetrics on V_n, the polytope of all semimetrics, and the polytope, generated by all cut semimetrics on V_n.

Metrics and cuts are well-known and central objects in Graph Theory, Combinatorial Optimization and, more generally, Discrete Mathematics. They also occur in other areas of Mathematics and its Applications such as Distance Geometry, Geometry of Numbers, Combinatorial Matrix Theory, Theory of Designs, Quantum Mechanics, Statistical Physics, Analysis and Probability Theory. Indeed, cuts are many faceted objects, and they can be interpreted as graph theoretic objects, as metrics, or as probabilistic pairwise correlations.

The knowledge of extreme rays of the metric cone MET_n for small n will allow to build a Theory of Multicommodity Flows on non-oriented graphs. The cut cone CUT_n plays a central role in the study of metric spaces that can be isometrically embedded into ℓ_1-spaces. See [DeLa97], [DDF96], [DeDe95], [Duto08] for details on MET_n and CUT_n.

Using computer search we list facets and generators for these cones and tables of their adjacencies and incidences. The different orbits were determined manually, using symmetries. We study two graphs, the 1-skeleton graph G_C and the ridge graph G_C^* of these polyhedra: the number of their nodes and edges, their diameters, conditions of adjacency, inclusions among the graphs and their restrictions on some orbits of nodes. In fact, we would like to describe two graphs G_C, G_C^* for our cones as fully as possible, but in the cases, when it is too difficult, we will give some partial information on adjacencies in those graphs. Especially we are interested in the diameters of the graphs, in a good criterion of adjacency in their local graphs (i.e., in the subgraphs induced by all neighbors of a given vertex) and in their restrictions on some orbits.

The *Adjacency Decomposition Method* of Christof and Reinelt (1996) [ChRe96] and the program *cdd* from Fukuda (1995) [Fuku95] were used to find the extreme rays and facets of the polyhedra. We were able also to test adjacency of facets or extreme rays of a cone without knowledge of extreme rays or of facets of this cone; namely, we used a Linear Programming Method.

The following polyhedra will be studied below:

- The *metric cone* MET_n: the set of all *semimetrics* on n points, generated by all *triangle inequalities* $d_{ik} \leq d_{ij} + d_{jk}$ on V_n;
- the *cut cone* CUT_n: the conic hull of all $2^{n-1} - 1$ non-zero *cut semimetrics* on n points;
- the *multicut cone* $MCUT_n$: the conic hull of all non-zero *multicut semimetrics* on n points;
- the *metric polytope* $MET_n^{\square}$: the set of all $d \in MET_n$, satisfying, in addition, all *perimeter inequalities* $d_{ik} + d_{ij} + d_{jk} \leq 2$ on V_n;
- the *cut polytope* $CUT_n^{\square}$: the convex hull of all 2^{n-1} *cut semimetrics* on n points;
- the *multicut polytope* $MCUT_n^{\square}$: the convex hull of all *multicut semimetrics* on n points.

9.2 Cones and polytopes of semimetrics and cuts

Let $V_n = \{1, 2, ..., n\}$; let $E_n = |\{(i,j) \mid i, j \in V_n, i \neq j\}|$, where (i,j) denotes the unordered pair of the integers i, j, i.e., $E_n = \binom{n}{2} = \frac{n(n-1)}{2}$.

Let d be a *semimetric* on the set V_n, i.e., a function $d : V_n \times V_n \to \mathbb{R}_{\geq 0}$, such that for any $x, y, \in V_n$ it holds $d(x,x) = 0$; $d(x,y) = d(y,x)$; $d(x,y) \leq d(x,z) + d(z,y)$ (in fact, in the symmetric case the non-negativity $d(x,y) \geq 0$ follows from the conditions $d(x,x) = 0$ and $d(x,y) \leq d(x,z) + d(z,y)$).

Because of the symmetry and since $d(i,i) = 0$ for $i \in V_n$, we can view a semimetric d as a vector $(d_{ij})_{1 \leq i < j \leq n} \in \mathbb{R}^{E_n}$, $d_{ij} = d(i,j)$. Clearly, one can also view a semimetric as a symmetric $n \times n$ matrix $((d_{ij}))$ with zeros on the main diagonal.

Hence, a semimetric on V_n can be viewed alternatively as a function on $V_n \times V_n$, as a symmetric $n \times n$ matrix $((d_{ij}))$, or as a vector in $\mathbb{R}^{E_n}$. We will use all three these representations for a semimetric on V_n. Moreover, we will use both symbols $d(i,j)$ and d_{ij} for the values of the semimetric d between points i and j.

Denote by MET_n the set of all semimetrics on n points. Then MET_n is a full-dimensional cone in $\mathbb{R}^{E_n}$, defined by the $\binom{n}{3} = \frac{n(n-1)(n-2)}{2}$ *triangle inequalities* $T_{ij,k}$ on V_n:

- $T_{ij,k} : d_{ik} + d_{kj} - d_{ij} \geq 0$.

In the symmetric case, the non-negativity condition $d_{ij} \geq 0$ follows from the triangle inequality and condition $d_{ii} = 0$, as was shown above.

For example, the cone MET_3 is a full-dimensional cone in $\mathbb{R}^3$, generated by 3 triangle inequalities $T_{12,3} : d_{13} + d_{32} - d_{12} \geq 0$, $T_{13,2} : d_{12} + d_{23} - d_{13} \geq 0$, $T_{23,1} : d_{21} + d_{13} - d_{23} \geq 0$. The cone MET_4 is a full-dimensional cone in $\mathbb{R}^6$, generated by 12 triangle inequalities $T_{12,3} : d_{13} + d_{32} - d_{12} \geq 0$, $T_{13,2} : d_{12} + d_{23} - d_{13} \geq 0$, $T_{23,1} : d_{21} + d_{13} - d_{23} \geq 0$, $T_{12,4} : d_{14} + d_{42} - d_{12} \geq 0$, $T_{14,2} : d_{12} + d_{24} - d_{14} \geq 0$, $T_{24,1} : d_{21} + d_{14} - d_{24} \geq 0$, $T_{14,3} : d_{13} + d_{34} - d_{14} \geq 0$, $T_{13,4} : d_{14} + d_{43} - d_{13} \geq 0$, $T_{34,1} : d_{31} + d_{14} - d_{34} \geq 0$, $T_{23,4} : d_{24} + d_{43} - d_{23} \geq 0$, $T_{24,3} : d_{23} + d_{34} - d_{24} \geq 0$, $T_{34,2} : d_{32} + d_{24} - d_{34} \geq 0$.

Note, that any triangle equality for the cone MET_n can be written as a vector in $\mathbb{R}^{E_n}$. So, for the cone MET_3 in $\mathbb{R}^3$ 3 corresponding triangle inequalities can be written as $T_{12,3} : (-1, 1, 1)$, $T_{13,2} : (1, -1, 1)$, and $T_{23,1} : (1, 1, -1)$. For the cone MET_4 in $\mathbb{R}^6$, the generating triangle inequalities have the form $T_{12,3} : (-1, 1, 0, 1, 0, 0)$, $T_{13,2} : (1, -1, 0, 1, 0, 0)$, $T_{23,1} : (1, 1, 0, -1, 0, 0)$, $T_{12,4} : (-1, 0, 1, 0, 1, 0)$, $T_{14,2} : (1, 0, -1, 0, 1, 0)$, $T_{24,1} : (1, 0, 1, 0, -1, 0)$, $T_{14,3} : (0, 1, -1, 0, 0, 1)$, $T_{13,4} : (0, -1, 1, 0, 0, 1)$, $T_{34,1} : (0, 1, 1, 0, 0, -1)$, $T_{23,4} : (0, 0, 0, -1, 1, 1)$, $T_{24,3} : (0, 0, 0, 1, -1, 1)$, $T_{34,2} : (0, 0, 0, 1, 1, -1)$.

The cone, generated by all $2^{n-1} - 1$ non-zero *cut semimetrics* δ_S for $S \subseteq V_n$, where $\delta_S(i, j) = 1$ if $i \neq j$, $|S \cap \{i, j\}| = 1$, and is equal to 0, otherwise, is called the *cut cone* and denoted by CUT_n. It is a full-dimensional cone in $\mathbb{R}^{E_n}$.

For example, the cut cone CUT_3 is a simplicial cone in $\mathbb{R}^3$, generated by 3 non-zero cuts $\delta_{\{1\}} = (1, 1, 0)$, $\delta_{\{2\}} = (1, 0, 1)$, and $\delta_{\{3\}} = (0, 1, 1)$. The cut cone CUT_4 is a full-dimensional cone in $\mathbb{R}^6$, generated by 7 non-zero cuts $\delta_{\{1\}} = (1, 1, 1, 0, 0, 0)$, $\delta_{\{2\}} = (1, 0, 0, 1, 1, 0)$, $\delta_{\{3\}} = (0, 1, 0, 1, 0, 1)$, $\delta_{\{4\}} = (0, 0, 1, 0, 1, 1)$, $\delta_{\{1,2\}} = (0, 1, 1, 1, 1, 0)$, $\delta_{\{1,3\}} = (1, 0, 1, 1, 0, 1)$, $\delta_{\{2,3\}} = (1, 1, 0, 0, 1, 1)$. Obviously, δ_S corresponds to a ray of MET_n, which is, moreover, an extreme ray.

The full-dimensional cone in $\mathbb{R}^{E_n}$, generated by all non-zero multicut semimetrics $\delta_{S_1,...,S_q}$ on V_n, where $\{S_1, ..., S_q\}$, $q \geq 2$, is a partition of the set $V_n = \{1, 2, ..., n\}$, and $\delta_{S_1,...,S_q}(i, j) = 0$, if $i, j \in S_\alpha$ for some α, $1 \leq \alpha \leq q$, and is equal to 1, otherwise, is called the *multicut cone* and denoted by $MCUT_n$.

It is easy to check, that

$$\delta_{S_1,...,S_q} = \frac{1}{2} \sum_{1 \leq i \leq q} \delta_{S_i},$$

i.e., *the multicuts with* $q > 2$ *are interior points of* CUT_n. So, $MCUT_n = CUT_n$ for any $n \geq 3$ (see [DeLa97], Proposition 4.2.9).

Among the facets of CUT_n the most simple ones are the triangle facets, i.e., those defined by triangle inequalities. Hence,

$$CUT_n = MCUT_n \subseteq MET_n \subseteq \mathbb{R}^{E_n}_{\geq 0}.$$

Clearly, all faces of MET_n and CUT_n are preserved by any permutation of the nodes. It means, that every permutation of V_n induces a symmetry of the cones MET_n and CUT_n. So, the group *Sym*(n) is a symmetry group of those cones.

It is proved (see [DGL91], [Duto02]), that *Sym*(n) is the full symmetry group of MET_n and CUT_n for $n \neq 4$. For $n = 4$ there are some additional symmetries: the full symmetry group $Is(CUT_4 = MET_4)$ of the cones CUT_4 and MET_4 is $Sym(4) \times Sym(3)$. All orbits of facets considered below, are under action of the group $Sym(n)$.

The triangle inequalities are sufficient for describing the cut polyhedra for $n \leq 4$, i.e., $MET_3 = CUT_3$, and $MET_4 = CUT_4$, but $CUT_n \subset MET_n$ (strictly) for $n \geq 5$. The complete description of all the facets of the cut polyhedra CUT_n is known for $n \leq 8$, the complete description of the semimetric polyhedra MET_n is known for $n \leq 8$. Here the "combinatorial explosion" starts from $n = 8$: for example, CUT_8 has 49604520 facets. See [DeLa97]; see also [Gris90], [Gris92], [ChRe01], [DDF96], [DFM03], [DeDu13] for a detailed description of CUT_7, MET_7, CUT_8, MET_8.

Denote by $MET_n^{\square}$ the *semimetric polytope*, i.e., the set of all semimetrics on n points, satisfying to all $3\binom{n}{3} = \frac{n(n-1)(n-2)}{2}$ *triangle inequalities* $T_{ij,k}$, and to all $\binom{n}{3} = \frac{n(n-1)(n-2)}{6}$ *perimeter inequalities* (*non-homogeneous triangle inequalities*) G_{ij} on V_n:

- $T_{ij,k} : d_{jk} + d_{kj} - d_{ij} \geq 0$;
- $G_{ij} : d_{ij} + d_{ik} + d_{jk} \leq 2$.

(See [DeLa97], page 421).

For example, the polytope $MET_3^{\square}$ in $\mathbb{R}^3$ is generated by 3 triangle inequalities $T_{12,3} : d_{13} + d_{32} - d_{12} \geq 0$, $T_{13,2} : d_{12} + d_{23} - d_{13} \geq 0$, $T_{23,1} : d_{21} + d_{13} - d_{23} \geq 0$, and one perimeter inequality $G_{123} : d_{12} + d_{13} + d_{23} \leq 2$. The polytope $MET_4^{\square}$ is a full-dimensional cone in $\mathbb{R}^6$, generated by 12 triangle inequalities $T_{12,3} : d_{13} + d_{32} - d_{12} \geq 0$, $T_{13,2} : d_{12} + d_{23} - d_{13} \geq 0$, $T_{23,1} : d_{21} + d_{13} - d_{23} \geq 0$, $T_{12,4} : d_{14} + d_{42} - d_{12} \geq 0$, $T_{14,2} : d_{12} + d_{24} - d_{14} \geq 0$, $T_{24,1} : d_{21} + d_{14} - d_{24} \geq 0$, $T_{14,3} : d_{13} + d_{34} - d_{14} \geq 0$, $T_{13,4} : d_{14} + d_{43} - d_{13} \geq 0$, $T_{34,1} : d_{31} + d_{14} - d_{34} \geq 0$, $T_{23,4} : d_{24} + d_{43} - d_{23} \geq 0$, $T_{24,3} : d_{23} + d_{34} - d_{24} \geq 0$, $T_{34,2} : d_{32} + d_{24} - d_{34} \geq 0$, and 4 perimeter inequalities $G_{123} : d_{12} + d_{13} + d_{23} \leq 2$, $G_{124} : d_{12} + d_{14} + d_{24} \leq 2$, $G_{134} : d_{13} + d_{14} + d_{34} \leq 2$, $G_{234} : d_{23} + d_{24} + d_{34} \leq 2$.

The convex hull of all cut semimetrics on n points is called the *cut polytope* and is denoted by $CUT_n^{\square}$; the convex hull of all multicut semimetrics on n points is called the *multicut polytope* and is denoted by $MCUT_n^{\square}$; these two polytopes do not coincide.

The cut polytope $CUT_3^{\square}$ in $\mathbb{R}^3$ is generated by 4 cuts $\delta_{\emptyset} = (0,0,0)$, $\delta_{\{1\}} = (1,1,0)$, $\delta_{\{2\}} = (1,0,1)$ and $\delta_{\{3\}} = (0,1,1)$. The multicut polytope $MCUT_3^{\square}$ in $\mathbb{R}^3$ is generated by five multicuts: four cuts $\delta_{\emptyset} = (0,0,0)$, $\delta_{\{1\}} = (1,1,0)$, $\delta_{\{2\}} = (1,0,1)$, $\delta_{\{3\}} = (0,1,1)$, and one "proper" multicut $\delta_{\{1\},\{2\},\{3\}} = (1,1,1)$. The cut polytope $CUT_4^{\square}$ in $\mathbb{R}^6$ is generated by 8 cuts $\delta_{\emptyset} = (0,0,0,0,0,0)$, $\delta_{\{1\}} = (1,1,1,0,0,0)$, $\delta_{\{2\}} = (1,0,0,1,1,0)$, $\delta_{\{3\}} = (0,1,0,1,0,1)$, $\delta_{\{4\}} = (0,0,1,0,1,1)$, $\delta_{\{1,2\}} = (0,1,1,1,1,0)$, $\delta_{\{1,3\}} = (1,0,1,1,0,1)$, $\delta_{\{2,3\}} = (1,1,0,0,1,1)$, The multicut polytope $MCUT_4^{\square}$ in $\mathbb{R}^6$ is generated by 15 muticuts, including zero cut $\delta_{\emptyset}$, 4 2-cuts (ordinary cuts) of the form $\delta_{\{i_1\},\{i_2,i_3,i_4\}}$ (e.g., $\delta_{\{1\}} = (1,1,1,0,0,0)$), 3 2-cuts (ordinary cuts) of the form $\delta_{\{i_1,i_2\},\{i_3,i_4\}}$ (e.g., $\delta_{\{1,2\}} = (0,1,1,1,1,0)$), 6 3-cuts of the form $\delta_{\{i_1\},\{i_2\},\{i_3,i_4\}}$ (e.g., $\delta_{\{1\},\{2\},\{3,4\}} = (1,1,1,1,1,0)$) and one 4-cut $\delta_{\{1\},\{2\},\{3\},\{4\}} = (1,1,1,1,1,1)$.

Among the facets of $CUT_n^{\square}$ the most simple ones are the triangle facets, i.e., those defined by the triangle inequalities and the non-homogeneous triangle (perimeter) inequalities. Hence,

$$CUT_n^{\square} \subseteq MCUT_n^{\square} \subseteq MET_n^{\square} \subseteq [0,1]^{E_n}.$$

Clearly, all faces of $CUT_n^{\square}$ are preserved by any permutation of the nodes. But there exists an other symmetry, defined by so-called *switching operation.*

For a vector $v \in \mathbb{R}^{E_n}$ and a cut vector δ_S, let v^{δ_S} be defined by $v_{ij}^{\delta_S} = -v_{ij}$, if $\delta_S(i,j) = 1$, and $v_{ij}^{\delta_S} = v_{ij}$, if $\delta_S(i,j) = 0$. Consider the mapping $r_{\delta_S} : \mathbb{R}^{E_n} \longrightarrow \mathbb{R}^{E_n}$, defined by $r_{\delta_S}(v) = v^{\delta_S} + \delta_S$. The mapping r_{δ_S} is an affine bijection of the space $\mathbb{R}^{E_n}$, called a *switching* mapping. The facets of $CUT_n^{\square}$ are preserved under switching operation (see [DeLa97], pages 403–409): a consequence of the simple fact, that the symmetric difference of two cuts is again a cut.

In other words, the polytopes $CUT_n^{\square}$ and $MET_n^{\square}$ are invariant, besides permutations, under the operation U_S on semimetrics, such that, for a given subset S of V_n, $U_S(d) = d'$, where

$$d'_{ij} = \begin{cases} 1 - d_{ij}, & \text{if } |S \cap \{i,j\}| = 1, \\ d_{ij}, & \text{otherwise.} \end{cases}$$

Moreover, it is shown in [DGL91], that for $n \neq 4$ switchings and permutations are the only symmetries of $CUT_n^{\square}$. For $n = 4$ there are some additional symmetries. Together, permutations and switching form a group of order $2^{n-1} \times n!$. For $n \neq 4$, this is the full symmetry group $Is(CUT_n^{\square})$ of the cut polytope on n points. For $n = 4$, the full symmetry group $Is(CUT_4^{\square})$ is $Aut(K_{4,4})$, and so, its order is $2 \times (4!)^2 = 2^3 \times 144 = 2^7 \cdot 3^2$.

It is shown in [Laur96] that the semimetric polytope $MET_n^{\square}$ has the same group of symmetries as $CUT_n^{\square}$; that is, $Is(MET_n^{\square}) = Is(CUT_n^{\square})$ for any $n \geq 3$.

The triangle inequalities are sufficient for describing the cut polyhedra for $n \leq 4$, but $CUT_n^{\square} \subset MET_n^{\square}$ (strictly) for $n \geq 5$. The complete description of all the facets of $CUT_n^{\square}$ is known for $n \leq 8$, the complete description of $MET_n^{\square}$ is known for $n \leq 8$. Here the "combinatorial explosion" starts from $n = 8$.

In the table 9.1 we give a summary of the most important information concerning the considered cones of semimetrics and cuts. The column 2 indicates the dimension of the cone, the columns 3 and 4 give the number of extreme rays and facets, respectively; in parenthesis are given the numbers of their orbits. The column 5 gives the diameter of the 1-skeleton graph and of the ridge graph of the cone.

Cone	Dimension	# of ext. rays (orbits)	# of facets (orbits)	Diameters
CUT_3=MET_3	3	3(1)	3(1)	1; 1
CUT_4=MET_4	6	7(2)	12(1)	1; 2
CUT_5	10	15(2)	40(2)	1; 2
MET_5	10	25(3)	30(1)	2; 2
CUT_6	15	31(3)	210(4)	1; 3
MET_6	15	296(7)	60(1)	2; 2
CUT_7	21	63(3)	38780(36)	1; 3
MET_7	21	55226(46)	105(1)	3; 2
CUT_8	28	127(4)	49604520(2169)	1; 3 or 4?
MET_8	28	119269588(3918)	168(1)	3?; 2

Table 9.1: Small cones of semimetrics

9.3 Small cones and polytopes of semimetrics and cuts

The enumeration of orbits of facets of CUT_n for $n \leq 7$ was done in [Deza60], [AvMu89], [Gris90], [Bara99] for $n = 5, 6$ and 7, respectively. For CUT_8 and $CUT_8^\square$, sets of facets were found in [ChRe01]; completeness of these sets was shown in [DeDu13]. The enumeration of orbits of extreme rays of MET_n for $n \leq 8$ was done in [Gris92], [DDF96], [DFM03].

Non-oriented case. The case of 3 points

In the symmetric case $CUT_3 = MET_3$, $CUT_3^\square = MET_3^\square$ and, hence, the only facet-defining inequalities for CUT_3 and $CUT_3^\square$ are the triangle inequalities, 3 inequalities (from one orbit, obtained by permutations) for CUT_3 and 4 inequalities (from one orbit, obtained by permutations and switchings) for $CUT_3^\square$. All the extreme rays of the cone $CUT_3 = MET_3$ correspond to non-zero cut vectors, 3 cut vectors (from one orbit, obtained by permutations); all the vertices of $CUT_3^\square = MET_3^\square$ are 4 cuts (from one orbit, obtained by permutations and switchings).

Non-oriented case. The case of 4 points

In the same way, $CUT_4 = MET_4$, and $CUT_4^\square = MET_4^\square$. The only facet-defining inequalities for CUT_4 and $CUT_4^\square$ are the triangle inequalities, 12 inequalities (from one orbit, obtained by permutations) for CUT_4 and 16 inequalities (from one orbit, obtained by permutations and switchings) for $CUT_4^\square$. All the extreme rays of the cone $CUT_4 = MET_4$ correspond to non-zero cut vectors, 7 cut vectors (from two orbits, represented by cuts $\delta_{\{1\}}$ and $\delta_{\{1,2\}}$, respectively, and obtained by permutations); all the vertices of $CUT_4^\square = MET_4^\square$ are 8 cuts (from one orbit, obtained by permutations and switchings).

Non-oriented case. The case of 5 points

In the symmetric case $CUT_5 \subset MET_5$ and $CUT_5^\square \subset MET_5^\square$ strictly, and in general, $CUT_n^\square \subset MET_n^\square$, $n \geq 5$. To see it, observe that the vector $(\frac{2}{3}, \frac{2}{3}, ..., \frac{2}{3}) \in \mathbb{R}^{E_n}$ belongs to $MET_n^\square$, but not to $CUT_n^\square$, $n \geq 5$. Alternatively, the pentagonal inequality defines a non-triangle facet of $CUT_n^\square$, $n \geq 5$.

For $b \in \mathbb{R}^n$ with $b_i \in \mathbb{Z}$, let $Q_n(b)$ denote the vector from $\mathbb{R}^{E_n}$, defined by

$$Q_n(b)_{ij} = b_i b_j \text{ for } 1 \leq i < j \leq n.$$

When the parameter n is clear from the context, we also denote $Q_n(b)$ by $Q(b)$. In this case, the inequality $\sum_{1 \leq i<j \leq n} b_i b_j d_{ij} \leq 0$ in the variable $d = (d_{ij}) \in \mathbb{R}^{E_n}$ can be rewritten as

$$Q_n(b)^T d \leq 0.$$

When $\sum_{i=1}^n b_i = 1$, the inequality

$$Q_n(b)^T d = \sum_{1 \leq i<j \leq n} b_i b_j d_{ij} \leq 0$$

is called an *hypermetric inequality* and is denoted by Hyp_b.

In fact, all the facets of the cut cone on 5 points have the form

$$Q_n(b)^T d = \sum_{ij \in E_n} b_i b_j d_{ij} \leq 0.$$

The cone CUT_5 has 40 facets from 2 orbits (under permutations): 30 triangle inequalities

$$Hyp_{(1,1,-1,0,0)} : Q_3(1,1,-1,0,0)^T d \leq 0,$$

and 10 *pentagonal inequalities*

$$Hyp_{(1,1,1,-1,-1)} : Q_5(1,1,1,-1,-1)^T d \leq 0.$$

The polytope $CUT_5^\square$ has 56 facets from 2 orbits (under permutations and switchings): 40 facets, induced by the triangle inequality

$$Hyp_{(1,1,-1,0,0)} : Q_3(1,1,-1,0,0)^T d \leq 0,$$

and 16 facets, induced by the pentagonal inequality

$$Hyp_{(1,1,1,-1,-1)} : Q_5(1,1,1,-1,-1)^T d \leq 0.$$

The extreme rays of CUT_5 correspond to 15 non-zero cuts (from two orbits, represented by cuts $\delta_{\{1\}}$ and $\delta_{\{1,2\}}$, respectively); all the vertices of $CUT_5^\square$ are 16 cuts (from one orbit, obtained by permutations and switchings).

The facets of MET_5 and of $MET_5^\square$ correspond to the triangle inequalities: 30 facets (from one orbit under permutations) for MET_5, and 40 facets (from one orbit under permutations and switchings) for $MET_5^\square$. The extreme rays of MET_5 and the vertices of $MET_5^\square$ are also known; namely, besides the cut vectors, all of them arise by a switching of the vector $(\frac{2}{3}, \frac{2}{3}, ..., \frac{2}{3})$. So, MET_5 has 25 extreme rays from 3 orbits, while $MET_5^\square$ has 32 vertices from 2 orbits.

Non-oriented case. The case of 6 points

The cut cone CUT_6 has $60 + 60 + 90 = 210$ facets from 4 orbits; the cut polytope $CUT_6^\square$ has $80 + 96 + 192 = 368$ facets from 3 orbits. All the facets, up to permutations and switchings, are induced by one of the following inequalities:

- $Hyp_{(1,1,-1,0,0,0)} : Q_6(1,1,-1,0,0,0)^T d \leq 0$ (triangle inequalities);
- $Hyp_{(1,1,1,-1,-1,0)} : Q_6(1,1,1,-1,-1,0)^T d \leq 0$ (pentagonal inequalities);
- $Hyp_{(2,1,1,-1,-1,-1)} : Q_6(2,1,1,-1,-1,-1)^T d \leq 0$ (7-gonal inequalities).

The extreme rays of CUT_6 correspond to 31 non-zero cuts (from three orbits, represented by the cuts $\delta_{\{1\}}$, $\delta_{\{1,2\}}$, $\delta_{\{1,2,3\}}$, respectively); all the vertices of $CUT_6^\square$ are 32 cuts (from one orbit, obtained by permutations and switchings).

The facets of MET_6 and of $MET_6^\square$ correspond to the triangle inequalities: 60 facets (from one orbit under permutations) for MET_6, and 140 facets (from one orbit under permutations and switchings) for $MET_6^\square$. The metric cone MET_6 has 296 extreme rays from 7 orbits. The metric polytope $MET_6^\square$ has 544 vertices from 3 orbits.

Non-oriented case. The case of 7 points

CUT_7 has 38780 facets from 36 orbits; $CUT_7^{\square}$ has 116764 facets from 11 orbits. All the facets of these polyhedra, up to permutations and switchings, induced by one of the following eleven inequalities:

- $Hyp_{(1,1,-1,0,0,0,0)} : Q_7(1,1,-1,0,0,0,0)^T d \leq 0$ (triangle inequalities);
- $Hyp_{(1,1,1,-1,-1,0,0)} : Q_7(1,1,1,-1,-1,0,0)^T d \leq 0$ (pentagonal inequalities);
- $Hyp_{(2,1,1,-1,-1,-1,0)} : Q_7(2,1,1,-1,-1,-1,0)^T d \leq 0$;
- $Hyp_{(1,1,1,1,-1,-1,-1)} : Q_7(1,1,1,1,-1,-1,-1)^T d \leq 0$;
- $Hyp_{(2,2,1,-1,-1,-1,-1)} : Q_7(2,2,1,-1,-1,-1,-1)^T d \leq 0$;
- $Hyp_{(3,1,1,-1,-1,-1,-1)} : Q_7(3,1,1,-1,-1,-1,-1)^T d \leq 0$;
- $CW_7^1(1,1,1,1,1,-1,-1)^T d \leq 0$;
- $CW_7^1(3,2,2,-1,-1,-1,-1)^T d \leq 0$;
- $(Par_7)^T d \leq 0$;
- $(Cr_7)^T d \leq 0$.

Here $CW_n^r(b)^T d \leq 0$, $(Par_n)^T d \leq 0$ and $(Cr_n)^T d \leq 0$ denote *Clique-Web inequality* (see [DeLa97], p. 467), *Parachute inequality* (see [DeLa97], p. 497), and *Grishukhin inequality* (see [DeLa97], p. 502). The extreme rays of CUT_7 correspond to 63 non-zero cuts (from three orbits, represented by the cuts $\delta_{\{1\}}$, $\delta_{\{1,2\}}$, $\delta_{\{1,2,3\}}$, respectively); all the vertices of $CUT_7^{\square}$ are 64 cuts (from one orbit, obtained by permutations and switchings).

The facets of MET_7 and of $MET_7^{\square}$ correspond to the triangle inequalities: 105 facets (from one orbit under permutations) for MET_7, and 140 facets (from one orbit under permutations and switchings) for $MET_7^{\square}$. MET_7 has 55226 extreme rays from 46 orbits. So, the cone MET_7 has 46 orbits of extreme rays and not 41 as, by a technical mistake, was given in Grishukhin (1992), as well as in Deza and Laurent (1997). $MET_7^{\square}$ has 275840 vertices from 13 orbits. The diameter of G_{MET_7} is 3.

Non-oriented case. The case of 8 points

CUT_8 has 49604520 facets from 2169 orbits; $CUT_8^{\square}$ has 217093472 facets from 147 orbits (see [ChRe96]). In [ChRe96], the Adjacency Decomposition Method was introduced and was applied to the Transporting Salesman polytope, the Linear Ordering polytope and the cut polytope. For the cut polytope $CUT_8^{\square}$, the authors found 147 orbits, consisting of $217,093,472$ facets, but this list was potentially incomplete, since they were not able to the treat the triangle, pentagonal and 7-gonal inequalities at that time. Therefore, they only prove that the number of orbits is at least 147. Sometimes ([DeLa97]) this is incorrectly understood and it is reported that the number of orbits is exactly 147 with [ChRe96] as a reference. We showed that Christof-Reinelt's list is complete.

The facets of MET_8 and of $MET_8^{\square}$ correspond to the triangle inequalities: 168 facets (from one orbit under permutations) for MET_8, and 224 facets (from one orbit under permutations and switchings) for $MET_8^{\square}$.

Deza, Fukuda, Pasechnik and Sato (2001) [DFPS01] obtained a list of 1,550,825,600 rays of MET_8; we computed (see [DDD15]) that this list consists of 3918 orbits of extreme rays under the symmetry group $Sym(8)$ of this cone.

The lower bound for the number of extreme rays of the cone MET_8, was obtained, using the computation in [DFPS01] of the vertices of the polytope $MET_8^{\square}$ (see [DeLa97], page 421). Note also that we were able to test adjacency of facets or extreme rays of a cone, without knowledge of extreme rays or of facets of this cone; namely, we used a linear programming method.

Table 9.2 gives known results of the number of facets (f) and extreme rays or vertices (r or v, respectively) for small cut and metric cones/polytopes. The numbers of orbits, given there in parentheses, is under $Sym(n)$ for cones and under $Sym(n)$ and 2^{n-1} switchings for polytopes.

Cone/Polytope	$n=3$	$n=4$	$n=5$	$n=6$	$n=7$	$n=8$
CUT_n, r	3(1)	7(2)	15(2)	31(3)	63(3)	127(4)
CUT_n, f	3(1)	12(1)	40(2)	210(4)	38,780(36)	49,604,520(2,169)
MET_n, r	3(1)	7(2)	25(3)	296(7)	55,226(46)	119,269,588(3,918)
MET_n, f	3(1)	12(1)	30(1)	60(1)	105(1)	168(1)
$CUT_n^{\square}, v$	4(1)	8(1)	16(1)	32(1)	64(1)	128(1)
$CUT_n^{\square}, f$	4(1)	16(1)	56(2)	368(3)	116,764(11)	217,093,472(147)
$MET_n^{\square}, v$	4(1)	8(1)	32(2)	554(3)	275,840(13)	1,550,825,600(533)
$MET_n^{\square}, f$	4(1)	16(1)	40(1)	80(1)	140(1)	224(1)

Table 9.2: The number of extreme rays and facets of CUT_n, MET_n, $CUT_n^{\square}$ and $MET_n^{\square}$ for $3 \leq n \leq 8$

9.4 Theorems and conjectures for general case

The metric cone $MET_n \in \mathbb{R}^{\binom{n}{2}}$ has an unique orbit of $3\binom{n}{3}$ facets $Tr_{ij,k}$.

Its full symmetry group $Is(MET_n)$ is $Sym(n)$ if $n \neq 4$, and is $Sym(4) \times Sym(3)$ for $n = 4$.

The number of extreme rays (orbits) of MET_n is 3 (1), 7 (2), 25 (3), 296 (7), 55226 (46) 119269588 (3918), and $> 10^{14}$ for $3 \leq n \leq 9$.

In the ridge graph $G^*_{MET_n}$ with $n > 3$, each facet $Tr_{ij,k}$ is not adjacent only to $4n - 10$ others: $Tr_{ik,t}, Tr_{kj,t}, Tr_{it,j}, Tr_{tj,i}$, where $t \neq i, j$. In other words, $Tr_{ij,k} \not\sim Tr_{i'j',k'}$ whenever they are *conflicting*, i.e., have non-zero values of different sign at some position pq, $p, q \in \{i, j, k\} \cap \{i', j', k'\}$. Clearly, $|\{i, j, k\} \cap \{i', j', k'\}|$ should be 3 or 2, and $Tr_{ij,k}$ conflicts with 2 and $4(n-3)$ $Tr_{i'j',k'}$'s, respectively.

Similarly, two triangle inequalities are adjacent in the ridge graph of $MET_n^{\square}$ if and only if they are non-conflicting (see [DeDe94]).

So, $d(G^*_{MET_n}) = 2$ for $n > 3$, while $G^*_{MET_3} = G_{MET_3} = K_3$. $d(G_{MET_n})$ is 1 for $n = 4$, 2 for $5 \leq n \leq 6$, 3 for $n = 7$; it is conjectured to stay 3 for $n > 7$.

For the cone CUT_n, generated by all $2^{n-1} - 1$ non-zero cuts δ_S on n points, the equality $CUT_n = MET_n$ holds for $n \leq 4$, and the equality $Is(CUT_n) = Is(MET_n)$ holds for any $n \geq 3$.

The number of facets (orbits) of CUT_n is 3 (1), 12 (1), 40 (2), 210 (4), 38780 (36), 49604520 (2169) for $3 \leq n \leq 8$. $d(G_{CUT_n}) = 1$, and $d(G^*_{CUT_n}) = 2, 3, 3$ for $n = 5, 6, 7$.

All triangle inequalities are facet-inducing in CUT_n and in $CUT_n^{\square}$ for any $n \geq 3$; the cut vectors form a single switching class, which is a clique in the 1-skeleton graph of

$MET_n^\square$. So, 1-skeleton graph $G_{CUT_n} = K_{S(n,2)}$, where $S(n,2) = 2^{n-1} - 1$ is the Stirling number of the second kind. On the other hand, it is shown in Laurent [Laur96] that every other switching class is a stable set in the 1-skeleton graph of $MET_n^\square$, that is, no two non-integral switching equivalent vertices of $MET_n^\square$ form an edge on $MET_n^\square$.

The $\{0,1\}$-valued elements $d \in MET_n$ are all $B(n)$ multicuts $\delta_{S_1,\dots,S_q}$ of V_n, where $B(n)$ is the n-th *Bell number*. (It follows by induction, using that $d_{1i} = d_{1j} = 0$ implies $d_{ij} = 0$, and $d_{1i} \neq d_{1j}$ implies $d_{ij} = 1$.) In fact, $S_1, \dots, S_q$ are the equivalence classes of the equivalence $\sim$ on V_n, defined by $i \sim j$ if $d_{ij} = 0$.

The set of extreme rays of the cone MET_n, containing a non-zero $\{0,1\}$-valued point, consists of all $S(n,2) = 2^{n-1} - 1$ non-zero cuts; so, $\{0,1\}$-$MET_n = CUT_n$: the cone $\{0,1\}$-MET_n, generated by all extreme rays of MET_n, containing a non-zero $\{0,1\}$-valued point, coincides with cut cone CUT_n.

In fact, CUT_n is the set of all n-vertex semimetrics, which embed isometrically into some metric space l_1, and rational-valued elements of CUT_n correspond exactly to the n-vertex semimetrics, which embed isometrically, *up to a scale* $\lambda \in \mathbb{N}$, into the path metric of some N-cube $H(N,2)$. It shows importance of this cone in Analysis and Combinatorics.

Conjecture 9.1 *([DeLa97]) Any facet of $CUT_n^\square$ is adjacent to a triangle inequality facet.*

The conjecture was checked for $n \leq 7$; we confirm it for $n = 8$.

Looking for a counterexample to this conjecture, we applied our sampling framework to $CUT_n^\square$ for $n = 10$, 11 and 12. We got initial facets of low incidence and then we complemented this with random walks in the set of all facets. This allowed us to find many simplicial facets (more than $10,000$ for each) of these $CUT_n^\square$, but all of them were adjacent to at least one triangle inequality facet.

Chapter 10

Oriented case: quasi-semimetrics and oriented cuts

10.1 Preliminaries

For given two quasi-semimetrics d_1 and d_2 on a set X their non-negative linear combination $d = \alpha d_1 + \beta d_2, \alpha, \beta \geq 0$, is a quasi-semimetric on X. Here, as usual, for all $x, y \in X$ it holds

$$(\alpha d_1 + \beta d_2)(x, y) = \alpha d_1(x, y) + \beta d_2(x, y).$$

Then we can speak about the cone of all quasi-semimetrics on n points, in fact, on the set $V_n = \{1, 2, ..., n\}$. We can consider already the cones, generated by all oriented cuts and oriented multicuts on n points, and similar polytopes.

In this chapter we consider, for small values of n, the cone of all quasi-semimetrics on V_n, the cone, generated by all oriented cut semimetrics on V_n, the cone, generated by all oriented multicut semimetrics on V_n, the polytope of all quasi-semimetrics on V_n, the polytope, generated by all oriented cut semimetrics on V_n, and the polytope, generated by all oriented multicut semimetrics on V_n.

The cone $QMET_n$ was introduced and studied, for small n, in [DePa99] and [DDP03].

The knowledge of extreme rays of $QMET_n$ for small n will allow to build a Theory of Multicommodity Flows on oriented graphs, as well as it was done for non-oriented graphs using dual MET_n.

In this chapter, after a short review of general quasi-semimetrics, we consider for small values of n (for $n = 3, 4, 5$ and, partially, for $n = 6$) the cone and the polytope of all quasi-semimetrics on $V_n = \{1, 2, ..., n\}$, and the cone and the polytope, generated by all oriented multicuts on V_n. We list the facets and generators for these polyhedra and tables of their adjacencies and incidences. We study the 1-skeleton graphs and the ridge graphs of these polyhedra: the number of nodes and edges of these graphs, their diameters, adjacency conditions, inclusions among these graphs and their restrictions on some orbits of nodes.

Finally, we compare the results obtained for oriented case (see [DePa99], [DDP03]) with similar results for the symmetric case (see [DeDe94], [DeDe95], [DDF96], [DeLa97]).

All computations used the programs *cdd* (Fukuda (1995) [Fuku95]) and an adaptation, by Dutour, of Adjacency Decomposition Method from Christof and Reinelt (1996) [ChRe96].

Comparing with the cones of semimetrics, the amount of computation and memory is much bigger in the oriented case, because the dimension of the cones $OMCUT_n$ and

$QMET_n$ is twice those of CUT_n and MET_n, and because o-multicuts with $q > 2$ are not interior points of the cone $OCUT_n$. The *combinatorial explosion* starts from $n = 5$, while for corresponding semimetric cones it starts from $n = 8$.

The following polyhedra will be studied below:

- the *quasi-semimetric cone* $QMET_n$: the set of all *quasi-semimetrics* on n points, defined by all *oriented triangle inequalities* $q_{ik} \le q_{ij} + q_{jk}$, and all *non-negativity inequalities* $q_{ij} \ge 0$ on V_n;

- the *oriented cut cone* $OCUT_n$: the conic hull of all non-zero *oriented cuts semimetrics* on n points;

- the *oriented multicut cone* $OMCUT_n$: the conic hull of all non-zero *oriented multicut semimeitrics* on n points;

- the *quasi-semimetric polytope* $QMET_n^\square$: the set of all $q \in QMET_n$, satisfying, in addition, all *perimeter inequalities* $q_{ij} + q_{ji} \le 2$ on V_n;

- the *oriented cut polytope* $OCUT_n^\square$: the convex hull of all *oriented cut semimetrics*, including zero o-cut, on V_n;

- the *oriented multicut polytope* $OMCUT_n^\square$: the convex hull of all *oriented multicut semimetrics*, including zero o-multicut, on V_n.

10.2 Cones and polytopes of quasi-semimetrics and oriented multicuts

Let $V_n = \{1, 2, ..., n\}$; let $I_n = |\{\langle i,j\rangle \,|\, i,j \in V_n, i \ne j\}|$, where $\langle i,j\rangle$ denotes the ordered pair of the integers i, j, i.e., $I_n = 2\binom{n}{2} = n(n-1)$.

Let q be a *quasi-semimetric* on the set V_n, i.e., a function $q : V_n \times V_n \to \mathbb{R}_{\ge 0}$, such that for any $x, y, \in V_n$ it holds $q(x,x) = 0$, and $q(x,y) \le q(x,z) + q(z,y)$. In the oriented case the non-negativity condition $q(x,y) \ge 0$ does not follow from other conditions.

Since $q(i,i) = 0$ for $i \in V_n$, we can view a quasi-semimetric q on the set V_n as a vector $(q_{ij})_{1 \le i \ne j \le n} \in \mathbb{R}^{I_n}$, where $q_{ij} = q(i,j)$. Clearly, one can also view a quasi-semimetric as an (in general, non-symmetric) $n \times n$ matrix $((q_{ij}))$ with zeros on the main diagonal.

Hence, a quasi-semimetric on V_n can be viewed alternatively as a function on $V_n \times V_n$, as a vector in $\mathbb{R}^{I_n}$ or as an (in general, non-symmetric) $n \times n$ matrix with zeros on the main diagonal. We will use all these representations for a quasi-semimetric on V_n. Moreover, we will use both symbols $q(i,j)$ and q_{ij} for the values of the quasi-semimetric between points i and j.

Denote by $QMET_n$ the set of all quasi-semimetrics on n points. Then $QMET_n$ is a full-dimensional cone in $\mathbb{R}^{I_n}$, defined by the $6\binom{n}{3} = n(n-1)(n-2)$ *oriented triangle inequalities* $OT_{ij,k}$, and $2\binom{n}{2} = n(n-1)$ *non-negativity inequalities* NN_{ij} on V_n:

- $NN_{ij} : d_{ij} \ge 0$;

- $OT_{ij,k} : d_{ik} + d_{kj} - d_{ij} \ge 0$.

If in the symmetric case the non-negativity condition follows from triangle inequality and condition $q_{ii} = 0$, in oriented case, as was shown above, it is not the case.

Note, that without the non-negativity condition we have in the cone $QMET_n$ the subspace of all mappings d^* satisfying $d^*(i,j) = -d^*(j,i)$ and $d^*(i,j) + d^*(j,n) = d^*(i,n)$ for all $1 \leq i,j \leq n$. The dimension of this subspace is $n(n-1) - (\binom{n}{2} + \binom{n-1}{2}) = n-1$.

For example, the cone $QMET_3$ is a full-dimensional cone in $\mathbb{R}^6$, generated by 6 oriented triangle inequalities $OT_{12,3} : d_{13}+d_{32}-d_{12} \geq 0$, $OT_{13,2} : d_{12}+d_{23}-d_{13} \geq 0$, $T_{23,1} : d_{21}+d_{13}-d_{23} \geq 0$, $OT_{21,3} : d_{23}+d_{31}-d_{21} \geq 0$, $OT_{31,2} : d_{32}+d_{21}-d_{31} \geq 0$, $T_{32,1} : d_{31}+d_{12}-d_{32} \geq 0$, and 6 non-negativity inequalities $N_{12} : d_{12} \geq 0$, $N_{21} : d_{21} \geq 0$, $N_{13} : d_{13} \geq 0$, $N_{31} : d_{31} \geq 0$, $N_{23} : d_{23} \geq 0$,$N_{32} : d_{32} \geq 0$.

The cone $QMET_4$ is a full-dimensional cone in $\mathbb{R}^{12}$, generated by 24 oriented triangle inequalities $OT_{12,4}$, $OT_{21,4}$, $OT_{23,4}$, $OT_{32,4}$, $OT_{13,4}$, $OT_{31,4}$, $OT_{12,3}$, $OT_{21,3}$, $OT_{14,3}$, $OT_{41,3}$, $OT_{24,3}$, $OT_{42,3}$, $OT_{13,2}$, $OT_{31,2}$, $OT_{14,2}$, $OT_{41,2}$, $OT_{34,2}$, $OT_{43,2}$, $OT_{23,1}$, $OT_{32,1}$, $OT_{24,1}$, $OT_{42,1}$, $OT_{34,1}$, $OT_{43,1}$, and 12 non-negativity inequalities N_{12}, N_{21}, N_{13}, N_{31}, N_{14}, N_{41}, N_{23}, N_{32}, N_{24}, N_{42}, N_{34}, N_{43}.

Any oriented triangle inequality and any non-negativity inequality for the cone $QMET_n$ can be written as an $\{0,1\}$-vector in $\mathbb{R}^{I_n}$. So, for the cone $QMET_3$ in $\mathbb{R}^6$ 6 corresponding oriented triangle inequalities can be written as

$$T_{12,3} : (-1,1,0,0,0,1),\ T_{21,3} : (0,1,-1,1,1,0),\ T_{13,2} : (1,-1,0,1,0,0),$$

$$T_{31,2} : (0,0,1,0,-1,1),\ T_{23,1} : (0,3,1,-1,0,0),\ T_{32,1} : (1,0,0,0,1,-1),$$

while 6 non-negativity inequalities have the form

$$N_{12} : (1,0,0,0,0,0),\ N_{21} : (0,0,1,0,0,0),\ N_{13} : (0,0,1,0,0,0),$$

$$N_{31} : (0,0,0,0,1,0),\ N_{23} : (0,0,0,1,0,0),\ N_{32} : (0,0,0,0,0,1).$$

For the cone $QMET_4$ in $\mathbb{R}^{12}$ we have, for example, that $OT_{12,4}$ triangle facet corresponds to the vector

$$OT_{12,4} : (-1,0,1,0,0,0,0,0,0,0,1,0),$$

and N_{12} non-negativity facet corresponds to the vector

$$N_{12} : (1,0,0,0,0,0,0,0,0,0,0,0,0).$$

The cone, generated by all $2^n - 2$ non-zero oriented cut semimetrics δ_S^O for $S \subseteq V_n$, where $\delta_S^O(i,j) = 1$, if $i \in S, j \notin S$, and is equal to 0, otherwise, is called the *oriented cut cone* and denoted by $OCUT_n$. It belongs to $\mathbb{R}^{I_n}$, and is of dimension $\binom{n+1}{2} - 1$.

For example, $OCUT_3$ is an 5-dimensional cone in $\mathbb{R}^6$, generated by 6 oriented cuts $\delta^O_{\{1\}} = (1,1,0,0,0,0)$, $\delta^O_{\{2\}} = (0,0,1,1,0,0)$, $\delta^O_{\{3\}} = (0,0,0,0,1,1)$, $\delta^O_{\{1,2\}} = (0,1,0,1,0,0)$, $\delta^O_{\{1,3\}} = (1,0,0,0,0,1)$, $\delta^O_{\{2,3\}} = (0,0,1,0,1,0)$.

The cone $OCUT_4$ is an 9-dimensional cone in $\mathbb{R}^{12}$, generated by all 14 non-zero oriented cuts $\delta^O_{\{1\}} = (1,1,1,0,0,0,0,0,0,0,0,0)$, $\delta^O_{\{2\}}$, $\delta^O_{\{3\}}$, $\delta^O_{\{4\}}$, $\delta^O_{\{2,3,4\}}$, $\delta^O_{\{1,3,4\}}$, $\delta^O_{\{1,2,4\}}$, $\delta^O_{\{1,2,3\}}$, $\delta^O_{\{1,2\}}$, $\delta^O_{\{1,3\}}$, $\delta^O_{\{1,4\}}$, $\delta^O_{\{2,3\}}$, $\delta^O_{\{2,4\}}$ $\delta^O_{\{3,4\}}$.

The full-dimensional cone in $\mathbb{R}^{I_n}$, generated by all non-zero oriented multicut semimetrics $\delta^O_{S_1,\ldots,S_q}$ on V_n, where $\langle S_1,\ldots,S_q\rangle$, $q \geq 2$, is an *ordered partition* of the set $V_n =$

$\{1, 2, ..., n\}$, and $\delta^O_{S_1,...,S_q}(i, j) = 1$, if $i \in S_\alpha, j \in S_\beta, \alpha < \beta$, and is equal to 0, otherwise, is called *oriented multicut cone* and denoted by $OMCUT_n$.

The number of generators of $OMCUT_n$ is $p^O(n) - 1$, where $p^O(n)$ is the number of all ordered partitions of n. In fact, $p^O(3) - 1 = 12$, $p^O(4) - 1 = 74$, $p^O(5) - 1 = 540$, $p^O(6) - 1 = 4682$, $p^O(7) - 1 = 47292$, $p^O(8) - 1 = 545834$ (see Chapter 7).

For example, $OMCUT_3$ is the simplicial cone in $\mathbb{R}^6$, generated by 12 non-zero oriented multicuts: 6 non-zero o-cuts (in fact, all 2-cuts) $\delta^O_{\{1\}} = (1, 1, 0, 0, 0, 0)$, $\delta^O_{\{2\}} = (0, 0, 1, 1, 0, 0)$, $\delta^O_{\{3\}} = (0, 0, 0, 0, 1, 1)$, $\delta^O_{\{1,2\}} = (0, 1, 0, 1, 0, 0)$, $\delta^O_{\{1,3\}} = (1, 0, 0, 0, 0, 1)$, $\delta^O_{\{2,3\}} = (0, 0, 1, 0, 1, 0)$, and 6 other multicuts (in fact, 3-cuts), obtained from $\delta^O_{\{1\},\{2\},\{3\}} = (1, 1, 0, 1, 0, 0)$ by permutation of subsets $\{1\}, \{2\}, \{3\}$.

The cone $OCUT_4$ is a full-dimensional cone in $\mathbb{R}^{12}$, generated by all 74 non-zero o-multicuts on 4 points: 14 o-cuts (in fact, 2-cuts) of the forms $\delta^O_{\{i_1\},\{i_2,i_3\}}$, $\delta^O_{\{i_1,i_2\},\{i_3,i_4\}}$, $\delta^O_{\{i_1,i_2,i_3\}}$ (and those obtained from them by permutations of given subsets), 36 3-cuts of the forms $\delta^O_{\{i_1\},\{1_2\},\{i_3,i_4\}}$ (and those obtained from them by permutations of given subsets), and 24 4-cuts, obtained from $\delta^O_{\{1\},\{2\},\{3\},\{4\}}$ by permutation of subsets $\{1\}, \{2\}, \{3\}, \{4\}$.

As every oriented cut is an oriented multicut and as every oriented multicut is a quasi-semimetric, we have, that

$$OCUT_n \subseteq OMCUT_n \subseteq OMET_n \subseteq \mathbb{R}^{I_n}_{\geq 0}.$$

In non-oriented case any multicut is a non-negative linear combination of cuts. Similar property fails for oriented multicuts. For example, $\delta^O_{\{1\},\{2\},\{3\}}$ is not a linear combination of oriented cuts on 3 points. So, $OCUT_n \subset OMCUT_n$ strictly for any $n \geq 3$.

It is easy to see that

$$MET_n = \{q + q^T \mid a \in QMET_n\},$$

where q^T denotes the transposed matrix. Moreover, any extreme ray r of MET_n has a form $r = g + g^T$, where g is an extreme ray of $QMET_n$ (in fact, r is a sum of extreme rays of $QMET_n$ with non-negative coefficients, and $r = \frac{1}{2}(r + r^T)$).

By the formula

$$\delta_{S_1,...,S_q} = \delta^O_{S_1,...,S_q} + \delta^O_{S_q,...,S_1}$$

we have, that

$$CUT_n = \{q + q^T \mid q \in OMCUT_n\}.$$

Clearly, all faces of $QMET_n$ and $OMCUT_n$ are preserved by any permutation of the nodes. It means, that every permutation of V_n induce a symmetry of the cones $QMET_n$ and $OMCUT_n$, so the group $Sym(n)$ of all permutations is a symmetry group of those cones.

But another symmetry, called *reversal*, exists (see [DDP03]): associate to each ray d the ray d^r, defined by $d^r_{ij} = d_{ji}$, i.e., in matrix terms, the reversal corresponds to the transposition of matrices: $((d^r_{ij})) = ((d_{ij}))^T$. This yields that the group $Z_2 \times Sym(n)$ is a symmetry group of the cones $QMET_n$ and $OMCUT_n$. It is proved in Dutour (2002) [Duto02] that this group is the full symmetry group of those cones:

$$Is(OMCUT_n) = Is(QMET_n) = Z_2 \times Sym(n).$$

It is so also ([DGD12]) for $OCUT_n$:

$$Is(OCUT_n) = Z_2 \times Sym(n).$$

All orbits of facets, considered below, are under action of this group.

In the oriented case, $OCUT_3 \subset QMET_3$, while the oriented triangle inequalities and non-negativity inequalities are sufficient for describing the oriented multicut cone for $n = 3$: $OMCUT_n = QMET_n$ for $n = 3$, but $OMCUT_n \subset MET_n$ strictly for $n \geq 4$. The complete description of all the facets of the oriented polyhedra $QMET_n$ and $OMCUT_n$ is known for $n \leq 5$ only (see, for example, [DDP03]). The *combinatorial explosion* starts from $n = 5$ (for instance, $QMET_5$ has 43590 extreme rays), while for corresponding semimetric cones it starts from $n = 8$. Comparing with the cones of semimetrics, the amount of computation and memory is much bigger in the oriented case, because the dimension of the cones $OMCUT_n$ and $QMET_n$ is twice those of CUT_n and MET_n, and because oriented multicuts with $q > 2$ are not interior points of the cone $OCUT_n$.

Denote by $QMET_n^\square$ the *quasi-semimetric polytope*, i.e., the set of all quasi-semimetrics on n points, generated by the $6\binom{n}{3} = n(n-1)(n-2)$ *oriented triangle inequalities* $OT_{ij,k}$, by the $\binom{n}{2} = \frac{n(n-1)}{2}$ *perimeter inequalities (non-homogeneous triangle inequalities)* OG_{ij}, and $2\binom{n}{2} = n(n-1)$ *non-negativity inequalities* NN_{ij} on V_n:

- $OT_{ij,k} : q_{jk} + q_{kj} - q_{ij} \geq 0$;
- $OG_{ij} : q_{ij} + q_{ji} \leq 2$;
- $NN_{ij} : q_{ij} \geq 0$.

For example, the polytope $QMET_3^\square$ is a full-dimensional cone in $\mathbb{R}^6$, generated by 6 oriented triangle inequalities $OT_{12,3} : d_{13} + d_{32} - d_{12} \geq 0$, $OT_{13,2} : d_{12} + d_{23} - d_{13} \geq 0$, $T_{23,1} : d_{21} + d_{13} - d_{23} \geq 0$, $OT_{21,3} : d_{23} + d_{31} - d_{21} \geq 0$, $OT_{31,2} : d_{32} + d_{21} - d_{31} \geq 0$, $T_{32,1} : d_{31} + d_{12} - d_{32} \geq 0$, 6 non-negativity inequalities $N_{12} : d_{12} \geq 0$, $N_{21} : d_{21} \geq 0$, $N_{13} : d_{13} \geq 0$, $N_{31} : d_{31} \geq 0$, $N_{23} : d_{23} \geq 0$,$N_{32} : d_{32} \geq 0$, and 3 non-homogeneous oriented triangle inequalities (perimeter inequalities) $OG_{12} : d_{12} + d_{21} \leq 2$, $OG_{13} : d_{13} + d_{31} \leq 2$, $OG_{23} : d_{23} + d_{32} \leq 2$.

In the same way, denote by $OCUT_n^\square$ the *oriented cut polytope* in $\mathbb{R}^{I_n}$: the convex hull of all *oriented cut semimetrics* on V_n; denote by $OMCUT_n^\square$ the *oriented multicut polytope* in $\mathbb{R}^{I_n}$: the convex hull of all *oriented multicut semimetrics* on V_n.

For example, $OCUT_3^\square$ in $\mathbb{R}^6$ is obtained as convex hull of all 7 o-cuts on 3 points: $\delta_\emptyset^O = (0,0,0,0,0,0)$, $\delta_{\{1\}}^O = (1,1,0,0,0,0)$, $\delta_{\{2\}}^O = (0,0,1,1,0,0)$, $\delta_{\{3\}}^O = (0,0,0,0,1,1)$, $\delta_{\{1,2\}}^O = (0,1,0,1,0,0)$, $\delta_{\{1,3\}}^O = (1,0,0,0,0,1)$, $\delta_{\{2,3\}}^O = (0,0,1,0,1,0)$, while $OMCUT_3^\square$ in $\mathbb{R}^6$ is obtained as convex hull of all 13 o-multicuts on 3 points (add to 7 o-cuts above 6 other multicuts, obtained from $\delta_{\{1\},\{2\},\{3\}}^O = (1,1,0,1,0,0)$ by permutations of subsets $\{1\},\{2\},\{3\}$).

The polytope $OCUT_4^\square$ in $\mathbb{R}^{12}$ is obtained as convex hull of all 15 o-cuts on 4 points, while $OMCUT_4^\square$ in $\mathbb{R}^{12}$ is obtained as convex hull of all 75 non-zero o-multicuts on 4 points (add to the above lists of generators of corresponding cones the zero o-cut $\delta_\emptyset^O = (0,0,0,0,0,0,0,0,0,0,0,0)$.)

Among the facets of $OMCUT_n^{\square}$ the most simple ones are the *triangle facets*, induced by oriented triangle inequalities, and by perimeter (non-homogenous oriented triangle) inequalities. Hence,

$$OCUT_n^{\square} \subset OMCUT_n^{\square} \subseteq QMET_n^{\square} \subseteq [0,1]^{I_n}.$$

Obviously, $OMCUT_n^{\square} \subset QMET_n^{\square}$ strictly for $n \geq 4$.

In the oriented case all orbits of facets of quasi-semimetric polyhedra on V_n are preserved under any permutation of the set $V_n = \{1, 2, ..., n\}$, but a switching is not a symmetry of $OMCUT_n^{\square}$, because the set of o-multicuts is not closed under the symmetric difference of sets. But the orbits of faces of $OMCUT_n^{\square}$ are preserved under the *reversal* operation.

For an o-multicut $\delta^O_{S_1,...,S_q}$ on V_n, the *reversal* of $\delta^O_{S_1,...,S_q}$ is the o-multicut $\delta^O_{S_q,...,S_1}$ (in the symmetric case the reversal of a multicut is the same multicut). We conjecture, that the symmetry group of $OMCUT_n^{\square}$ consists only of permutations and reversals, i.e., $Is(OMCUT_n^{\square})$ is the group $Z_2 \times Sym(n)$ of signed permutations, and that the full symmetry group $Is(QMET_n^{\square})$ of $QMET_n^{\square}$ is $Sym(n)$.

For oriented case, we give the complete description of oriented semimetric polytopes only for $n \leq 5$, and consider some information for $n = 6$. The oriented triangle inequalities are sufficient for describing the oriented multicut polyhedra for $n = 3$ only, and $OMCUT_n^{\square} \subset QMET_n^{\square}$ strictly for $n \geq 4$. Here the "combinatorial explosion" starts from $n = 5$.

In Table 10.1 there is a summary of the most important information concerning the cones of quasi-semimetrics,oriented cuts and oriented multicuts.

The column 2 indicates the dimension of the cone, the columns 3 and 4 give the number of extreme rays and facets, respectively; in parenthesis are given the numbers of their orbits. The column 5 gives the diameter of the 1-skeleton graph and of the ridge graph of the cone.

Cone	Dimension	# of ext. rays (orbits)	# of facets (orbits)	Diameters
$OCUT_3$	5	6(2)	9(2)	1; 2
$OMCUT_3$=$QMET_3$	6	12(2)	12(2)	2; 2
$OCUT_4$	9	14(3)	30(3)	1; 2
$OMCUT_4$	12	74(5)	72(4)	2; 2
$QMET_4$	12	164(10)	36(2)	3; 2
$OCUT_5$	14	30(4)	130(6)	1; 3
$OMCUT_5$	20	540(9)	35320(194)	2; 3
$QMET_5$	20	43590(229)	80(2)	3; 2
$OCUT_6$	20	62(5)	16,460(61)	1; 3
$OMCUT_6$	30	4682(19)	$\geq$ 217847040($\geq$ 163822)	2; ?
$QMET_6$	30	$\geq$ 182403032($\geq$ 127779)	150(2)	?; 2

Table 10.1: Small cones of quasi-semimetrics, oriented cuts and oriented multicuts

In Appendix 1 we represent also all 229 orbits of extreme rays for $QMET_5$ with adjacency, incidence of their representatives and their size. In Appendix 2 we give the same information for facets of $OMCUT_5$. For representation matrices detailing orbit-wise adjacencies and additional information on those cones see http://www.liga.ens.fr/~dutour/.

10.3 Small cones and polytopes of quasi-semimetrics and oriented multicuts

Oriented case. The case of 3 points

Recall, that in the symmetric case $CUT_3 = MET_3$ ($CUT_3^{\square} = MET_3^{\square}$) and, hence, the only facet-defining inequalities for CUT_3 and $CUT_3^{\square}$ are the triangle inequalities, 3 inequalities (from one orbit, obtained by permutations) for CUT_3 and 4 inequalities (from one orbit, obtained by permutations and switchings) for $CUT_3^{\square}$.

In the oriented case, $OCUT_3 \subset QMET_3$, and $OCUT_3^{\square} \subset QMET_3^{\square}$, while $OMCUT_3 = QMET_3$, and $OMCUT_3^{\square} = QMET_3^{\square}$.

We present here the complete linear description for case $n = 3$.

Clearly, $OCUT_3 \subset QMET_3$ strictly, but for $n = 3$ the oriented triangle inequalities with the non-negativity inequalities describe $OMCUT_3$: $OMCUT_3 = QMET_3$.

There are 12 non-zero o-multicuts, including 6 o-cuts (see Table 10.2 below) on V_3, which form 2 orbits: the orbit O_1 of o-cuts and the orbit O_2 of other o-multicuts.

O-multicut	$(v_{12}, v_{13}, v_{21}, v_{23}, v_{31}, v_{32})$	Orbit Number
$\delta^O_{\{1\}}$	$(1,1,0,0,0,0)$	O_1
$\delta^O_{\{2\}}$	$(0,0,1,1,0,0)$	O_1
$\delta^O_{\{3\}}$	$(0,0,0,0,1,1)$	O_1
$\delta^O_{\{1,2\}}$	$(0,1,0,1,0,0)$	O_1
$\delta^O_{\{1,3\}}$	$(1,0,0,0,0,1)$	O_1
$\delta^O_{\{2,3\}}$	$(0,1,0,0,1,0)$	O_1
$\delta^O_{\{1\},\{2\},\{3\}}$	$(1,1,0,1,0,0)$	O_2
$\delta^O_{\{1\},\{3\},\{2\}}$	$(1,1,0,0,0,1)$	O_2
$\delta^O_{\{2\},\{1\},\{3\}}$	$(0,1,1,1,0,0)$	O_2
$\delta^O_{\{2\},\{3\},\{1\}}$	$(0,0,1,1,1,0)$	O_2
$\delta^O_{\{3\},\{1\},\{2\}}$	$(1,0,0,0,1,1)$	O_2
$\delta^O_{\{3\},\{2\},\{1\}}$	$(0,0,1,0,1,1)$	O_2

Table 10.2: Non-zero o-multicuts on 3 points

All o-cuts above can be obtained from $\delta^O_{\{1\}}$ by a permutation ($\delta^O_{\{2\}}$ and $\delta^O_{\{3\}}$) or by a reversal and a permutation ($\delta^O_{\{1,2\}}$, $\delta^O_{\{1,3\}}$, and $\delta^O_{\{2,3\}}$); all o-multicuts above can be obtained from $\delta^O_{\{1\},\{2\},\{3\}}$ by some permutation.

The only facet-defining inequalities of $OMCUT_3$ are the 6 *oriented triangle inequalities* $OT_{ij,k}$, and 6 *non-negativity inequalities* NN_{ij}, which form two orbits $F_1 = OT$ and $F_2 = NN$, respectively:

- $OT_{ij,k} : q_{ik} + q_{kj} - q_{ij} \geq 0$;
- $NN_{ij} : q_{ij} \geq 0$.

The list of representatives of the orbits of facets in $OMCUT_3$ is given in Table 10.3.

O/F	Representative	v_{12}	v_{13}	v_{21}	v_{23}	v_{31}	v_{32}
O_1	$\delta^O_{\{1\},\{2,3\}}$:	1	1	0	0	0	0
O_2	$\delta^O_{\{1\},\{2\},\{3\}}$:	1	1	0	1	0	0
F_1	$OT_{12,3}$	-1	1	0	0	0	1
F_2	NN_{12}	1	0	0	0	0	0

Table 10.3: The representatives of orbits of extreme rays and facets in $OMCUT_3$

Adjacencies of facets (extreme rays) of $OMCUT_3$ are shown in Tables 10.4 and 10.5. For each orbit a representative and a number of adjacent ones from other orbits are given, as well as the total number of adjacent ones and the cardinality of orbits.

Orbit	Representative	F_1	F_2	Total adjacency	$\|F_i\|$
F_1	$OT_{12,3}$	3	5	8	6
F_2	NN_{12}	5	2	7	6

Table 10.4: The adjacencies of facets in $OMCUT_3$

Orbit	Representative	O_1	O_2	Total adjacency	$\|O_i\|$
O_1	$\delta^O_{\{1\}}$	5	4	9	6
O_2	$\delta^O_{\{1\},\{2\},\{3\}}$	4	2	6	6

Table 10.5: The adjacencies of extreme rays in $OMCUT_3$

The incidences of facets and extreme rays for $OMCUT_3$ are shown in Table 10.6. Namely, for each orbit F_i we give the number of extreme rays from orbits O_j, which belong to a representative of F_i, and the total number of extreme rays, which belong to it. Remind that, in general, the table of incidences of extreme rays and facets can be obtained from the table of incidences of facets and extreme rays by the formulas $F_{ij} \times |O_j| = O_{ji} \times |F_i|$, where $|O_j|$ and $|F_i|$ are the orbit sizes, F_{ij} is the number of elements from orbit O_j, which are incident to a representative of F_i, and O_{ji} is the number of elements from orbit F_i, to which is incident a representative from O_j.

Orbit	O_1	O_2	Total incidence
F_1	4	3	7
F_2	4	3	7

Table 10.6: The incidences of facets and extreme rays in $OMCUT_3$

The 1-skeleton graph G_{OMCUT_3} has 12 nodes and $45 = \frac{1}{2}(9 \times 6 + 6 \times 6)$ edges. Figure 10.1 shows the complement $\overline{G}_{OMCUT_3}$ of this graph (here $a_1 = \delta^O_{\{2,3\}}$, $a_2 = \delta^O_{\{1,3\}}$, $a_3 = \delta^O_{\{1,2\}}$, $a_1^* = \delta^O_{\{1\}}$, $a_2^* = \delta^O_{\{2\}}$, $a_3^* = \delta^O_{\{3\}}$, $b_1 = \delta^O_{\{2\},\{3\},\{1\}}$, $b_2 = \delta^O_{\{1\},\{3\},\{2\}}$, $b_3 = \delta^O_{\{1\},\{2\},\{3\}}$, $b_1^* = \delta^O_{\{3\},\{2\},\{1\}}$, $b_2^* = \delta^O_{\{3\},\{1\},\{2\}}$, $b_3^* = \delta^O_{\{2\},\{1\},\{3\}}$).

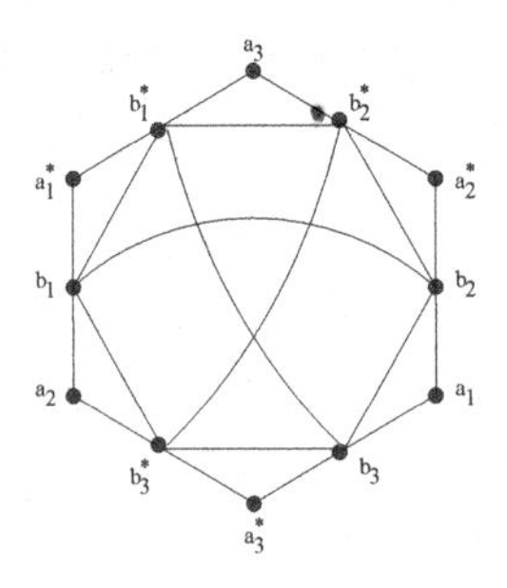

Figure 10.1: Graph $\overline{G}_{OMCUT_3}$

As any two nodes of G_{OMCUT_3} have at least 3 common neighbors, we obtain

Proposition 10.1 *The diameter of the 1-skeleton graph G_{OMCUT_3} is 2.*

The graph $G^*_{OMCUT_3}$ has also 12 nodes and 45 edges. Figure 10.2 shows its complement.

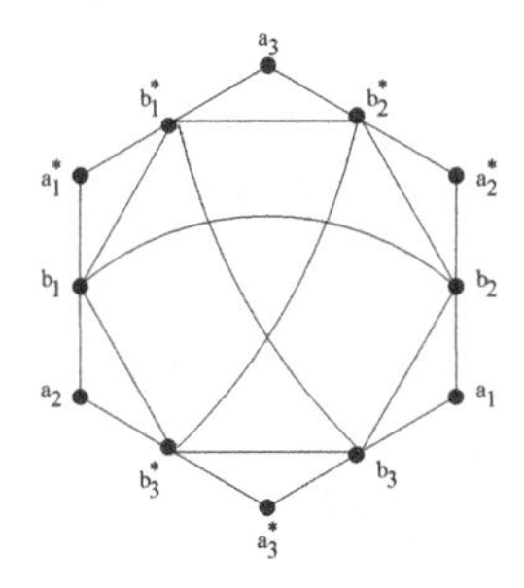

Figure 10.2: Graph $\overline{G}^*_{OMCUT_3}$

As any two nodes of $G^*_{OMCUT_3}$ have at least 3 common neighbors, we obtain

Proposition 10.2 *The diameter of the ridge graph $G^*_{OMCUT_3}$ is* 2.

It is easy to see, that in $G^*_{OMCUT_3}$ a triangle facet is adjacent to some other facet if and only if they are non-conflicting. (Remaind,that two vectors from $\{0,1,-1\}^n$ are said to be *conflicting* if there exists a pair ij such that the two vectors have non-zero coordinates of distinct signs at the position ij.) More exactly, we obtain the following result.

Proposition 10.3 *For the ridge graph $G^*_{QMET_3}$ it holds:*
(i) The triangle facet $OT_{ij,k}$ is adjacent to a facet if and only if they are non-conflicting;
(ii) The non-negativity facet NN_{ij} is adjacent also to the facets NN_{im}, NN_{kj} ($m \neq j$, $k \neq i$).

Table 10.7 gives the full representation matrices of G_{OMCUT_3} and $G^*_{OMCUT_3}$.

O	Representative	O_1	O_2	Adjacency	Incidence	$\|O_i\|$
O_1	$\delta^O_{\{1\},\{2,3\}}$	5	4	9	8	6
O_2	$\delta^O_{\{1\},\{2\},\{3\}}$	4	2	6	6	6

O	Representative	F_1	F_2	Adjacency	Incidence	$\|F_i\|$
F_1	$OT_{12,3}$	3	5	8	7	6
F_2	NN_{12}	5	2	7	7	6

Table 10.7: Representation matrices of G_{OMCUT_3} and $G^*_{OMCUT_3}$

Consider now polytope $OMCUT_3^\square$: the convex hull of all 13 o-multicuts on V_3.

This polytope has 13 vertices, which form 3 orbits (the orbit O_1 of o-cuts, the orbit O_2 of other o-multicuts, and the new orbit O_1^p, consisting of only $\delta^O_\emptyset$). $OMCUT_3^\square$ has 15 facets: 6 facets of type $OT_{ij,k}$ (orbit $F_1 = OT$), 6 facets of type NN_{ij} (orbit $F_2 = NN$) and 3 new facets (orbit $F_1^p = OG$), which are induced by the perimeter inequalities OG_{ij}:

- $OG_{ij} : d_{ij} + d_{ji} \leq 2$.

As the cone $OMCUT_3$ coincides with the cone $QMET_3$, we define $QMET_3^\square$ by all inequalities of types $OT_{ij,k}$ (oriented triangle inequalities), NN_{ij} (non-negativity inequalities), and OG_{ij} (perimeter, or oriented non-homogeneous triangle inequalities). Hence, by definition, $OMCUT_3^\square = QMET_3^\square$.

Connections between facets and vertices of the o-multicut polytope $OMCUT_3^\square$ are shown in Tables 10.8 – 10.10, which are constructed in the same way, as Tables 10.4 – 10.6.

Orbit	Representative	F_1	F_2	F_1^p	Total adjacency	$\|F_i\|$
F_1	$OT_{12,3}$	3	5	1	9	6
F_2	NN_{12}	5	2	3	10	6
F_1^p	OG_{12}	2	6	2	10	3

Table 10.8: The adjacencies of facets in $OMCUT_3^\square$

Orbit	Representative	O_1	O_2	O_1^p	Total adjacency	$\|O_i\|$
O_1	$\delta^O_{\{1\}}$	5	4	1	10	6
O_2	$\delta^O_{\{1\},\{2\},\{3\}}$	4	2	1	7	6
O_1^p	$\delta^O_\emptyset$	6	6	0	12	1

Table 10.9: The adjacencies of vertices in $OMCUT_3^\square$

Orbit	O_1	O_2	O_1^p	Total incidence
F_1	4	3	1	8
F_2	4	3	1	8
F_1^p	4	6	0	10

Table 10.10: The incidences of facets and vertices in $OMCUT_3^\square$

$G_{OMCUT_3^\square}$ has 13 nodes and 57 edges. Figure 10.3 shows the complement of it. Here the points $a_1, \ldots, b_3^*$ are the same as in Figure 10.1, and $a_0 = \delta^O_\emptyset$.

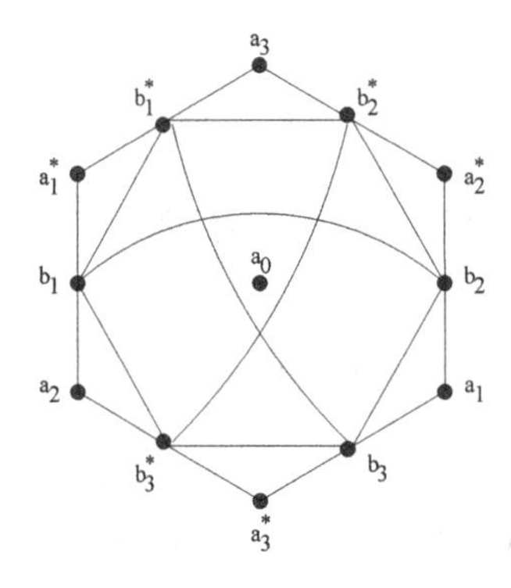

Figure 10.3: Graph $\overline{G}_{OMCUT_3^\square}$

Since any two nodes of $G_{OMCUT_3^\square}$ have $\delta^O_\emptyset$ as a common neighbor, we obtain

Proposition 10.4 *The diameter of the* 1*-skeleton graph* $G_{OMCUT_3^\square}$ *is* 2.

$G^*_{OMCUT_3^\square}$ has 15 nodes and 72 edges; Figure 10.4 shows the complement of it.

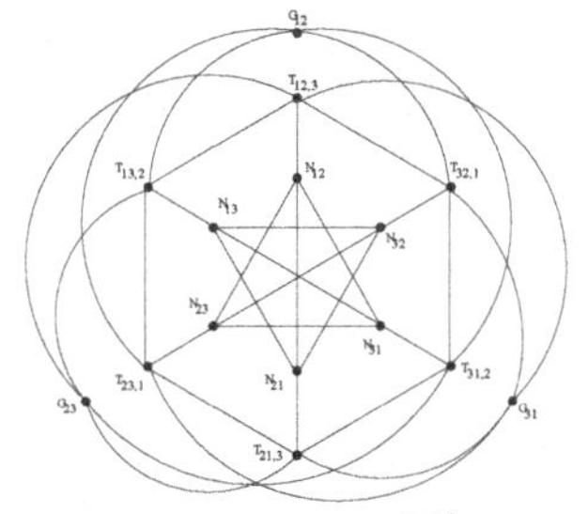

Figure 10.4: Graph $\overline{G}^*_{OMCUT_3^\square}$

As any two nodes of $G^*_{OMCUT_3^\square}$ have at least 3 common neighbors, we obtain

Proposition 10.5 *The diameter of ridge graph* $G^*_{OMCUT_3^\square}$ *is* 2.

It is easy to check the following

Proposition 10.6 *For the ridge graph* $G^*_{QMET_3^\square}$ *it holds:*
(i) the triangle facet $OT_{ij,k}$ *is adjacent to a facet if and only if they are non-conflicting;*
(ii) the facet NN_{ij} *is adjacent also to* NN_{im}, NN_{kj} $(m \neq j,\ k \neq i)$, *and to all* OG_{mk}*;*
(iii) the facet OG_{ij} *is adjacent also to all non-triangle facets.*

It turns out, that the 1-skeleton graph G_{OMCUT_3}, and the ridge graph $G^*_{OMCUT_3}$ are induced subgraphs of $G_{OMCUT_3^\square}$ and $G^*_{OMCUT_3^\square}$, respectively.

Oriented case. The case of 4 points

Recall, that in the symmetric case $CUT_4 = MET_4$, and $CUT_4^\square = MET_4^\square$. The only facet-defining inequalities for CUT_4 and $CUT_4^\square$ are the triangle inequalities, 12 inequalities (from one orbit, obtained by permutations) for CUT_4, and 16 inequalities (from one orbit, obtained by permutations and switchings) for $CUT_4^\square$.

We present here the complete linear description of $OMCUT_4$, $QMET_4$, $OMCUT_4^\square$, and $QMET_4^\square$.

The *oriented multicut cone* $OMCUT_4$ has 74 extreme rays (all non-zero o-multicuts on V_4), which form 5 orbits with the representatives $\delta^O_{\{1\}}$ (orbit O_1), $\delta^O_{\{1,2\}}$ (orbit O_2), $\delta^O_{\{1\},\{2\},\{3,4\}}$ (orbit O_3), $\delta^O_{\{1\},\{2,3\},\{4\}}$ (orbit O_4), and $\delta^O_{\{1\},\{2\},\{3\},\{4\}}$ (orbit O_5). The list of representatives is given in Table 10.11.

O/F	Representative	v_{12}	v_{13}	v_{14}	v_{21}	v_{23}	v_{24}	v_{31}	v_{32}	v_{34}	v_{41}	v_{42}	v_{43}
O_1	$\delta^O(\{4\},\{1,2,3\})$	0	0	0	0	0	0	0	0	0	1	1	1
O_2	$\delta^O(\{3,4\},\{1,2\})$	0	0	0	0	0	0	1	1	0	1	1	0
O_3	$\delta^O(\{4\},\{3\},\{1,2\})$	0	0	0	0	0	0	1	1	0	1	1	1
O_4	$\delta^O(\{4\},\{2,3\},\{1\})$	0	0	0	1	0	0	1	0	0	1	1	1
O_5	$\delta^O(\{4\},\{3\},\{2\},\{1\})$	0	0	0	1	0	0	1	1	0	1	1	1
$F_1 = OT$	$OT_{12,3}$	-1	1	0	0	0	0	0	1	0	0	0	0
$F_2 = NN$	NN_{12}	1	0	0	0	0	0	0	0	0	0	0	0
$F_3 = A^4$	$A^4_{1,2;3,4}$	-1	0	1	-1	0	1	1	1	-1	0	0	1
$F_4 = B^4$	$B^4_{1,2;3,4}$	-1	1	1	-1	1	1	1	1	-1	0	0	0

Table 10.11: The representatives of orbits of extreme rays and facets in $OMCUT_4$

$OMCUT_4$ has 72 facets from 4 orbits, which are induced by 24 *oriented triangle inequalities* (orbit $F_1 = OT$) $OT_{ij,k}$, 12 *non-negativity inequalities* (orbit $F_2 = ON$) NN_{ij}, 12 inequalities (orbit $F_3 = A^4$) $A^4_{i,j;k,m}$, and 24 inequalities (orbit $F_4 = B^4$) $B^4_{i,j;k,m}$ and $(B^4)^r_{i,j;k,m}$:

- $OT_{ij,k} : q_{ik} + q_{kj} - q_{ij} \geq 0$;
- $NN_{ij} : q_{ij} \geq 0$;
- $A^4_{i,j;k,m} : q_{ij} + q_{ji} + q_{km} \leq q_{im} + q_{jm} + q_{ki} + q_{kj} + q_{mk}$;
- $B^4_{i,j;k,m} : q_{ij} + q_{ji} + q_{km} \leq q_{ik} + q_{im} + q_{jk} + q_{jm} + q_{ki} + q_{kj}$;
- $(B^4)^r_{i,j;k,m} : q_{ij} + q_{ji} + q_{mk} \leq q_{ik} + q_{jk} + q_{ki} + q_{kj} + q_{mi} + q_{mj}$.

Note, that for the orbits F_1, F_2, F_3, but not for F_4, the reversal operation coincides with some permutation; the reversal of $B^4_{i,j;k,m}$ is $(B^4)^r_{i,j;k,m}$. See Table 10.11 for representatives of the orbits.

Tables 10.12 – 10.14 show connections between the facets and the extreme rays of $OMCUT_4$.

Orbit	Representative	F_1	F_2	F_3	F_4	Total adjacency	$\|F_i\|$
F_1	$OT_{12,3}$	17	11	5	8	41	24
F_2	NN_{12}	22	6	12	8	48	12
F_3	$A^4_{1,2;3,4}$	10	12	0	2	24	12
F_4	$B^4_{1,2;3,4}$	8	4	1	3	16	24

Table 10.12: The adjacencies of facets in $OMCUT_4$

Orbit	Representative	O_1	O_2	O_3	O_4	O_5	Total adjacency	$\|O_i\|$
O_1	$\delta^O_{\{1\}}$	7	6	21	9	18	61	8
O_2	$\delta^O_{\{1,2\}}$	8	5	20	12	8	53	6
O_3	$\delta^O_{\{1\},\{2\},\{3,4\}}$	7	5	15	7	10	44	24
O_4	$\delta^O_{\{1\},\{2,3\},\{4\}}$	6	6	14	6	8	40	12
O_5	$\delta^O_{\{1\},\{2\},\{3\},\{4\}}$	6	2	10	4	12	34	24

Table 10.13: The adjacencies of extreme rays in $OMCUT_4$

Orbit	O_1	O_2	O_3	O_4	O_5	Total incidence
F_1	6	4	14	7	12	43
F_2	6	4	14	7	12	43
F_3	4	4	8	4	8	28
F_4	3	4	4	4	2	17

Table 10.14: The incidences of facets and extreme rays in $OMCUT_4$

The 1-skeleton graph G_{OMCUT_4} of $OMCUT_4$ has 74 nodes and 1479 edges. As 14 o-cuts (orbits O_1 and O_2 together) form a dominating clique, we obtain

Proposition 10.7 *The diameter of the* 1*-skeleton graph* G_{OMCUT_4} *is* 2 *or* 3.

The ridge graph $G^*_{OMCUT_4}$ has 72 nodes and 1404 edges. The full representation matrices of G_{OMCUT_4} and G_{QMET_4} are given in Tables 10.15 and 10.16.

O	Representative	O_1	O_2	O_3	O_4	O_5	Adjacency	Incidence	$\|O_i\|$
O_1	$\delta^O_{\{4\},\{1,2,3\}}$	7	6	21	9	18	61	42	8
O_2	$\delta^O_{\{4,3\},\{1,2\}}$	8	5	20	12	8	53	48	6
O_3	$\delta^O_{\{4\},\{3\},\{2,1\}}$	7	5	15	7	10	44	29	24
O_4	$\delta^O_{\{4\},\{3,2\},\{4\}}$	6	6	14	6	8	40	33	12
O_5	$\delta^O_{\{4\},\{3\},\{2\},\{1\}}$	6	2	10	4	12	34	24	24

Table 10.15: Representation matrix of G_{OMCUT_4}

F	Representative	F_1	F_2	F_3	F_4	Adjacency	Incidence	$\|F_i\|$
F_1	$OT_{12,3}$	17	11	5	8	41	43	24
F_2	NN_{12}	22	6	12	8	48	43	12
F_3	$A^4_{12,34}$	10	12	0	2	24	28	12
F_4	$B^4_{12,34}$	8	4	1	3	16	17	24

Table 10.16: Representation matrix of $G^*_{OMCUT_4}$

The *quasi-semimetric cone* $QMET_4$ has 36 facets, distributed into two orbits: 24 triangle facets (orbit $F_1 = OT$) and 12 non-negativity facets (orbit $F_2 = NN$). In Table 10.17 there are representatives of these orbits.

O/F	Representative	v_{12}	v_{13}	v_{14}	v_{21}	v_{23}	v_{24}	v_{31}	v_{32}	v_{34}	v_{41}	v_{42}	v_{43}
O_1	$\delta^O(\{1\},\{2,3,4\})$	0	0	0	0	0	0	0	0	0	1	1	1
O_2	$\delta^O(\{1,2\},\{3,4\})$	0	0	0	0	0	0	1	1	0	1	1	0
O_3	$\delta^O(\{1\},\{2\},\{3,4\})$	0	0	0	0	0	0	1	1	0	1	1	1
O_4	$\delta^O(\{1\},\{2,3\},\{4\})$	0	0	0	1	0	0	1	0	0	1	1	1
O_5	$\delta^O(\{1\},\{2\},\{3\},\{4\})$	0	0	0	1	0	0	1	1	0	1	1	1
O_6	$\delta^O(\{1\},\{2\},\{3\},\{4\}) + e_{1,4}$	0	0	0	1	0	0	1	1	0	2	1	1
O_7	$\delta^O(\{1\},\{2\},\{3\},\{4\}) + e_{4,3}$	0	0	0	1	0	0	1	1	1	1	1	1
O_8	$\delta^O(\{1\},\{2\},\{3\},\{4\}) + e_{3,2}$	0	0	0	1	0	1	1	1	1	1	1	0
O_9	$\delta^O(\{1\},\{2\},\{3\},\{4\}) + e_{2,1} + e_{4,3}$	0	0	1	1	1	1	1	1	1	1	0	0
O_{10}	$\delta^O(\{1\},\{2\},\{3\},\{4\}) + e_{1,4} + e_{2,1} + e_{4,3}$	0	0	1	1	1	1	1	1	2	1	0	0
F_1	$OT_{12,3}$	-1	1	0	0	0	0	0	1	0	0	0	0
F_2	NN_{12}	1	0	0	0	0	0	0	0	0	0	0	0

Table 10.17: The representatives of orbits of extreme rays and facets in $QMET_4$

There are 164 extreme rays in $QMET_4$, which form 10 orbits: orbits $O_1 - O_5$ with the same representatives as in $OMCUT_4$, and 5 other orbits.

Denote by $e_{i,j}$ the $\{0,1\}$-vector in $\mathbb{R}^{12}$ with 1 only on the place ij. Then we can write representatives of the orbits $O_6 - O_{10}$ using oriented multicut $\delta^O_{\{1\},\{2\},\{3\},\{4\}}$ and vectors $e_{i,j}$. In Table 10.17 there is the list of 10 orbits of extreme rays of $QMET_4$, given in such form.

The adjacencies and incidences of the facets and extreme rays of $QMET_4$ are given in Tables 10.18 – 10.20.

Orbit	Representative	F_1	F_2	Total adjacency	$\|F_i\|$
F_1	$OT_{12,3}$	17	11	28	24
F_2	NN_{12}	22	6	28	12

Table 10.18: The adjacencies of facets in $QMET_4$

O	O_1	O_2	O_3	O_4	O_5	O_6	O_7	O_8	O_9	O_{10}	Ad.	$\|O_i\|$
O_1	7	6	21	9	18	6	9	6	3	6	91	8
O_2	8	5	20	12	8	12	16	4	4	8	97	6
O_3	7	5	7	5	10	4	4	2	0	2	46	24
O_4	6	6	10	2	8	4	4	0	2	4	46	12
O_5	6	2	10	4	3	4	2	1	0	1	33	24
O_6	2	3	4	2	4	0	2	1	0	0	18	24
O_7	3	4	4	2	2	2	0	1	1	2	21	24
O_8	4	2	4	0	2	2	2	0	0	0	16	12
O_9	4	4	0	4	0	0	4	0	0	4	20	6
O_{10}	2	2	2	2	1	0	2	0	1	0	12	24

Table 10.19: The adjacencies of extreme rays in $QMET_4$

Orbit	F_1	F_2	Total incidence
O_1	18	9	27
O_2	16	8	24
O_3	14	7	21
O_4	14	7	21
O_5	12	6	18
O_6	10	6	16
O_7	10	5	15
O_8	10	5	15
O_9	8	4	12
O_{10}	8	4	12

Table 10.20: The incidences of extreme rays and facets in $QMET_4$

The 1-skeleton graph G_{QMET_4} has 164 nodes and 2647 edges. As 14 o-cuts (orbits O_1 and O_2 together) form a dominating clique, and as there are two representatives from orbit O_{10}, which have not common neighbors, we obtain

Proposition 10.8 *The diameter of the* 1*-skeleton graph* G_{QMET_4} *is* 3.

Note, that the complement of the graph of neighbors for a representative of O_{10} is $4K_1 + P_3 + P_2 + K_{1,3}$.

The ridge graph $G^*_{QMET_4}$ has 36 nodes and 504 edges. One can check (case by case)

Proposition 10.9 *For the ridge graph* $G^*_{QMET_4}$ *it holds:*

(i) the triangle facet $OT_{ij,k}$ *is adjacent to a facet if and only if they are non-conflicting;*

(ii) the non-negativity facet NN_{ij} *is adjacent also to the facets* NN_{im}*,* NN_{kj}*,* NN_{km} $(m \neq j,\ k \neq i)$*;*

(iii) the diameter of the graph $G^*_{QMET_4}$ *is* 2*.*

In Tables 10.21 – 10.22 are given full representation matrices for G_{QMET_4} and $G^*_{QMET_4}$.

O	O_1	O_2	O_3	O_4	O_5	O_6	O_7	O_8	O_9	O_{10}	Adjacency	Incidence	$\|O_i\|$
O_1	7	6	21	9	18	6	9	6	3	6	91	27	8
O_2	8	5	20	12	8	12	16	4	4	8	97	24	6
O_3	7	5	7	5	10	4	4	2	0	2	46	21	24
O_4	6	6	10	2	8	4	4	0	2	4	46	21	12
O_5	6	2	10	4	3	4	2	1	0	1	33	18	24
O_6	2	3	4	2	4	0	2	1	0	0	18	16	24
O_7	3	4	4	2	2	2	0	1	1	2	21	15	24
O_8	4	2	4	0	2	2	2	0	0	0	16	15	12
O_9	4	4	0	4	0	0	4	0	0	4	20	12	6
O_{10}	2	2	2	2	1	0	2	0	1	0	12	12	24

Table 10.21: Representation matrix of G_{QMET_4}

F	Representative	F_1	F_2	Adjacency	Incidence	$\|F_i\|$
F_1	$OT_{12,3}$	17	11	28	78	24
F_2	NN_{12}	22	6	28	80	12

Table 10.22: Representation matrix of $G^*_{QMET_4}$

Note, that in $QMET_4$ the adjacencies of facets $OT_{ij,k}$ and NN_{ij} are the same as in $OMCUT_4$ (see Tables 10.12 and 10.18); hence, $G^*_{QMET_4}$ is an induced subgraph of $G^*_{OMCUT_4}$. But the adjacencies of o-multicuts from orbits O_3, O_4 and O_5 are decreased in the cone $QMET_4$ (see Tables 10.13 and 10.19); hence, G_{OMCUT_4} is not an induced subgraph of G_{QMET_4}.

Consider now $OMCUT_4^{\square}$: the convex hull of all 75 o-multicuts on V_4 (with $\delta^O_\emptyset$). It has 75 vertices, which belong to 6 orbits (orbits $O_1 - O_5$ with the same representatives as in $OMCUT_4$, and a new orbit O_1^p, which has only one element $\delta^O_\emptyset$). $OMCUT_4^{\square}$ has 106 facets from 7 orbits: 72 facets from the orbits $F_1 - F_4$, induced by the inequalities of the types $OT_{ij,k}$, NN_{ij}, $A^4_{ij,km}$ and $B^4_{ij,km}$, respectively, 6 facets (orbit $F_1^p = OG$), induced by the inequalities OG_{ij}, 4 facets (orbit $F_2^p = M$), induced by the inequalities M_{ijkm}, and 24 facets (orbit $F_3^p = R$), induced by the inequalities R_{ijkm}:

- $OG_{ij} : q_{ij} + q_{ji} \leq 2;$
- $M_{ijkm} : q_{ik} + q_{im} + q_{ki} + q_{km} + q_{mi} + q_{mk} - q_{ij} - q_{ji} - q_{jk} - q_{jm} - q_{kj} - q_{mj} \leq 1;$
- $R_{ijkm} : q_{jk} + q_{jm} + q_{km} + q_{mk} - q_{ik} - q_{im} - q_{ji} - q_{ki} - q_{mi} \leq 1.$

Tables 10.23 – 10.25 give connections between the facets and the vertices of this polytope.

Orbit	Representative	F_1	F_2	F_3	F_4	F_1^p	F_2^p	F_3^p	Total adjacency	$\|F_i\|$
F_1	$OT_{12,3}$	17	11	5	8	4	1	4	50	24
F_2	NN_{12}	22	6	12	8	6	4	8	66	12
F_3	$A^4_{12,34}$	10	12	0	2	1	0	0	25	12
F_4	$B^4_{12,34}$	8	4	1	3	4	0	1	21	24
F_1^p	G_{12}	16	12	2	6	5	2	12	65	6
F_2^p	M_{1234}	6	12	0	0	3	0	6	27	4
F_3^p	R_{1234}	4	4	0	1	3	1	0	13	24

Table 10.23: The adjacencies of facets in $OMCUT_4^{\square}$

Orbit	Representative	O_1	O_2	O_3	O_4	O_5	O_1^p	Total adjacency	$\|O_i\|$
O_1	$\delta^O_{\{1\}}$	7	6	21	9	18	1	62	8
O_2	$\delta^O_{\{1,2\}}$	8	5	20	12	8	1	54	6
O_3	$\delta^O_{\{1\},\{2\},\{3,4\}}$	7	5	16	8	10	1	47	24
O_4	$\delta^O_{\{1\},\{2,3\},\{4\}}$	6	6	16	6	10	1	45	12
O_5	$\delta^O_{\{1\},\{2\},\{3\},\{4\}}$	6	2	10	5	12	1	36	24
O_1^p	$\delta^O_\emptyset$	8	6	24	12	24	0	74	1

Table 10.24: The adjacencies of vertices in $OMCUT_4^{\square}$

Orbit	O_1	O_2	O_3	O_4	O_5	O_1^p	Total incidence
F_1	6	4	14	7	12	1	34
F_2	6	4	14	7	12	1	34
F_3	4	4	8	4	8	1	29
F_4	3	4	4	4	2	1	18
F_1^p	4	2	4	2	0	1	13
F_2^p	6	0	12	6	0	0	24
F_3^p	1	0	8	1	3	0	13

Table 10.25: The incidences of facets and extreme rays in $OMCUT_4^\square$

The 1-skeleton graph $G_{OMCUT_4^\square}$ has 75 nodes and 1604 edges. As $\delta_\emptyset^O$ is adjacent to all other vertices, we obtain

Proposition 10.10 *The diameter of the* 1*-skeleton graph* $G_{OMCUT_4^\square}$ *is* 2.

The ridge graph of $G^*_{OMCUT_4^\square}$ has 72 nodes and 1683 edges.

It turns out, that $G^*_{OMCUT_4}$ is an induced subgraph of $G^*_{OMCUT_4^\square}$ (see Tables 10.12 and 10.23), and $G^*_{QMET_4}$ is the induced subgraph of $G^*_{OMCUT_4^\square}$ (see Tables 10.18 and 10.23), but G_{OMCUT_4} is not an induced subgraph of $G_{OMCUT_4^\square}$ (see Tables 10.13 and 10.24).

Similarly to $QMET_3^\square$, we define $QMET_4^\square$ by the inequalities of the types $OT_{ij,k}$, NN_{ij}, and G_{ij}. Hence, this polytope has 42 facets from 3 orbits: 24 triangle facets (orbit $F_1 = OT$), 12 non-negativity facets (orbit $F_2 = NN$), and 6 facets, induced by inequalities OG_{ij} (orbit $F_3 = OG$).

$QMET_4^\square$ has 221 vertices, which belong to 14 orbits: 10 orbits $O_1 - O_{10}$ with the same representatives as in $QMET_4$, orbit O_1^p (see $OMCUT_4^\square$), and 3 new orbits $O_2^p - O_4^p$ with the representatives $(1,1,1,1,2,2,0,0,1,0,0,1)$ (orbit O_2^p), $(1,1,2,1,1,2,0,0,1,0,0,0)$ (orbit O_3^p), and $(1,1,1,0,1,1,0,1,1,0,1,1)$ (orbit O_4^p). Connections of the facets and the vertices of $QMET_4^\square$ are given in Tables 10.26 – 10.28.

Orbit	Representative	F_1	F_2	F_1^p	Total adjacency	$\lvert F_i \rvert$
F_1	$OT_{12,3}$	17	11	4	32	24
F_2	NN_{12}	22	6	6	34	12
F_1^p	OG_{12}	16	12	5	33	6

Table 10.26: The adjacencies of facets in $QMET_4^\square$

O	O_1	O_2	O_3	O_4	O_5	O_6	O_7	O_8	O_9	O_{10}	O_1^p	O_2^p	O_3^p	O_4^p	Adj.	$\lvert O_i \rvert$
O_1	7	6	21	9	18	6	6	6	3	6	1	9	9	3	110	8
O_2	8	5	20	12	8	12	16	4	4	8	1	0	0	0	98	6
O_3	7	5	16	8	10	4	4	2	0	2	1	4	6	2	71	24
O_4	6	6	16	6	10	4	4	0	2	6	1	4	4	0	69	12
O_5	6	2	10	5	12	4	2	1	0	1	1	2	0	0	46	24
O_6	2	3	4	2	4	0	2	1	0	0	1	0	0	0	19	24
O_7	2	4	4	2	2	2	0	1	1	2	1	1	0	0	22	24
O_8	4	2	4	0	2	2	2	0	0	0	1	0	0	0	17	12
O_9	4	4	0	4	0	0	4	0	0	4	1	0	0	0	21	6
O_{10}	2	2	2	3	1	0	2	0	1	0	1	2	0	0	16	24
O_1^p	8	6	24	12	24	24	24	12	6	24	0	0	0	0	164	1
O_2^p	3	0	4	2	2	0	1	0	0	2	0	0	1	0	15	24
O_3^p	3	0	6	2	0	0	0	0	0	0	0	1	2	1	15	24
O_4^p	3	0	6	0	0	0	0	0	0	0	0	0	3	0	12	8

Table 10.27: The adjacencies of vertices in $QMET_4^\square$

Orbit	F_1	F_2	F_1^p	Total incidence
O_1	18	9	3	30
O_2	16	8	4	28
O_3	14	7	5	26
O_4	14	7	5	26
O_5	12	6	6	24
O_6	10	6	1	17
O_7	10	5	1	16
O_8	10	5	1	16
O_9	8	4	2	14
O_{10}	8	4	3	15
O_1^p	24	12	0	36
O_2^p	6	4	4	14
O_3^p	6	5	3	14
O_4^p	6	3	3	12

Table 10.28: The incidences of vertices and facets in $QMET_4^{\square}$

The 1-skeleton graph $G_{QMET_4^{\square}}$ has 221 nodes and 3534 edges. As the orbits O_1 and O_2 together form a dominating clique, and as there are two representatives from the orbit O_4^p, which have no common neighbors, we obtain

Proposition 10.11 *The diameter of the 1-skeleton graph $G_{QMET_4^{\square}}$ is 3.*

The ridge graph $G^*_{QMET_4^{\square}}$ has 80 nodes and 686 edges. It is easy to check the following

Proposition 10.12 *For the ridge graph $G^*_{QMET_4^{\square}}$ it holds:*

(i) the triangle facet $OT_{ij,k}$ is adjacent to a facet if and only if they are non-conflicting;

(ii) the facet NN_{ij} is adjacent also to NN_{im}, NN_{kj}, NN_{km} ($m \neq j$, $k \neq i$), and to all G_{mk};

(iii) the facet G_{ij} is adjacent also to all non-triangle facets.

Figures 10.5 and 10.6 show some subgraphs of $G_{QMET_4^{\square}}$. The complement of the graph of neighbors for a representative of the orbit O_4^p is given in Figure 10.5. The restriction of $G_{QMET_4^{\square}}$ on the union of orbits O_1 and O_4^p consists of two disjoint cube graphs; see Figure 10.6: here the black (white) points are the elements from O_1 (O_4^p).

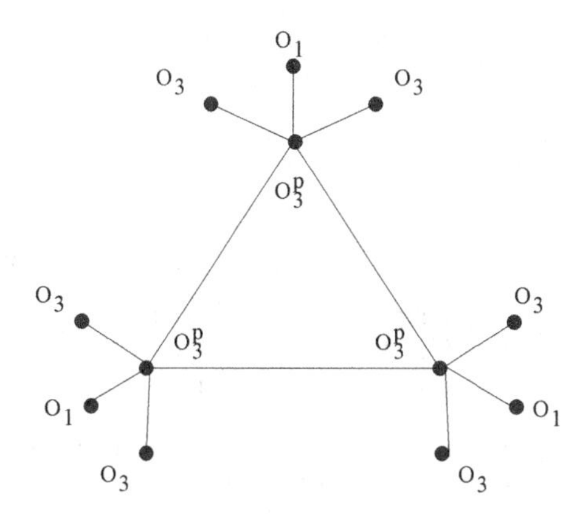

Figure 10.5: The complement of the graph of neighbors for a representative of O_4^p

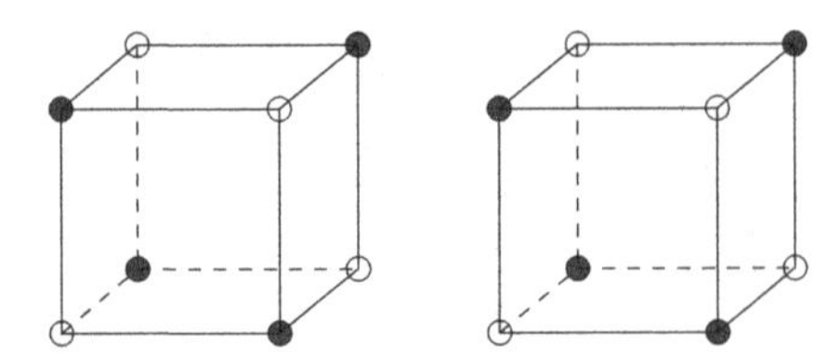

Figure 10.6: The restriction of $G_{QMET_4^\square}$ on $O_1 \cup O_4^p$

Note, that in $QMET_4^\square$ the adjacencies of facets $OT_{ij,k}$ and NN_{ij} are the same as in $QMET_4$ (see Tables 10.18 and 10.26), and the ridge graph $G^*_{QMET_4}$ is an induced subgraph of $G^*_{QMET_4^\square}$. Similarly (see Tables 10.23 and 10.26), the graph $G^*_{QMET_4^\square}$ is an induced subgraph of $G^*_{OMCUT_4^\square}$. $G_{OMCUT_4^\square}$ is the induced subgraph of $G_{QMET_4^\square}$ (see Tables 10.24 and 10.27), but G_{OMCUT_4} is not an induced subgraph of $G_{QMET_4^\square}$ (see Tables 10.13 and 10.27).

Oriented case. The case of 5 points

Recall, that for the symmetric case $CUT_5 \subset MET_5$, and $CUT_5^\square \subset MET_5^\square$ strictly. While CUT_5 has 40 facets from 2 orbits, $CUT_5^\square$ has 56 facets from 2 orbits. The extreme rays of MET_5 and the vertices of $MET_5^\square$ are also known; namely, besides the cut vectors, all of them arise by a switching of the vector $(\frac{2}{3}, \frac{2}{3}, ..., \frac{2}{3})$. So, MET_5 has 25 extreme rays from 3 orbits, $MET_5^\square$ has 32 vertices from 2 orbits.

The *quasi-semimetric cone* $QMET_5$ has 80 facets, distributed in two orbits: 60 triangle facets (orbit $F_1 = OT$) and 20 non-negativity facets (orbit $F_2 = NN$). The list of representatives of the orbits is given in Table 10.29.

F	Representative	v_{12}	v_{13}	v_{14}	v_{15}	v_{21}	v_{23}	v_{24}	v_{25}	v_{31}	v_{32}	v_{34}	v_{35}	v_{41}	v_{42}	v_{43}	v_{45}	v_{51}	v_{52}	v_{53}	v_{54}
F_1	$OT_{1,2;3}$	-1	1	0	0	0	0	0	0	0	1	0	0	0	0	0	0	0	0	0	0
F_2	NN_{12}	1	0	0	0	0	0	0	0	0	0	0	0	0	0	0	0	0	0	0	0

Table 10.29: The representatives of orbits of facets in $QMET_5$

There are 43590 extreme rays in $QMET_5$. Under the group action we have 229 orbits to consider (see Appendix 1 for the full list of representatives); amongst them, nine orbits of 540 oriented multicuts on V_5 are given in Table 10.31 with the representatives $\delta^O_{\{5\}}$ (orbit O_1), $\delta^O_{\{5,4\}}$ (orbit O_2) – they are two orbits of oriented cuts, $\delta^O_{\{5,4\},\{3\},\{2,1\}}$ (orbit O_3), $\delta^O_{\{5\},\{4,3\},\{2,1\}}$ (orbit O_4), $\delta^O_{\{5\},\{4\},\{3,2,1\}}$ (orbit O_5), $\delta^O_{\{5\},\{4,3,2\},\{1\}}$ (orbit O_6), $\delta^O_{\{5\},\{4\},\{3\},\{2,1\}}$ (orbit O_7), $\delta^O_{\{5\},\{4\},\{3,2\},\{1\}}$ (orbit O_8) and $\delta^O_{\{5\},\{4\},\{3\},\{2\},\{1\}}$ (orbit O_9).

Besides of o-multicut orbits, there are, for example, orbits with the representatives

$(0,0,0,1,1,0,0,1,1,0,0,1,1,1,1,2,0,0,0,0)$ (the adjacency 404, the incidence 41),
$(0,1,1,1,1,1,1,1,1,0,1,1,0,0,1,1,0,0,0,1)$ (the adjacency 58, the incidence 30),
$(0,0,1,1,1,1,2,1,2,1,2,1,1,1,0,0,1,1,1,1)$ (the adjacency 20, the incidence 20).

The representatives of some orbits of the cone $QMET_5$, which are not those of oriented multicuts, have, however, a form, similar to the corresponding orbits of the cone $QMET_4$.

For example, the representative of orbit O_{18} is similar to the representative $\delta^O_{\{1\},\{2\},\{3\},\{4\}} + e_{1,4}$ of the orbit O_6 in the cone $QMET_4$: it is $\delta^O_{\{1\},\{2\},\{3\},\{4\},\{5\}} + e_{1,5}$, and the representative of orbit O_{11} is similar to the representative $\delta^O_{\{1\},\{2\},\{3\},\{4\}} + e_{4,3}$ of the orbit O_7 in the cone $QMET_4$: it is $\delta^O_{\{1\},\{2\},\{3\},\{4\},\{5\}} + e_{5,4}$.

We give below more such examples. If v_i is the representative of orbit O_i, then

$v_{11} = \delta^O_{\{1\},\{2\},\{3\},\{4\},\{5\}} + e_{5,4}$,
$v_{18} = \delta^O_{\{1\},\{2\},\{3\},\{4\},\{5\}} + e_{1,5}$,
$v_{25} = \delta^O_{\{1\},\{2\},\{3\},\{4\},\{5\}} + e_{5,3} + e_{5,4}$,
$v_{26} = \delta^O_{\{1\},\{2\},\{3\},\{4\},\{5\}} + e_{1,5} + e_{5,4}$,
$v_{32} = \delta^O_{\{1\},\{2\},\{3\},\{4\},\{5\}} + e_{1,5} + e_{2,5}$,
$v_{48} = \delta^O_{\{1\},\{2\},\{3\},\{4\},\{5\}} + e_{1,4} + e_{2,4} + e_{5,3} + e_{5,4}$,
$v_{49} = \delta^O_{\{1\},\{2\},\{3\},\{4\},\{5\}} + e_{1,4} + e_{1,5} + e_{5,4}$,
$v_{52} = \delta^O_{\{1\},\{2\},\{3\},\{4\},\{5\}} + e_{5,3} + e_{5,4} + e_{4,3}$,
$v_{53} = \delta^O_{\{1\},\{2\},\{3\},\{4\},\{5\}} + e_{1,4} + e_{5,3} + e_{5,4}$,
$v_{55} = \delta^O_{\{1\},\{2\},\{3\},\{4\},\{5\}} + e_{1,4} + e_{1,5} + e_{2,5} + e_{3,5}$,
$v_{56} = \delta^O_{\{1\},\{2\},\{3\},\{4\},\{5\}} + e_{1,4} + e_{1,5} + e_{2,5} + e_{3,5} + e_{4,5}$,
$v_{72} = \delta^O_{\{1\},\{2\},\{3\},\{4\},\{5\}} + e_{1,4} + e_{5,2} + e_{5,3} + e_{5,4}$,
$v_{84} = \delta^O_{\{1\},\{2\},\{3\},\{4\},\{5\}} + e_{1,4} + e_{1,5} + e_{2,5}$,
$v_{94} = \delta^O_{\{1\},\{2\},\{3\},\{4\},\{5\}} + e_{1,2} + e_{1,3} + e_{1,4} + e_{1,5} + e_{2,3} + e_{2,4} + e_{2,5} + e_{3,4} + e_{4,3} + e_{4,5} + 2e_{5,3} + e_{5,4}$,
$v_{99} = \delta^O_{\{1\},\{2\},\{3\},\{4\},\{5\}} + e_{1,3} + e_{1,4} + 2e_{1,5} + e_{2,4} + e_{2,5} + e_{3,5}$,
$v_{129} = \delta^O_{\{1\},\{2\},\{3\},\{4\},\{5\}} + e_{1,3} + e_{1,54} + e_{2,4} + e_{5,2} + e_{5,3} + 2e_{5,4}$,
$v_{174} = \delta^O_{\{1\},\{2\},\{3\},\{4\},\{5\}} + e_{1,3} + e_{1,5} + e_{2,5} + e_{4,2} + e_{4,3} + e_{4,5} + e_{5,2} + e_{5,3}$.

But many other representatives have distinct form; for example,

$v_{10} = \delta^O_{\{1\},\{2\},\{3\},\{4\},\{5\}} + e_{5,3} + e_{5,4} - e_{3,4}$,
$v_{12} = \delta^O_{\{1\},\{2\},\{3\},\{4\},\{5\}} + e_{5,4} - e_{2,3}$.

In the 1-skeleton graph G_{QMET_5}, the oriented cuts (orbits O_1 and O_2 together) form a clique, but, distinctly from cases $n = 3, 4$, not a dominating clique. For example, a representative from orbit O_{95} (and the same holds for orbits O_{96}, O_{97}, O_{98}, O_{198}, O_{204}, O_{210}, O_{212} – O_{223}, O_{228}, O_{229}) is not adjacent to any oriented cut.

The representation matrix of the ridge graph $G^*_{QMET_5}$ of $QMET_5$ is given in Table 10.30.

F	$F_1 = OT$	$F_2 = NN$	Adjacency	Incidence	$\|F_i\|$
$F_1 = OT$	49	19	68	13590	60
$F_2 = NN$	57	12	69	14359	20

Table 10.30: The representation matrix of $G^*_{QMET_5}$

The cone $OMCUT_5$ has 540 extreme rays (all non-zero o-multicuts on V_5), which form 9 orbits. See Table 10.31 for representatives of the orbits and Table 10.32 for the representation matrix of the 1-skeleton graph G_{OMCUT_5}.

There are 35320 facets in $OMCUT_5$, which form 194 orbits. The list of the representatives of these orbits is given in Appendix 2.

Representative	v_{12}	v_{13}	v_{14}	v_{15}	v_{21}	v_{23}	v_{24}	v_{25}	v_{31}	v_{32}	v_{34}	v_{35}	v_{41}	v_{42}	v_{43}	v_{45}	v_{51}	v_{52}	v_{53}	v_{54}
$\delta^O_{\{1\},\{2,3,4,5\}}$	1	1	1	1	0	0	0	0	0	0	0	0	0	0	0	0	0	0	0	0
$\delta^O_{\{1,5\},\{2,3,4\}}$	1	1	1	0	0	0	0	0	0	0	0	0	0	0	0	0	0	1	1	1
$\delta^O_{\{1\},\{2\},\{3,4,5\}}$	1	1	1	1	0	1	1	1	0	0	0	0	0	0	0	0	0	0	0	0
$\delta^O_{\{1,2\},\{3\},\{4,5\}}$	0	1	1	1	0	1	1	1	0	0	1	1	0	0	0	0	0	0	0	0
$\delta^O_{\{1\},\{2,3,4\},\{5\}}$	1	1	1	1	0	0	0	1	0	0	0	1	0	0	0	1	0	0	0	0
$\delta^O_{\{1\},\{2,3\},\{4,5\}}$	1	1	1	1	0	0	1	1	0	0	1	1	0	0	0	0	0	0	0	0
$\delta^O_{\{1\},\{2\},\{3\},\{4,5\}}$	1	1	1	1	0	1	1	1	0	0	1	1	0	0	0	0	0	0	0	0
$\delta^O_{\{1\},\{2\},\{3,4\},\{5\}}$	1	1	1	1	0	1	1	1	0	0	0	1	0	0	0	1	0	0	0	0
$\delta^O_{\{1\},\{2\},\{3\},\{4\},\{5\}}$	1	1	1	1	0	1	1	1	0	0	1	1	0	0	0	1	0	0	0	0

Table 10.31: The representatives of orbits of extreme rays in $OMCUT_5$

O	Representative	O_1	O_2	O_3	O_4	O_5	O_6	O_7	O_8	O_9	Adj.	Inc.	$\|O_i\|$
O_1	$\delta^O_{\{1\},\{2,3,4,5\}}$	9	20	36	30	16	54	108	96	96	465	8840	10
O_2	$\delta^O_{\{1,5\},\{2,3,4\}}$	10	19	38	27	20	57	90	96	60	417	10562	20
O_3	$\delta^O_{\{1\},\{2\},\{3,4,5\}}$	9	19	34	24	16	42	84	72	66	366	3106	40
O_4	$\delta^O_{\{1,2\},\{3\},\{4,5\}}$	10	18	32	24	16	54	72	76	40	342	3172	30
O_5	$\delta^O_{\{1\},\{2,3,4\},\{5\}}$	8	20	32	24	12	42	72	72	54	336	4372	20
O_6	$\delta^O_{\{1\},\{2,3\},\{4,5\}}$	9	19	28	27	14	51	64	66	36	314	3576	60
O_7	$\delta^O_{\{1\},\{2\},\{3\},\{4,5\}}$	9	15	28	18	12	32	57	55	36	262	1598	120
O_8	$\delta^O_{\{1\},\{2\},\{3,4\},\{5\}}$	8	16	24	19	12	33	55	49	36	252	1930	120
O_9	$\delta^O_{\{1\},\{2\},\{3\},\{4\},\{5\}}$	8	10	22	10	9	18	36	36	38	187	1123	120

Table 10.32: The representation matrix of orbits of extreme rays in $OMCUT_5$

Besides of zero-extensions of all facets of $OMCUT_4$, the cone $OMCUT_5$ has, for example, three facets A^5, B^5, C^5, given below; here $S_{a...x}$ denotes the sum of distances along oriented cycle $a \dots x$ (for example, $S_{axb} = q_{ax} + q_{xb} + q_{ba}$, and $S_{ab} = q_{ab} + q_{ba}$):

- $A^5_{xyz,ab} : S_{axb} + S_{ayb} + S_{azb} \geq S_{ab} + S_{xyz}$;
- $B^5_{xyz,ab} : S_{abx} + S_{zay} + S_{zby} = S_{xayzabyzb} \geq S_{ab} + (S_{xz} + S_{yz})$;
- $C^5_{xyz,ab} : S_{axb} + S_{ayx} + S_{byz} + S_{az} \geq S_{ab} + (S_{xyz} + S_{xzy})$.

An example of a quasi-metric violating inequality A^5 is provided by the oriented graph on $\{a, b, x, y, z\}$, having ai, ib, ba, xy, zy (for $i = x, y, z$) as the set of arcs. If we replace all (oriented) cycles in A^5 by non-oriented ones, then it will became a *pentagonal inequality*; remind, that 10 pentagonal inequalities, together with 30 triangle inequalities, give all facets of the cone CUT_5.

Besides these facets, the next simplest facets of $OMCUT_5$, i.e. having no more than 8 non-zeros components, are:

- $M^5 : (q_{15} - q_{25}) + q_{54} - q_{14} + q_{34} + q_{31} + q_{23} + q_{12} \geq 0$;
- $R^5 : (-q_{51} + q_{52}) + q_{54} - q_{14} + q_{34} + q_{31} + q_{23} + q_{12} \geq 0$.

$QMET_5^{\square}$ has 90 facets (3 orbits) and 79391 vertices (more than 113 orbits).
$OMCUT_5^{\square}$ has 541 vertices (11 orbits) and more than 128 orbits of facets.

10.4 Theorems and conjectures for general case

It is known (see Proposition 8 in [LLR94]), that any quasi-metric on n points embeds isometrically into $\mathbb{R}^n$ equipped with some directed norm.

Let us define *oriented l_p-norm* for $1 \leq p < \infty$ by

$$||x - y||_{p,or} = \sqrt[p]{\sum_{k=1}^{n} [\max\{x_k - y_k, 0\}]^p},$$

and, for $p = \infty$, by

$$||x - y||_{\infty,or} = \max_{1 \leq k \leq n} \max\{x_k - y_k, 0\}.$$

Theorem 10.1 *Let q be a quasi-semimetric on n points. Then it holds:*
(i) q is embeddable in $l^n_{\infty,or}$;
(ii) q is embeddable in $l^m_{1,or}$ for some m if and only if $q \in OCUT_n$.

Proof. (i) Let $v_1, ..., v_n$ in $\mathbb{R}^n$ defined as $v_i = (q_{i1}, q_{i2}, ..., q_{in})$.

Then $||v_i - v_j||_{\infty,or} = \max\{q_{ik} - q_{jk}, 0\}$ (it is $\leq q_{ij}$ from oriented triangle inequality), and $q_{ij} - q_{jj} = q_{ij}$, so, $||v_i - v_j||_{\infty,or} = q_{ij}$.

(ii) The proof is the same as in Proposition 4.2.2 of Deza and Laurent (1997) [DeLa97] for non-oriented case.

□

The extreme rays of $QMET_n$ have been studied in [DDP03]. It was proved that they are not symmetric and have at least $n-1$ zeros, implying that no one is the directed path distance of an directed graph. The oriented multicuts define extreme rays of $QMET_n$. Also, the *vertex-splitting* of an extreme ray is still an extreme ray.

Theorem 10.2 *Every extreme ray of $QMET_n$ has at least $n - 1$ coordinates equal to zero. This lower bound is met for $n = 4, 5, 6$.*

Proof. The rank of the system $(q_{ij} = q_{ik} + q_{kj})_{1 \leq i,j,k \leq n}$ is $(n-1)^2$ (see Deza and Panteleeva (2000) [DePa99]). Let q be an extreme ray of $QMET_n$; let $NN = (NN_\alpha)_{\alpha \in A}$ be the set of all non-negativity facets, to which q is incident; let $OT = (OT_\beta)_{\beta \in B}$ be the set of all oriented triangle facets, to which it is incident. So, the rank of $NN \cup OT$ is $n(n-1) - 1$.

If $rank\,(OT) = (n-1)^2$, then the vector q is incident to all oriented triangle inequalities. So, $q_{ij} + q_{ji} = 0$, and since q belongs to $QMET_n$, the equalities $q_{ij} = q_{ji} = 0$ hold. So, we have $rank\,(OT) \leq (n-1)^2 - 1$, and

$$n(n-1) - 1 = rank\,(NN \cup OT) \leq rank\,(OT) + rank\,(NN) \leq (n-1)^2 - 1 + |A|,$$

implying $n - 1 \leq |A|$.

□

Above Theorem implies that any extreme ray of $QMET_n$ is not the directed path distance of an directed graph (see, for example, Chartrand, Johns, Tian and Winters (1993) [CJTW93], for those notions).

The *vertex-splitting* of a vector $(q_{ij})_{1\le i\neq j\le n}$ is denoted by $(q^{vs}_{ij})_{1\le i\neq j\le n+1}$ and defined by

$$q^{vs}_{n\,n+1} = q^{vs}_{n+1\,n} = 0,\; q^{vs}_{i\,n+1} = q_{i\,n},\ \text{and}\ q^{vs}_{n+1\,i} = q_{n\,i}\ .$$

The vertex-splitting of an oriented multicut $q = \delta^O_{S_1,\dots,S_q}$ is $q^{vs} = \delta^O_{S_1,\dots,S_l\cup\{n+1\},\dots,S_q}$, if $n \in S_l$.

Theorem 10.3 *The vextex-splitting of an extreme ray of $QMET_n$ is an extreme ray of $QMET_{n+1}$.*

Proof. If q lies on an extreme ray of $QMET_n$, then one can check easily the validity of oriented triangle and non-negativity inequalities for q^{vs}.
The ray q^{vs} is incident to the oriented triangle inequalities $OT_{ni,n+1}$, $OT_{in,n+1}$, $OT_{(n+1)i,n}$, and $OT_{i(n+1),n}$, as well as to the non-negativity inequalities $NN_{n,n+1}$, and $NN_{n+1,n}$.
Assume now that e is a vector of $QMET_{n+1}$, which is incident to all facets incident to q^{vs}. Then we obtain

$$e_{n+1,i} = e_{n,i},\quad e_{i,n+1} = e_{i,n},\ \text{and}\ e_{n,n+1} = e_{n+1,n} = 0\ .$$

Since the restriction of e on V_n yields a function that satisfies to all inequalities, which are satisfied by q, we obtain (since q is an extreme ray), that the restriction $e_{|V_n}$ is a multiple of q. So, we get $e_{|V_n} = \lambda q$ with $\lambda \ge 0$. Above equalities yield $e = \lambda q^{vs}$.

□

Theorem 10.4 *There is no symmetric extreme rays in $QMET_n$.*

Proof. If q lies on an extreme ray, then by Theorem 10.2 this vector has a zero, say $q_{n,n+1} = 0$. By symmetry one obtains $q_{n+1,n} = 0$, then the validity of triangle inequalities $OT_{ni,n+1}$, $OT_{in,n+1}$ gives $q_{i,n} = q_{i,n+1}$, and then again by symmetry $q_{n,i} = q_{n+1,i}$. So, q lies on the vertex-splitting ray of an $QMET_{n-1}$, which will be again a symmetric extreme ray. One can then conclude by induction.

□

Proposition 10.13 *The oriented multicut $\delta^O_{\{1\},\{2\},\dots,\{n\}}$ is an extreme ray of $QMET_n$.*

Proof. Let $q = \delta^O_{\{1\},\{2\},\dots,\{n\}}$. Then we get that $q_{ij} = 1$, if $j > i$, and $q_{ij} = 0$, if $j < i$. The vector q satisfy to all oriented triangle and non-negativity inequalities; so, it lies on a ray of $QMET_n$. Now, let us prove that it defines an extreme ray of $QMET_n$. The vector q is incident to:

- non-negativity inequalities NN_{ij} with $j < i$;
- oriented triangle inequalities $OT_{ik,j}$ with any (i, j, k), except those with $i < j < k$.

Let e be a vector of $QMET_n$, which is incident to the same inequalities of $QMET_n$. We will prove now that e is a non-negative multiple of d. Since e is incident to the non-negativity inequalities NN_{ij} with $j < i$, one has $e_{ij} = 0$, if $j < i$. It remains to prove that the numbers e_{ij} with $j > i$ are all equal. One has e incident to $OT_{ij,k}$ with $i < k < j$. So, $e_{ik} = e_{ij}$; in the same way $e_{ik} = e_{jk}$, if $j < k$, so we get the result, i.e., $e = \lambda q$, and λ is positive, since the inequality $e_{12} \ge 0$ is valid.

□

Theorem 10.5 *Oriented multicuts are extreme rays of $QMET_n$.*

Proof. By repeated applications of above Proposition, Theorem 10.3, and using group of symmetries, the Theorem follows.

□

Two vectors are said to be *conflicting* if there exist a component, on which they have non-zero values of different sign.

Theorem 10.6 *For the ridge graph $G^*_{QMET_n}$ it holds:*
(i) a triangle facet is non-adjacent to any facet, to which it conflicts;
(ii) the non-negativity facets NN_{ij} and $NN_{i'j'}$ are non-adjacent if either $i' = j$, or $j' = i$.

Proof. (i) If q is a quasi-semimetric vector, which is incident to both, $OT_{ij,k}$ and NN_{ij}, then we have $0 = q_{ik}+q_{kj}$. But since q is a quasi-semimetric, we have $q_{ik} \geq 0$, and $q_{kj} \geq 0$; so, $q_{ki} = q_{kj} = 0$. The vector q must lie in a space of dimension $n(n-1)-3$. So, the facets $OT_{ij,k}$ and NN_{ij} are non-adjacent.
If q is a quasi-semimetric vector, incident to $OT_{ij,k}$ and $OT_{ik,l}$, then we will have

$$q_{ij} = q_{ik} + q_{kj} = q_{il} + q_{lk} + q_{kj} = q_{il} + (q_{lk} + q_{kj}) .$$

Since q is a quasi-semimetric, we have $q_{ij} \leq q_{il}+q_{lj}$, and $q_{lj} \leq q_{lk}+q_{kj}$. These inequalities must be equalities in order to meet above equality. So, the vector q belongs to a space of dimension $n(n-1)-4$, and the facets are not adjacent.

(ii) is obvious, since a vector q, incident to NN_{ij} and NN_{ki}, is also incident to NN_{kj}, and gives a lower than expected rank. Similarly, if q satisfies $q_{ij} = 0$ and $q_{ji} = 0$, then we have the equality $q_{ik} = q_{jk}$ for all k, and we will get again a too low rank.

□

We think that the necessary condition (i) and (ii) are also sufficient; it will imply that the diameter of $G^*_{QMET_n}$ is two. Moreover, we expect that the ridge graph $G^*_{QMET_n}$ is induced subgraph of $G^*_{OMCUT_n}$. Those suppositions have been computer checked up to $n = 7$. Apropos, any two facets of $QMET_n$ will be not adjacent if, instead of all oriented multicuts, we restrict ourself only to oriented cuts.

Conjecture 10.1 *For the ridge graph $G^*_{QMET_n}$ it holds:*
(i) a triangle facet $OT_{ij,k}$ is adjacent to a facet if and only if they are non-conflicting;
(ii) a non-negativity facet NN_{ij} is adjacent to $NN_{i'j'}$ if and only if neither $i' = j$, nor $j' = i$.

Conjecture 10.2 *For the ridge graph $G^*_{QMET^\square_n}$ it holds:*
(i) a triangle facet $OT_{ij,k}$ is adjacent to some facet if and only if they are non-conflicting;
(ii) a non-negativity facet NN_{ij} is adjacent to $NN_{i'j'}$ if and only if neither $i' = j$, nor $j' = i$; it is adjacent also to all facets OG_{mk};
(iii) a facet OG_{ij} is adjacent also to all non-triangle facets.

Conjectures 10.1, 10.2 would imply that the ridge graphs of $QMET_n$ and $QMET_n^{\square}$ have diameter 2. Let us see it for $QMET_n$.

The "non-conflicting" graph, restricted on the orbit of all triangle facets, has diameter 2. It follows (case by case check) from the fact that the complement of its local graph is $2K_1 + (2n-6)K_2$.

For example, $OT_{ij,k}$ conflicts with $4n-10$ facets: $OT_{ik,j}$; $OT_{kj,i}$; $OT_{ik,l}$; $OT_{il,j}$; $OT_{kj,l}$; $OT_{lj,i}$ for all l different of i,j,k. All other possible pairs of non-adjacent facets are $(OT_{ij,k}, NN_{ij})$, (NN_{ij}, NN_{ji}), (NN_{ij}, NN_{ki}), (NN_{ij}, NN_{jk}). Examples of common neighbors for those pairs are, respectively, NN_{ik}, $OT_{ik,j}$, NN_{kj}, NN_{ik}.

Conjecture 10.2 implies that if we take a triangle facet $OT_{ij,k}$ in $QMET_n^{\square}$, then the "conflicting" graph (the graph of "non-neighbors") of $OT_{ij,k}$ has $4(n-2)+1$ nodes (facets NN_{ij}, OG_{jk}, OG_{ik}, $OT_{ik,j}$, $OT_{kj,i}$, $OT_{il,j}$, $OT_{ik,l}$, $OT_{li,j}$, $OT_{kj,l}$, where $l \neq i,j,k$), and $2(n-2)$ edges (between facets OG_{jk} and $OT_{ik,j}$; OG_{ik} and $OT_{kj,i}$; $OT_{il,j}$ and $OT_{ik,l}$; $OT_{lj,i}$ and $OT_{kj,l}$). So, it is the graph $2(n-2)K_2 + K_1$.

Remark. The set

$$E_n = \{e + e^T \mid e \text{ is an extreme ray of } QMET_n\}$$

consists, for $n = 3,4,5$ of $1,7,79$ orbits (amongst of $2,10,229$), including $0,3,10$ orbits of path metrics of graphs.

More exactly, a path metric d_G belongs to E_4 for the graphs $G \in \{K_4, P_2, C_4, P_4\}$. Now, $d_G \in E_5$ for $G \in \{K_{2,3}, K_5 - K_3, K_5 - P_2 - P_3, K_5, C_5, \overline{P_2}, \overline{P_3}, \overline{P_4}, \overline{P_5}, \overline{2P_2}\}$, where $d_{K_{2,3}}$ is an extreme ray of MET_5; $d_{K_{2,3}}$, $d_{K_5-K_3}$ and $d_{K_5-P_2-P_3}$ do not belong to CUT_5, and the path metrics of the remaining seven graphs belong to CUT_5.

In fact, those seven path metrics d_G are all of form $e + e^T$, were e is an extreme ray of the *oriented hypermetric cone* $QHYP_5$ (see Chapters 8, 12).

Now consider some general results about $OMCUT_n$.

Conjecture 10.3 *The oriented cuts form a dominating clique (i.e., a complete subgraph such that any vertex is adjacent to one of its elements) in G_{OMCUT_n}; so, the diameter of G_{OMCUT_n} is 2 or 3 for any $n \geq 4$.*

We checked this conjecture by computer for $n \leq 7$, using only non-negativity facets and oriented triangle facets. The adjacency between oriented cuts and any other oriented multicut are preserved even if we consider only oriented triangle and non-negativity facets, but already for $n = 4$ (when $OMCUT_n$ has other facets) the adjacency is diminished for any oriented multicut orbit different from oriented cut.

More general, we have

Conjecture 10.4 *For $OMCUT_n$, $QMET_n$, $OMCUT_n^{\square}$ and $QMET_n^{\square}$ it holds:*
(i) every o-cut is adjacent to all other o-cuts;
(ii) every extreme ray (vertex) is adjacent to some o-cut;
(iii) the diameter of the 1-skeleton graph is equal to 2 or 3.

It is conjectured that the diameters of $OCUT_n$ and $OMCUT_n$ are 1 and 2, respectively.

Let us consider the following inequalities:

- the *zero-extension* of an inequality $\sum_{1\le i\ne j\le n-1} f_{ij}q_{ij} \ge 0$: an inequality
$$\sum_{1\le i\ne j\le n} f'_{ij}q_{ij} \ge 0 \text{ with } f'_{ni} = f'_{in} = 0, \text{ and } f'_{ij} = f_{ij}, \text{ otherwise;}$$
- the inequality
$$A^n_{c_1\dots c_{n-2},ab} : \sum_{i=1}^{n-2}(q_{ac_i} + q_{c_ib}) + q_{ba} \ge S_{ab} + S_{c_1\dots c_{n-2}},$$
where $S_{c_1\dots c_{n-2}}$ denotes the sum of distances along oriented cycle $c_1, ..., c_{n-2}$;
- the inequality
$$B^n_{c_1\dots c_{n-2},ab} : \sum_{i=1}^{n-2}(q_{c_ia} + q_{ac_i} + q_{c_ib}) \ge q_{ab} + S_{c_1\dots c_{n-2}};$$
- the *oriented hypermetric inequality*
$$Hyp^O_b : Hyp^O_b(q) = \sum_{1\le i\ne j\le n} b_ib_jq_{ij} \le 0,$$
where $b = (b_1, ..., b_n) \in \mathbb{Z}^n$, and $\sum_{i=1}^n b_i = 1$.

Call a face of $OMCUT_n$ *symmetric* if, in matrix terms, it is preserved by the transposition.

Theorem 10.7 *The following inequalities are valid on $OMCUT_n$:*
(i) zero-extensions of valid faces of $OMCUT_{n-1}$;
(ii) symmetric faces coming from CUT_n (so, including any inequality Hyp^O_b);
(iii) $A^n_{c_1\dots c_n,ab}$ and $B^n_{c_1\dots c_n,ab}$.

Proof. (i) is obvious.

(ii) Since we have $CUT_n = \{q + q^T \mid d \in OMCUT_n\}$, the validity of symmetric faces coming from CUT_n is obvious, too.

(iii) Consider the inequality A^n on an oriented multicut $\delta^O_{S_1,\dots,S_t}$. In this case $S_{c_1\dots c_{n-2}} \le n-3$, and $S_{c_1\dots c_{n-2}} = n-3$ if and only if $c_1 \in S_{\alpha_1}, ..., c_{n-2} \in S_{\alpha_{n-2}}$, where $1 \le \alpha_1 < ... < \alpha_{n-2} \le n-2$.

Let $b \in S_\alpha, a \in S_\beta$, where $\alpha < \beta$. Then $S_{ab} = q_{ba} = 1$, and $S_{ab} + S_{c_1\dots c_{n-2}} \le n-2$. But $\sum_{i=1}^{n-2}(q_{ac_i} + q_{c_ib} + q_{ba}) \ge \sum_{i=1}^{n-2} q_{ba} = n-2$, and A^n holds.

Let now $a \in S_\alpha, b \in S_\beta$, where $\alpha < \beta$. Then $S_{ab} = q_{ab} = 1$, and $S_{ab} + S_{c_1\dots c_{n-2}} \le n-2$. But in this case $q_{ac_i} + q_{c_ib} \ge 1$; hence, $\sum_{i=1}^{n-2}(q_{ac_i} + q_{c_ib} + q_{ba}) = \sum_{i=1}^{n-2}(q_{ac_i} + q_{c_ib}) \ge n-2$ and A^n holds.

If $a, b \in S_\alpha$, then $S_{ab} = 0$ and $S_{ab} + S_{c_1\dots c_{n-2}} \le n-3$. But in this case $q_{ac_i} + q_{c_ib} = 1$ for $c_i \notin S_\alpha$ and $q_{ac_i} + q_{c_ib} = 0$ for $c_i \in S_\alpha$; hence, $\sum_{i=1}^{n-2}(q_{ac_i} + q_{c_ib} + q_{ba}) = \sum_{i=1}^{n-2}(q_{ac_i} + q_{c_ib}) \ge S_{c_1\dots c_{n-2}}$, and A^n holds.

Consider the inequality B^n on an oriented multicut $\delta^O_{S_1,\dots,S_t}$. In this case $S_{c_1\dots c_{n-2}} \le n-3$, and $S_{c_1\dots c_{n-2}} \le n-3$ if and only if $c_1 \in S_{\alpha_1}, ..., c_{n-2} \in S_{\alpha_{n-2}}$, where $1 \le \alpha_1 < ... < \alpha_{n-2} \le t$.

Let $b \in S_\alpha, a \in S_\beta$, where $\alpha < \beta$. Then $q_{ab} = 0$, and $q_{ab} + S_{c_1 \dots c_{n-2}} \le n-3$. But in this case $q_{ac_i} + q_{c_i b} \ge 1$; hence, $\sum_{i=1}^{n-2}(q_{ac_i} + q_{c_i a} + q_{c_i b}) \ge \sum_{i=1}^{n-2}(q_{ac_i} + q_{c_i b}) \ge n-2$, and B^n holds.

Let $a \in S_\alpha, b \in S_\beta$, where $\alpha < \beta$. Then $q_{ab} = 1$, and $q_{ab} + S_{c_1 \dots c_{n-2}} \le n-2$. In this case $q_{ac_i} + q_{c_i b} \ge 1$; hence, $\sum_{i=1}^{n-2}(q_{ac_i} + q_{c_i a} + q_{c_i b}) \ge \sum_{i=1}^{n-2}(q_{ac_i} + q_{c_i b}) \ge n-2$, and B^n holds.

If $a, b \in S_\alpha$, then $q_{ab} = 0$, and $q_{ab} + S_{c_1 \dots c_{n-2}} \le n-3$. But in this case $q_{ac_i} + q_{c_i a} + q_{c_i b} \ge 1$ for $c_i \notin S_\alpha$, and $q_{ac_i} + q_{c_i a} + q_{c_i b} = 0$ for $c_i \in S_\alpha$; hence, $\sum_{i=1}^{n-2}(q_{ac_i} + q_{c_i b} + q_{c_i b}) \ge S_{c_1 \dots c_{n-2}}$, and B^n holds.

□

Conjecture 10.5 *The following inequalities correspond to facets of $OMCUT_n$:*

(i) zero-extensions of any facet of $OMCUT_{n-1}$ (so, including all oriented triangle and non-negativity inequalities);

(ii) any oriented hypermetric inequality Hyp_b^O, except of non-oriented triangle inequalities;

(iii) $A^n_{c_1 \dots c_{n-2}, ab}$ and $B^n_{c_1 \dots c_{n-2}, ab}$.

We checked this conjecture by computer for $n \le 7$.

Any oriented triangle inequality is a zero extension of A^3, while any non-negativity inequality is a zero-extension of a degenerated B^3 with $b = c_1$.

Furthermore, if $f \ge 0$ defines a facet of $OMCUT_n$, then the zero-extension of f to $OMCUT_{n+1}$ is still a facet, just as in the case of the cone CUT_n.

Theorem 10.8 *For the cone $OMCUT_n$ it holds:*

(i) any symmetric facet of the cone $OMCUT_n$ corresponds to a facet of the cone CUT_n;

(ii) all orbits of symmetric facets of $OMCUT_n$ with $n \le 6$ are represented by Hyp_b^O, where $b = (1,1,1,-1,-1)$, $(1,1,1,-1,-1,0)$, $(2,1,1,-1,-1,-1)$, and $(1,1,1,1,-1,-2)$;

(iii) all orbits of symmetric facets of $OMCUT_7$ are all 9 orbits of hypermetric facets of CUT_7, which are different from the orbit of triangle facets, and 18 out of 26 non-hypermetric ones (namely, all but O_6, O_{13}, O_{22}, O_{18}, O_{20}, O_{24}, O_{25}, and O_{26} in terms of Deza and Dutour (2001) [DeDu03]).

Proof. (ii) and (iii) were obtained by computer using (i). In order to prove (i), let us fix a symmetric facet F of $OMCUT_n$. Set $U_F = \{q \in OMCUT_n \mid F(q) = 0\}$. If F is a symmetric facet, then U_F is a set, invariant by the reversal (transposition).
Denote $SU(X) = \{q + q^r \mid q \in X\}$ for any $X \subset OMCUT_n$. Then $SU(U_F)$ is a set of semimetrics, which are incident to F. Moreover, we have $SU(U_F) \subset CUT_n$, since $CUT_n = SU(OMCUT_n)$. By hypothesis F is a facet, so, U_F has dimension $n(n-1)-1$. The mapping $q \to q + q^r$ decrease dimension by at most $\frac{n(n-1)}{2}$; so, we get that $SU(U_F)$ has dimension $\frac{n(n-1)}{2} - 1$, i.e., F is a facet of CUT_n.

□

Given two ordered partitions A and B, we will write $A < B$, if A is a proper refinement of B. We will write $A \dot{<} B$ if, moreover, each part of A is a proper subset of a part of B. Moreover, for a given ordered partition A we will denote by A^r the *reversal* of A, i.e. the same sets, but written in reversed order.

Conjecture 10.6 *For the cone $OMCUT_n$ it holds:*

(i) an oriented multicut δ_A^O is not adjacent to all oriented multicuts δ_B^O such that $B < A^r$;

(ii) the orbit of extreme rays of oriented cut $\delta^O_{\{1\},\{2,\dots,n\}}$ is unique orbit, such that extreme rays in this orbit is not adjacent only to oriented multicuts described in (i) above; the total adjacency is $p^O(n) - p^O(n-1) - 1$, and it is the maximal total adjacency;

(iii) an extreme ray from the orbit of the oriented cut $\delta^O_{\{1,2\},\{3,\dots,n\}}$ is not adjacent only to oriented multicuts δ_B^O, such that $B < \langle\{1,2\},\{3,...,n\}\rangle^r$, or B is any cyclic shift of C with $C \dot{<} \langle\{1,2\},\{3,...,n\}\rangle$.

(iv) the diameter of G_{OMCUT_n} is 2.

We checked this conjecture by computer for $n \leq 6$.

Conjecture 10.7 *The ridge graphs $G^*_{QMET_n}$ and $G^*_{QMET_n^\square}$ are induced subgraphs of $G^*_{OMCUT_n}$ and $G^*_{OMCUT_n^\square}$, respectively.*

It is conjectured that the diameters of $OCUT_n$ and $OMCUT_n$ are 1 and 2, respectively. Furthermore, if f defines a facet of $OMCUT_n$, then the zero-extension of f to $OMCUT_{n+1}$ is still a facet, just as in the case of the cone CUT_n.

Remind, that in the symmetric case all triangle inequalities are facet-inducing in CUT_n and in $CUT_n^\square$ for any $n \geq 3$; the cut vectors form a single switching class, which is a clique in the 1-skeleton graph of $MET_n^\square$ (on the other hand, it is shown in Laurent (1996) [Laur96] that every other switching class is a stable set in the 1-skeleton graph of $MET_n^\square$, that is, no two non-integral switching equivalent vertices of $MET_n^\square$ form an edge on $MET_n^\square$); two triangle inequalities are adjacent in the ridge graph of $MET_n^\square$ if and only if they are non-conflicting (see [DeDe94]).

10.5 Other constructions of quasi-semimetric polyhedra

Define the *quasi-semimetric polytope* $QMETP_n$ as the set of all $q \in QMET_n$, satisfying all

$$q_{ki} + q_{ij} + q_{jk} \leq 2.$$

For $QMETP_n$, one can define the following analog of the switching operation of $MET_n^\square$: given $S \subset \{1,...,n\}$, call *oriented switching* the operation $U_S(q) = q'$, such that

$$q'_{ij} = \begin{cases} 1 - q_{ji}, & \text{if } |S \cap \{i,j\}| = 1, \\ q_{ij}, & \text{otherwise.} \end{cases}$$

This, together with reversals and $Sym(n)$, gives a group of order $2^n \cdot n!$, expected to be the full symmetry group of $QMETP_n$ (checked for $n \leq 9$).

Define *oriented cut polytope* $OCUTP_n$ as the convex hull of all the oriented cuts and their images under oriented switchings. Define *oriented multicut polytope* $OMCUTP_n$ as the convex hull of all the oriented multicuts and their images under oriented switchings. We expect, that $OCUTP_n$ have exactly 2^{2n-2} vertices, which is much higher than $2^n - 2$, the number of extreme rays of $OCUT_n$. Both polytopes have the same symmetry group as $QMETP_n$.

In Table 10.33 data for small quasi-semimetric polytopes of this type is given.

Polytope	Dimension	# of vertices (orbits)	# of facets (orbits)
$OCUTP_3$	5	16(2)	16(2)
$OCUTP_4$	9	64(3)	40(2)
$OCUTP_5$	14	256(3)	1,056(5)
$OCUTP_6$	20	1,024(4)	1,625,068(97)
$QMETP_3$	6	22(3)	20(2)
$QMETP_4$	12	544(8)	56(2)
$QMETP_5$	20	1,155,136(392)	120(2)

Table 10.33: The number of vertices and facets in some quasi-metric polytopes for $3 \leq n \leq 6$

Chapter 11

Multidimensional case: m-hemimetrics

11.1 Preliminaries

For given two m-hemimetrics d_1 and d_2 on a set X, their non-negative linear combination $d = \alpha d_1 + \beta d_2, \alpha, \beta \geq 0$, is a m-hemimetric on X. Here, as usual, for all $x_1, ..., x_{m+1} \in X$ it holds

$$(\alpha d_1 + \beta d_2)(x_1, ..., x_{m+1}) = \alpha d_1(x_1, ..., x_{m+1}) + \beta d_2(x_1, ..., x_{m+1}).$$

Then we can speak about the cone of all m-hemimetrics on n points, in fact, on the set $V_n = \{1, 2, ..., n\}$, which generalize the notion of m-dimensional volumes, and, more generally, of the cone of (m, s)-supermetrics. We can consider already the cones, generated by all partition m-hemimetrics on n points - multidimensional analogues of multicut semimetrics.

In this chapter we introduce, for small values of n and m, the polyhedral cones, associated with m-hemimetrics on V_n, in particular, the cone, generated by all partition m-hemimetrics on V_n, coming from partitions of an n-set into $m+1$ blocks; we construct the cone of all (m, s)-supermetrics on V_n, as well as the cone, generated by all $\{0, 1\}$-valued extreme rays of it.

The notions of m-metrics, m-hemimetrics and partition m-hemimetrics are $(m+1)$-ary generalizations of the binary notions of semimetrics and cuts, which are well-known and central objects in Graph Theory and Combinatorial Optimization.

The notion of m-hemimetric on n points was introduced in [DeRo00] and studied, using the program *cdd* [Fuku95], for small parameters m, n, in [DeRo02]. The notion of (m, s)-supermetric was introduced in [DeDu03]. (See also [DeDe10]).

Using computer search, we list facets and generators for these cones and give tables of their adjacencies and incidences. The different orbits are determined manually, using symmetries. We study two graphs, the 1-skeleton graph and the ridge graph, of these polyhedra: the number of their nodes and edges, their diameters, conditions of adjacency, inclusions among the graphs and their restrictions on some orbits of nodes.

Finally, we compare obtained results with similar results for classical semimetric case (see [DeDe95], [DDF96], [DeLa97]), and for asymmetric quasi-semimetric case (see [DePa99], [DDP03]).

All computation was done using the programs *cdd* of Fukuda (1995) [Fuku95]. The calculations are rather complex due to the high dimension, $\binom{n}{m+1}$, of the cones even for small n; moreover, matrices can no longer be used.

The following structures will be studied below:

- the *m-hemimetric cone* $HMET_n^m$: the set of all *m-hemimetrics* on n points, defined by all *m-simplex* inequalities

$$d_{i_1 \dots i_{m+1}} \leq \sum_{i=1}^{m+1} d_{i_1 \dots i_{k-1} i_{k+1} \dots i_{m+2}},$$

  and all *non-negativity inequalities* $d_{i_1 \dots i_{m+1}} \geq 0$ on V_n;

- the *non-positive m-hemimetric cone* HM_n^m: the set of all *non-positive m-hemimetrics* on n points, defined only by all m-simplex inequalities

$$d_{i_1 \dots i_{m+1}} \leq \sum_{i=1}^{m+1} d_{i_1 \dots i_{k-1} i_{k+1} \dots i_{m+2}}$$

  on V_n (so, negative values of functions are allowed);

- the *partition m-hemimetric cone* $HCUT_n^m$: the conic hull of all non-zero *partition m-hemimetrics* on n points;

- the *(m, s)-supermetric cone* $SMET_n^{m,s}$: the set of all *(m, s)-supermetrics* on n points, defined by all *(m, s)-simplex inequalities*

$$s \times d_{i_1 \dots i_{m+1}} \leq \sum_{i=1}^{m+1} d_{i_1 \dots i_{k-1} i_{k+1} \dots i_{m+2}},$$

  and all non-negativity inequalities $d_{i_1 \dots i_{m+1}} \geq 0$ on V_n;

- the *binary (m, s)-supermetric cone* $SCUT_n^{m,s}$: the conic hull of all $\{0, 1\}$-valued extreme rays of $SMET_n^{m,s}$.

All these cones lie in the space $\mathbb{R}^{\binom{n}{m+1}}$, and are of full dimension $\binom{n}{m+1}$ each. For $m = 1$, the last two cones coincide and the first two cones are the *semimetric cone* MET_n and the *cut cone* CUT_n, considered in detail in [DeLa97] and in the references listed there.

11.2 Cones and polytopes of m-hemimetrics and (m, s)-supermetrics

Let $V_n = \{1, 2, \dots, n\}$; let $E_n^k = |\{(i_1, \dots, i_k) \,|\, i_1, \dots, i_k \in V_n, i \neq j\}|$, where $(i_1, \dots, i_k)$ denotes the unordered set of different integers $i_1, \dots, i_k$, i.e., E_n^k is the amount of the elements in the family of all k-element subsets of V_n $(k = 1, 2, \dots, n)$. In fact, $E_n^k = \binom{n}{k} = \frac{n!}{(n-k)!k!}$. Moreover, $E_n^2 = E_n$.

Let d be an *m-hemimetric* on the set V_n, i.e., a function $d : V_n^{m+1} \longrightarrow \mathbb{R}$, such that for all $x_i \in V_n$ it holds $d(x_1, \dots, x_{m+1}) \geq 0$; $d(x_1, \dots, x_{m+1}) = 0$ whenever $x_1, \dots, x_{m+1}$ are not pairwise distinct; $d(x_1, \dots, x_{m+1}) = d(\pi(x_1), \dots, \pi(x_{m+1}))$ for any permutation π

of the set $\{x_1, ..., x_{m+1}\}$; $d(x_1, ..., x_{m+1}) \leq \sum_{i=1}^{m+1} d(x_1, ..., x_{i-1}, x_{i+1}, ..., x_{m+2})$ (*m-simplex inequality*).

Then we can view an m-hemimetric d on the set V_n as a vector

$$(d_{i_1 \dots i_{m+1}})_{1 \leq i_1 < \dots < i_{m+1} \leq n} \in \mathbb{R}^{E_n^{m+1}} \ : \ d_{i_1 \dots i_{m+1}} = d(i_1, ..., i_{m+1}).$$

Hence, any m-semimetric on V_n can be viewed alternatively as a function on V_n^{m+1} or as a vector in $\mathbb{R}^{E_n^{m+1}}$. But the matrix representation can no longer be used. We will use both these representations. Moreover, we will use both symbols $d(i_1, ..., i_{m+1})$ and $d_{i_1 \dots i_{m+1}}$ for the values of an m-semimetric between points i_1,..., i_{m+1}.

On the other hand, any vector $v = (v_{i_1 \dots i_{m+1}}) \in \mathbb{R}^{E_n^{m+1}}$, indexed by $(m+1)$-subsets of the set V_n, can be seen as vertex-labeled subgraph of the *Johnson graph* $J(n, m+1)$. We consider the restriction of $J(n, m+1)$ on the *support* of v (i.e., the set of all indices of the vector, on which its components are non-zero): the vertices of this graph are all unordered $(m+1)$-tuples $(i_1, ..., i_{m+1})$, such that $v_{i_1 \dots i_{m+1}}$ is non-zero; this value will be the label of the vertex; we will omit the label when it is 1; two vertices of the Johnson graph (and also of its induced subgraph) are adjacent if the corresponding $(m+1)$-tuples have m common elements. Call this restriction *labeled representation graph of v* (in Johnson graph $J(n, m+1)$) and denote it by G_v. In the special case, when v is an $\{0,1\}$-valued vector, G_v is just usual graph.

In Section 11.4 we will introduce another graph H_v for the special case $n = m+3$.

As any extreme ray v (or facet f) of the cones under consideration can be given by a vector from $\mathbb{R}^{E_n^{m+1}}$, we can speak about their labeled representation graphs G_v and G_f.

For example, given a partition m-hemimetric $\alpha_{S_1,...,S_{m+1}}$, its labeled representation graph $G_{\alpha_{S_1,...,S_{m+1}}}$ is the *Hamming graph* $H(|S_1|, \dots, |S_{m+1}|)$, i.e., the direct (Cartesian) product $K_{|S_1|} \times K_{|S_2|} \times ... \times K_{|S_{m+1}|}$ of the cliques $K_{|S_i|}$, $1 \leq i \leq m+1$.

Another examples: the graph G_v of the vector v, defining a non-negativity facet is a vertex and, for an m-simplex facet, it is the complete graph K_{m+2} with one vertex, labeled -1.

Denote by $HMET_n^m$ the set of all m-hemimetrics on n points. Then $HMET_n^m$ is a full-dimensional cone in $\mathbb{R}_n^{E^{m+1}}$, defined by all $(m+2)\binom{n}{m+2}$ *simplex inequalities* $ST_{i_1 \dots i_{m+1}, i_{m+2}}$ and all $\binom{n}{m+1}$ *non-negativity inequalities* $NN_{i_1 \dots i_{m+1}}$ on V_n:

- $ST_{i_1 \dots i_{m+1}, i_{m+2}} : \sum_{i=1}^{m+1} d_{i_1 \dots i_{i-1} i_{i+1} \dots i_{m+2}} - d_{i_1 \dots i_{m+1}} \geq 0$;
- $NN_{i_1 \dots i_{m+1}} : d_{i_1 \dots i_{m+1}} \geq 0$.

For example, the cone $HMET_4^2$ is a full-dimensional cone in $\mathbb{R}^4$, generated by 4 *tetrahedron inequalities* $ST_{123,4} : d_{423} + d_{143} + d_{124} - d_{123} \geq 0$, $ST_{124,3} : d_{124} + d_{134} + d_{123} - d_{124} \geq 0$, $ST_{134,2} : d_{234} + d_{124} + d_{132} - d_{134} \geq 0$, $ST_{234,1} : d_{134} + d_{214} + d_{231} - d_{234} \geq 0$, and 4 non-negativity inequalities $NN_{123} : d_{123} \geq 0$, $NN_{124} : d_{124} \geq 0$, $NN_{134} : d_{124} \geq 0$, $NN_{234} : d_{234} \geq 0$.

Any inequality, defining the cone $HMET_4^2$, can be written as a vector in $\mathbb{R}^4$; for example, $ST_{123,4} : d_{423} + d_{143} + d_{124} - d_{123} \geq 0$ corresponds to the vector $(-1, 1, 1, 1)$, while $NN_{123} : d_{123} \geq 0$ corresponds to the vector $(1, 0, 0, 0)$.

The cone, generated by all non-zero partition m-hemimetrics $\alpha_{S_1,...,S_{m+1}}$, is called the *partition m-hemimetric cone* and denoted by $HCUT_n^m$. It is also the full-dimensional cone in $\mathbb{R}^{E_n^{m+1}}$.

For example, the cone $HCUT_4^2$ is a full-dimensional cone in $\mathbb{R}^4$, generated by 6 partition 2-hemimetrics α_{S_1,S_2,S_3} for the 3-partitions of V_4:

$$\{1\},\{2\},\{3,4\};\{1\},\{3\},\{2,4\};\{1\},\{4\},\{2,3\};$$

$$\{2\},\{3\},\{1,4\};\{2\},\{4\},\{1,3\};\{3\},\{4\},\{1,2\}.$$

Any partition 2-hemimetric α_{S_1,S_2,S_3} in the cone $HCUT_4^2$ can be written as a vector in $\mathbb{R}^4$:

$$\alpha_{\{1\},\{2\},\{3,4\}} = (1,1,0,0),\ \alpha_{\{1\},\{3\},\{2,4\}} = (1,0,1,0),$$

$$\alpha_{\{1\},\{4\},\{2,3\}} = (0,1,1,0),\ \alpha_{\{2\},\{3\},\{1,4\}} = (1,0,0,1),$$

$$\alpha_{\{2\},\{4\},\{1,3\}} = (0,1,0,0),\ \alpha_{\{3\},\{4\},\{1,2\}} = (0,0,1,1).$$

As every m-partition hemimetric is an m-hemimetric, we have, that

$$HCUT_n^m \subseteq HMET_n^m \subseteq \mathbb{R}_{\geq 0}^{E_n^{m+1}}.$$

In [DeRo02] the *non-positive m-hemimetric cone* HM_n^m, defined only by all *m-simplex inequalities* on V_n, was considered. These generalized m-hemimetrics are not, in general, non-negative. Clearly, HM_n^m is of full dimension $\binom{n}{m+1}$, too. For $m = 1$ the cones $HMET_n^m$ and HM_n^m coincide.

Denote by $SHMET_n^{m,s}$ the set of all (m,s)-hemimetrics on n points. Then $SHMET_n^{m,s}$ is a full-dimensional cone in $\mathbb{R}^{E_n^{m+1}}$, defined by all $(m+2)\binom{n}{m+2}$ *(m,s)-simplex inequalities* $sST_{i_1\dots i_{m+1},i_{m+2}}$, and all $\binom{n}{m+1}$ *non-negativity inequalities* $NN_{i_1\dots i_{m+1}}$ on V_n:

- $sST_{i_1\dots i_{m+1},i_{m+2}} : \sum_{j=1}^{m+1} d_{i_1\dots i_{j-1}i_{j+1}\dots i_{m+2}} - s \times d_{i_1\dots i_{m+1}} \geq 0;$
- $NN_{i_1\dots i_{m+1}} : d_{i_1\dots i_{m+1}} \geq 0.$

The cone, generated by all $\{0,1\}$-valued extreme rays of $SHMET_n^m$, is denoted by $SHCUT_n^{m,s}$. It is also the full-dimensional cone in $\mathbb{R}^{E_n^{m+1}}$.

By definition, we have, that

$$SHCUT_n^{m,s} \subseteq SHMET_n^{m,s} \subseteq \mathbb{R}_{\geq 0}^{E_n^{m+1}}.$$

Remark. We can define the *m-hemimetric polytope* by adding to the definition of $HMET_n^m$ the following *m-perimeter inequalities*:

$$d_{i_2\dots i_{m+2}} + d_{i_1i_3\dots i_{m+2}} + \dots + d_{i_1\dots i_{m+1}} \leq 2.$$

However, there is no equivalent of the *switching* operation in that case and so, the studies have been limited to cones so far.

Clearly, all faces of $HMET_n^m$ and $HCUT_n^m$, as well as HM_n^m, $SHMET_n^{m,s}$ and $SHCUT_n^{m,s}$, are preserved by any permutation of the nodes. It means, that every permutation of V_n induce a symmetry of the cones $HMET_n^m$, $HCUT_n^m$, HM_n^m, $SHMET_n^{m,s}$ and $SHCUT_n^{m,s}$, so, the group $Sym(n)$ is a symmetry group of those cones.

In order to find the full symmetry group of a given cone, for example, the cone $SMET_n^{m,s}$, one should consider the group of automorphisms of the graphs $G_{SMET_n^{m,s}}$

and $G^*_{SMET_n^{m,s}}$. If one of those groups is equal to $Sym(n)$, then the full symmetry group of the cone $SMET_n^{m,s}$ is $Sym(n)$. If those groups are bigger than $Sym(n)$, then their realizability, as group of linear symmetries, can be checked by elementary Linear Algebra.

Using this method, we checked by computer, that the symmetry group of $SMET_n^{m,s}$ is $Sym(n)$, the only exception being $MET_4 = SMET_4^{1,1}$, for which it is $Sym(4) \times Sym(3)$, and that the symmetry group of $SCUT_6^{3,3}$ has size 518400.

We conjecture, that in general case, the full symmetry group of our cones consists only of permutations of V_n, i.e., it is the group $Sym(n)$ of all permutations on V_n (see Theorem 3.3 in [DGL91] stating that the symmetry group of a truncated multicut polytope is $Sym(n)$).

The minimal n for which the hemimetric cones are non-trivial, is $m+2$; the dimension of the cones is also $m+2$ for $n = m+2$.

The m-simplex inequalities and non-negativity inequalities are sufficient for describing the m-hemimetric cone on n points for $n = m+2$: $HCUT^m_{m+2} = HMET^m_{m+2}$ for any $m \geq 1$.

For $m = 1$, the cones $HCUT^1_n$ and $HMET^1_n$ are the *cut cone* CUT_n and the *metric cone* MET_n, considered in detail in [DeLa97] (cf. Chapter 9).

The full description of the listed cones of m-hemimetrics for $n \leq 7$ is given below. In general, the calculations are rather complex due to the high dimension, $\binom{n}{m+1}$, of the cones even for small n; moreover, matrices can no longer be used.

In Table 11.1 we summarize the most important numeric information on m-hemimetric cones.

Cone	Dimension	# of ext. rays (orbits)	# of facets (orbits)	Diameters
$HCUT^m_{m+2}$=$HMET^m_{m+2}$ for $m \geq 2$	m+2	$\binom{m+2}{2}$ (1)	$2m+4$ (2)	2; 3,2 for m=2, >2
$HCUT^m_{m+3}$ for $m \geq 2$	$\binom{m+3}{2}$	$\binom{m+3}{3}+3\binom{m+3}{4}$ (2)	?	?;?
$HMET_{m+3}$ for $m \geq 2$	$\binom{m+3}{2}$	?	$3\binom{m+3}{2}$(2)	?;?
$HCUT^2_5$	10	25(2)	120(4)	2; 3
$HMET^2_5$	10	37(3)	30(2)	2; 2
$HCUT^3_6$	15	65(2)	4065(16)	2; 3
$HMET^3_6$	15	287(5)	45(2)	3; 2
$HCUT^4_7$	21	140(2)	474390(153)	2; 3
$HMET^4_7$	21	3692(8)	63(2)	3; 2
$HCUT^5_8$	28	266(2)	$\geq$ 409893148($\geq$ 11274)	2; ?
$HMET^5_8$	28	55898(13)	84(2)	3; 2
$HMET^6_9$	36	864174(20)	108(2)	?; 2
$HCUT^2_6$	20	90(3)	2095154(3086)	2; ?
$HMET^2_6$	20	12492(41)	80(2)	3; 2
$HMET^2_7$	35	$\geq$ 454191608($\geq$ 91836)	175(2)	?; 2
$HMET^3_7$	35	$\geq$ 551467967($\geq$ 110782)	140(2)	?; 2

Table 11.1: Small cones of m-hemimetrics

The column 2 indicates the dimension of the cone, the columns 3 and 4 give the number of extreme rays and facets, respectively; in parenthesis are given the numbers of their orbits. The column 5 gives the diameters of the 1-skeleton graph and of the ridge graph.

In Table 11.2 we give the similar information for cones of (m,s)-supermetrics.

Cone	Dimension	# of ext. rays (orbits)	# of facets (orbits)	Diameters
$SMET^{m,s}_{m+2}$, $1 \leq s \leq m-1$	m+2	$\binom{m+2}{s+1}(1)$	2m+4(2)	min(s+1,m-s+1); 2 but 2; 3 if m=2,s=1
$SMET^{m,m}_{m+2}$	m+2	m+2(1)	m+2(1)	1; 1
$SMET^{2,2}_5$	10	132(6)	20(1)	2; 1
$SCUT^{2,2}_5$	10	20(2)	220(6)	1; 3
$SMET^{3,3/2}_6$	15	331989(596)	45(2)	6; 2
$SMET^{3,2}_6$	15	12670(40)	45(2)	4; 2
$SCUT^{3,2}_6$	15	247(5)	866745(1345) (*conj.*)	2; ?
$SMET^{3,5/2}_6$	15	85504(201)	45(2)	6; 2
$SMET^{3,3}_6$	15	1138(12)	30(1)	3; 1
$SCUT^{3,3}_6$	15	21(2)	150(3)	1; 3
$SMET^{4,2}_7$	21	2561166(661) (a.d.m.)	63(2)	?; 2
$SMET^{4,3}_7$	21	838729(274) (a.d.m.)	63(2)	?; 2
$SMET^{4,4}_7$	21	39406(37)	42(1)	3; 1
$SCUT^{4,4}_7$	21	112(2)	148554(114) (a.d.m.)	1; 4
$SMET^{5,2}_8$	28	$\geq$ 222891598($\geq$ 6228)	84(2)	?; 2
$SMET^{5,3}_8$	28	$\geq$ 881351739($\geq$ 23722)	84(2)	?; 2
$SMET^{5,4}_8$	28	$\geq$ 136793411($\geq$ 4562)	84(2)	?; 2
$SMET^{5,5}_8$	28	775807(92) (a.d.m.)	56(1)	?; 1
$SMET^{6,6}_9$	36	30058078(335) (*conj.*)	72(1)	?;1
$SMET^{7,7}_{10}$	45	923072558(1067) (*conj.*)	90(1)	?;1
$SMET^{2,2}_6$	20	21775425(30827) (*conj.*)	60(1)	?; 1
$SCUT^{2,2}_6$	20	96(3)	$\geq$ 243692840($\geq$ 341551)	1; ?
$SMET^{3,3}_7$	35	$\geq$ 594481939($\geq$ 119732)	105(1)	?; 1
$SMET^{2,2}_7$	35	$\geq$ 465468248($\geq$ 93128)	140(1)	?; 1

Table 11.2: Small cones of (m, s)-supermetrics

11.3 Small cones of m-hemimetrics and (m, s)-supermetrics

Small 2-hemimetrics. The case of 4 points

As was noted above, the minimal n, for which the m-hemimetric cones are non-trivial, is $m+2$; the dimension of the cones is also $m+2$ for $n = m+2$. So, the smallest non-trivial 2-hemimetric cone can be defined on 4 points.

In this section we present a complete linear description for case $(m, n) = (2, 4)$.

It turns out that $HCUT^2_4 = HMET^2_4$. This cone has 6 extreme rays (all in the same orbit under $Sym(4)$): α_{S_1,S_2,S_3} for the 3-partitions

$$(1, 2, 34), (1, 3, 24), (1, 4, 23), (2, 3, 14), (2, 4, 13), (3, 4, 12).$$

There are 8 facets, which form 2 orbits: the orbit F_1 of 4 *tetrahedron facets* and the orbit F_2 of 4 non-negativity facets.

Adjacencies of facets of $HMET^2_4$ are shown in Table 11.3. For each orbit a representative and the number of adjacent facets from other orbits are given, as well as the total number of adjacent ones, the number of incident extreme rays and the cardinality of orbits.

Orbit	Representative	F_1	F_2	Adj.	Inc.	$\|F_i\|$
F_1	$ST_{123,4}$	0	3	3	3	4
F_2	NN_{123}	3	0	3	3	4

Table 11.3: The adjacencies of facets in the cone $HMET^2_4$

The 1-skeleton graph $G_{HMET_4^2}$ is $K_6 - 3K_2$ (the octahedron); the 3 pairs of non-adjacent rays are of the form $\alpha_{\{a\},\{b\},\{c,d\}}, \alpha_{\{c\},\{d\},\{a,b\}}$. Each extreme ray (say, $\alpha_{\{1\},\{2\},\{3,4\}}$) is incident to 2 tetrahedron and to 2 non-negativity facets (namely, to $ST_{123,4}, ST_{124,3}$, and NN_{134}, NN_{234}).

The ridge graph $G^*_{HMET_4^2}$ is the cube.

Proposition 11.1 *For the ridge graph of $HMET_4^2$ it holds:*
(i) the tetrahedron facet $ST_{ijk,l}$ is adjacent only to the facets NN_{ijl}, N_{ikl}, N_{jkl};
(ii) the non-negativity facet NN_{ijk} is adjacent only to the facets $ST_{ijl,k}, T_{ikl,j}, T_{jkl,i}$.

Small 2-hemimetrics. The case of 5 points

We present here the complete linear description of $HCUT_5^2$, $HMET_5^2$, and HM_5^2.

The cone $HCUT_5^2$ has 25 extreme rays, which form 2 orbits with representatives $\alpha_{\{1\},\{2\},\{3,4,5\}}$ (orbit O_1) and $\alpha_{\{1\},\{2,3\},\{4,5\}}$ (orbit O_2). Clearly, the graphs G_α for partition 2-hemimetrics $\alpha_{\{1\},\{2\},\{3,4,5\}}$ and $\alpha_{\{1\},\{2,3\},\{4,5\}}$ are the cycles C_3 and C_4, respectively.

The cone $HCUT_5^2$ has 120 facets divided into 4 orbits, induced by 20 tetrahedron inequalities (orbit $F_1 = ST$), 10 *non-negativity inequalities* (orbit $F_2 = NN$), 60 inequalities (orbit $F_3 = A$), represented by

- $A : 2d_{123} - (d_{124} + d_{135}) + (d_{134} + d_{125} + d_{245} + d_{345}) \geq 0$;

and the 30 inequalities (orbit $F_4 = B$), represented by

- $B : 2(d_{123} + d_{145} + d_{245} - d_{345}) + (d_{134} + d_{135} + d_{234} + d_{235} - d_{124} - d_{125}) \geq 0.$

The above two inequalities are the 2-hemimetric analogs of the *5-gonal inequality* (the simplest inequality, different from the triangle inequality), appearing in the cone CUT_n for $n \geq 5$:

$$(d_{13} + d_{14} + d_{15} + d_{23} + d_{24} + d_{25}) - (d_{12} + d_{34} + d_{35} + d_{45}) \geq 0.$$

This last facet and facet B have both the *Petersen graph* as their $\overline{G}_F$, $\overline{G}_B$ (i.e. the complement of their graphs G_F, G_B).

The graphs $\overline{G}_A$, $\overline{G}_B$ are given in Figure 11.1.

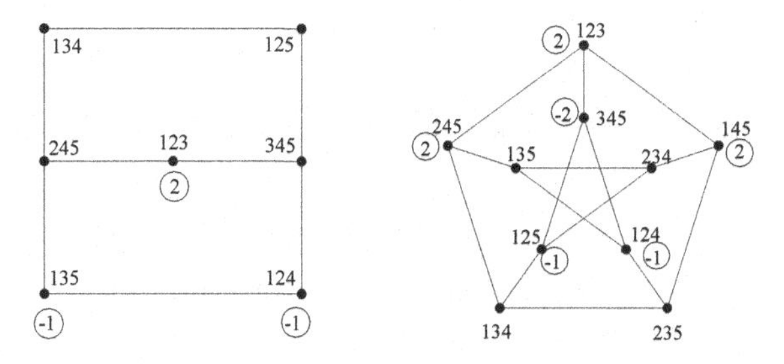

Figure 11.1: $\overline{G}_A$, $\overline{G}_B$ in the cone $HCUT_5^2$

The facets from the orbit F_4 are simplex cones, i.e., the extreme rays on them are linearly independent.

The 1-skeleton graph and the ridge graph of $HCUT_5^2$ have 270 and 1185 edges, respectively.

Among 9 neighbors of a facet B, 4 are from the orbit F_1, 4 are from the orbit F_3, and exactly one (actually, NN_{123}) is from the orbit F_2.

The local graph of a facet from F_4 (i.e., the subgraph of the ridge graph of $HCUT_5^2$, induced by all neighbors of a facet B) is $K_9 - C_4$. In fact, all non-adjacencies in this local graph are the four edges of the 4-cycle of the 4 neighbors of B from the orbit F_3 (see Tables 11.4, 11.5).

Orbit	Representative	O_1	O_2	Adjacency	Incidence	$\vert O_i\vert$
O_1	$\alpha_{\{1\},\{2\},\{3,4,5\}}$	9	12	21	54	10
O_2	$\alpha_{\{1\},\{2,3\},\{4,5\}}$	8	14	22	54	15

Table 11.4: The adjacencies of extreme rays in the cone $HCUT_5^2$

Orbit	Representative	F_1	F_2	F_3	F_4	Adjacency	Incidence	$\vert F_i\vert$
F_1	$ST_{123,4}$	16	9	18	6	49	16	20
F_2	NN_{123}	18	3	18	3	42	16	10
F_3	A	6	3	4	2	15	10	60
F_4	B	4	1	4	0	9	9	30

Table 11.5: The adjacencies of facets in the cone $HCUT_5^2$

Facets from orbits F_1, F_2, F_3, F_4 are incident, respectively, to 7,9; 7,9; 4,6; 3,6 extreme rays from orbits O_1, O_2 of $HCUT_5^2$.

The adjacencies of 37 extreme rays and of 30 facets of the cone $HMET_5^2$ are given in Tables 11.6, 11.7.

Orbit	Representative	O_1	O_2	O_3	Adjacency	Incidence	$\vert O_i\vert$
O_1	$\alpha_{\{1\},\{2\},\{3,4,5\}}$	9	12	6	27	21	10
O_2	$\alpha_{\{1\},\{2,3\},\{4,5\}}$	8	6	8	22	18	15
O_3	r_3	5	10	5	20	15	12

Table 11.6: The adjacencies of extreme rays in the cone $HMET_5^2$

Orbit	Representative	F_1	F_2	Adjacency	Incidence	$\vert F_i\vert$
F_1	$ST_{123,4}$	16	9	25	22	20
F_2	NN_{123}	18	3	21	22	10

Table 11.7: The adjacencies of facets in the cone $HMET_5^2$

The extreme rays of $HMET_5^2$ are divided into 3 orbits O_1, O_2, O_3, represented by $\{0,1\}$-valued vectors

$$r_1 = \alpha_{\{1\},\{2\},\{3,4,5\}} = (1,1,1,0,0,0,0,0,0,0),$$

$$r_2 = \alpha_{\{1\},\{2,3\},\{4,5\}} = (0,1,1,1,1,0,0,0,0,0),$$

$$r_3 = (1,0,1,0,0,1,1,0,0,1);$$

their G-graphs are C_3, C_4, C_5, respectively.

Each facet (from both orbits) of $HMET_5^2$ is incident to 7,9,6 extreme rays from orbits O_1, O_2, O_3, respectively.

The 1-skeleton graph and the ridge graph of $HMET_5^2$ have 420 and 355 edges, respectively.

Proposition 11.2 *The diameters of the* 1*-skeleton graphs of* $HCUT_5^2$ *and of* $HMET_5^2$ *are* 2.

Proof. In fact, each of the orbits O_1, O_2 of $HCUT_5^2$ is a dominating clique. There is only one type of a non-edge, represented by $\alpha_{\{1\},\{2,3\},\{4,5\}}, \alpha_{\{2\},\{3\},\{1,4,5\}}$, but $\alpha_{\{1\},\{3\},\{2,4,5\}}$ is one of common neighbors. The complement of the 1-skeleton graph of $HCUT_5^2$ turns out to be the Petersen graph with a new vertex (corresponding to a member of the orbit O_2) on each of 15 edges. The result for $HMET_5^2$ comes also by finding out a common neighbor to each possible non-edge.

□

Proposition 11.3 *For the ridge graph of* $HMET_5^2$ *it holds:*
(i) the diameter of the ridge graph of $HMET_5^2$ *is* 2;
(ii) its restrictions on the orbits F_1 *and* F_2 *are* $K_{4,4,4,4,4}$ *and the Petersen graph, respectively;*

The cone HM_5^2 has 92 extreme rays divided into 6 orbits. Below we give some representatives $r_1, ..., r_6$ of those orbits $O_1, ..., O_6$:
$r_1 = (1, 1, 1, 0, 0, 0, 0, 0, 0, 0)$;
$r_2 = (0, 1, 1, 1, 1, 0, 0, 0, 0, 0)$;
$r_3 = (1, 0, 1, 0, 0, 1, 1, 0, 0, 1)$;
$r_4 = (1, 1, 1, -1, 0, 0, 1, 0, 1, 1)$;
$r_5 = (1, 1, 1, -1, -1, 1, 1, 1, 1, 1)$;
$r_6 = (1, 0, 1, 0, 1, -1, 1, 1, 2, 1)$;
here $r \in \mathbb{R}^{10}$ has the form $r = (d_{123}, d_{124}, d_{125}, d_{134}, d_{135}, d_{145}, d_{234}, d_{235}, d_{245}, d_{345})$. The first two represent both orbits of $HCUT_5^2$, the first 3 represent the 3 orbits of $HMET_5^2$ (see Table 11.8).

Orbit	Representative	O_1	O_2	O_3	O_4	O_5	O_6	Adj.	Inc.	$\lvert O_i \rvert$
O_1	$r_1 = \alpha_{\{1\},\{2,3\},\{3,4,5\}}$	6	12	6	6	9	9	48	14	10
O_2	$r_2 = \alpha_{\{1\},\{2,3\},\{4,5\}}$	8	2	4	2	4	8	28	12	15
O_3	r_3	5	5	0	5	5	5	25	10	12
O_4	r_4	6	3	6	0	3	6	24	12	10
O_5	r_5	6	4	4	2	0	4	20	12	15
O_6	r_6	3	4	2	2	2	0	13	10	30

Table 11.8: The adjacencies of extreme rays in the cone HM_5^2

Each (tetrahedron) facet of HM_5^2 is incident to 7, 9, 6, 6, 9, 15 extreme rays from orbits $O_1, ..., O_6$, respectively.

Proposition 11.4 *The ridge graph of* HM_5^2 *is* $K_{4,4,4,4,4}$ *(of diameter* 2*).*

Small 2-hemimetrics. The case of 6 points

The cone $HMET_6^2$ has exactly 12492 extreme rays from 41 orbits O_1 – O_{41}. We list in Table 11.9 the representatives of these orbits with their adjacency numbers, sizes and incidence numbers.

Three of the orbits, O_1, O_2 and O_4, are represented by 3-partition 2-hemimetrics $\alpha_{\{1\},\{2\},\{3,4,5,6\}}, \alpha_{\{1\},\{2,3\},\{4,5,6\}}, \alpha_{\{1,2\},\{3,4\},\{5,6\}}$ and have size 15, 60, 15, respectively.

O	123	124	125	126	134	135	136	145	146	156	234	235	236	245	246	256	345	346	356	456	Adj.	Size	Inc.
O_1	0	0	0	0	0	0	0	0	0	1	0	0	0	0	0	1	0	0	1	1	2778	15	64
O_2	0	0	0	0	0	0	0	0	1	1	0	0	0	0	1	1	0	1	1	0	1321	60	56
O_3	0	0	1	1	1	0	1	1	0	0	1	1	0	0	1	0	0	0	1	1	1030	12	40
O_4	0	0	0	0	0	1	1	1	1	0	0	1	1	1	1	0	0	0	0	0	818	15	48
O_5	0	0	0	0	0	0	1	1	0	1	0	0	1	1	0	1	1	1	0	0	731	180	48
O_6	0	0	0	1	0	1	0	1	1	0	0	1	1	1	0	1	0	1	1	0	358	180	40
O_7	0	0	1	1	1	0	1	1	0	0	1	1	0	0	1	0	0	2	1	1	270	120	36
O_8	0	0	1	1	1	0	1	1	0	2	1	1	2	2	1	0	0	0	1	1	93	120	28
O_9	0	0	1	1	1	0	1	1	0	2	1	1	0	2	1	0	0	2	1	1	66	240	28
O_{10}	0	0	1	1	1	0	1	1	0	2	1	1	0	0	1	2	0	2	1	1	51	360	28
O_{11}	0	0	1	1	1	0	1	0	1	1	1	1	0	1	0	1	1	1	0	2	47	120	28
O_{12}	0	0	0	1	0	0	1	0	1	2	0	0	2	0	2	1	0	2	1	1	46	60	39
O_{13}	0	0	0	1	0	0	1	1	1	2	0	0	2	1	2	1	1	2	1	0	37	360	31
O_{14}	0	0	0	1	0	0	1	2	2	2	0	0	2	2	1	1	2	1	1	0	37	180	31
O_{15}	0	0	0	2	0	0	2	2	1	1	0	0	2	2	1	1	2	1	1	0	37	60	31
O_{16}	0	0	0	2	0	3	1	3	1	1	0	3	1	3	1	1	0	2	1	1	32	90	28
O_{17}	0	0	0	1	0	1	1	2	2	2	0	1	2	2	1	1	1	1	0	0	30	720	27
O_{18}	0	0	0	2	0	1	2	2	1	1	0	1	2	2	1	1	1	1	0	0	30	360	27
O_{19}	0	0	0	1	0	1	1	1	1	2	0	1	2	1	2	1	2	2	0	0	30	360	27
O_{20}	0	0	0	2	0	2	1	2	1	1	0	2	1	2	1	1	2	2	0	0	30	180	26
O_{21}	0	0	0	2	0	1	1	1	1	2	0	1	1	1	1	2	2	2	0	0	30	180	27
O_{22}	0	0	0	2	0	2	1	3	1	1	0	2	1	3	1	1	1	2	0	1	29	360	26
O_{23}	0	0	0	2	0	1	1	2	1	1	0	1	1	2	1	1	1	2	3	0	29	360	26
O_{24}	0	0	1	1	1	0	1	0	2	2	1	1	2	1	1	0	1	0	1	2	27	360	23
O_{25}	0	0	0	1	1	1	1	2	2	2	1	1	2	2	1	1	0	0	0	0	27	360	27
O_{26}	0	0	1	1	2	0	1	0	1	2	2	1	2	1	2	0	2	0	1	1	27	360	23
O_{27}	0	0	1	2	3	0	1	3	1	1	3	1	1	2	1	0	0	1	2	1	27	360	23
O_{28}	0	0	1	2	2	1	2	1	2	0	2	2	0	2	0	1	0	2	1	1	27	180	23
O_{29}	0	0	0	2	1	1	2	2	1	1	1	1	2	2	1	1	0	0	0	0	27	180	27
O_{30}	0	0	0	2	2	2	1	2	1	1	2	2	1	2	1	1	0	0	0	0	27	60	26
O_{31}	0	0	1	2	1	1	1	2	1	0	1	2	1	1	1	1	0	3	0	3	26	720	23
O_{32}	0	0	0	2	1	1	1	2	1	1	1	1	1	2	1	1	0	3	3	0	26	180	24
O_{33}	0	0	1	1	2	0	1	1	1	2	2	1	2	2	2	0	1	0	1	0	25	720	23
O_{34}	0	0	1	1	1	0	2	1	1	2	1	1	1	2	2	0	2	0	2	0	25	720	23
O_{35}	0	0	1	1	0	1	2	2	1	0	0	2	1	1	2	0	1	3	1	1	25	360	25
O_{36}	0	0	1	1	0	1	2	2	1	0	0	2	1	1	2	0	1	1	1	3	25	360	25
O_{37}	0	0	1	1	1	0	1	2	2	2	1	1	2	1	1	0	1	0	1	0	25	360	23
O_{38}	0	0	1	1	1	0	2	2	1	1	1	1	1	1	2	3	1	0	1	0	23	720	22
O_{39}	0	0	1	2	1	1	1	2	0	1	1	2	1	1	2	0	0	2	3	1	22	720	21
O_{40}	0	0	1	2	1	1	2	2	1	0	1	2	0	1	3	1	0	2	1	1	22	720	21
O_{41}	0	0	1	2	2	1	1	1	1	1	2	2	1	2	1	0	0	0	3	3	21	360	21

Table 11.9: Representatives of orbits of extreme rays of $HMET_6^2$

Orbits $O_1, ..., O_6$ have representation graph K_4 (the skeleton of the tetrahedron), $\overline{C}_6 = K_6 - C_6 = K_3 \times K_2$ (i.e., $Prism_3$), Petersen graph, the skeleton of 3-cube, 2-truncated tetrahedron and an 10-vertex graph (with 4 vertices of degree five and all others of degree three), respectively. The orbit O_7 has the graph G (non-planar, non-regular), given in Figure 11.2, together with one for the orbit O_5. All these orbits consist of $\{0, 1\}$-valued extreme rays. The extreme rays of the remaining orbits are $\{0, 1, 2\}$-valued and $\{0, 1, 2, 3\}$-valued vectors.

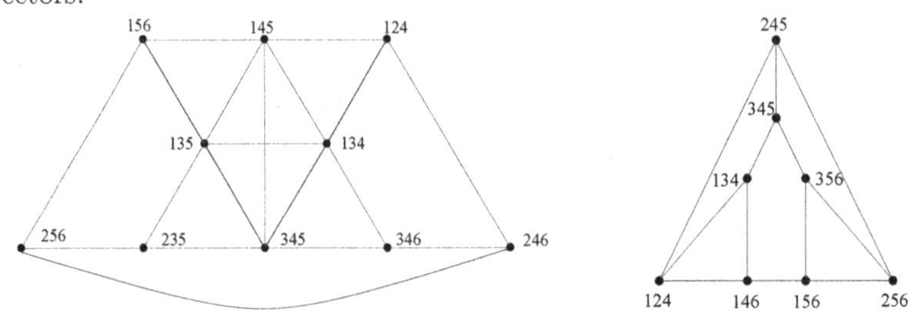

Figure 11.2: The graphs G of extreme rays from orbits O_7, O_5 of the cone $HMET_6^2$

Table 11.10 gives representatives of two facet orbits of this cone.

F	123	124	125	126	134	135	136	145	146	156	234	235	236	245	246	256	345	346	356	456	Adj.	Size	Inc.
F_1	−1	0	0	1	0	0	1	0	0	0	0	0	1	0	0	0	0	0	0	0	75	60	4001
F_2	0	0	0	0	0	0	0	0	0	0	0	0	0	0	0	0	0	0	0	1	67	20	3939

Table 11.10: Representatives of orbits of facets of $HMET_6^2$

Table 11.11 shows the adjacencies of these facets.

Orbit	Representative	F_1	F_2	Adjacency	Incidence	$\lvert F_i \rvert$
F_1	$ST_{123,4}$	56	19	75	4001	60
F_2	NN_{123}	57	10	67	3939	20

Table 11.11: The adjacencies of facets in the cone $HMET_6^2$

The cone $HCUT_6^2$ has 2095154 facets which form 3086 orbits. Here are two examples of an $\{0,1,-1\}$-valued facets of $HCUT_6^2$ (see also Figure 11.3 for the facet W):

$W : (-d_{145}+d_{146}+d_{136}+d_{123}+d_{125})+(d_{245}+d_{234}+d_{346}+d_{356}+d_{256})-(d_{235}+d_{236}) \geq 0;$

$Z : \sum d_{ijk}-(d_{124}+d_{125}+d_{145})-(d_{234}+d_{235}+d_{345})-2(d_{146}+d_{156}+d_{456})-2d_{236} \geq 0.$

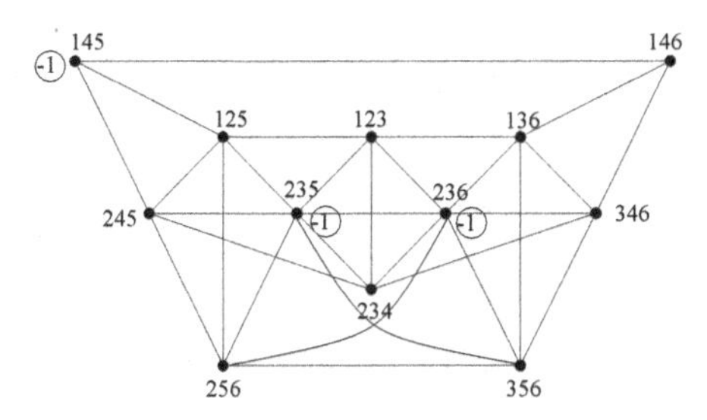

Figure 11.3: G_W in the cone $HCUT_6^2$

Remark, that the triples with coefficients zero, in W and Z, form the skeleton of 1- and 2-truncated tetrahedron, respectively; the triples with coefficient -1 form K_3+K_1 and K_2+K_1, respectively.

We list in Table 11.12 only the representatives of the orbits of facets in $HCUT_6^2$ with incidence greater or equal to 29, besides the non-negativity and 3-simplex facets.

123	124	125	126	134	135	136	145	146	156	234	235	236	245	246	256	345	346	356	456	Adj.	Size	Inc.
−1	0	0	1	0	1	0	0	0	1	0	1	2	0	0	−1	0	0	0	0	15270	360	45
−1	−1	0	2	2	0	1	0	1	0	2	0	1	0	1	0	0	−2	0	0	5908	180	42
−1	−1	−1	3	−1	1	1	1	1	1	−1	1	1	1	1	1	1	1	−1	−1	600	180	33
−1	−1	−1	3	1	1	1	1	1	1	1	1	1	1	1	1	−1	−1	−1	−1	939	60	33
−1	−1	1	1	0	0	1	1	0	0	0	0	1	2	1	−1	1	−1	1	0	1579	720	33
−1	−1	0	2	1	−1	1	1	1	0	1	−1	1	1	1	0	3	−1	1	−1	496	360	32
−1	−1	1	1	2	0	1	1	0	0	2	1	0	0	1	0	−1	−1	0	0	1856	180	32
−1	−1	1	1	0	0	1	0	1	1	1	1	1	1	1	−1	−1	0	0	0	515	360	31
−1	0	0	1	0	0	1	1	−1	1	0	0	1	1	1	−1	2	0	0	0	629	360	31
−1	−1	0	2	1	−1	1	1	1	0	2	−1	2	0	1	1	1	−2	1	0	404	720	30
−1	−1	1	1	0	0	1	1	0	0	0	1	0	2	1	0	0	0	1	−1	458	720	30
−1	−1	1	1	2	0	1	0	1	1	3	1	1	1	1	−1	−1	−2	0	0	558	360	30
−1	−1	2	2	2	−1	2	2	−1	−1	2	2	−1	−1	2	−1	−1	−1	2	2	2265	12	30
−1	0	0	1	0	0	1	1	−1	1	0	1	2	0	0	−1	1	1	0	0	441	720	30
−1	0	0	1	0	1	0	1	1	0	1	0	0	1	0	1	0	1	1	−2	1867	120	30
−1	−1	−2	4	−2	2	1	2	1	2	−2	2	1	2	1	2	4	2	−2	−2	149	180	29
−1	−1	0	2	1	−1	1	1	1	0	1	−1	1	1	1	0	2	−2	2	0	473	360	29
−1	−1	0	2	1	2	0	2	0	2	2	1	2	1	2	−2	−1	−2	0	0	323	360	29
−1	−1	0	2	3	−1	1	0	2	1	3	1	1	−1	1	0	1	−3	1	0	288	720	29
−1	−1	1	1	−1	1	1	1	1	−1	0	0	1	0	1	1	1	0	0	0	283	360	29

Table 11.12: Representatives of orbits of facets of $HCUT_6^2$ with incidence at least 29

Small 3-hemimetrics

The cone $HMET_6^3$ has 287 extreme rays divided into 5 orbits. Below we give representatives $r_1, ..., r_5$ of the orbits $O_1, ..., O_5$ in the notation

$$(d_{\overline{12}}, d_{\overline{13}}, d_{\overline{14}}, d_{\overline{15}}, d_{\overline{16}}, d_{\overline{23}}, d_{\overline{24}}, d_{\overline{25}}, d_{\overline{26}}, d_{\overline{34}}, d_{\overline{35}}, d_{\overline{36}}, d_{\overline{45}}, d_{\overline{46}}, d_{\overline{56}})$$

(these vectors are indexed by 4-subsets of the set $\{1, ..., 6\}$; in our representation the 4-subsets are given as the complements of corresponding 2-subsets):

$r_1 = (0, 0, 1, 1, 0, 0, 0, 0, 0, 0, 0, 0, 1, 0, 0)$;
$r_2 = (0, 0, 1, 1, 0, 0, 0, 0, 0, 1, 1, 0, 0, 0, 0)$;
$r_3 = (0, 0, 1, 1, 0, 0, 0, 0, 0, 1, 0, 1, 0, 0, 1)$;
$r_4 = (0, 0, 1, 1, 0, 0, 0, 1, 1, 1, 0, 1, 0, 0, 0)$;
$r_5 = (0, 0, 1, 1, 0, 1, 0, 0, 1, 0, 0, 1, 1, 2, 0)$.

The first four extreme rays are $\{0, 1\}$-valued; their G-graphs (in the Johnson graph $J(6, 4)$ of all 4-tuples) are the cycles C_3, C_4, C_5, C_6, respectively. The first two are partition 3-hemimetrics; they represent both orbits of $HCUT_6^3$. The graphs G_{r_4} and G_{r_5} are given in Figure 11.4.

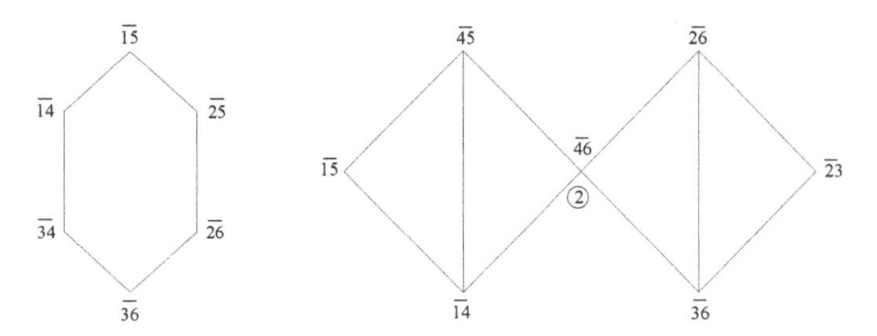

Figure 11.4: G_{r_4}, G_{r_5} in the cone $HMET_6^3$

Table 11.13 shows the adjacencies of extreme rays in the cone $HMET_6^3$.

Orbit	Representative	O_1	O_2	O_3	O_4	O_5	Adjacency	Incidence	$\|O_i\|$
O_1	$\alpha_{\{1\},\{2\},\{3\},\{4,5,6\}}$	19	36	36	18	27	136	21, 12	20
O_2	$\alpha_{\{1\},\{2\},\{3,4\},\{5,6\}}$	16	18	24	20	16	94	18, 11	45
O_3	r_3	10	15	20	15	10	70	15, 10	72
O_4	r_4	6	15	18	9	6	54	12, 9	60
O_5	r_5	6	8	8	4	0	26	10, 8	90

Table 11.13: The adjacencies of extreme rays in the cone $HMET_6^3$

In Table 11.14 the adjacencies of facets in the cone $HMET_6^3$ are given.

Orbit	Representative	F_1	F_2	Adjacency	Incidence	$\|F_i\|$
F_1	$ST_{1234,5}$	25	14	39	131	30
F_2	NN_{1234}	28	14	42	181	15

Table 11.14: The adjacencies of facets in the cone $HMET_6^3$

The cone $HCUT_6^3$ has 4065 facets divided into 16 orbits. The representatives of corresponding orbits F_i, $1 \le i \le 16$, and their representation matrices are given in Tables 11.15 - 11.16, where $\overline{ij}$ means the complement of an 2-subset of V_6.

The facets from the orbits F_1 and F_2 are non-negativity and 3-simplex facets, respectively. Any facet f_{11} from the orbit F_{11} is a simplex cone. The G-graphs of the facets f_3 and f_4 from the orbits F_3 and F_4 are given in Figure 11.5.

The orbit F_1 of the non-negativity facets form a dominating clique in the ridge graph of the cone $HCUT_6^3$. Also, the ridge graph of the cone $HCUT_6^3$, restricted on the orbits F_1

	$\overline{56}$	$\overline{46}$	$\overline{45}$	$\overline{36}$	$\overline{35}$	$\overline{34}$	$\overline{26}$	$\overline{25}$	$\overline{24}$	$\overline{23}$	$\overline{16}$	$\overline{15}$	$\overline{14}$	$\overline{13}$	$\overline{12}$	Adj.	Size	Inc.
F_1	0	0	0	0	0	0	0	0	0	0	0	0	0	0	1	1526	15	49
F_2	-1	0	1	0	1	0	0	1	0	0	0	1	0	0	0	703	30	41
F_3	-1	0	1	0	1	0	1	0	1	1	2	1	0	0	-1	100	180	23
F_4	-1	0	1	0	1	0	1	0	1	1	1	2	-1	1	0	37	360	19
F_5	-1	-1	2	0	1	1	2	1	1	0	2	1	1	2	-2	31	360	18
F_6	-1	-1	2	1	0	2	1	2	0	2	2	1	1	-1	-1	30	180	18
F_7	-1	-1	2	1	0	2	2	1	1	-1	2	3	-1	1	0	23	360	16
F_8	-1	-2	3	1	2	1	2	1	2	-1	2	1	2	3	-2	23	360	17
F_9	-1	-2	3	2	1	2	2	1	2	-2	3	4	-1	1	1	23	180	15
F_{10}	-1	-1	2	-1	2	2	2	1	1	1	2	1	1	1	-2	22	60	18
F_{11}	-1	-1	2	0	1	1	1	2	2	-1	2	1	1	2	-1	18	360	16
F_{12}	-1	-1	2	1	1	1	1	1	1	-1	1	1	1	-1	2	18	90	16
F_{13}	-1	-1	2	1	0	2	1	2	0	0	2	1	1	-1	1	14	720	14
F_{14}	-1	0	1	0	1	2	2	1	0	2	2	1	2	0	-2	14	360	14
F_{15}	-1	-1	2	1	0	2	2	1	1	1	2	1	1	1	-2	14	360	14
F_{16}	-1	0	1	0	1	0	1	0	1	1	1	0	1	1	0	14	90	14

Table 11.15: Representatives of orbits of facets of $HCUT_6^3$

	F_1	F_2	F_3	F_4	F_5	F_6	F_7	F_8	F_9	F_{10}	F_{11}	F_{12}	F_{13}	F_{14}	F_{15}	F_{16}	Adj.	Size
F_1	14	28	144	144	192	96	120	168	36	20	120	36	240	72	72	24	1526	15
F_2	14	25	72	96	60	24	72	48	36	10	48	12	72	48	48	18	703	30
F_3	12	12	14	10	10	6	10	0	0	2	2	0	12	4	4	2	100	180
F_4	6	8	5	2	2	2	2	2	2	0	2	0	0	2	2	0	37	360
F_5	8	5	5	2	3	0	0	2	0	1	1	0	0	2	2	0	31	360
F_6	8	4	6	4	0	2	4	2	0	0	0	0	0	0	0	0	30	180
F_7	5	6	5	2	0	2	0	0	1	0	0	0	2	0	0	0	23	360
F_8	7	4	0	2	2	1	0	0	2	0	2	1	2	0	0	0	23	360
F_9	3	6	0	4	0	0	2	4	0	0	0	0	0	2	2	0	23	180
F_{10}	5	5	6	0	6	0	0	0	0	0	0	0	0	0	0	0	22	60
F_{11}	5	4	1	2	1	0	0	2	0	0	0	1	2	0	0	0	18	360
F_{12}	6	4	0	0	0	0	0	4	0	0	4	0	0	0	0	0	18	90
F_{13}	5	3	3	0	0	0	1	1	0	0	1	0	0	0	0	0	14	720
F_{14}	3	4	2	2	2	0	0	0	1	0	0	0	0	0	0	0	14	360
F_{15}	3	4	2	2	2	0	0	0	1	0	0	0	0	0	0	0	14	360
F_{16}	4	6	4	0	0	0	0	0	0	0	0	0	0	0	0	0	14	90
Size	15	30	180	360	360	180	360	360	180	60	360	90	720	360	360	90		4065

Table 11.16: Representation matrix of the ridge graph of $HCUT_6^3$

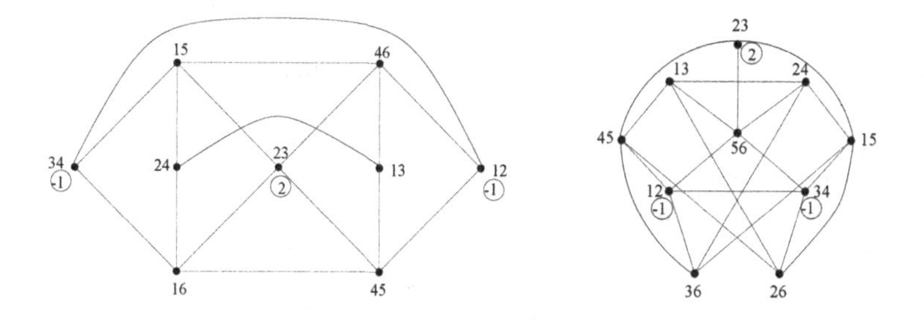

Figure 11.5: $\overline{G}_{f_3}, \overline{G}_{f_4}$ in the cone $HCUT_6^3$

and F_2, coincides with the ridge graph of the cone $HMET_6^3$. The complement of the *local graph* (i.e., of the graph induced by all neighbors of an representative of the orbit) for the orbits $F_{16}, F_{15}, F_{14}, F_{13}, F_{12}$ of small adjacency have, respectively $8, 8, 9, 7, 12$ vertices. It is $4K_2$ for F_{16} and some connected planar graphs for F_{15}, F_{14}, F_{13}; for F_{12} it is $K_8 - C_8$ on the set V_8 plus four pendent edges, which are incident with vertices $1, 3, 5, 7$, respectively.

There are two orbits of extreme rays in $HCUT_6^3$, corresponding to partition 3-hemimetrics on V_6 and giving us the total of 65 extreme rays in $HCUT_6^3$. The representatives of two orbits of extreme rays of this cone are given in Table 11.17.

The adjacencies of extreme rays in the cone $HCUT_6^3$ are given in Table 11.18.

O	$\overline{56}$	$\overline{46}$	$\overline{45}$	$\overline{36}$	$\overline{35}$	$\overline{34}$	$\overline{26}$	$\overline{25}$	$\overline{24}$	$\overline{23}$	$\overline{16}$	$\overline{15}$	$\overline{14}$	$\overline{13}$	$\overline{12}$	Adj.	Size	Inc.
O_1	0	0	0	0	0	0	0	0	1	1	0	0	1	1	0	58	45	993
O_2	0	0	0	0	0	0	0	0	0	1	0	0	0	1	1	55	20	1113

Table 11.17: Representatives of orbits of extreme rays of $HCUT_6^3$

Orbit	Representative	O_1	O_2	Adjacency	Incidence	$\lvert O_i \rvert$
O_1	$\alpha_{\{1\},\{2\},\{3\},\{4,5,6\}}$	19	36	55	1113	20
O_2	$\alpha_{\{1\},\{2\},\{3,4\},\{5,6\}}$	16	42	58	993	45

Table 11.18: The adjacencies of extreme rays in the cone $HCUT_6^3$

Proposition 11.5 *The diameter of the* 1*-skeleton graph of* $HCUT_6^3$ *is* 2*. Moreover, it holds:*

(i) $G_{O_1} = K_{20}$, $G_{O_2} = K_{45} - 15K_3$;

(ii) all non-edges are represented by $\alpha_{12,34,5,6}$ *that are non-adjacent:*

- *to* $\alpha_{\{1,2\},\{3\},\{4\},\{5,6\}}, \alpha_{\{1\},\{2\},\{3,4\},\{5,6\}}$ *(from the same orbit* O_2*);*
- *to* $\alpha_{\{1,2,5\},\{3\},\{4\},\{6\}}$, $\alpha_{\{1,2,6\},\{3\},\{4\},\{5\}}$, $\alpha_{\{3,4,5\},\{1\},\{2\},\{6\}}$, $\alpha_{\{3,4,6\},\{1\},\{2\},\{5\}}$.

In fact, both non-neighbors of $\alpha_{\{1,2\},\{3,4\},\{5\},\{6\}}$ are in O_2. For both types of non-edges - $\alpha_{\{1,2\},\{3,4\},\{5\},\{6\}}$ with $\alpha_{\{1,2\},\{3\},\{4\},\{5,6\}}$ and $\alpha_{\{1,2,5\},\{3\},\{4\},\{6\}}$ - the ray $\alpha_{\{1,3\},\{2,4\},\{5\},\{6\}}$ is a common neighbor. Also, all 9 non-neighbors of a ray from O_1, form K_9 in the 1-skeleton graph.

Notice, that the 1-skeleton of $HCUT_6^3$ is not an induced subgraph of the 1-skeleton graph of $HMET_6^3$; the only difference is in their restriction G_{O_2} to the orbit of rays, represented by r_2.

One can check that all neighbors of a partition hemimetric $\alpha_{\{a_1\},\{a_2\},\{b_1,b_2\},\{c_1,c_2\}}$ from the same orbit O_2 of $HMET_6^3$ are the 10 rays obtained by a transposition (xy) and the 8 rays obtained by a product $(a_1b_i)(a_2c_j)$ or $(a_1c_i)(a_2b_j)$ of two transpositions.

But in the 1-skeleton graph of $HCUT_6^3$, the ray $\alpha_{\{a_1\},\{a_2\},\{b_1,b_2\},\{c_1,c_2\}}$ is adjacent to all other members of O_2, except for the two rays, obtained from it by $(a_1b_1)(a_2b_2)$ or $(a_1c_1)(a_2c_2)$.

The complement of the graph, induced by all 18 neighbors of the ray $\alpha_{\{a_1\},\{a_2\},\{b_1,b_2\},\{c_1,c_2\}}$ from the same orbit O_2 of $HMET_6^3$, is $C_4 + C_4$ on 8 rays, obtained by a product $(a_1b_i)(a_2c_j)$ or by a product $(a_1c_i)(a_2b_j)$, the skeleton of the cube on 8 rays, obtained by (a_ib_j) or (a_ic_j), and it is $\overline{K}_2$ on two rays obtained by (b_ic_j).

The 1-skeleton graphs of $HCUT_6^3$ and $HMET_6^3$ both contain a dominating clique O_1; so, their diameters are 2 or 3. In order to see closer the 1-skeleton graph of $HMET_6^3$, we now describe the local graph, denoted by H, of the ray r_5.

All 26 neighbors are in orbits O_1, O_2, O_3, O_4 only. It will be easier to describe $\overline{H}$. The restrictions of $\overline{H}$ on them are $\overline{K}_6$, C_8, the skeleton of the cube and $2K_2$, respectively.

Two vertices from O_1 (say, 15 and 16) are isolated; so the diameter of H is 2. Here we denote by ij the j-th member of the orbit O_i in H. All edges of $\overline{H}$ (without isolated vertices 15 and 16) are presented in Figure 11.6.

On the right picture the members of O_1 are excluded, while on the left one the members of O_2 are excluded. $\overline{H}$ does not contain cross-edges among orbits O_1 and O_2 (see Tables 11.18, 11.11).

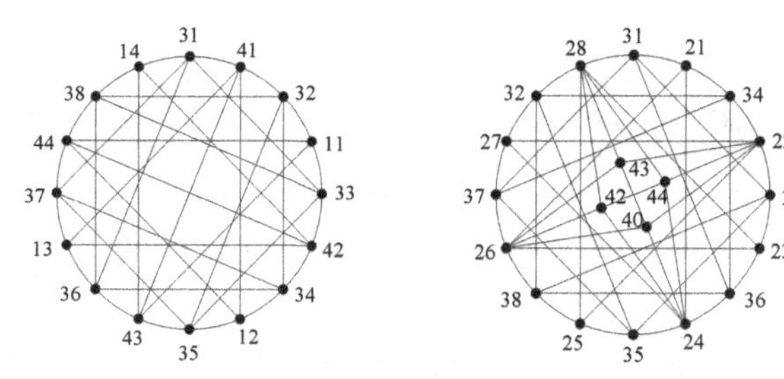

Figure 11.6: A presentation of the local graph of a ray of the orbit O_5 in $HMET_6^3$

Proposition 11.6 *The ridge graph of $HMET_6^3$ has diameter 2. Moreover, it holds:*

(i) any 3-simplex facet $ST_{ijkl,m}$ is adjacent to all but 5 facets: NN_{ijkl} and all 4 other 3-simplex facets with the same support;

(ii) the restrictions of the ridge graph of $HMET_6^3$ to the orbits F_1 and F_2 are $K_{5,5,5,5,5,5}$ and K_{15}, respectively.

Small 4-hemimetrics

The cone $HMET_7^4$ has 3692 extreme rays divided into 8 orbits. We give below representatives $w_1, ..., w_8$ of their orbits $O_1, ..., O_8$ in the notation

$$(\overline{12}, \overline{13}, \overline{14}, \overline{15}, \overline{16}, \overline{17}, \overline{23}, \overline{24}, \overline{25}, \overline{26}, \overline{27}, \overline{34}, \overline{35}, \overline{36}, \overline{37}, \overline{45}, \overline{46}, \overline{47}, \overline{56}, \overline{57}, \overline{67})$$

(these vectors are indexed by 5-subsets of the set $V_7 = \{1, 2, ..., 7\}$; the 5-subsets are given as the complements of the corresponding 2-subsets):

$w_1 = (0,0,0,1,1,0,0,0,0,0,0,0,0,0,0,0,0,1,0,0)$;
$w_2 = (0,0,0,1,1,0,0,0,1,1,0,0,0,0,0,0,0,0,0,0,0)$;
$w_3 = (0,0,0,1,1,0,1,0,0,1,0,0,1,0,0,0,0,0,0,0,0)$;
$w_4 = (0,0,0,1,1,0,1,0,0,1,0,1,0,0,0,1,0,0,0,0,0)$;
$w_5 = (0,0,0,1,1,0,1,0,0,1,0,1,0,0,0,0,0,1,0,1,0)$;
$w_6 = (0,0,0,1,1,0,0,0,0,0,0,1,0,2,1,0,0,1,1,0,0)$;
$w_7 = (0,0,0,2,2,0,0,0,0,1,1,1,1,0,0,1,0,0,0,0,1)$;
$w_8 = (0,0,0,1,1,0,0,1,0,0,1,2,1,1,0,0,0,1,0,0,0)$.

The (*adjacency, incidence*) pairs of those rays are, respectively, (985,48), (535,43), (315,38), (192,33), (126,28), (67,30), (43,25), (42,25).

The first five vectors are $\{0, 1\}$-valued; their graphs G are C_3, C_4, C_5, C_6, C_7, respectively. The first two are partition 4-hemimetrics; they represent both orbits of $HCUT_7^4$. The vectors $w_i, 1 \leq i \leq 4$, and w_6 have the same G-graphs as the members of orbits $O_i, 1 \leq i \leq 5$, of $HMET_6^3$, respectively; so, the graphs of Figure 11.4 represent also w_4 and w_6. The graphs G_{w_6}, G_{w_7} and G_{w_8} are given in Figure 11.7.

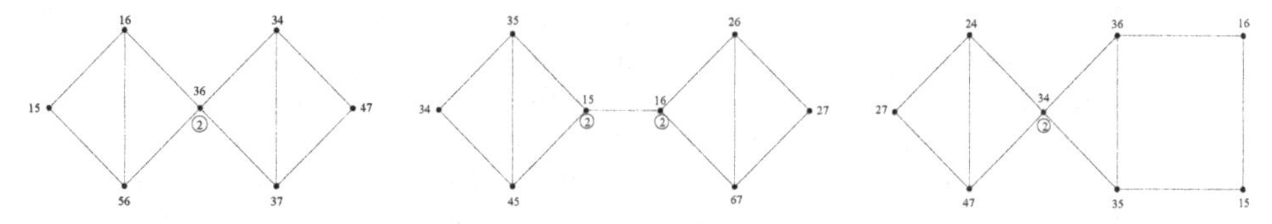

Figure 11.7: $G_{w_6}, G_{w_7}, G_{w_8}$ in the cone $HMET_7^4$

The adjacencies of facets in the cone $HMET_7^4$ are given in Table 11.19.

Orbit	Representative	F_1	F_2	Adjacency	Incidence	$\|F_i\|$
F_1	$ST_{12345,6}$	36	20	56	1302	42
F_2	NN_{12345}	40	20	60	2437	21

Table 11.19: The adjacencies of facets in the cone $HMET_7^4$

It is easy to check that the ridge graph of $HMET_7^4$ is $K_{6,6,6,6,6,6,6}$ on F_1, and K_{21} on F_2. All non-edges among F_1, F_2 are of the form $ST_{i_1 \dots i_5, i_6}$ and $NN_{i_1 \dots i_5}$.

The information on the facets of the cone $HCUT_7^4$ is given in Table 11.20. We list here only the orbits with incidence greater or equal to 29, besides the non-negativity and 4-simplex facets.

$\overline{67}$	$\overline{57}$	$\overline{56}$	$\overline{47}$	$\overline{46}$	$\overline{45}$	$\overline{37}$	$\overline{36}$	$\overline{35}$	$\overline{34}$	$\overline{27}$	$\overline{26}$	$\overline{25}$	$\overline{24}$	$\overline{23}$	$\overline{17}$	$\overline{16}$	$\overline{15}$	$\overline{14}$	$\overline{13}$	$\overline{12}$	Adj.	Size	Inc.
−1	0	1	0	1	0	0	1	0	0	1	0	1	1	1	2	1	0	0	0	−1	3490	420	50
−1	0	1	0	1	0	0	1	0	0	1	0	1	1	1	1	2	−1	1	1	0	343	1260	37
−1	−1	2	0	1	1	0	1	1	0	2	1	1	0	0	2	1	1	2	2	−2	320	1260	36
−1	−1	2	1	0	2	1	0	2	0	1	2	0	0	0	1	2	0	0	0	0	353	630	34
−1	0	1	0	1	0	1	0	1	1	1	0	1	1	0	1	2	−1	1	0	0	429	1260	34
−1	0	1	0	1	0	1	0	1	1	1	2	−1	1	0	1	2	1	−1	0	2	405	840	34
−1	−1	2	−1	2	2	0	1	1	1	2	1	1	1	0	2	1	1	1	2	−2	66	840	32
−1	−1	2	0	1	1	1	0	2	1	1	2	0	1	2	2	1	1	0	−1	−1	213	1260	32
−1	0	1	0	1	0	1	0	1	1	1	2	−1	1	0	2	1	2	0	−1	1	394	2520	32
−1	−1	2	−1	2	2	−1	2	2	2	2	1	1	1	1	2	1	1	1	1	−2	30	105	30
−1	−1	2	−1	2	2	1	1	1	1	1	1	1	1	−1	1	1	1	1	−1	2	33	420	29
−1	−1	2	−2	3	3	1	2	2	1	2	1	1	2	−1	2	1	1	2	3	−2	44	2520	29
−1	−2	3	0	1	2	1	2	1	1	2	1	2	0	−1	2	1	2	2	3	−2	89	2520	29

Table 11.20: Representatives of orbits of facets of $HCUT_7^4$ with incidence at least 29

Comparison of the small cones

Now we compare some semimetric, quasi-semimetric and m-hemimetric cones on n points for small n.

The triangle inequalities suffice to describe the cut cones for $n \leq 4$, but $CUT_n \subset MET_n$ (strictly) for $n \geq 5$. The complete description of all the facets of the cut cone CUT_n is known for $n \leq 8$, the complete description of the metric cone MET_n is known for $n \leq 8$, too (see Chapter 9). Here the "combinatorial explosion" starts from $n = 8$. The number of orbits of facets and of extreme rays of those and other cones, when it is known, is given in Table 11.21.

For the 3-dimensional case, the equality $HCUT_n^2 = HMET_n^2$ holds only for the smallest value $n = 4$. For $n = 4, 5$ we computed all facets, extreme rays and their adjacencies and incidences for the three cones $HCUT_n^2, HMET_n^2, HM_n^2$. For 2-hemimetrics the "combinatorial explosion" (in terms of the amount of computation and memory) starts already for the cone $HCUT_6^2$.

In Table 11.21 we compare the small 2-hemimetric cones $HCUT_n^2, HMET_n^2$ with the 1-hemimetric cones CUT_n, MET_n and their generalization in another direction: the cones $OMCUT_n, QMET_n$. Last two cones consist of all *quasi-semimetrics* on V_n and of those obtained from *oriented multicuts* (see [DePa99] for the notions; see also Chapter 10).

In Table 11.21, the column 2 gives the dimension of a cone under consideration, the columns 3 and 4 give the number of its extreme rays and facets, respectively; in parenthesis are given the numbers of their orbits. In column 5 are given the diameters of the 1-skeleton graph and the ridge graph of the cone, specified in the row.

Cone	Dimension	# of ext. rays (orbits)	# of facets (orbits)	Diameters
$HCUT^m_{m+2}$=$HMET^m_{m+2}$ $m \geq 3$	m+2	$\binom{m+2}{2}$ (1)	$2m+4$ (2)	2; 2
$CUT_3 = MET_3$	3	3(1)	3(1)	1; 1
$HCUT^2_4$=$HMET^2_4$	4	6(1)	8(2)	2; 3
$CUT_4 = MET_4$	6	7(2)	12(1)	1; 2
$OMCUT_3 = QMET_3$	6	12(2)	12(2)	2; 2
CUT_5	10	15(2)	40(2)	1; 2
MET_5	10	25(3)	30(1)	2; 2
$HCUT^2_5$	10	25(2)	120(4)	2; 3
$HMET^2_5$	10	37(3)	30(2)	2; 2
$OMCUT_4$	12	74(5)	72(4)	2; 2
$QMET_4$	12	164(10)	36(2)	3; 2
CUT_6	15	31(3)	210(4)	1; 3
MET_6	15	296(7)	60(1)	2; 2
$HCUT^3_6$	15	65(2)	4065(16)	2; 3
$HMET^3_6$	15	287(5)	45(2)	3; 2
$HCUT^2_6$	20	90(3)	$\geq$ 2095154($\geq$ 3086)	2; ?
$HMET^2_6$	20	12492(41)	80(2)	3; 2
$OMCUT_5$	20	540(10)	35320(194)	2; 3
$QMET_5$	20	43590(229)	80(2)	3; 2
$HCUT^4_7$	21	140(2)	474390(153)	2; 3
$HMET^4_7$	21	3692(8)	63(2)	3; 2
CUT_7	21	63(3)	38780(36)	1; 3
MET_7	21	55226(46)	105(1)	3; 2
CUT_8	28	127(4)	49604520(2169)	1; ?
$HCUT^5_8$	28	266(2)	$\geq$ 322416108($\geq$ 8792)	?; ?
$HMET^5_8$	28	55898(13)	84(2)	3; 2

Table 11.21: Some parameters of metric cones and their generalizations for small n

Note, that the number of orbits of extreme rays and the diameter of the 1-skeleton graph for the cones $QMET_5$, $HMET^2_6$, $HCUT^2_6$, $HCUT^4_7$ and for the duals of $HCUT^3_6$, $OMCUT_5$ are taken from the work [DeDu03], as well as the exact value of the diameter for $HMET^3_6$, $HMET^4_7$ and for the duals of $HCUT^2_5$, CUT_7.

The cones $HMET^2_n$ and $QMET_n$ have, besides of generalizations of the usual triangle inequalities, only non-negativity facets.

Incidences (to the extreme rays) of facets $ST_{ijk,l}$ and NN_{ijk} on the cones $HCUT^2_4 = HMET^2_4$, $HCUT^2_5, HMET^2_5$ amount to 3, 14 and 22, respectively, but they are different (4001 and 3939) on HM^2_6. Incidences of similar facets $ST_{ij,k}$ (*oriented triangular inequality*, i.e., $q_{ij} \leq q_{ik} + q_{kj}$ for a quasimetric q) and NN_{ij} (*non-negativity inequality* $q_{ij} \geq 0$) are equal (to 7, 43) on cones $OMCUT_3 = QMET_3, OMCUT_4$, but they are different (78 and 80) on $QMET_4$.

For $n = 4, 5$ we observe that the ridge graphs of HM^2_n and $HMET^2_n$ are induced subgraphs of the ridge graphs of $HMET^2_n$ and $HCUT^2_n$, respectively. The similar property does *not* hold for the 1-skeletons of those cones.

For example, any extreme ray of the orbit O_2 is adjacent to 14, 6, 2 members of the same orbit in the cones $HCUT^2_n, HMET^2_n, HM^2_n$, respectively.

Also, the ridge graph of $QMET_4$ is an induced subgraph of the ridge graph of $OMCUT_4$, but the 1-skeleton graph of $OMCUT_4$ is not an induced subgraph of the 1-skeleton graph of $QMET_4$ (see [DePa99]).

On the other hand, the ridge graph of MET_n and the 1-skeleton graph of CUT_n (for any n) have diameters 2 and 1, respectively, and those graphs are induced subgraphs of the ridge graph of CUT_n, and of the 1-skeleton graph of MET_n, respectively (see Lemma 2.1 and Theorem 3.5 in [DeDe94]).

Small supermetrics

We compute all information about small supermetrics for the case $n = m+3$. In fact, in this case the dimension of the supermetric cone $SMET^{m,s}_{m+3}$ is $\binom{n}{n-2} = \binom{n}{2}$, i.e., the same as of MET_n.

This correspondence allows us to replace the representation graph G_v by a simpler graph: any facet or extreme ray of the cone $SMET^{m,s}_{m+3}$ is given by a vector, say, v, indexed by $(m+1)$-subsets of V_n. It can be seen also as a function on 2-subsets of V_n, which are complements of $(m+1)$-subsets of V_{m+3}. So, to every $\{0,1\}$-valued extreme ray of $SMET^{m,s}_{m+3}$ one can associate a set of pairs (ij), and this set of pairs is the edge set of a graph H_v, such that the graph G_v is the line graph of H_v. If some vertices are isolated, then we remove them.

For example, if v is a cut semimetric $\delta_{\{1,2,3\},\{4\}}$, then v has support $\{\{1,4\},\{2,4\},\{3,4\}\}$, i.e. the complements are $\{\{2,3\},\{1,3\},\{1,2\}\}$; so, H_v is the complete graph on $\{1,2,3\}$.

If $v = \alpha_{\{1,2,3\},\{4\},\{5\}}$ (an extreme ray of $HMET^2_5$), then its support is

$$\{\{1,4,5\},\{2,4,5\},\{3,4,5\}\},$$

i.e., the complements are $\{\{2,3\},\{1,3\},\{1,2\}\}$; so, H_v is the complete graph on vertices $\{1,2,3\}$ again.

It follows from the fact, that the vector $\alpha_{\{1,2,3\},\{4\},\{5\}}$ is a zero-extension of the vector $\delta_{\{1,2,3\},\{4\}}$: for any $\{0,1\}$-valued vector d, the graph $H_{d^{ze}}$ of a zero-extension of d is equal to H_d.

In these terms, we have the following results.

1. All extreme rays of $SMET^{1,1}_4 = MET_4$ have $H_v = K_3$ or $K_{2,2}(= C_4)$.
2. All extreme rays of $HMET^2_5$ have $H_v = K_3(= C_3)$, $K_{2,2}(= C_4)$; $\overline{C}_5(= C_5)$.
3. All $\{0,1\}$-valued extreme rays of $SMET^{2,2}_5$ have $H_v = K_4$ or $K_{2,2,1}$.
4. All $\{0,1\}$-valued extreme rays of $HMET^3_6$ have $H_v = K_3(= C_3)$, $K_{2,2}(= C_4)$; $\overline{C}_5(= C_5)$; C_6.
5. All $\{0,1\}$-valued extreme rays of $SMET^{3,2}_6$ have $H_v = K_4$, $K_{2,2,1}$; $\overline{C}_6(=Prism_3)$, $\overline{C_1+C_5}$, $\overline{C_3+C_3}$.
6. All $\{0,1\}$-valued extreme rays of $SMET^{3,3}_6$ have $H_v = K_5$ or $K_{2,2,2}$.
7. All $\{0,1\}$-valued extreme rays of $HMET^4_7$ have $H_v = K_3(= C_3)$, $K_{2,2}(= C_4)$; C_5, C_6, C_7.
8. All $\{0,1\}$-valued extreme rays of $SMET^{4,2}_7$ have $H_v = K_4$, $K_{2,2,1}$; $\overline{C}_6$, $\overline{C_1+C_5}$, $\overline{C_3+C_3}$; ∇C_6, or the graphs in Figure 11.8.

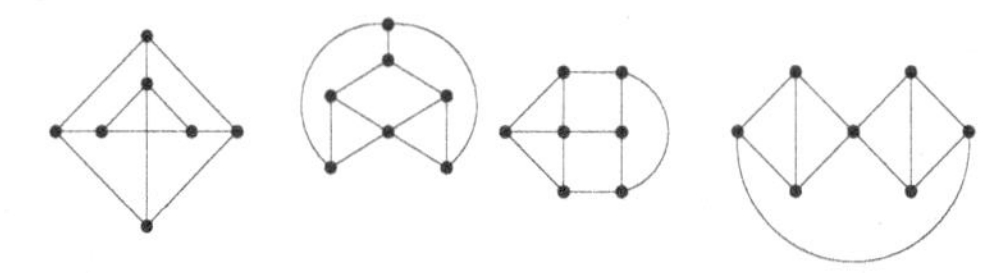

Figure 11.8: Graphs from the cone $SMET^{4,2}_7$

9. All $\{0,1\}$-valued extreme rays of $SMET_7^{4,3}$ have $H_v = K_5$, $K_{2,2,2}$; $\overline{C_7}$, $\overline{C_1+C_6}$, $\overline{C_2+C_5}$, $\overline{C_3+C_4}$, $\overline{C_1+C_3+C_3}$.

10. All $\{0,1\}$-valued extreme rays of $SMET_7^{4,4}$ have $H_v = K_6$ or $K_{2,2,2,1}$.

11. All $\{0,1\}$-valued extreme rays of $HMET_8^5$ have $H_v = K_3 (= C_3)$, $K_{2,2} (= C_4)$; C_5, C_6, C_7, C_8.

12. Some $\{0,1\}$-valued extreme rays of $SMET_8^{5,2}$ have $H_v = K_4$, $K_{2,2,1}$; $\overline{C_6}$, $\overline{C_1+C_5}$, $\overline{C_3+C_3}$, ∇C_7; four graphs depicted above for $SMET_7^{4,2}$, ∇C_6, 3-cube, or the graphs in Figure 11.9.

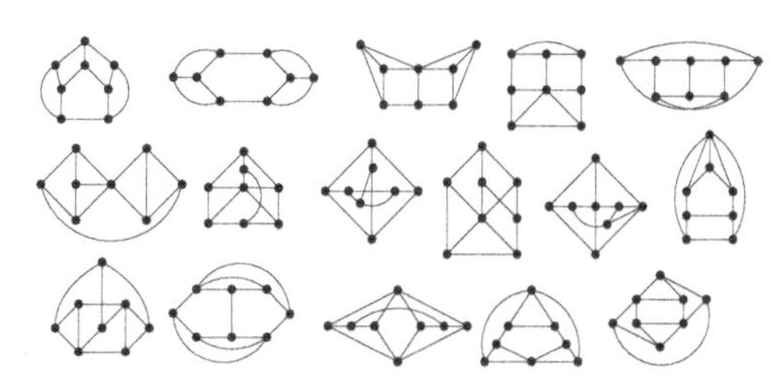

Figure 11.9: Graphs from the cone $SMET_8^{5,2}$

13. Some $\{0,1\}$-extreme rays of $SMET_8^{5,3}$ have $H_v = K_5$, $K_{2,2,2}$; $K_{4,4}$, $\overline{C_7}$, $\overline{C_1+C_6}$, $\overline{C_2+C_5}$, $\overline{C_3+C_4}$, $\overline{C_1+C_3+C_3}$; complement of 3-cube or the complement of the graphs in Figure 11.10.

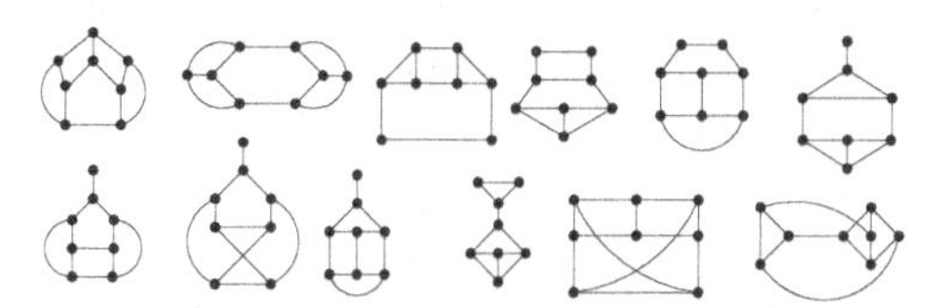

Figure 11.10: Graphs from the cone $SMET_8^{5,3}$

14. Some $\{0,1\}$-valued extreme rays of $SMET_8^{5,4}$ have $H_v = K_6$, $K_{2,2,2,1}$; $\overline{C_8}$, $\overline{C_1+C_7}$, $\overline{C_2+C_6}$, $\overline{C_3+C_5}$, $\overline{C_4+C_4}$, $\overline{C_1+C_3+C_4}$, $\overline{C_2+C_3+C_3}$.

15. All $\{0,1\}$-valued extreme rays of $SMET_8^{5,5}$ have $H_v = K_7$ or $K_{2,2,2,2}$.

The representatives of orbits of facets and extreme rays of the cones $SMET_5^{2,2}$, $SCUT_5^{2,2}$, $SMET_6^{3,3}$, $SCUT_6^{3,3}$, $SMET_6^{3,2}$, $SMET_7^{4,4}$ are presented in Tables 11.22, 11.23, 11.24, 11.25, 11.26, 11.27, respectively.

Proposition 11.7 *The number of orbits of $\{0,1\}$-valued extreme rays and the minimal number of zeros of an extreme ray are:*

$(3,5)$ *for* $HMET_5^2$	$(2,0)$ *for* $SMET_5^{2,2}$		
$(4,8)$ *for* $HMET_6^3$	$(5,3)$ *for* $SMET_6^{3,2}$	$(2,0)$ *for* $SMET_6^{3,3}$	
$(5,13)$ *for* $HMET_7^4$	$(10,7)$ *for* $SMET_7^{4,2}$	$(7,3)$ *for* $SMET_7^{4,3}$	$(2,0)$ *for* $SMET_7^{4,4}$
$(6,9)$ *for* $HMET_6^2$	$(3,0)$ *for* $SMET_6^{2,2}$		

F/O	$\overline{45}$	$\overline{35}$	$\overline{34}$	$\overline{25}$	$\overline{24}$	$\overline{23}$	$\overline{15}$	$\overline{14}$	$\overline{13}$	$\overline{12}$	Adjacency	Size	Incidence
F_1	-2	0	1	0	1	0	0	1	0	0	19	20	70
O_1	0	0	1	0	1	1	0	1	1	1	92	5	16
O_2	0	2	2	2	2	1	2	2	1	1	32	10	12
O_3	0	1	1	1	1	0	1	1	1	1	28	15	12
O_4	1	1	2	2	1	2	2	2	1	1	25	12	10
O_5	1	2	4	3	3	2	3	4	4	1	13	60	10
O_6	0	4	4	4	4	2	4	4	2	5	13	30	10

O	O_1	O_2	O_3	O_4	O_5	O_6	Adjacency	Size
O_1	4	10	12	12	36	18	92	5
O_2	5	0	3	6	12	6	32	10
O_3	4	2	2	4	12	4	28	15
O_4	5	5	5	0	5	5	25	12
O_5	3	2	3	1	2	2	13	60
O_6	3	2	2	2	4	0	13	30
Size	5	10	15	12	60	30		132

Table 11.22: Representatives of orbits of facets and extreme rays of $SMET_5^{2,2}$, followed by the representation matrix of 1-skeleton graph of $SMET_5^{2,2}$

O/F	$\overline{45}$	$\overline{35}$	$\overline{34}$	$\overline{25}$	$\overline{24}$	$\overline{23}$	$\overline{15}$	$\overline{14}$	$\overline{13}$	$\overline{12}$	Adjacency	Size	Incidence
O_1	0	1	1	1	1	0	1	1	1	1	19	15	104
O_2	0	0	1	0	1	1	0	1	1	1	19	5	134
F_1	-2	0	1	0	1	0	0	1	0	0	61	20	13
F_2	-1	-3	1	1	1	1	1	1	1	-1	22	60	11
F_3	-3	-3	3	-3	3	3	1	1	1	1	15	20	10
F_4	-1	-1	1	1	1	1	1	1	1	-3	13	30	10
F_5	-3	-3	3	1	1	1	1	3	3	-3	9	60	9
F_6	-1	-1	3	5	3	3	5	3	3	-15	9	30	9

F	F_1	F_2	F_3	F_4	F_5	F_6	Adj.	Size
F_1	13	21	6	6	9	6	61	20
F_2	7	6	2	3	2	2	22	60
F_3	6	6	0	0	3	0	15	20
F_4	4	6	0	0	2	1	13	30
F_5	3	2	1	1	2	0	9	60
F_6	4	4	0	1	0	0	9	30
Size	20	60	20	30	60	30		220

Table 11.23: Representatives of orbits of extreme rays and facets of $SCUT_5^{2,2}$, followed by the representation matrix of the ridge graph of $SCUT_5^{2,2}$

F/O	$\overline{56}$	$\overline{46}$	$\overline{45}$	$\overline{36}$	$\overline{35}$	$\overline{34}$	$\overline{26}$	$\overline{25}$	$\overline{24}$	$\overline{23}$	$\overline{16}$	$\overline{15}$	$\overline{14}$	$\overline{13}$	$\overline{12}$	Adj.	Size	Inc.
F_1	-3	0	1	0	1	0	0	1	0	0	0	1	0	0	0	29	30	594
O_1	0	0	1	0	1	1	0	1	1	1	0	1	1	1	1	650	6	25
O_2	0	1	1	1	1	0	1	1	1	1	1	1	1	1	0	449	15	24
O_3	0	3	3	3	3	0	3	3	3	3	3	3	3	3	4	93	45	18
O_4	1	1	1	2	2	2	2	2	2	1	2	2	2	1	1	57	10	18
O_5	0	2	2	2	2	1	2	2	1	1	2	2	2	2	2	56	60	17
O_6	1	1	3	2	2	2	2	3	3	1	2	3	3	1	3	51	90	18
O_7	1	1	2	2	1	2	2	2	1	1	2	2	2	2	2	30	72	15
O_8	0	3	3	3	3	2	3	3	2	4	3	3	2	4	4	27	60	16
O_9	0	4	4	4	4	2	4	4	2	5	4	4	4	5	5	23	180	15
O_{10}	1	2	4	3	3	2	3	4	4	1	3	4	4	3	4	18	360	15
O_{11}	1	3	3	4	4	4	4	4	5	2	4	4	5	2	5	18	180	15
O_{12}	3	3	3	6	6	6	6	6	6	3	6	6	6	3	7	14	60	14

O	O_1	O_2	O_3	O_4	O_5	O_6	O_7	O_8	O_9	O_{10}	O_{11}	O_{12}	Adj.	Size
O_1	5	15	30	10	50	90	60	30	90	180	60	30	650	6
O_2	6	14	39	6	36	48	24	36	72	72	72	24	449	15
O_3	4	13	2	2	4	16	8	4	8	24	8	0	93	45
O_4	6	9	9	0	6	9	0	0	0	0	18	0	57	10
O_5	5	9	3	1	0	6	6	0	6	12	6	2	56	60
O_6	6	8	8	1	4	4	0	4	4	8	4	0	51	90
O_7	5	5	5	0	5	0	0	0	5	5	0	0	30	72
O_8	3	9	3	0	0	6	0	0	3	0	3	0	27	60
O_9	3	6	2	0	2	2	2	1	0	4	1	0	23	180
O_{10}	3	3	3	0	2	2	1	0	2	2	0	0	18	360
O_{11}	2	6	2	1	2	2	0	1	1	0	0	1	18	180
O_{12}	3	6	0	0	2	0	0	0	0	0	3	0	14	60
Size	6	15	45	10	60	90	72	60	180	360	180	60		1138

Table 11.24: Representatives of orbits of facets and extreme rays of $SMET_6^{3,3}$, followed by the representation matrix of the 1-skeleton graph of $SMET_6^{3,3}$

O/F	$\overline{56}$	$\overline{46}$	$\overline{45}$	$\overline{36}$	$\overline{35}$	$\overline{34}$	$\overline{26}$	$\overline{25}$	$\overline{24}$	$\overline{23}$	$\overline{16}$	$\overline{15}$	$\overline{14}$	$\overline{13}$	$\overline{12}$	Adj.	Size	Inc.
O_1	0	0	1	0	1	1	0	1	1	1	0	1	1	1	1	20	6	125
O_2	0	1	1	1	1	0	1	1	1	1	1	1	1	1	0	20	15	108
F_1	-1	-1	0	-1	0	0	0	1	1	1	0	1	1	1	-2	25	60	17
F_2	-3	0	1	0	1	0	0	1	0	0	0	1	0	0	0	25	30	17
F_3	-1	-1	1	-1	1	1	0	0	0	0	0	0	0	0	1	14	60	14

F	F_1	F_2	F_3	Adj.	Size
F_1	12	7	6	25	60
F_2	14	5	6	25	30
F_3	6	3	5	14	60
Size	60	30	60		150

Table 11.25: Representatives of orbits of extreme rays and facets of $SCUT_6^{3,3}$, followed by the representation matrix of the ridge graph of $SCUT_6^{3,3}$

F/O	$\overline{56}$	$\overline{46}$	$\overline{45}$	$\overline{36}$	$\overline{35}$	$\overline{34}$	$\overline{26}$	$\overline{25}$	$\overline{24}$	$\overline{23}$	$\overline{16}$	$\overline{15}$	$\overline{14}$	$\overline{13}$	$\overline{12}$	Adj.	Size	Inc.
F_1	-2	0	1	0	1	0	0	1	0	0	0	1	0	0	0	43	30	5404
F_2	0	0	0	0	0	0	0	0	0	0	0	0	0	0	1	34	15	3195
O_1	0	0	0	0	0	1	0	0	1	1	0	0	1	1	1	1642	15	31
O_2	0	0	1	1	0	1	1	1	0	1	1	1	1	0	0	953	60	24
O_3	0	0	0	1	1	1	1	1	1	0	1	1	1	0	0	696	10	24
O_4	0	0	0	0	1	1	0	1	1	0	0	1	1	1	1	274	90	24
O_5	0	0	1	2	0	0	2	1	1	2	2	1	1	2	0	248	90	21
O_6	0	0	0	0	2	2	0	2	2	1	0	2	2	1	1	183	60	23
O_7	0	0	1	1	0	1	1	1	0	0	1	1	1	1	1	125	72	20
O_8	0	0	0	2	2	2	2	2	2	0	2	2	2	0	3	103	60	19
O_9	0	0	2	2	0	2	2	2	0	2	2	2	2	0	3	92	360	19
O_{10}	0	0	1	1	0	1	1	1	2	2	1	1	2	2	0	84	180	18
O_{11}	0	0	1	2	0	3	2	1	2	1	2	1	3	3	0	73	720	18
O_{12}	0	0	0	0	4	4	0	4	4	2	0	4	4	2	5	59	180	21
O_{13}	0	0	1	3	0	2	3	1	3	2	3	1	3	2	0	46	360	17
O_{14}	0	1	2	1	2	1	2	1	0	2	2	1	2	0	1	39	60	15
O_{15}	0	0	1	0	1	2	0	2	1	2	0	2	2	1	1	35	72	20
O_{16}	0	0	3	1	1	1	1	2	3	3	1	3	2	3	0	32	360	16
O_{17}	0	1	2	3	4	1	4	2	0	4	4	4	2	0	2	30	360	15
O_{18}	0	0	1	0	2	4	0	3	3	2	0	3	4	4	1	23	360	20
O_{19}	0	1	1	4	4	2	5	5	0	2	5	5	2	0	3	23	360	15
O_{20}	0	1	2	1	2	3	2	1	0	3	2	1	3	0	3	23	180	15
O_{21}	0	1	1	2	2	0	3	3	2	1	3	3	2	1	0	23	90	15
O_{22}	0	0	1	2	0	1	2	1	2	1	2	1	2	2	0	22	720	16
O_{23}	0	0	2	6	0	4	6	2	6	4	6	2	6	7	0	22	720	16
O_{24}	0	0	1	2	0	2	2	1	2	1	2	1	2	1	0	22	360	16
O_{25}	0	0	2	2	0	3	2	2	1	1	2	2	3	3	0	22	360	16
O_{26}	0	0	2	3	0	2	3	2	1	3	3	2	1	3	0	22	360	16
O_{27}	0	0	2	3	2	5	3	2	5	5	3	3	3	0	0	22	360	17
O_{28}	0	0	2	2	0	1	2	2	1	1	2	2	2	2	0	22	360	16
O_{29}	0	0	2	6	0	7	6	2	6	4	6	2	6	4	0	22	360	16
O_{30}	0	0	2	2	0	5	2	2	4	4	2	2	4	4	0	22	180	16
O_{31}	0	0	5	1	1	2	1	4	3	4	1	5	5	4	0	20	720	16
O_{32}	0	0	1	1	1	3	1	2	2	1	1	2	3	3	0	20	720	16
O_{33}	0	0	3	3	1	5	3	2	5	5	3	3	2	1	0	20	720	16
O_{34}	0	0	5	3	2	3	3	3	2	0	3	5	5	1	1	20	360	16
O_{35}	0	0	1	2	1	3	2	2	2	0	2	2	3	3	0	18	720	16
O_{36}	0	0	2	2	1	3	2	1	3	3	2	2	2	0	0	18	360	16
O_{37}	0	0	1	2	1	1	2	2	2	0	2	2	2	2	0	18	360	16
O_{38}	0	1	2	5	5	2	6	3	0	3	6	5	1	0	6	18	360	15
O_{39}	0	1	4	3	6	3	4	2	4	0	4	6	0	6	2	18	360	15
O_{40}	0	1	2	1	2	1	2	4	0	2	2	4	2	0	4	18	120	15

Table 11.26: Representatives of orbits of facets and extreme rays of $SMET_6^{3,2}$

F/O	$\overline{67}$	$\overline{57}$	$\overline{56}$	$\overline{47}$	$\overline{46}$	$\overline{45}$	$\overline{37}$	$\overline{36}$	$\overline{35}$	$\overline{34}$	$\overline{27}$	$\overline{26}$	$\overline{25}$	$\overline{24}$	$\overline{23}$	$\overline{17}$	$\overline{16}$	$\overline{15}$	$\overline{14}$	$\overline{13}$	$\overline{12}$	Adj.	Size	Inc.
F_1	-4	0	1	0	1	0	0	1	0	0	0	1	0	0	0	0	1	0	0	0	0	41	42	21363
O_1	0	0	1	0	1	1	0	1	1	1	0	1	1	1	1	0	1	1	1	1	1	17163	7	36
O_2	0	1	1	1	1	0	1	1	1	1	1	1	1	1	0	1	1	1	1	1	1	1486	105	30
O_3	0	3	3	3	3	0	3	3	3	3	3	3	3	3	4	3	3	3	3	4	4	1314	105	26
O_4	0	2	2	2	2	0	2	2	2	2	2	2	2	2	1	2	2	2	2	1	1	1228	105	32
O_5	0	2	2	2	2	1	2	2	1	2	2	2	2	1	2	2	2	2	2	1	1	343	252	30
O_6	0	8	8	8	8	0	8	8	8	8	8	8	8	8	4	8	8	8	8	4	9	294	315	26
O_7	0	4	4	4	4	2	4	4	2	2	4	4	4	4	4	4	4	4	4	4	5	238	210	24
O_8	0	4	4	4	4	2	4	4	2	5	4	4	4	5	5	4	4	4	5	5	3	153	630	24
O_9	1	2	4	3	3	2	3	4	4	1	3	4	4	3	4	3	4	4	3	4	2	120	1260	26
O_{10}	1	1	2	2	1	2	2	2	1	2	2	2	2	1	2	2	2	2	2	1	1	112	360	28
O_{11}	2	2	4	4	2	5	4	4	4	4	4	4	5	5	2	4	4	5	5	2	5	111	1260	24
O_{12}	1	1	3	2	2	2	2	3	3	1	2	3	3	1	3	2	3	3	2	3	3	108	630	24
O_{13}	0	8	8	8	8	4	8	8	4	8	8	8	8	4	8	8	8	8	8	4	9	95	1260	24
O_{14}	0	8	8	8	8	10	8	8	10	10	8	8	10	10	11	8	8	4	4	8	8	93	630	22
O_{15}	0	9	9	9	9	10	9	9	10	10	9	9	3	6	9	9	9	9	6	3	9	70	1260	24
O_{16}	1	1	1	2	2	2	2	2	2	1	2	2	2	1	1	2	2	2	2	2	2	63	70	24
O_{17}	2	2	4	4	2	4	4	4	2	2	4	4	4	4	4	4	4	4	4	4	5	61	252	22
O_{18}	2	4	8	6	6	4	6	8	8	2	6	8	8	6	8	6	8	8	6	8	9	49	1260	22
O_{19}	0	8	8	8	8	11	8	8	11	11	8	8	11	11	11	8	8	6	6	6	6	47	105	24
O_{20}	3	3	8	4	8	8	5	5	5	4	5	8	8	8	3	5	8	8	8	3	8	46	630	26
O_{21}	2	2	6	4	4	4	4	6	6	2	4	6	6	3	6	4	6	6	3	6	5	44	1260	25
O_{22}	1	2	4	3	3	3	3	4	3	3	3	4	4	1	4	3	4	4	2	3	4	40	2520	24
O_{23}	0	3	3	3	3	2	3	3	2	4	3	3	2	4	4	3	3	3	4	4	4	40	420	22
O_{24}	4	4	8	8	4	8	8	8	4	8	8	8	8	4	8	8	8	8	8	4	9	36	2520	22
O_{25}	2	6	9	8	8	4	8	8	8	8	8	9	9	4	6	8	9	9	8	2	9	35	2520	24
O_{26}	2	6	9	8	8	6	8	8	6	8	8	9	9	2	8	8	9	9	8	2	9	35	840	24
O_{27}	2	4	8	6	6	6	6	8	9	4	6	8	9	4	9	6	8	9	4	9	9	31	840	23
O_{28}	1	5	5	6	6	6	6	6	8	4	6	6	8	4	8	6	6	8	4	8	8	31	420	23
O_{29}	10	10	10	3	6	9	9	6	9	9	9	9	3	9	9	9	9	9	9	3	6	28	5040	22
O_{30}	12	12	12	12	12	8	12	16	16	4	4	16	16	12	16	8	17	16	12	16	16	24	5040	21
O_{31}	12	12	12	12	15	3	12	9	9	15	6	16	12	15	15	6	16	12	15	15	16	24	2520	21
O_{32}	1	3	3	4	4	4	4	4	5	2	4	4	5	2	5	4	4	5	4	5	5	24	1260	21
O_{33}	12	12	12	12	12	3	12	6	9	9	6	14	12	12	12	6	14	12	12	12	14	24	1260	21
O_{34}	12	12	25	16	16	16	16	24	24	8	16	24	24	8	24	8	24	24	16	24	24	24	1260	21
O_{35}	12	12	28	12	28	28	18	18	18	18	18	27	27	27	9	18	27	27	27	9	27	24	420	21
O_{36}	12	12	22	12	22	22	18	18	18	18	18	18	18	18	9	18	18	18	18	9	9	24	140	21
O_{37}	3	3	3	6	6	6	6	6	6	3	6	6	6	3	7	6	6	6	6	7	7	20	420	20

Table 11.27: Representatives of orbits of facets and extreme rays of $SMET_7^{4,4}$

11.4 Some special cases of parameters

The case $s \geq m+1$

Proposition 11.8 *For $SMET_n^{m,s}$ it holds:*
(i) if $s > m+1$, the cone $SMET_n^{m,s}$ collapses to 0;
(ii) if $s = m+1$, the cone $SMET_n^{m,m+1}$ collapses to the half-line of all non-negative multiples of the vector of all ones;
(iii) if $m \leq s < m+1$, the non-negativity inequalities are implied by the (m,s)-simplex inequalities.

Proof. The (m,s)-simplex inequality $sST_{x_1 \dots x_{m+1}, x_{m+2}}$ has the form

$$\sum_{j=1}^{m+1} d(x_1, \dots, x_{j-1}, x_{j+1}, \dots, x_{m+2}) - s \times d(x_1, \dots, x_{m+1}) \geq 0.$$

If we denote $T = \{x_1, \dots, x_{m+2}\}$, $d_{x_i} = d(x_1, \dots, x_{i-1}, x_{i+1}, \dots, x_{m+2})$, and $\Sigma_T = \sum_{i=1}^{m+2} d_{x_i}$, then the (m,s)-simplex inequality $sST_{T;x_{m+2}}$ can be rewritten as $(s+1)d_{x_{m+2}} \leq \Sigma_T$.

Let take the fixed support $T = \{x_1, \dots, x_{m+2}\}$; by summing all (m,s)-simplex inequalities $(s+1)d_{x_i} \leq \Sigma_T$, one obtain that

$$(s+1)\Sigma_T \leq (m+2)\Sigma_T.$$

By non-negativity inequalities it holds $\Sigma_T \geq 0$, and we have that, for $s > m+1$, only the vector of all zero is possible as d, i.e., the cone collapses to zero and one obtain (i).

Now, summing the (m, s)-inequalities over all $i \neq k$, we obtain

$$(s+1)(\Sigma_T - d_{x_k}) \leq (m+1)\Sigma_T,$$

i.e., $(s+1)d_{x_k} \geq (s-m)\Sigma_T$. This inequality and the (m, s)-simplex equality gives

$$(s-m)\Sigma_T \leq (s+1)d_{x_k} \leq \Sigma_T. \qquad (*)$$

For $s = m+1$, this inequality imply $(m+2)d_{x_k} = \Sigma_T$, i.e., d is a positive multiple of the vector of all ones and one obtain (ii).

Let $m \leq s \leq m+1$. Then, the inequalities $(*)$ imply the inequality $\Sigma_T \geq 0$ and, therefore, the non-negativity inequalities. This gives (iii).

□

We will now assume that $0 < s < m+1$.

The case $n = m+2$

In general, $HCUT^m_{m+2} = HMET^m_{m+2}$ for any $m \geq 2$. This cone has $\binom{m+2}{2}$ extreme rays, all in the same orbit, represented by $\alpha_{\{1,2\},\{3\},\dots,\{m+2\}}$, i.e., by any vector of length $m+2$, consisting of two ones and m zeros.

The 1-skeleton graph of $HCUT^m_{m+2}$ is the Johnson graph $J(m+2, 2)$, called also the *triangular graph* $T(m+2)$, which is is the line graph $L(K_{m+2})$ of the complete graph K_{m+2}. It is also the 1-skeleton graph of the $(m+1)$-polytope (called *ambo-*α_{m+1}), obtained from the $(m+1)$-simplex as the convex hull of the mid-points of all its edges; e.g., $T(4)$ is the skeleton of the octahedron, $T(5)$ is the complement of the Petersen graph. In general, $T(m), m \geq 2$, has diameter 2; moreover, it is a strongly regular graph.

The cone $HCUT^m_{m+2}$ has two orbits, F_1 and F_2, of facets, containing $m+2$ facets each and represented by the m-simplex facet $T_{1\dots(m+1),(m+2)}$ and by the non-negativity facet $NN_{1\dots(m+1)}$. The orbit F_1 consists of simplex cones, i.e., facets from this orbit are incident to $m+1$ linearly independent extreme rays.

Any non-negativity inequality defines the cone $HCUT^{m-1}_{m+1} = HMET^{m-1}_{m+1}$, i.e., it becomes equality on this smaller cone. So, a non-negativity inequality is non-facet only for $m = 1$, and it is a simplex cone only for $m = 2$; in general, a non-negativity inequality is incident to $\binom{m+1}{2}$ extreme rays.

The ridge graph is $\overline{K}_{m+2}$ on F_1; on F_2 it is $\overline{K}_4$ for $m = 2$, and K_{m+2} for $m \geq 3$. Finally, the $m+2$ pairs $(T_{\bar{i},i}, N_{\bar{i}})$ of m-simplex and non-negativity facets are the only non-edges for pairs of facets from different orbits.

The cone HM^m_{m+2} is a simplex cone of dimension $(m+2)$; so $G_{HM^m_{m+2}} = G^*_{HM^m_{m+2}} = K_{m+2}$. Its facets are all $\{1, -1\}$-valued $(m+2)$-vectors with only one -1; its generators are all $\{1-m, 1\}$-valued $(m+2)$-vectors with only one $1-m$.

Notice, that $HCUT^1_3 = HM^1_3 = CUT_3 = MET_3$.

Conjecture 11.1 *Let $m > 1$ and $n \geq m+2$; then $HCUT^m_n = HMET^m_n$ only if $n = m+2$.*

Theorem 11.1 *For the cone $SMET^{m,s}_{m+2}$ it holds:*

(i) it has only one orbit of extreme rays; each extreme ray contains a vector with $\lfloor s \rfloor + 1$ components 1, one component $s - \lfloor s \rfloor$, and the other ones 0; all such vectors appear on different extreme rays;

(ii) if s is integer, the 1-skeleton graph of $SMET^{m,s}_{m+2}$ is the Johnson graph $J(m+2, s+1)$; if s is not integer, then two extreme rays are adjacent if and only if they either have the same support, or they differ only by the position of the value $s - \lfloor s \rfloor$ in the associated vector;

(iii) the ridge graph of $SMET^{m,s}_{m+2}$ has the following form:

$$G^*_{SMET^{m,s}_{m+2}} = \begin{cases} K_{m+2}, & \textit{if} \quad m \le s < m+1; \\ K_{(m+2)\times 2}, & \textit{if} \quad 1 < s < m-1; \\ K_{(m+2)\times 2} - K_{m+2}, & \textit{if} \quad s = 1 < m-1 \textit{ or } 1 < s = m-1; \\ K_{(2+2)\times 2} - 2K_{2+2}, & \textit{if} \quad s = 1, \textit{ and } m = 2. \end{cases}$$

Proof. (i) The cone is defined by $\sum d_i \ge (s+1)d_k \ge 0$ for all $1 \le k \le m+2$.

In fact, $SMET^{m,s}_{m+2}$ has dimension $m+2$ and two orbits of facets, each consisting of $m+2$ linearly independent members: (m,s)-simplex facets - all $(1,-s)$-vectors of length $m+2$ with exactly one $-s$, and non-negativity facets - all $\{0,1\}$-vectors of length $m+2$ with exactly one 1.

Fix an extreme ray d of the cone. It lies on $m+1$ linearly independent facets. Without loss of generality, one can suppose that it lies on (m,s)-simplex facets with $-s$ on positions $1, ..., p$ only, and on non-negativity facets with 1 on positions $p+2, ..., m+2$ only. So, $d_1 = ... = d_p = t$, $d_{p+2} = ... = d_{m+2} = 0$, and $(-s+p-1)t + d_{p+1} = 0$.

The validity of the (m,s)-simplex facet with $-s$ on position $p+1$ and of the non-negativity facets imply $t > 0$, $d_{p+1} \ge 0$, and $pt - sd_{p+1} \ge 0$. This yield the inequalities $p-1 \le s \le p$; if s is integer, then the values $p = s+1$ and $p = s$ yield the same vector d; so, one can assume, in general, that $p = \lfloor s \rfloor + 1$.

(ii) If s is integer, the vertices of the 1-skeleton graph are the same as for the Johnson graph $J(m+2, s+1)$. Let us see now that it is this Johnson graph, i.e., two extreme rays are adjacent if and only if the rank of the set of facets, containing them both, is m. If corresponding vectors have, say, i common ones, then this rank is $m + 2i - 2s$ (namely, i (m,s)-simplex facets, and $m+i-2s$ non-negativity facets). The maximum of this number is m and is attained exactly for $i = s$. The diameter of 1-skeleton graph is $\min\{s+1, m-s+1\}$.

If s is non-integer, then each extreme ray belongs to exactly $m+1$ (linearly independent) facets: $\lfloor s \rfloor + 1$ (m,s)-simplex facets with $-s$ on position, where the ray has 1, $m - \lfloor s \rfloor$ non-negativity facets with 1 on position, where the ray has 0. So, the adjacency of extreme rays follows.

(iii) First, non-negativity inequalities are not facets if and only if $m \le s < m+1$. Two (m,s)-simplex facets are adjacent, unless $s = 1 < m$. Two non-negativity facets are adjacent, unless $s \ge m-1$. Fixed (m,s)-simplex and non-negativity facets are adjacent if and only if there is no position, in which the first has $-s$ and the second has 1. So, the ridge graph follows.

□

Remark. Any $SMET^{1,s}_3$ has only one orbit of extreme rays, represented by $(0, s, 1)$ for $0 < s \le 1$, and by $(s-1, 1, 1)$ for $1 \le s \le m+1 = 2$. Both, the 1-skeleton graph and the ridge graph of this cone, are C_6 for $0 < s < 1$, and K_3 for $1 \le s < 2$.

The number of extreme rays (number of orbits) of $SMET^{1,s}_n$ is 54(5), 2900(35), 988105(1567) for $s = \frac{1}{2}$ and $n = 4, 5, 6$; it is 25(4), 1235(24), 530143(890) for $s = \frac{3}{2}$ and $n = 4, 5, 6$.

The case $n = m + 3$

The dimension of the supermetric cone $SMET^{m,s}_{m+3}$ is $\binom{n}{n-2}$, i.e., the same as of MET_n.

This correspondence allows us to replace the representation graph G_v by a simpler graph: any facet or extreme ray of the cone $SMET^{m,s}_n$ is given by a vector, say, v, indexed by $(m+1)$-subsets of V_n. It can be seen also as a function on 2-subsets of V_n, which are complements of $(m+1)$-subsets of V_{m+3}. So, to every $\{0,1\}$-valued extreme ray of $SMET^{m,s}_n$ one can associate a set of pairs (ij), and this set of pairs is edge set of a graph H_v, such that the graph G_v is the line graph of H_v. If some vertices are isolated, then we remove them.

Conjecture 11.2 *In terms of graph H_v, associated to $\{0,1\}$-valued vectors, we have for the cone $SMET^{m,s}_{n=m+3}$:*

(i) for $s = m$, there are two orbits of $\{0,1\}$-valued extreme rays: K_{n-1}, and $K_n - \lfloor\frac{n}{2}\rfloor K_2$;

(ii) for $s = m-1$, besides zero-extensions from $SMET^{m-1,m-1}_{n-1}$, all $\{0,1\}$-valued extreme rays for $m = 2, 3, 4$ and some for $m = 5$ have H_v being the complement of the union of disjoint circuits with lengths partitioning $n = m+3$, but lengths-vectors (the lengths are not increasing) $(..., 1, 1)$, $(..., 2, 1)$, $(..., 2, 2)$ and $(n \leq 4)$, $(n-1 \leq 4, 1)$, $(n-2 \leq 4, 2)$ are excluded.

We checked this conjecture up to $n = 12$.

Conjecture 11.3 *The complete list of extreme rays of $HMET^m_{m+3}$ consists of:*

(i) $\{0,1\}$-valued extreme rays v with $H_v = C_i$ for $3 \leq i \leq m+3$;

(ii) $\{0,1,2\}$-valued extreme rays v, represented by $H_v = C_{1,2,\dots,i} + P_{i,i+1,\dots,k} + C_{k,k+1,\dots,j}$ (with $3 \leq i < k$, $j \geq k+2$, and $i+k-1 \leq j \leq n$), having value 2 on edges of the path (so, besides zero-extensions, they are those with $j = n$, i.e., with all $3 \leq i < k \leq n-i+1$).

Another interesting fact is that the minimal incidence number for extreme rays of $SMET^{m,2}_{m+3}$ is $\binom{m+3}{2}$ and not (a priori possible) minimum $\binom{m+3}{2} - 1$, which occurs for other supermetric cones.

Conjecture 11.4 *The number of extreme rays, the number of orbits, and the diameter of $G_{SMET^{2,s}_5}$ are:*

$(1170, 16, 5)$, *if* $s \in]0, 1[$	$(37, 3, 2)$, *if* $s = 1$	$(2462, 35, 5)$, *if* $s \in]1, \frac{3}{2}[$	$(1442, 25, 4)$, *if* $s = \frac{3}{2}$
$(2102, 31, 5)$, *if* $s \in]\frac{3}{2}, \frac{5}{3}[$	$(1742, 28, 5)$, *if* $s = \frac{5}{3}$	$(1862, 30, 5)$, *if* $s \in]\frac{5}{3}, 2[$	$(132, 6, 2)$, *if* $s = 2$

This conjecture was verified for about one hundred values of s.

The case $m = s$ and $n \geq m + 4$

As we saw above, the case $n = m+2$ was completely solvable and the case $n = m+3$ was computationally reasonable. But already for $n = m+4$ the combinatorial explosion happens (see, for example, linear descriptions of the cones $HCUT^2_6$, $SMET^{2,2}_6$, and $SCUT^{2,2}_6$).

A remarkable feature of the cone $SCUT^{1,1}_n = CUT_n$ is that its 1-skeleton graph is a complete graph. We checked that this happens also for $SCUT^{m,s}_n$ with (m, s, n)=$(2, 2, 5)$, $(3, 3, 6)$, $(4, 4, 7)$, $(2, 2, 6)$, $(2, 2, 7)$, but not for $(5, 5, 8)$, $(3, 3, 7)$, and $(3, 2, 6)$.

It is conjectured (see [DeLa97]) that every facet of CUT_n is adjacent to a triangle facet; this conjecture, if true, would imply that the diameter of $G^*_{CUT_n}$ is 3 or 4. The cone $SCUT_7^{4,4}$ is the smallest example, when $SCUT_n^{m,s}$ has facets, which are *not adjacent* to any of (m,s)-simplex facets.

The cone $SMET_n^{m,m}$ shares with $SMET_n^{1,1} = MET_n$ the property that the non-negativity inequality is redundant. We expect that it shares also the following conjectured property.

Conjecture 11.5 *The cone $SMET_n^{m,m}$ has extreme rays without components zeros.*

This conjecture holds for the cones $MET_n = SMET_n^{1,1}$, $SMET_5^{2,2}$, $SMET_6^{3,3}$, $SMET_7^{4,4}$, $SMET_6^{2,2}$, $SMET_7^{3,3}$, and $SMET_7^{2,2}$.

Proposition 11.9 *Given an $(m+1)$-subset of $V_n = \{1,2,...,n\}$, the vector v_S, defined by*

$$v_S = \begin{cases} 1, & \text{if} \quad 1 \in S, \\ 0, & \text{if} \quad 1 \notin S, \end{cases}$$

is an extreme ray of the cone $SMET_n^{m,m}$. Its representation graph G_{v_S} is the Johnson graph $J(n-1,m)$, and the orbit, represented by v_S, has size n.

Proof. If $T = \{x_1, ..., x_{m+2}\}$ is an $(m+2)$-set, then v_S is incident to the (m,m)-simplex facet $(s+1)d_{x_i} \leq \Sigma_T$ if and only if $1 \notin T$, or $1 \in T$, and $x_i \neq 1$. Let $d \in SMET_n^{m,m}$ be a ray, incident to all those facets.

If $1 \notin T$, then, by summing all (m,m)-simplex equalities with support in T, one obtains $(m+1)\Sigma_T = (m+2)\Sigma_T$, i.e. $\Sigma_T = 0$. So, $d(S) = 0$, if $1 \notin S$.

If $1 \in T$, we get $(m+1)d_x = \Sigma_T$ for all $x \in T \backslash \{1\}$. Considering all $1 \in T$, we obtain $d(S) = \lambda$ for all sets S with $1 \in S$. So, $d = \lambda v$, which proves the result.

□

11.5 Theorems and conjectures for general case

The group $Sym(n)$ acts in an obvious way on V_n and so, on $SMET_n^{m,s}$, which proves that $Sym(n)$ is a subgroup of the full symmetry group of $SMET_n^{m,s}$.

In order to find the symmetry group of $SMET_n^{m,s}$, one should consider the group of automorphisms of the graphs $G_{SMET_n^{m,s}}$ and $G^*_{SMET_n^{m,s}}$. If one of those groups is equal to $Sym(n)$, then the full symmetry group of the cone $SMET_n^{m,s}$ is $Sym(n)$. If those groups are bigger than $Sym(n)$, then their realizability, as group of linear symmetries, can be checked by elementary Linear Algebra.

Using this method, we checked by computer, that the symmetry group of MET_4 is $Sym(4) \times Sym(3)$.

We checked by computer, using the *nauty* program ([McKa15]), for $(m,s,n) = (2,2,5)$, $(2,1,1)$, $(2,1,6)$, $(3,2,5)$, $(3,3,6)$, $(3,1,6)$, $(3,2,6)$, $(4,1,7)$ and for MET_n with $5 \leq n \leq 14$, the validity of the following conjecture.

Conjecture 11.6 *The symmetry group of $SMET_n^{m,s}$ is $Sym(n)$, the only exception being $MET_4 = SMET_4^{1,1}$, for which it is $Sym(4) \times Sym(3)$.*

By analogy with the binomial coefficient $\binom{n}{m}$, let us denote by $\binom{A}{m}$ the *set of all m-subsets of the set A*.

Definition 11.1 *(i) For any function d on the set* $\binom{V_n}{m+1}$ *one can define a function* d^{ze} *on the set* $\binom{V_{n+1}}{m+2}$, *called zero-extension, by*

$$d^{ze}(S) = \begin{cases} 0, & \text{if } S \subset V_n, \\ d(S \cap V_n), & \text{otherwise,} \end{cases}$$

for any $(m+2)$*-subset S of* V_{n+1}*;*

(ii) For any function d on the set $\binom{V_n}{m+1}$ *one can define a function* d^{vs} *on the set* $\binom{V_{n+1}}{m+1}$, *called vertex-splitting (of the vertex n into two vertices n and* $n+1$*), by*

$$d^{vs}(S) = \begin{cases} 0, & \text{if } \{n, n+1\} \subset S, \\ d(S), & \text{if } S \subset V_n, \\ d((S \backslash \{n+1\}) \cup \{n\}), & \text{if } n+1 \in S \text{ and } n \notin S, \end{cases}$$

for any $(m+1)$*-subset S of* V_{n+1}*.*

For example, for the cut semimetrics $\delta_{\{1\},\{2,3,4\}}$ and $\delta_{\{1,2\},\{3,4\}}$ (which are representatives of two orbits of extreme rays of MET_4), their zero-extensions are vectors

$\overline{45}$	$\overline{35}$	$\overline{34}$	$\overline{25}$	$\overline{24}$	$\overline{23}$	$\overline{15}$	$\overline{14}$	$\overline{13}$	$\overline{12}$
0	0	1	0	1	1	0	0	0	0
0	0	0	0	1	1	0	1	1	0

(which are extreme rays of $HMET_5^2$), and their vertex-splittings are vectors

12	13	14	15	23	24	25	34	35	45
1	1	1	1	0	0	0	0	0	0
0	1	1	1	1	1	1	0	0	0

i.e., $\delta_{\{1\},\{2,3,4,5\}}$ and $\delta_{\{1,2\},\{3,4,5\}}$ (which are extreme rays of MET_5).

The representation graph $G_{d^{ze}}$ is equal to G_d for any $\{0,1\}$-valued vector. The representation graph $G_{d^{vs}}$ is obtained from G_d by splitting each vertex, associated to a set $S \subset V_n$, with $n \in S$, in two vertices: one for the subset S, and another one for the subset $(S \backslash \{n\}) \cup \{n+1\}$.

The zero-extension of the partition m-hemimetric $\alpha_{S_1,\dots,S_{m+1}}$ is the partition $(m+1)$-hemimetric $\alpha_{S_1,\dots,S_{m+1},\{n+1\}}$, while its vertex-splitting is the partition m-hemimetric $\alpha_{S_1,\dots,S_p \cup \{n+1\},\dots,S_{m+1}}$, if $n \in S_p$.

Theorem 11.2 *For the cones* $SMET_n^{m,s}$ *and* $HMET_n^m$ *it holds:*

(i) zero-extension of any extreme ray of $SMET_n^{m,s}$ *is an extreme ray of* $SMET_{n+1}^{m+1,s}$*;*

(ii) vertex-splitting of any extreme ray of $HMET_n^m$ *is an extreme ray of* $HMET_{n+1}^m$*.*

Proof. (i) If d is an extreme ray of $SMET_n^{m,s}$, then one can check easily the validity of (m,s)-simplex inequalities and non-negativity inequalities for d^{ze}. The ray d^{ze} is incident to the non-negativity facet NN_A, if $A \subset V_n$, and to the (m,s)-simplex facet $sST_{A,i}$, if $n+1 \in A$, and d is incident to $sST_{A \backslash \{n+1\},i}$.

Assume now that e belongs to $SMET_{n+1}^{m+1,s}$, and is incident to all facets incident to d^{ze}. Then one obtain $e(A) = 0$, if $A \subset V_n$.

So, the restriction of e to the subsets, containing $n+1$, is identified to a function on $\binom{V_n}{m+1}$, which is incident to the facets, incident to d. We get $e = \lambda d^{ze}$ with $\lambda \geq 0$ by non-negativity.

(ii) If d is an extreme ray of $HMET_n^{m,s}$, then one can check easily the validity of (m,s)-simplex inequalities and non-negativity inequalities for d^{vs}. The ray d^{vs} is incident to the non-negativity facet NN_A, if $\{n, n+1\} \subset A$, and to the (m,s)-simplex facet $sST_{A,i}$, if $A \subset V_n$, and d is incident to $sST_{A,i}$.

Assume now, that e belongs to $HMET_{n+1}^{m,s}$, and is incident to all facets incident to d^{vs}. Then one obtain $e(A) = 0$, if $\{n, n+1\} \subset A$. Now, if $\{n, n+1\} \subset T$, then, applying m-simplex inequalities $ST_{T,n+1}$ and $ST_{T,n}$, we get $e(T \backslash \{n\}) = e(T \backslash \{n+1\})$.

Since the restriction of e on V_n yields a vector, which is incident to all facets, incident to d, one obtain (since d is an extreme ray), that the restriction $e_{|V_n}$ is a multiple of d. So, we get $e_{|V_n} = \lambda d$ with $\lambda \geq 0$. Above equalities yield $e = \lambda d^{vs}$.

□

Remind, that the same result holds for the similar notion of vertex-splitting of an ray in $QMET_n$ (see [DDP03]).

We will suppose from now that $n \geq m+3$.

Theorem 11.3 *Partition m-hemimetrics are extreme rays of $HMET_n^m$.*

Proof. Any ray $\alpha_{S_1, S_2, \ldots, S_{m+1}}$ is a function on $\binom{V_n}{m+1}$. Using symmetry and above remark, it can be viewed as a vertex-splitting of the ray $\alpha_{\{1\},\{2\},\ldots,\{m+1\}}$ on $\binom{V_{m+1}}{m+1}$. The hemimetric cone $HMET_{m+1}^m$ has dimension one and the vector $\alpha_{\{1\},\{2\},\ldots,\{m+1\}}$ generates it; so, applying Theorem 11.2, we obtain the result.

□

Conjecture 11.7 *In the 1-skeleton graph of $HCUT_n^m$, two partition m-hemimetrics are non-adjacent if and only if, up to permutation, corresponding $(m+1)$-partitions can be written as $\langle S_1, \ldots, S_{m+1} \rangle$ and $\langle S'_1, \ldots, S'_{m+1} \rangle$, which differ only by $S_1, S_2, S_3 = A \cup B, C, D$ and $S'_1, S'_2, S'_3 = A, B, C \cup D$ for some disjoints sets A, B, C, D.*

This conjecture holds for $n = m+2$ and it was checked also for $(m,n) = (2,5)$, $(3,6)$, $(4,7)$, $(5,8)$, $(6,9)$, $(2,6)$, $(3,7)$, $(4,8)$, $(2,7)$.

If this conjecture holds, it will imply that the 1-skeleton graph of $HCUT_n^m$ has diameter 2, since any two non-adjacent partitions m-hemimetrics are both adjacent to partition m-hemimetric $\alpha_{A \cup C, B, D, S_4, \ldots, S_{m+1}}$.

Conjecture 11.8 *The two partition m-hemimetrics $\alpha_{S_1, \ldots, S_{m+1}}$ and $\alpha_{T_1, \ldots, T_{m+1}}$ on V_n are non-adjacent in the 1-skeleton graph of $HCUT_n^m$ if and only if there exist six different subsets S_i, S_j, S_k and $T_{i'}, T_{j'}, T_{k'}$, such that $S_i \cup S_j = T_{k'}$, and $S_k = T_{i'} \cup T_{j'}$.*

The conjecture holds for $m = 1$: all cut semimetrics are adjacent. It holds for $n - m = 2$: we have the graph $J(m+2, 2)$. It also holds for $(m,n) = (2,5)$ and $(3,6)$.

Conjecture 11.9 *The two partition m-hemimetrics $\alpha(S_1, \ldots, S_{m+1})$ and $\alpha(T_1, \ldots, T_{m+1})$ on V_n are non-adjacent in the 1-skeleton graph of $HCUT_n^m$ if and only if there exist six different subsets S_i, S_j, S_k and $T_{i'}, T_{j'}, T_{k'}$, such that $S_i \cup S_j = T_{k'}$, and $S_k = T_{i'} \cup T_{j'}$.*

The conjecture holds for $m = 1$: all cut semimetrics are adjacent. It holds for $n - m = 2$: we have the graph $J(m+2, 2)$. It also holds for $(m, n) = (2, 5)$ and $(3, 6)$.

Conjecture 11.10 *The extreme rays of $HMET_n^m$ include:*
(i) any ray whose G-graph is an G-graph of an extreme ray of $HMET_{n-1}^{m-1}$;
(ii) every $\{0,1\}$-valued extreme ray of $HMET_{m+3}^m$ with G-graph C_i, $3 \leq i \leq m+3$.

Conjecture 11.11 *For the extreme rays of $HMET_n^m$ it holds:*
(i) adjacency rule in $HMET_n^m$ is the same as above for the first orbit, represented by $\alpha_{\{1\},\{2\},\dots,\{m\},\{m+1,\dots,n\}}$;
(ii) two elements of the second orbit, represented by $\alpha_{\{1\},\{2\},\dots,\{m-1\},\{m,m+1\},\{m+2,\dots,n\}}$, are non-adjacent if and only if the rule as above is completed by the condition

$$rank(\inf\{a,b\}) - (m+1) = (m+1) - rank(\sup\{a,b\}) > 1;$$

rank here is the number of parts, while $\inf\{a,b\}$ *and* $\sup\{a,b\}$ *are join and union operation in the lattice of partitions.*

Theorem 11.4 *For the facets of $SMET_n^{m,s}$ it holds:*
(i) the non-negativity facet $NN_{A\setminus\{i\}}$ is non-adjacent to the (m,s)-simplex facet $sST_{A,i}$ in the cone $SMET_n^{m,s}$;
(ii) two (m,s)-simplex facets are non-adjacent if they have the same support;
(iii) for $m-1 \leq s < m$, the non-negativity facets NN_A and NN_B are non-adjacent, if $|A \cap B| = m$.

Proof. (i) If d is an m-hemimetric, satisfying $d(A\setminus\{i\}) = 0$, and

$$s \times d(A\setminus\{i\}) = \sum_{k \in A\setminus\{i\}} d(A\setminus\{k\}),$$

then we have $d(A\setminus\{k\}) = 0$ for all $k \in A\setminus\{i\}$, and d lies in a space of insufficient rank.

(ii) Assume that d is incident to the facets $sST_{T,i}$ and $sST_{T,j}$ with T being a subset of V_n of size $m+2$, and $i, j \in T$. Then we have $2d_i = \Sigma_T$, and $2d_j = \Sigma_T$, which implies $d_i = d_j$, and $0 = \sum_{k \in T\setminus\{i,j\}} d(T\setminus\{k\})$. So, $d(T\setminus\{k\}) = 0$, and the rank is again too low.

(iii) If $|A \cap B| = m$, then write $A \cup B = (A \cap B) \cup \{i, j\}$, and set $d((A \cup B)\setminus\{k\}) = d_k$ for $k \in A \cup B$. The non-negativity inequalities NN_A and NN_B have the form $d_i \geq 0$ and $d_j \geq 0$. Let d lie on NN_A and NN_B. Then $d_i = d_j = 0$. By summing (m,s)-simplex inequalities $(s+1)d_k \leq \Sigma_{A\cup B}$ over all $k \neq i, j$, and using $d_i = d_j = 0$, one obtain

$$(s+1)\Sigma_{A\cup B} \leq m\Sigma_{A\cup B}.$$

If $m-1 < s < m$, this inequality holds only if $\Sigma_{A\cup B} = 0$. Hence, $d_k = 0$ for all $k \in A \cup B$, i.e. the rank is too low.

If $s = m-1$, the last inequality is equality. This means that all summed inequalities are equalities, too. Hence, $md_k = \Sigma_{A\cup B}$ for all $k \neq i, j$. This implies, that the intersection of NN_A and NN_B is the ray $d_i = d_j = 0$, $d_k = const \geq 0$, $k \neq i, j$. Hence, codimension of this intersection is $m+1$, which is strictly greater than two, needed for adjacency.

□

Conjecture 11.12 *For the facets of $SMET_n^{m,s}$ it holds:*

(i) two (m,s)-simplex facets are not adjacent if and only if $s = 1$, and they have the same support;

(ii) a non-negativity and an (m,s)-simplex facets are not adjacent if and only if they are conflicting (i.e., there exist a position, for which they have non-zero values of different signs);

(iii) two non-negativity facets, say, NN_A and NN_B, are not adjacent if and only if $|A \cap B| = m$, and $m - 1 \leq s < m$;

(iv) the ridge graph is complete if $s \geq m$; otherwise, it has diameter two and, for $1 < s < m - 1$, it is $K_{(n-m)\binom{n}{m+1}} - \binom{n}{m+1}K_2$.

This conjecture was checked for all cases of Table 11.1.

Conjecture 11.13 *The ridge graphs of HM_n^m and of $HMET_n^m$ are induced subgraphs of the ridge graphs of $HMET_n^m$ and $HCUT_n^m$, respectively.*

Recall, that the ridge graph of $HMET_n^m$ has two orbits of vertices, F_1 and F_2, consisting of $(n-m-1)\binom{n}{m+1}$ m-simplex inequalities, and $\binom{n}{m+1}$ non-negativity inequalities.

Conjecture 11.14 *The non-negativity inequalities and the m-simplex inequalities are facets of $HCUT_n^m$.*

This conjecture was verified for $(m, n) = (3, 6)$, $(4, 7)$, $(5, 8)$, and $(2, 6)$.

Conjecture 11.15 *For the ridge graph $HMET_n^m$ it holds:*

(i) the m-simplex facet $ST_{i_1 \ldots i_{m+1}, i_{m+2}}$ is adjacent to all other facets, except the following $m + 2$ facets: all other m-simplex facets with the same support, and $NN_{i_1 \ldots i_{m+1}}$;

(ii) $G_{F_2} = \overline{J(n,3)}$ for $m = 2$, and $G_{F_2} = K_{\binom{n}{m+1}}$ for $m \geq 3$.

Clearly, (i) implies that the restriction of the ridge graph on F_1 is $G_{F_1} = K_{m+2,\ldots,m+2}$. It is easy to see, that this Conjecture would imply that the diameter of the ridge graph of $HMET_n^m$ is 2. It was proved in [DeDe94] for $m = 1$: the diameter of the ridge graph of $HMET_n^1 = MET_n$ is 2. To see it for $m = 2$, consider all 3 types of pairs of non-adjacent vertices:

(a) let $x, y \in F_1$ have the same support, say, 1234. Suppose that $x_{124} = y_{124} = -1$. Then NN_{123} is a common neighbor for x and y;

(b) for NN_{123} and NN_{124}, any tetrahedron facet $ST_{134,2}$ is their common neighbor;

(c) for NN_{123} and $ST_{123,4}$, the facet NN_{345} is their common neighbor.

Part V

Important cases of polyhedra of generalized finite semimetrics

Chapter 12

Cones of partial semimetrics and weightable quasi-semimetrics

12.1 Preliminaries

For given two partial semimetrics p_1 and p_2 on a set X their non-negative linear combination $d = \alpha p_1 + \beta p_2, \alpha, \beta \geq 0$, is a partial semimetric on X. Here, as usual, for all $x, y \in X$ it holds

$$(\alpha p_1 + \beta p_2)(x, y) = \alpha p_1(x, y) + \beta p_2(x, y).$$

Similarly, for given two weightable quasi-semimetrics q_1 and q_2 on a set X their non-negative linear combination $q = \alpha q_1 + \beta q_2, \alpha, \beta \geq 0$, is a weightable quasi-semimetric on X.

Then we can speak about the cones of all partial semimetrics and all weightable quasi-semimetrics on n points, in fact, on the set $V_n = \{1, 2, ..., n\}$. We can consider already the similar cones, related to cuts semimetrics, and some corresponding polytopes.

In this chapter we consider, for small values of n, the cone of all partial semimetrics on V_n (as well as the cones of weak and strong partial semimetrics on V_n), the cone of all weightable quasi-semimetrics on V_n (including weak and strong weightable quasi-semimetrics on V_n), the cone of all weighted semimetrics on V_n (together with down-weighted and strong-weighted semimetrics on V_n). For any cone C under consideration we construct its $\{0,1\}$-C cone, generated by all extreme rays of C, containing a non-zero $\{0,1\}$-valued point. In some cases we try to construct similar polytopes.

Partial semimetrics are generalization of semimetrics, having important applications in Computer Science (Domain Theory, Analysis of Data Flow Deadlock, Complexity Analysis of Programs, etc.). They are used for treatment of partially defined/computed objects in Semantics of Computation.

Partial semimetrics were introduced by Matthews in [Matt92] for treatment of partially defined objects in Computer Science. Weak partial semimetrics were introduced in [Heck99] as a generalization of partial semimetrics, introduced in [Matt92]. The cone $PMET_n$ of partial semimetrics was considered in [DeDe10] and [DDV11]; see also [Hitz01], [Seda97].

We will consider in this chapter the following convex polyhedra:

- the *weak partial semimetric cone* $wPMET_n$: the set of all *weak partial semimetrics* on n points, generated by all *non-negativity inequalities* $p_{ii} \geq 0$, and all *sharp triangle inequalities* $p_{ik} + p_{kj} - p_{ij} - p_{ii} \geq 0$ on V_n;

- the *partial semimetric cone* $PMET_n$: the set of all *partial semimetrics* on n points, i.e., all $p \in wPMET_n$, satisfying, in addition, all *small self-distance conditions* $p_{ij} \geq p_{ii}$ on V_n;

- the *strong partial semimetric cone* $sPMET_n$: the set of al *strong partial semimetrics* on n points, i.e., all $p \in PMET_n$, satisfying, in addition, all *large self-distance conditions* $p_{ii} + p_{jj} - p_{ij} \geq 0$ on V_n;

- the *weak weightable quasi-semimetric cone* $wWQMET_n$: the set of all *weak weightable quasi-semimetrics* on n points, generated by all *weak non-negativity inequalities* $q_{ij} + q_{ji} \geq 0$, and all $3\binom{n}{3}$ *semi-oriented triangle inequalities* $q_{ik} + q_{kj} - q_{ij} \geq 0$ on V_n (since the condition $q_{ik} + q_{kj} - q_{ij} \geq 0$ is equivalent to the condition $q_{jk} + q_{ki} - q_{ji} \geq 0$ for weak weightable quasi-semimetric, satisfying $q_{ij} + w_i = q_{ji} + w_j$);

- the *weightable quasi-semimetric cone* $WQMET_n$: the set of all *weightable quasi-semimetrics* on n points, defined by all *non-negativity inequalities* $q_{ij} \geq 0$, and all $3\binom{n}{3}$ *semi-oriented triangle inequalities* $q_{ik} + q_{kj} - q_{ij} \geq 0$ on V_n (since the condition $q_{ik} + q_{kj} - q_{ij} \geq 0$ is equivalent to the condition $q_{jk} + q_{ki} - q_{ji} \geq 0$ for weightable quasi-semimetric, satisfying $q_{ij} + w_i = q_{ji} + w_j$);

- the *strong weightable quasi-semimetric cone* $sWQMET_n$: the set of all *strong weightable quasi-semimetrics* on n points, i.e., all $q \in WQMET_n$, satisfying, in addition, all *large weight conditions* $q_{ij} \leq w_j$ on V_n;

- the *weighted semimetric cone* $WMET_n$: the set of all *weighted semimetrics* on n points, defined by all *weight non-negativity conditions* $w_i \geq 0$, and all *triangle inequalities* $d_{ik} + d_{kj} - d_{ij} \geq 0$ on V_n;

- the *down-weighted semimetric cone* $dWMET_n$: the set of all *down-weighted semimetrics* on n points, i.e., all $d \in WMET_n$, satisfying, in addition, all *down-weighted conditions* $d_{ij} \geq w_i - w_j$ on V_n;

- the *strongly weighted semimetric cone* $sWMET_n$: the set of all *strongly weighted semimetrics* on n points, i.e., all $d \in WMET_n$, satisfying, in addition, all *down-weighted conditions* $d_{ij} \geq w_i - w_j$, and all *up-weighted conditions* $d_{ij} \geq w_i + w_j$ on V_n;

- the cones $\{0,1\}$-$wPMET_n$, $\{0,1\}$-$PMET_n$, $\{0,1\}$-$sPMET_n$, $\{0,1\}$-$wWQMET_n$, $\{0,1\}$-$WQMET_n$, $\{0,1\}$-$sWQMET_n$, $\{0,1\}$-$WMET_n$, $\{0,1\}$-$dWMET_n$, $\{0,1\}$-$sWMET_n$, generated by all $\{0,1\}$-valued extreme rays of the cones $wPMET_n$, $PMET_n$, $sPMET_n$, $wWQMET_n$, $WQMET_n$, $sWQMET_n$, $WMET_n$, $dWMET_n$, $sWMET_n$, respectively.

12.2 Polyhedra of partial semimetrics and weightable quasi-semimetrics

Let $V_n = \{1, 2, ..., n\}$; let $E_n^{'} = |\{(i,j) \,|\, i,j \in V_n\}|$, where (i,j) denotes the unordered pair of the integers i,j, not necessary distinct, i.e., $E_n^{'} = E_n + n = \binom{n+1}{2} = \frac{n(n+1)}{2}$.

Let p be a *weak partial semimetric* on the set V_n, i.e., a function $p : V_n \times V_n \to \mathbb{R}_{\geq 0}$, such that for any $i,j,k \in V_n$ it holds $p(i,k) + p(k,j) - p(i,j) - p(k,k) \geq 0$, and $p(i,j) = p(j,i)$.

Because of the symmetry, we can view a weak partial semimetric p as a vector $(p_{ij})_{1 \leq i \leq j \leq n} \in \mathbb{R}^{E_n^{'}}$: $p_{ij} = p(i,j)$. Clearly, one can also view a weak partial semimetric as a symmetric $n \times n$ matrix $((p_{ij}))$, having, in general, non-zero entries on the main diagonal.

Hence, a weak partial semimetric on V_n can be viewed alternatively as a function on $V_n \times V_n$, as a symmetric $n \times n$ matrix $((p_{ij}))$, or as a vector in $\mathbb{R}^{E_n^{'}}$. We will use all three these representations for a partial semimetric on V_n. Moreover, we will use both symbols $p(i,j)$ and p_{ij} for the values of the weak partial semimetric p between points i and j.

Denote by $wPMET_n$ the set of all weak partial semimetrics on n points. The weak partial semimetrics on V_n form an $\binom{n+1}{2}$-dimensional convex cone, i.e., full-dimensional cone in $\mathbb{R}^{E_n^{'}}$, defined by n *non-negativity inequalities* PW_{ii}, and by $3\binom{n}{3}$ *sharp triangle inequalities* $ShT_{ij,k}$:

- $PW_{ii} : p_{ii} \geq 0$;
- $ShT_{ij,k} : p_{ik} + p_{kj} - p_{ij} - p_{kk} \geq 0$.

Clearly, all $ShT_{ij,i} = 0$, and $ShT_{ii,k} = 2p_{ik} - p_{ii} - p_{kk} = ShT_{ij,k} + ShT_{kj,i}$. So, it is sufficient to require $p_{ij} \geq 0$ only for $i = j$, and $p_{ik} + p_{kj} - p_{ij} - p_{kk} \geq 0$ only for different $i,j,k \in V_n$.

A *partial semimetric* on the set V_n is a function $p : V_n \times V_n \to \mathbb{R}_{\geq 0}$, such that for any $i,j,k \in V_n$ it holds $p(i,k) + p(k,j) - p(i,j) - p(k,k) \geq 0$, $p(i,j) = p(j,i)$, and $p(i,j) \geq p(i,i)$, i.e., it is a weak partial semimetric with an additional condition $p(i,j) \geq p(i,i)$(*small self-distances*).

Denote the set of all partial semimetrics on V_n by $PMET_n$. It forms an $\binom{n+1}{2}$-dimensional cone, a subcone of $wPMET_n$, generated by n *non-negativity inequalities* PW_{ii}, $2\binom{n}{2}$ *small self-distance conditions* PN_{ij}, and $3\binom{n}{3}$ *sharp triangle inequalities* $ShT_{ij,k}$:

- $PW_{ii} : p_{ii} \geq 0$;
- $PN_{ij} : p_{ij} - p_{ii} \geq 0$;
- $ShT_{ij,k} : p_{ik} + p_{kj} - p_{ij} - p_{kk} \geq 0$.

A *strong partial semimetric* on the set V_n is a function $p : V_n \times V_n \to \mathbb{R}_{\geq 0}$, such that for any $i,j,k \in V_n$ it holds $p(i,k) + p(k,j) - p(i,j) - p(k,k) \geq 0$, $p(i,j) = p(j,i)$, $p(i,j) \geq p(i,i)$, and $p(i,i) + p(j,j) - p(i,j) \geq 0$ i.e., it is a partial semimetric with an additional condition $p(i,i) + p(j,j) - p(i,j) \geq 0$ (*large self-distances*).

Denote the set of all strong partial semimetrics on V_n by $sPMET_n$. It form an $\binom{n+1}{2}$-dimensional cone, a subcone of $PMET_n$, generated by $2\binom{n}{2}$ *small self-distance conditions* PN_{ij}, $\binom{n}{2}$ *large self-distance conditions* PS_{ij}, and $3\binom{n}{3}$ *sharp triangle inequalities* $ShT_{ij,k}$ (so, alltogether, $3\binom{n+1}{3}$ facets):

- $PN_{ij} : p_{ij} - p_{ii} \geq 0$;
- $PS_{ij} : p_{ii} + p_{jj} - p_{ij} \geq 0$;
- $ShT_{ij,k} : p_{ik} + p_{kj} - p_{ij} - p_{kk} \geq 0$.

Here, $p_{ii} = PS_{ij} + PN_{ji} \geq 0$, i.e., $PS_{ij} : p_{ii} + p_{jj} - p_{ij} \geq 0$ and $PN_{ij} : p_{ij} - p_{ii} \geq 0$ imply $PW_{ii} : p_{ii} \geq 0$ for all $i \in V_n$.

For example, the cone $wPMET_3$ is a full-dimensional cone in $\mathbb{R}^6$, generated by 3 non-negativity inequalities $PW_{11} : p_{11} \geq 0$, $PW_{22} : p_{22} \geq 0$, $PW_{33} : p_{33} \geq 0$, and 3 sharp triangle inequalities $ShT_{23,1} : p_{21} + p_{13} - p_{23} - p_{11} \geq 0$, $ShT_{13,2} : p_{12} + p_{32} - p_{13} - p_{22} \geq 0$, $ShT_{12,3} : p_{13} + p_{32} - p_{12} - p_{33} \geq 0$.

In order to obtain the cone $PMET_3$, we should add 6 small self-distance conditions $PN_{12} : p_{12} \geq p_{11}$, $PN_{21} : p_{21} \geq p_{22}$, $PN_{13} : p_{13} \geq p_{11}$, $PN_{31} : p_{31} \geq p_{33}$, $PN_{23} : p_{23} \geq p_{22}$, $PN_{32} : p_{32} \geq p_{33}$.

In order to obtain the cone $PMET_3$, we should change 3 non-negativity inequalities to 3 large self-distance conditions $PS_{12} : p_{11} + p_{22} \geq p_{12}$, $PS_{13} : p_{11} + p_{33} \geq p_{13}$, $PS_{23} : p_{22} + p_{33} \geq p_{23}$.

Any above equality can be written as a vector $p = (p_{11}, p_{12}, p_{13}, p_{22}, p_{23}, p_{33})$ in $\mathbb{R}^6$. For example:

$PW_{11} : p_{11} \geq 0$ corresponds to the vector $(1, 0, 0, 0, 0, 0, 0)$;

$ShT_{23,1} : p_{12} + p_{13} - p_{23} - p_{11} \geq 0$ corresponds to the vector $(-1, 1, 1, 0, -1, 0)$;

$PN_{12} : p_{12} - p_{11} \geq 0$ corresponds to the vector $(-1, 1, 0, 0, 0, 0)$;

$PS_{13} : p_{11} + p_{33} - p_{13} \geq 0$ corresponds to the vector $(1, 0, -1, 0, 0, 1)$.

But in this chapter we will often use the matrix representation of considered objects. For the inequalities $PW_{11} : p_{11} \geq 0$, $ShT_{23,1} : p_{12} + p_{13} - p_{23} - p_{11} \geq 0$, $PN_{12} : p_{12} - p_{11} \geq 0$ the corresponding symmetric 3×3 matrices have the following forms:

$$PW_{11} : \begin{pmatrix} 1 & 0 & 0 \\ 0 & 0 & 0 \\ 0 & 0 & 0 \end{pmatrix}; \; ShT_{23,1} : \begin{pmatrix} -1 & 1 & 1 \\ 1 & 0 & -1 \\ 1 & -1 & 0 \end{pmatrix};$$

$$PN_{12} : \begin{pmatrix} -1 & 1 & 0 \\ 1 & 0 & 0 \\ 0 & 0 & 0 \end{pmatrix}; \; PS_{13} : \begin{pmatrix} 1 & 0 & -1 \\ 0 & 0 & 0 \\ -1 & -1 & 1 \end{pmatrix}.$$

As every strong partial semimetric is a partial semimetric, and every partial semimetric is a weak partial semimetric, we have

$$sPMET_n \subseteq PMET_n \subseteq wPMET_n \subseteq \mathbb{R}^{E'_n}_{\geq 0}.$$

For each cone of partial semimetrics consider the cone, generated by all its extreme rays, containing a non-zero $\{0,1\}$-valued point: let $\{0,1\}$-$wPMET_n$, $\{0,1\}$-$PMET_n$ and $\{0,1\}$-$sPMET_n$ be the cones, generated by all $\{0,1\}$-valued extreme rays of the cones $wPMET_n$, $PMET_n$, and $sPMET_n$, relatively.

The *partial semimetric convex body* $PMET_n^{\square}$ is the set of all $d \in PMET_n$, satisfying, in addition, all the conditions

$$p_{ij} \leq 1 + p_{ii},$$

and all the *perimeter inequalities*

$$p_{ij} + p_{jk} + p_{ki} \leq 2 + p_{ii} + p_{jj} + p_{kk}.$$

Remark. In $PMET_n^{\square}$, the entries p_{ij} are controlled by the p_{ii}, but the p_{ii} entries are not bounded; so, $PMET_n^{\square}$ is not a polytope. One may get a polytope by adding the inequalities $\sum_i p_{ii} \leq 1$; then the vertices of the obtained polytope are the vertices of $MET_n^{\square}$ together with the extreme rays of $PMET_n^{\square}$.

Let q be a *weightable quasi-semimetric* on the set V_n, i.e., a function $q : V_n \times V_n \to \mathbb{R}_{\geq 0}$, such that for any $x, y, z \in V_n$ it holds $q(x,x) = 0$; $q(x,y) \leq q(x,z) + q(z,y)$, and there exists a *weight function* $w = (w_i) : V_n \longrightarrow \mathbb{R}_{\geq 0}$, such that for any $i, j, k \in V_n$ it holds

$$q_{ij} + w_i = q_{ji} + w_j.$$

Define by $WQMET_n$ the set of all *weightable quasi-semimetrics* q (or, equivalently, pairs (q, w)) on V_n. It is a cone in $\mathbb{R}^{E_n'}$, generated by $2\binom{n}{2}$ *non-negativity inequalities* NN_{ij}, and $3\binom{n}{3}$ *semi-oriented triangle inequalities* $OT_{ij,k}$ (since, for a quasi-semimetric, $OT_{ij,k} = OT_{ji,k}$, if it is weightable):

- $NN_{ij} : q_{ij} \geq 0$;
- $OT_{ij,k} : p_{ik} + p_{kj} - p_{ij} \geq 0$.

It is easy to see, that the cone $WQMET_n$ is of dimension $\binom{n+1}{2} - 1$.

Clearly, $((2q_{ij} + w_i - w_j))$ is the *symmetrization semimetric* of q. Moreover, the function $q = ((q_{ij}))$ is a weightable quasi-semimetric if and only if the function $p = ((p_{ij}))$, where $p_{ii} = w_i$, and $p_{ij} = q_{ij} + p_{ii}$, is a partial semimetric. Easy to check, that a quasi-semimetric q is weightable if and only if it has *relaxed symmetry*, i.e., for distinct $i, j, k \in V_n$, it holds

$$q_{ij} + q_{jk} + q_{ki} = q_{ik} + q_{kj} + q_{ji}.$$

Also, $WQMET_n$ consists of all $((d_{ij} + d_{i0} - d_{j0}))$, where $((d_{ij}))$ is a semimetric on $\{0, 1, 2, ..., n\}$, but none of the inequalities $d_{i0} + d_{j0} - d_{ij} \geq 0$ is required.

A weightable quasi-semimetric (q, w) with all *large weight conditions*

- $QS_{ij} : q_{ij} \leq w_j$

is called a *strong weightable quasi-semimetric*. But if, on the contrary, the non-negativity condition $NN_{ij} : p_{ij} \geq 0$ is weakened to the *weak non-negativity condition*

- $QN_{ij} : q_{ij} + q_{ji} \geq 0$

(so, $q_{ij} < 0$ is allowed), (q, w) is called a *weak weightable quasi-semimetric*. Denote by $sWQMET_n$ and $wWQMET_n$ the corresponding cones.

As every strong weightable quasi-semimetric is an weightable quasi-semimetric, and as every weightable quasi-semimetric is a weak weightable quasi-semimetric, we have

$$wWQMET_n \subseteq WQMET_n \subseteq sWQMET_n \subseteq \mathbb{R}_{\geq 0}^{E_n'}.$$

Besides strict inclusion $WQMET_n \subset QMET_n$, we have, with equality only for $n = 3$, the following inclusion: $OCUT_n \subseteq WQMET_n$. So,

$$OCUT_n \subseteq WQMET_n \subset QMET_n.$$

Also, it is easy to check, that

$$MET_n = QMET_n \cap PMET_n.$$

As before, consider also the corresponding cones $\{0,1\}$-$wWQMET_n$, $\{0,1\}$-$WQMET_n$, and $\{0,1\}$-$sWQMET_n$, generated by all $\{0,1\}$-valued extreme rays of the cones $wWQMET_n$, $WQMET_n$, and $sWQMET_n$, respectively.

Define *weightable quasi-semimetric polytope* $WQMET_n^{\square}$ as the set of all weightable quasi-semimetrics, belonging to the *quasi-semimetric polytope* $QMET_n^{\square}$:

$$WQMET_n^{\square} = WQMET_n \cap QMET_n^{\square}.$$

A *weighted semimetric* $(d; w)$ on V_n is a semimetric d (i.e., a function $d : V_n \times V_n \to \mathbb{R}_{\geq 0}$, such that for any $x, y, z \in V_n$ it holds $d(x,x) = 0$; $d(x,y) = d(y,x)$, and $d(x,y) \leq d(x,z) + d(z,y)$) with a weight function $w : V_n \to \mathbb{R}_{\geq 0}$ on its points.

Denote by $WMET_n$ the set of all weighted semimetrics $(d; w)$ on V_n. It forms a convex cone in $\mathbb{R}^{E_n'}$ with n *weight non-negativity conditions* W_i, and $3\binom{n}{3}$ *triangle inequalities* $T_{ij,k}$:

- $W_i : w_i \geq 0$;
- $T_{ij,k} : d_{ik} + d_{kj} - d_{ij} \geq 0$.

Call a weighted semimetric $(d; w)$ *down-weighted*, if for all distinct $i, j \in V_n$ it holds the *down-weighted condition*:

- $D_{ij} : d_{ij} \geq w_i - w_j$.

Call a weighted semimetric $(d; w)$ *up-weighted*, if for all distinct $i, j \in V_n$ it holds the *up-weighted condition*:

- $U_{ij} : d_{ij} \geq w_i + w_j$.

Denote by $dWMET_n$ the cone of down-weighted semimetrics on V_n, and by $sWMET_n$ the cone of *strongly*, i.e., both, down-weighted and up-weighted, semimetrics.

Obviously, as every strongly weighted semimetric is an down weighted semimetric, and as every down-weighted semimetric is a weighted semimetric, we have

$$sWMET_n \subseteq dWMET_n \subseteq WMET_n \subseteq \mathbb{R}_{\geq 0}^{E_n'}.$$

Consider a weighted semimetric $(d; w)$ on V_n as the matrix $((d'_{ij}))$, $0 \leq i, j \leq n$, with $d'_{00} = 0$, $d'_{0i} = d'_{i0} = w_i$ for $i \in V_n$, and $d'_{ij} = d_{ij}$ for $i, j \in V_n$.

In this interpretation it is easy to see, that

$$MET_n \simeq \{(d; (k, ..., k)) \mid d \in WMET_n\}.$$

Moreover,

$$sWMET_n = MET_{n+1}.$$

In fact, any *pointed semimetric* (d, i_0), i.e., a semimetric d on $V_n \cup \{0\}$ with a selected base point 0, can be seen as a weighted semimetric $(d; w)$ on the set V_n with weight function $w : w_i = d_{i0}$.

Define by $\{0,1\}$-$WMET_n$, $\{0,1\}$-$dWMET_n$ and $\{0,1\}$-$sWMET_n$ the cones, generated by all $\{0,1\}$-valued extreme rays of the cones $WMET_n$, $dWMET_n$ and $sWMET_n$, respectively.

Every permutation of V_n induce a symmetry of the above polyhedra; so, $Sym(n)$ is a symmetry group of them.

In oriented case appears also a *reversal* symmetry (see [DDP03]), corresponding to transposition of matrix $((q_{ij}))$. We expect $Z_2 \times Sym(n)$ and $Sym(n)$ to be the full symmetry groups of $WQMET_n$, $\{0,1\}$-$WQMET_n$, and $PMET_n$, $\{0,1\}$-$PMET_n$, respectively.

It is checked for $n \leq 9$, that the full symmetry group of $PMET_n$ is $Sym(n)$.

However, $PMET_n^{\square}$ has additional symmetries. If for $S \subset \{1, 2, ..., n\}$ we define the *switching* operation $p' = U_S(p)$ as

$$p'_{ij} = \begin{cases} 1 + p_{ii} + p_{jj} - p_{ji}, & \text{if } |S \cap \{i, j\}| = 1, \\ p_{ij}, & \text{otherwise,} \end{cases}$$

we can check, that switchings are symmetries of $PMET_n^{\square}$. We expect that, together with $Sym(n)$, this defines the full symmetry group of $PMET_n^{\square}$ (checked for $n \leq 9$).

In Table 12.1 we summarize the most important numeric information on cones under consideration for $n \leq 6$. The column 2 indicates the dimension of the cone, the columns 3 and 4 give the number of extreme rays and facets, respectively; in parentheses are given the numbers of their orbits. All orbits are under $Sym(n)$. The columns 5 and 6 give the diameters of the 1-skeleton graph and the ridge graph of a given cone. The expanded version of the data can be found in [Vida15].

12.3 Maps P, Q and connections between considered polyhedra

Given a weighted semimetric $(d; w)$ on V_n, define the map P by the function $p = P(d; w)$ with

$$p_{ij} = \frac{d_{ij} + w_i + w_j}{2}.$$

Clearly, P is an *automorphism* (invertible linear operator) of the vector space $\mathbb{R}^{\binom{n+1}{2}}$, and $(d; w) = P^{-1}(p)$, where the inverse map P^{-1} is defined by

$$d_{ij} = 2p_{ij} - p_{ii} - p_{jj}, w_i = p_{ii}.$$

Define the map Q by the function $(q, w) = Q(d; w)$ with

$$q_{ij} = \frac{d_{ij} - w_i + w_j}{2}.$$

Cone	Dimension	# of ext. rays (orbits)	# of facets (orbits)	Diameter	Diameter dual
$wPMET_3$	6	6 (2)	6 (2)	1	1
$wPMET_4$	10	11 (3)	16 (2)	1	2
$wPMET_5$	15	30 (4)	35 (2)	2	2
$wPMET_6$	21	302 (8)	66 (2)	2	2
$PMET_3$={0, 1}-$PMET_3$	6	13 (5)	12 (3)	3	2
$PMET_4$	10	62 (11)	28 (3)	3	2
$PMET_5$	15	1696 (44)	55 (3)	3	2
$PMET_6$	21	337092 (734)	96 (3)	3	2
{0, 1}-$PMET_4$	10	44 (9)	46 (5)	3	2
{0, 1}-$PMET_5$	15	166 (14)	585 (15)	3	3
{0, 1}-$PMET_6$	21	705 (23)		3	
$sPMET_3$ = {0, 1}-$sPMET_3$	6	7 (2)	12 (1)	1	2
$sPMET_4$	10	25 (3)	30 (1)	2	2
$sPMET_5$	15	296 (7)	60 (1)	2	2
$sPMET_6$	21	55226 (46)	105 (1)	3	2
{0, 1}-$sPMET_4$	10	15 (2)	40 (2)	1	2
{0, 1}-$sPMET_5$	15	31 (3)	210 (4)	1	3
{0, 1}-$sPMET_6$	21	63 (3)	38780 (36)	1	3
{0, 1}-$dWMET_3$	6	10 (4)	15 (4)	2	2
{0, 1}-$dWMET_4$	10	22 (6)	62 (7)	2	3
{0, 1}-$dWMET_5$	15	46 (7)	1165 (27)	2	3
{0, 1}-$dWMET_6$	21	94 (9)	369401 (806)	2	
$WQMET_3 = OCUT_3$	5	6 (2)	9 (2)	1	2
$WQMET_4$={0, 1}-$WQMET_4$	9	20 (4)	24 (2)	2	2
$WQMET_5$	14	190 (11)	50 (2)	2	2
$WQMET_6$	20	18502 (77)	90 (2)		2
{0, 1}-$WQMET_5$	14	110 (8)	250 (5)	2	2
{0, 1}-$WQMET_6$	20	802 (17)			
$OCUT_4$	9	14 (3)	30 (3)	1	2
$OCUT_5$	14	30 (4)	130 (6)	1	3
$OCUT_6$	20	62 (5)	16460 (62)	1	

Table 12.1: Main parameters of considered cones with $n \leq 6$

So, $Q(d;w) = P(d;w) - ((1))w$, i.e.,

$$q_{ij} = p_{ij} - p_{ii},$$

and

$$d_{ij} = q_{ij} + q_{ji}$$

is the *symmetrization semimetric* of q.

In fact, given functions $d = (d;w)$, $p = P(d;w)$, and $q = (q;w) = Q(d;w)$, we have:

- $p_{ij} = q_{ij} + w_i = q_{ji} + w_j = \frac{d_{ij}+w_i+w_j}{2}$ with $p_{ii} = w_i$;
- $q_{ij} = \frac{d_{ij}-w_i+w_j}{2}$ with the weight $w = (w_i)$;
- $d_{ij} = q_{ij} + q_{ji} = 2p_{ij} - p_{ii} - p_{jj}$ with weight $w = (w_i)$.

Obviously, it holds $p = \mathcal{P}(q)$, where the map $\mathcal{P} = PQ^{-1}$ is defined by above conditions $p_{ij} = q_{ij} + w_i$.

Example. Three 6×6 matrices below represent:

- the weighted semimetric (d,w) on V_6, where $d = 2\delta_{\{5,6\},\{1\},\{2,3\},\{4\}} - \delta_{\{5,6\}} \in MET_6$, and the weight $w = (1_{i\in\{5,6\}}) = (0,0,0,0,1,1)$;
- the partial semimetric $P(d;w) = J(\{5,6\}) + \delta_{\{5,6\},\{1\},\{2,3\},\{4\}}$, where $J(\{5,6\})$ is denoted the $\{0,1\}$-valued function with $J(\{5,6\})_{ij} = 1$ exactly when $i,j \in \{5,6\}$ (its ray is extreme in $PMET_6$);

- the weightable quasi-semimetric $Q(d;w) = \delta^O_{\{1\}} + \delta^O_{\{2,3\}} + \delta^O_{\{4\}}$ (its ray is not extreme in $WQMET_6$).

$$\begin{array}{cccccc} \mathbf{0} & 2 & 2 & 2 & 1 & 1 \\ 2 & \mathbf{0} & 0 & 2 & 1 & 1 \\ 2 & 0 & \mathbf{0} & 2 & 1 & 1 \\ 2 & 2 & 2 & \mathbf{0} & 1 & 1 \\ 1 & 1 & 1 & 1 & \mathbf{1} & 0 \\ 1 & 1 & 1 & 1 & 0 & \mathbf{1} \end{array} \qquad \begin{array}{cccccc} \mathbf{0} & 1 & 1 & 1 & 1 & 1 \\ 1 & \mathbf{0} & 0 & 1 & 1 & 1 \\ 1 & 0 & \mathbf{0} & 1 & 1 & 1 \\ 1 & 1 & 1 & \mathbf{0} & 1 & 1 \\ 1 & 1 & 1 & 1 & \mathbf{1} & 1 \\ 1 & 1 & 1 & 1 & 1 & \mathbf{1} \end{array} \qquad \begin{array}{cccccc} \mathbf{0} & 1 & 1 & 1 & 1 & 1 \\ 1 & \mathbf{0} & 0 & 1 & 1 & 1 \\ 1 & 0 & \mathbf{0} & 1 & 1 & 1 \\ 1 & 1 & 1 & \mathbf{0} & 1 & 1 \\ 0 & 0 & 0 & 0 & \mathbf{1} & 0 \\ 0 & 0 & 0 & 0 & 0 & \mathbf{1} \end{array}$$

Clearly, it holds:

- $d_{ij} + d_{ik} - d_{jk} = p_{ij} + p_{ik} - p_{jk} - p_{ii} = q_{ji} + q_{ik} - q_{jk}$, i.e., the triangle inequalities $T_{jk,i}$, $ShT_{jk,i}$ and $OT_{jk,i}$ are equivalent on all three levels: d - of (weighted) semimetrics, p - of (would-be) partial semimetrics, and q - of (would-be) weightable quasi-semimetrics.

- $p_{ij} \geq p_{ii}$ if and only if $d_{ij} \geq w_i - w_j$ if and only if $q_{ij} \geq 0$; so, the *small self-distance condition* PN_{ij} for partial semimetrics is equivalent to the *down-weighted condition* D_{ij} for weighted semimetrics, and to the *non-negativity condition* NN_{ij} for weightable quasi-semimetrics;

- $p_{ij} \leq p_{ii} + p_{jj}$ if and only if $d_{ij} \leq w_i + w_j$ if and only if $q_{ij} \leq w_j$; so, the *large self-distance condition* PS_{ij} for partial semimetrics is equivalent to the *up-weighted condition* U_{ij} for weighted semimetrics, and to the *large weight condition* QS_{ij} for weightable quasi-semimetrics;

- $2p_{ij} \geq p_{ii} + p_{jj}$ if and only if $d_{ij} \geq 0$ if and only if $q_{ij} + q_{ji} \geq 0$; so, the *small semi-distance condition* $p_{ij} \geq p_{ii}$ for partial semimetrics is closely connected with the *non-negativity property* of weighted semimetrics, and with *weak non-negativity condition* for weak weightable quasi-semimetrics;

- moreover, as $p_{ii} = w_i$, then $p_{ii} \geq 0$ if and only if $w_i \geq 0$.

This implies the following important proposition.

Proposition 12.1 *The following statements hold:*

(i) there exists one-to-one correspondence between weighted semimetrics on V_n and weak partial semimetrics on V_n; between down-weighted semimetrics on V_n and partial semimetrics on V_n; between strongly weighted semimetrics on V_n and strong partial semimetrics on V_n:

$$wPMET_n = P(WMET_n),\ PMET_n = P(dWMET_n),\ sPMET_n = P(sWMET_n);$$

(ii) there exists one-to-one correspondence between weighted semimetrics on V_n and weak weightable quasi-semimetrics on V_n; between down-weighted semimetrics on V_n and weightable quasi-semimetrics on V_n; between strongly weighted semimetrics on V_n and strong weightable quasi-semimetrics on V_n:

$$wWQMET_n = Q(WMET_n), WQMET_n = Q(dWMET_n), sWQMET_n = Q(sWMET_n).$$

12.4 Small polyhedra of partial semimetrics and weightable quasi-semimetrics

In Table 12.1 we summarize the most important numeric information on cones under consideration for $n \leq 6$ from [DDV11]. The column 2 indicates the dimension of the cone, the columns 3 and 4 give the number of extreme rays and facets, respectively; in parentheses are given the numbers of their orbits. The columns 5 and 6 give the diameters of the 1-skeleton graph and the ridge graph of the considered cone. The expanded version of the data can be found on the Vidali's homepage ([Vida15]) http://lkrv.fri.uni-lj.si/~janos/cones/.

In Tables below $\gamma_{S_0;S_1,\dots,S_t}$ denote the *partial t-cut*, defined as $J(S_0)+\delta_{S_0,S_1,\dots,S_t}$, where, for $S_0 \subset V_n$, the value $J(S_0)$ is the $\{0,1\}$-valued function with $J(S_0)_{ij} = 1$ exactly when $i,j \in S_0$, and $S_1, \dots, S_t$ is a partition $\mathcal{S} = \{S_1, \dots, S_t\}$ of $\overline{S}_0$.

The case $n = 3$

In the simplest case $n = 3$, the partial semimetric cone $PMET_3$ is a full-dimensional cone in $\mathbb{R}^6$; in this case $\{0,1\}$-$PMET_3 = PMET_3$, so, there are 13 extreme rays in $PMET_3$, which form 5 orbits, and there are 12 facets from 3 orbits.

The representatives of orbits of extreme rays and facets in $PMET_3 = \{0,1\}$-$PMET_3$ are given in Table 12.2.

O/F	Representative	11	21	22	31	32	33	Inc.	Adj.	Size
O_1	$\gamma_{\{1,2,3\}}$	1	1	1	1	1	1	9	6	1
O_2	$\gamma_{\{1\};\{2,3\}}$	1	1	0	1	0	0	8	9	3
O_3	$\gamma_{\{2,3\};\{1\}}$	0	1	1	1	1	1	7	6	3
O_4	$\gamma_{\emptyset;\{1\},\{2,3\}}$	0	1	0	1	0	0	7	8	3
O_5	$\gamma_{\{3\};\{1\},\{2\}}$	0	1	0	1	1	1	5	5	3
F_1	$PW_{11} : p_{11} \geq 0$	1	0	0	0	0	0	8	9	3
F_2	$ShT_{12,3} : p_{13} + p_{23} - p_{12} - p_{33} \geq 0$	0	-1	0	1	1	-1	8	7	3
F_3	$PN_{12} : p_{12} - p_{11} \geq 0$	-1	1	0	0	0	0	7	6	6

Table 12.2: The representatives of orbits of extreme rays and facets in $PMET_3 = \{0,1\}$-$PMET_3$

Here, using cuts δ_S, o-cuts δ_S^O, and o-anti-cuts β_S^O, we obtain the partial cuts by the following equalities:

$$\gamma_{\{1,2,3\}} = \mathcal{P}(\delta_{\emptyset}^O),\ \gamma_{\{1\};\{2,3\}} = \mathcal{P}(\delta_{\{\overline{1}\}}^O),$$

$$\gamma_{\{2,3\};\{1\}} = \mathcal{P}(\delta_{\{1\}}^O),\ \gamma_{\emptyset;\{1\},\{2,3\}} = \mathcal{P}(\delta_{\{1\}}) = \delta_{\{1\}}^O + \delta_{\{\overline{1}\}}^O,$$

$$\gamma_{\{3\};\{1\},\{2\}} = \mathcal{P}(\beta_{\{3\}}^O) = \delta_{\{1\}}^O + \delta_{\{2\}}^O.$$

The 1-skeleton graph and the ridge graph of the cone $PMET_3 = \{0,1\}$-$PMET_3$ are represented in Figures 12.1 and 12.2. The extreme rays of the cone are denoted in Figure 12.1 by the symbols of black triangle (orbit O_1), white circle (orbit O_2), black circle (orbit O_3), white square (orbit O_4), black square (orbit O_5). The facets of the cone are denoted in Figure 12.2 by the symbols of white circle (orbit $F_1 = PW(L)$), black triangle (orbit $F_2 = ShT(Tr)$), black circle (orbit $F_3 = PN(M)$).

The cone $\{0,1\}$-$dWMET_3$ is also an full-dimensional cone in $\mathbb{R}^6$. It has 10 extreme rays, which form 4 orbits, and 15 facets from 4 orbits.

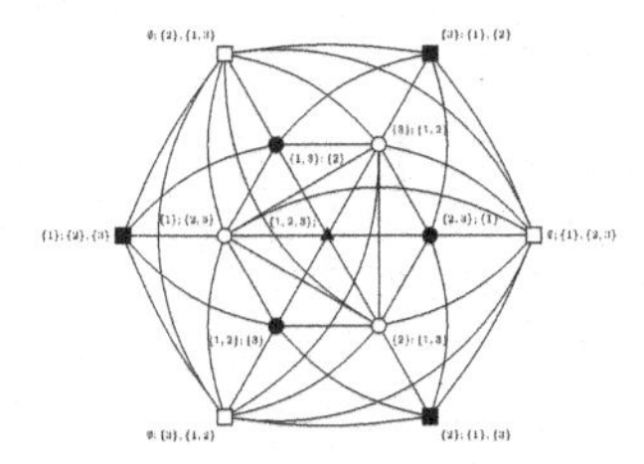

Figure 12.1: The 1-skeleton graph of $PMET_3 = \{0,1\}$-$PMET_3$

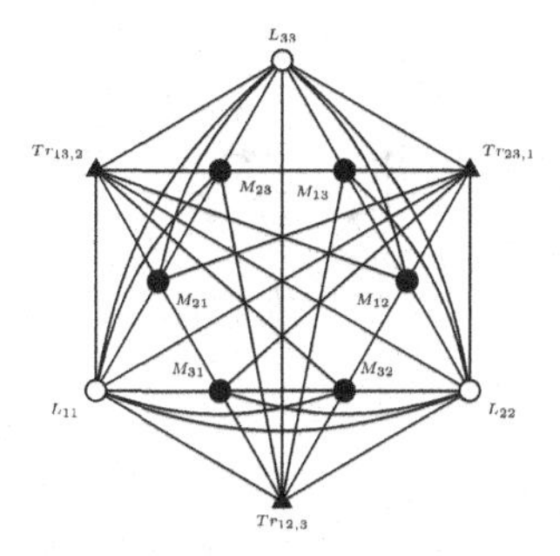

Figure 12.2: The ridge graph of $PMET_3 = \{0,1\}$-$PMET_3$

The representatives of orbits of extreme rays and facets in $\{0,1\}$-$dWMET_3$ are given in Table 12.3.

O/F	Representative	11	21	22	31	32	33	Inc.	Adj.	Size
O_1	$(\delta_\emptyset;(1))$	1	0	1	0	0	1	9	6	1
O_2	$(\delta_{\{1\}};w'')$	0	1	1	1	0	1	9	8	3
O_3	$(\delta_{\{1\}};w')$	1	1	0	1	0	0	9	8	3
O_4	$(\delta_{\{1\}};(0))$	0	1	0	1	0	0	9	8	3
F_1	$W_1 : w_1 \geq 0$	1	0	0	0	0	0	6	9	3
F_2	$T_{12,3} : d_{13} + d_{23} - d_{12} \geq 0$	0	-1	0	1	1	0	7	8	3
F_3	$D_{12} : d_{12} + (w_2 - w_1) \geq 0$	-1	1	1	0	0	0	6	6	6
F_4	$T'_{12,3} : (d_{13} + d_{23} - d_{12}) + 2(w_1 + w_2 - w_3) \geq 0$	2	-1	2	1	1	-2	5	5	3

Table 12.3: The orbits of extreme rays and facets for $\{0,1\}$-$dWMET_3$

In fact, the set of extreme rays of $\{0,1\}$-$dWMET_3$ consists of $(\delta_\emptyset;(1)) = (((0));(1))$, i.e., non-discrete semimeric $((0))$ with weight function $w = (1)$, and 2-cuts $(\delta_S; w)$ with weights (0), $w' = (1_{i \in S})$, or $w'' = (1_{i \notin S})$.

The 1-skeleton graph and the ridge graph of the cone $\{0,1\}$-$dWMET_3$ are represented in Figures 12.3 and 12.4. The extreme rays of the cone are denoted in Figure 12.3 by the symbols of black triangle (orbit O_1), white circle (orbit O_2), black circle (orbit O_3), white square (orbit O_4). The facets of the cone are denoted in Figure 12.4 by the symbols

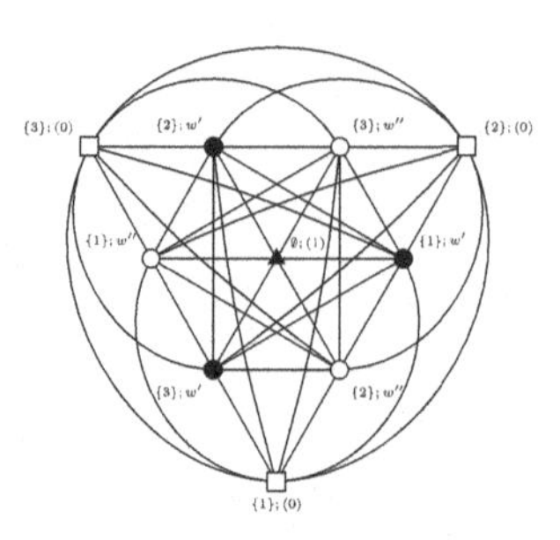

Figure 12.3: The 1-skeleton graph of $\{0,1\}$-$dWMET_3$

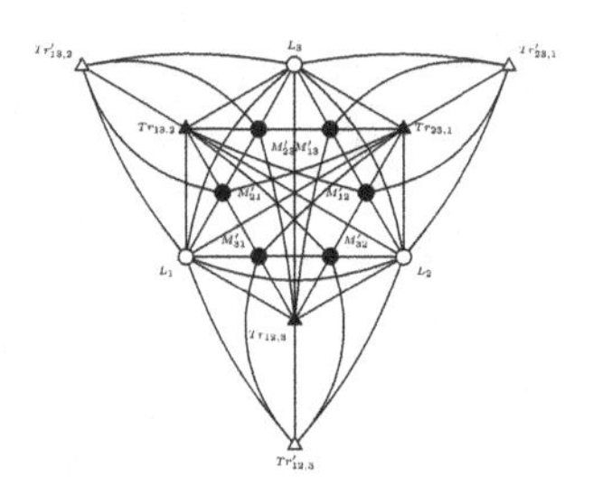

Figure 12.4: The ridge graph of $\{0,1\}$-$dWMET_3$

of white circle (orbit $F_1 = W(L)$), black triangle (orbit $F_2 = T(Tr)$), black circle (orbit $F_3 = D(M')$), white triangle (orbit $F_4 = T'(Tr')$).

The set $R(dWMET_3) \setminus R(\{0,1\}\text{-}dWMET_3)$ of extreme rays of $dWMET_3$, different from $\{0,1\}$-valued, and the set $F(\{0,1\}\text{-}dWMET_3) \setminus F(dWMET_3)$ of "new" facets of the cone $\{0,1\}$-$dWMET_3$ consist of 3 simplicial elements, forming $\overline{K}_3$ in the corresponding graph. But only $G^*_{\{0,1\}\text{-}dWMET_3}$ is an induced subgraph of $G^*_{dWMET_3}$.

Similarly, for $n = 3$ we get, that

$$wPMET_3 \simeq WMET_3 = \{0,1\}\text{-}WMET_3$$

is a simplicial cone in $\mathbb{R}^6$, having 6 extreme rays from 2 orbits, and 6 facets from 2 orbits; the simplicial cone $\{0,1\}$-$wPMET_3$ has 3 extreme rays and 3 facets.

On the other hand, it holds, that

$$sPMET_3 = \{0,1\}\text{-}sPMET_3 \simeq MET_4 = CUT_4 = \{0,1\}\text{-}WMET_3.$$

For this 6-dimensional cone we get 7 extreme rays from 2 orbits, and 12 facets from one orbit.

For the oriented case,

$$WQMET_3 = \{0,1\}\text{-}WQMET_3 = OCUT_3;$$

this cone has 6 extreme rays from 2 orbits, and 9 facets from 2 orbits. Note, that $QMET_3 = \{0,1\}\text{-}QMET_3 = OMCUT_3$ has 12 extreme rays from 2 orbits, and 12 facets from 2 orbits (see Chapter 10).

The case $n = 4$

The partial semimetric cone $PMET_4$ is an full-dimensional cone in $\mathbb{R}^{10}$. It has 28 facets, divided in 3 orbits, and 62 extreme rays, divided in 11 orbits.

The representatives of orbits of extreme rays and facets of the cone $PMET_4$ are given in Table 12.4.

F/O	Representative	11	21	22	31	32	33	41	42	43	44	Inc.	Adj.	Size
F_1	$PW_{11} : p_{11} \geq 0$	1	0	0	0	0	0	0	0	0	0	29	36	4
F_2	$ShT_{23,1} : p_{12} + p_{13} - p_{23} - p_{11} \geq 0$	-1	1	0	1	-1	0	0	0	0	0	26	24	12
F_3	$PN_{12} : p_{12} - p_{11} \geq 0$	-1	1	0	0	0	0	0	0	0	0	23	23	12
O_1	$\gamma_{\{1,2,3,4\}}$	1	1	1	1	1	1	1	1	1	1	24	20	1
O_2	$\gamma_{\{2\};\{\overline{2}\}}$	0	1	1	0	1	0	0	1	0	0	21	38	4
O_3	$\gamma_{\{\overline{3}\};\{3\}}$	1	1	1	1	1	0	1	1	1	1	19	17	4
O_4	$\gamma_{\emptyset;\{\overline{3}\},\{3\}}$	0	0	0	1	1	0	0	0	1	0	19	32	4
O_5	$\gamma_{\{1,2\};\{\overline{1,2}\}}$	1	1	1	1	1	0	1	1	0	0	18	31	6
O_6	$\gamma_{\emptyset;\{1,2\},\{\overline{1,2}\}}$	0	0	0	1	1	0	1	1	0	0	16	32	3
O_7	$\gamma_{\{1,4\};\{2\},\{3\}}$	1	1	0	1	1	0	1	1	1	1	14	14	6
O_8	$\gamma_{\{1\};\{2\},\{3,4\}}$	1	1	0	1	1	0	1	1	0	0	14	20	12
O_9	$\gamma_{\{4\};\{1\},\{2\},\{3\}}$	0	1	0	1	1	0	1	1	1	1	9	9	4
O_{10}		1	1	0	1	1	0	2	1	1	1	10	18	6
O_{11}		0	2	0	1	1	0	2	2	3	2	9	9	12

Table 12.4: The representatives of orbits of facets and extreme rays in $PMET_4$

In fact, the representatives of orbits O_i of extreme rays of the cone $PMET_4$, given in Table 12.4, are:

- $O_1 : \ \gamma_{\{1,2,3,4\}} = \mathcal{P}(\delta^O_\emptyset)$;
- $O_2 : \ \gamma_{\{2\};\{\overline{2}\}} = \mathcal{P}(\delta^O_{\{\overline{2}\}})$;
- $O_3 : \ \gamma_{\{\overline{3}\};\{3\}} = \mathcal{P}(\delta^O_{\{3\}})$;
- $O_4 : \ \gamma_{\emptyset;\{\overline{3}\}} = \mathcal{P}(\delta_{\{3\}})$;
- $O_5 : \ \gamma_{\{1,2\};\{\overline{1,2}\}} = \mathcal{P}(\delta^O_{\{3,4\}})$;
- $O_6 : \ \gamma_{\emptyset;\{1,2\},\{\overline{1,2}\}} = \mathcal{P}(\delta_{\{3,4\}})$;
- $O_7 : \ \gamma_{\{1,4\};\{2\},\{3\}} = \mathcal{P}(\delta^O_{\{2\}} + \delta^O_{\{3\}})$;
- $O_8 : \ \gamma_{\{1\};\{2\},\{3,4\}} = \mathcal{P}(\delta^O_{\{2\}} + \delta^O_{\{3,4\}})$;
- $O_9 : \ \gamma_{\{4\};\{1\},\{2\},\{3\}} = \mathcal{P}(\beta^O_{\{4\}})$;
- $O_{10} : \ \mathcal{P}(\beta^O_{\{1,4\}})$;
- $O_{11} : \ \mathcal{P}(\delta_{\{3\}} + 2\delta^O_{\{\overline{4}\}}) + 2d(K_{1,2})$.

The cone $\{0,1\}$-$PMET_4$ is also an full-dimensional cone in $\mathbb{R}^{10}$. It has 44 extreme rays, divided in 9 orbits, and 46 facets, divided in 5 orbits.

Given a sequence $b = (b_1, ..., b_n)$ of integers, satisfying the condition $\sum_{i=1}^{n} b_i \in \{0,1\}$, and an $p = ((p_{ij})) \in \{0,1\}$-$PMET_n$, consider the inequalities

$$Hyp_b(p) = -\sum_{1 \le i < j \le n} b_i b_j p_{ij} - \frac{1}{2}\sum_{i=1}^{n} b_i(b_i - 1)p_{ii} \ge 0,$$

and

$$Mod_b(p) = -2\sum_{1 \le i < j \le n} b_i b_j p_{ij} - \sum_{i=1}^{n} \max\{0, |b_i|(|b_i| + 1) - 2\}p_{ii} \ge 0.$$

In fact, the typical facet-defining inequalities $PN_{12} : p_{12} - p_{11} \ge 0$, and $ShT_{12,3} : p_{13} + p_{23} - p_{12} - p_{33} \ge 0$ of $PMET_n$, are simplest instances of $Hyp_b(p) \ge 0$ for $b = (-1, 1, 0, ..., 0)$ and $b = (1, 1, -1, 0, ..., 0)$, respectively.

The cone $\{0,1\}$-$PMET_4$, besides orbits F_1, F_2, F_3 of sizes 4, 12, 12 of facets of $PMET_4$, has orbits F_4, F_5 (of sizes 6, 12) of facets, represented by the following inequalities:

- $Hyp_{(1,1,-1,-1)}(p) = (p_{13} + p_{23} + p_{14} + p_{24}) - (p_{12} + p_{34}) - (p_{33} + p_{44}) \ge 0$;
- $Mod_{(2,1,-1,-1)}(p) = (2p_{13} + p_{23} + 2p_{14} + p_{24}) - (2p_{12} + p_{34}) - 2p_{11} \ge 0$.

O/F	Representative	11	21	22	31	32	33	41	42	43	44	Inc.	Adj.	Size
O_1	$\gamma_{\{1,2,3,4\}}$	1	1	1	1	1	1	1	1	1	1	24	20	1
O_2	$\gamma_{\{2\};\{\overline{2}\}}$	0	1	1	0	1	0	0	1	0	0	21	38	4
O_3	$\gamma_{\{\overline{3}\};\{3\}}$	1	1	1	1	1	0	1	1	1	1	19	17	4
O_4	$\gamma_{\emptyset;\{\overline{3}\},\{3\}}$	0	0	0	1	1	0	0	0	1	0	19	32	4
O_5	$\gamma_{\{1,2\};\{\overline{1,2}\}}$	1	1	1	1	1	0	1	1	0	0	18	31	6
O_6	$\gamma_{\emptyset;\{1,2\},\{\overline{1,2}\}}$	0	0	0	1	1	0	1	1	0	0	16	32	3
O_7	$\gamma_{\{1,4\};\{2\},\{3\}}$	1	1	0	1	1	0	1	1	1	1	14	14	6
O_8	$\gamma_{\{1\};\{2\},\{3,4\}}$	1	1	0	1	1	0	1	1	0	0	14	20	12
O_9	$\gamma_{\{4\};\{1\},\{2\},\{3\}}$	0	1	0	1	1	0	1	1	1	1	9	9	4
F_1	$PW_{11} : p_{11} \ge 0$	1	0	0	0	0	0	0	0	0	0	29	36	4
F_2	$ShT_{12,3} : Hyp_{(-1,1,1,0)}(p) \ge 0$	-1	1	0	1	-1	0	0	0	0	0	26	24	12
F_3	$PN_{12} : Hyp_{(-1,1,0,0)}(p) \ge 0$	-1	1	0	0	0	0	0	0	0	0	23	23	12
F_4	$Hyp_{(1,1,-1,-1)}(p) \ge 0$	0	-1	0	1	1	-1	1	1	-1	-1	16	12	6
F_5	$Mod_{(2,1,-1,-1)}(p) \ge 0$	-2	-2	0	2	1	0	2	1	-1	0	9	9	12

Table 12.5: The orbits of extreme rays and facets in $\{0,1\}$-$PMET_4$

Note, that the orbits O_{10} and O_{11} of extreme rays in $PMET_4$ are excluded in $\{0,1\}$-$PMET_4$ by the orbits F_4 and F_5 of facets, respectively. In fact, $P(\beta^{O}_{\{1,2\}})$ violates $Hyp_{(1,1,-1,-1)}(p) \ge 0$, while $P(\delta_{\{1\}} + 2\delta^{O}_{\{\overline{2}\}}) + 2d(K_{\{3,4\}})$ violates $Mod_{(2,1,-1,-1)}(p) \ge 0$.

The case $n = 5$

The partial semimetric cone $PMET_5$ is an full-dimensional cone in $\mathbb{R}^{15}$. It has 55 facets, divided in 3 orbits, and 1696 extreme rays, divided in 44 orbits.

The representatives of orbits of extreme rays of the cone $PMET_5$ are given in Table 12.6.

O_i	Representative	11	21	22	31	32	33	41	42	43	44	51	52	53	54	55	Inc.	$\vert O_i\vert$
O_1	$\mathcal{P}(\delta^O_\emptyset)$	1	1	1	1	1	1	1	1	1	1	1	1	1	1	1	90	1
O_2	$\mathcal{P}(\delta^O_{\{\overline{5}\}})$	0	0	0	0	0	0	0	0	0	0	1	1	1	1	1	44	5
O_3	$\mathcal{P}(\delta^O_{\{5\}})$	1	1	1	1	1	1	1	1	1	1	1	1	1	1	0	41	5
O_4	$\mathcal{P}(\delta_{\{1\}})$	0	1	0	1	0	0	1	0	0	0	1	0	0	0	0	41	5
O_5	$\mathcal{P}(\delta^O_{\{\overline{1,5}\}})$	1	1	0	1	0	0	1	0	0	0	1	1	1	1	1	38	10
O_6	$\mathcal{P}(\delta^O_{\{1,5\}})$	0	1	1	1	1	1	1	1	1	1	0	1	1	1	0	37	10
O_7	$\mathcal{P}(\delta_{\{1,5\}})$	0	1	0	1	0	0	1	0	0	0	0	1	1	1	0	34	10
O_8	$\mathcal{P}(\delta^O_{\{5\}} + \delta^O_{\{4\}})$	1	1	1	1	1	1	1	1	1	0	1	1	1	1	0	32	10
O_9	$\mathcal{P}(\delta^O_{\{1\}} + \delta^O_{\{\overline{1,2}\}})$	0	1	1	1	1	0	1	1	0	0	1	1	0	0	0	32	20
O_{10}	$\mathcal{P}(\delta^O_{\{1\}} + \delta^O_{\{1,5\}})$	0	1	0	1	0	0	1	0	0	0	1	0	0	0	0	29	30
O_{11}	$\mathcal{P}(\delta^O_{\{1,5\}} + \delta^O_{\{3,4\}})$	0	1	1	1	1	0	1	1	0	0	0	1	1	1	0	28	15
O_{12}	$\mathcal{P}(\delta^O_{\{5\}}+\delta^O_{\{4\}}+\delta^O_{\{1\}})$	0	1	1	1	1	1	1	1	1	0	1	1	1	1	0	23	10
O_{13}	$\mathcal{P}(\delta^O_{\{5\}}+\delta^O_{\{4\}}+\delta^O_{\{1,3\}})$	0	1	1	0	1	0	1	1	1	0	1	1	1	1	0	21	30
O_{14}	$\mathcal{P}(\beta^O_{\{1\}})$	1	1	0	1	1	0	1	1	1	0	1	1	1	1	0	14	5
O_{15}	$\mathcal{P}(\beta^O_{\{2,3\}})$	0	1	1	1	**2**	1	1	1	1	0	1	1	1	1	0	18	10
O_{16}	$\mathcal{P}(\beta^O_{\{\overline{4,5}\}})$	1	2	1	2	2	1	1	1	1	0	1	1	1	1	0	17	10
O_{17}	$\mathcal{P}(\beta_{\{4,5\}}) = d(K_{2,3})$	0	2	0	2	2	0	1	1	1	0	1	1	1	2	0	14	10
O_{18}	$\mathcal{P}(\beta^O_{\{23\}} + e_{14})$	0	1	1	1	2	1	1	1	1	0	0	1	1	1	0	24	30
O_{19}		1	1	1	2	2	1	1	1	1	0	1	1	1	1	0	23	30
O_{20}		2	2	1	2	2	1	2	1	1	0	2	1	1	1	0	19	30
O_{21}		1	2	2	1	2	0	2	2	1	1	2	2	1	2	1	17	20
O_{22}		0	1	1	2	2	1	1	1	1	0	1	1	1	1	0	17	60
O_{23}		1	2	1	1	2	0	2	1	1	0	1	1	1	1	0	16	60
O_{24}		1	2	2	1	2	0	2	2	1	1	2	2	1	2	0	16	60
O_{25}		1	2	2	1	2	0	2	2	1	0	2	2	1	2	0	15	60
O_{26}		0	2	2	1	2	0	2	2	1	0	2	2	1	2	0	14	20
O_{27}		0	2	2	1	**3**	0	2	2	1	0	2	2	**3**	2	2	22	30
O_{28}		0	2	2	1	3	0	2	2	1	0	0	2	1	2	0	22	60
O_{29}		0	3	2	0	3	0	1	2	1	0	1	2	1	2	0	21	30
O_{30}		0	2		2	3	3	2	1	2	2	0	1	2	2	0	18	30
O_{31}		0	2	2	1	3	0	2	2	1	0	2	2	1	2	0	16	20
O_{32}		0	2	2	3	3	0	2	2	1	0	2	2	1	2	0	16	60
O_{33}		3	3	2	3	3	0	3	2	1	0	3	2	1	2	0	16	60
O_{34}		1	3	2	1	3	0	2	2	1	0	1	2	1	2	0	16	120
O_{35}		0	3	2	2	3	0	1	2	1	0	1	2	1	2	0	15	30
O_{36}		2	3	2	2	3	0	3	2	1	0	3	2	1	2	0	15	60
O_{37}		0	2	2	2	3	1	1	2	1	0	1	2	2	2	0	15	120
O_{38}		0	2	2	2	3	0	1	2	1	0	1	2	1	2	0	14	60
O_{39}		0	2	2	3	3	2	1	2	2	0	1	2	2	2	0	14	60
O_{40}		2	3	2	2	3	0	3	2	1	0	1	1	1	1	0	14	120
O_{41}		2	**4**	2	3	3	0	2	2	1	0	2		1	2	0	16	30
O_{42}		0	2	2	3	4	2	1	2	2	0	1	2	2	2	0	15	60
O_{43}		2	3	2	2	3	0	3	2	1	0	4	3	2	3	2	14	60
O_{44}		0	4	4	3	**5**	0	2	4	1	0	2	4	3	4	2	15	120

Table 12.6: The representatives of orbits of extreme rays in $PMET_5$

The cone $\{0,1\}$-$PMET_5$ has 166 extreme rays from 14 orbits.

There are 585 facets in 15 orbits (up to a permutation), represented below (orbits F_1, F_2, F_3, F_5, F_7 consist of zero-extensions of facets of $\{0,1\}$-$PMET_4$):

- F_1 of size 5: $PW_{11} : p_{11} \geq 0$;
- F_2 of size 30: $ShT_{12,3} : Hyp_{(1,1,-1,0,0)}(p) \geq 0$;
- F_3 of size 20: $PN_{12} : Hyp_{(1,-1,0,0,0)}(p) \geq 0$;
- F_4 of size 10: $Hyp_{(1,1,1,-1,-1)}(p) \geq 0$;
- F_5 of size 30: $Hyp_{(1,1,-1,-1,0)}(p) \geq 0$;
- F_6 of size 20: $Hyp_{(1,1,1,-1,-2)}(p) \geq 0$;
- F_7 of size 20: $Mod_{(2,1,-1,-1,0)}(p) = Hyp_{(2,1,-1,-1,0)}(p) - 2p_{11} \geq 0$;
- F_8 of size 20: $Hyp_{(2,1,-1,-1,-1)}(p) \geq 0$;
- F_9 of size 20: $Mod_{(3,1,-1,-1,-1)}(p) = Hyp_{(3,1,-1,-1,-1)}(p) - 5p_{11} \geq 0$;
- F_{10} of size 10: $Mod_{(2,2,-1,-1,-1)}(p) = Hyp_{(2,2,-1,-1,-1)}(p) - 2(p_{11} + p_{22}) \geq 0$;

- F_{11} of size 60: $Mod_{(2,1,1,-1,-2)}(p) = Hyp_{(2,1,1,-1,-2)}(p) - 2(p_{11} + p_{55}) \geq 0$;
- F_{12} of size 60: $Hyp_{(3,2,-1,-1,-2)}(p) \geq -p_{12} + 5p_{11} + 2p_{22} + p_{15} + p_{25}$;
- F_{13} of size 60: $2p_{12} + 2p_{23} + p_{34} + 2p_{45} + 2p_{51} \geq 2p_{22} + p_{13} + 2p_{25} + p_{41} + 2p_{55}$;
- F_{14} of size 60: $(4p_{13} + 3p_{14} + 4p_{15}) + (2p_{23} + p_{24} + 2p_{25}) \geq 2(p_{34} + p_{35} + p_{45}) + 3p_{12} + 2(p_{11} + p_{33} + p_{55})$;
- F_{15} of size 120: $Hyp_{(-2,-1,1,2,1)}(p) \geq 2(p_{14} + p_{34} + p_{11} + p_{44})$, i.e.,

  $2(p_{13} + p_{14} + p_{15}) + (p_{23} + 2p_{24} + p_{25}) \geq (p_{34} + 2p_{45}) + 2p_{12} + 2(p_{11} + p_{44})$.

The representatives of orbits of facets in $\{0,1\}$-$PMET_5$ are given in Table 12.7.

F_i	Representative	11	21	22	31	32	33	41	42	43	44	51	52	53	54	55	Inc.	$\|F_i\|$
F_1	$PW_{11} : p_{11} \geq 0$	1	0	0	0	0	0	0	0	0	0	0	0	0	0	0	114	5
F_2	$Hyp_{(1,1,-1,0,0)}(p) \geq 0$	-1	1	0	1	-1	0	0	0	0	0	0	0	0	0	0	92	30
F_3	$Hyp_{(1,-1,0,0,0)}(p) \geq 0$	-1	1	0	0	0	0	0	0	0	0	0	0	0	0	0	81	20
F_4	$Hyp_{(1,1,1,-1,-1)}(p) \geq 0$	0	-1	0	-1	-1	0	1	1	1	-1	1	1	1	-1	-1	62	10
F_5	$Hyp_{(1,1,-1,-1,0)}(p) \geq 0$	0	-1	0	1	1	-1	1	1	-1	-1	0	0	0	0	0	54	30
F_6	$Hyp_{(1,1,1,-1,-2)}(p) \geq 0$	0	-1	0	-1	-1	0	1	1	1	-1	2	2	2	-2	-3	36	20
F_7	$Mod_{(2,1,-1,-1,0)}(p) \geq 0$	-2	-2	0	2	1	0	2	1	-1	0	0	0	0	0	0	31	60
F_8	$Hyp_{(2,1,-1,-1,-1)}(p) \geq 0$	-1	-2	0	2	1	-1	2	1	-1	-1	2	1	-1	-1	-1	29	20
F_9	$Mod_{(3,1,-1,-1,-1)}(p) \geq 0$	-5	-3	0	3	1	0	3	1	-1	0	3	1	-1	-1	0	23	20
F_{10}	$Mod_{(2,2,-1,-1,-1)}(p) \geq 0$	-2	-4	-2	2	2	0	2	2	-1	0	2	2	-1	-1	0	20	10
F_{11}	$Mod_{(2,1,1,-1,-2)}(p) \geq 0$	-2	0	-2	-1	0	2	1	1	0	4	2	2	-2	-2	2	20	60
F_{12}		-5	-5	-2	3	2	0	3	2	-1	0	5	3	-2	-2	0	19	60
F_{13}		0	2	-2	-1	2	0	-1	0	1	0	2	-2	0	2	-2	18	60
F_{14}		-2	-3	0	4	2	-2	3	1	-2	0	4	2	-2	-2	-2	18	60
F_{15}		-2	-2	2	1	0	2	2	-1	-2	2	1	0	-2	0	2	17	120

Table 12.7: The representatives of orbits of facets in $\{0,1\}$-$PMET_5$

Partial semimetric convex body for $3 \leq n \leq 6$

In Table 12.8 we present information on the structure of the partial semimetric cone $PMET_n$ and of the partial semimetric convex body $PMET_n^{\square}$ for $3 \leq n \leq 6$. The orbits of $PMET_n$ are given under the group $Sym(n)$ of order $n!$, while for $PMET_n^{\square}$ the orbits are under the group of order $n! \cdot 2^{n-1}$, generated by switchings and permutations. But both have 3 orbits of facets and, hopefully, ridge graphs of diameter 2.

Cone/Body	Dimension	# of ext. rays (orbits)	# of facets (orbits)	Diameters
$PMET_3$	6	13(5)	12(3)	3; 2
$PMET_4$	10	62(11)	28(3)	3; 2
$PMET_5$	15	1,696(44)	55(3)	3; 2
$PMET_6$	21	337,092(734)	96(3)	3; 2
$PMET_3^{\square}$	6	17(4)	19(3)	2; 2
$PMET_4^{\square}$	12	97(6)	44(3)	2; 2
$PMET_5^{\square}$	20	7953(24)	85(3)	3; 2
$PMET_6^{\square}$	12	5090337(427)	146(3)	?; 2

Table 12.8: The number of vertices and facets in $PMET_n$ and $PMET_n^{\square}$ for $3 \leq n \leq 6$

12.5 Theorems and conjectures for general case

Description of the cones $wPMET_n$ and $sPMET_n$

Remind, that for any $S \subset V_n$ we denote by $J(S)$ the $\{0,1\}$-valued function $J : V_n \times V_n \to \{0,1\}$ with $J(S)_{ij} = 1$ exactly when $i, j \in S$. So, $J(V_n)$ and $J(\emptyset)$ are all-ones and all-zeros partial semimetrics, respectively.

For any $S_0 \subset V_n$ and any partition $\mathcal{S} = \{S_1, \dots, S_t\}$ of $\overline{S_0}$, define the *partial multicut* or, specifically, the *partial t-cut* $\gamma_{S_0;S_1,\dots,S_t}$ by

$$\gamma_{S_0;S_1,\dots,S_t} = J(S_0) + \delta_{S_0,S_1,\dots,S_t}.$$

Clearly, $\gamma_{S_0;S_1,\dots,S_t} \in PMET_n$, and

$$\gamma_{S_0;S_1,\dots,S_t} = P(d;w),$$

where

$$d = 2\delta_{S_0,S_1,\dots,S_t} - \delta_{S_0} = \sum_{i=1}^{t} \delta_{S_i},\ w = (w_i = 1_{i \in S_0}).$$

Given integer $i \geq 0$, let $Q(n)$ denote the n-th *partition number*, i.e., the number of ways to write n as a sum of positive integers; $Q(n)$ form the sequence A000041 in [Sloa15]:

$$1, 1, 2, 3, 5, 7, 11, 15, \dots.$$

Given integer $n \geq 0$, let $B(n)$ denote the n-th *Bell number*, i.e., the number of partitions of $V_n = \{1, 2, \dots, n\}$; $B(n)$ form the sequence A000110 in [Sloa15]:

$$1, 1, 2, 5, 15, 52, 203, 877, \dots.$$

So, $B(n)$ is the number of all multicuts on V_n, while the number of all cuts on V_n is 2^{n-1} - the half of the number 2^n of all subsets of V_n.

The number of all oriented cuts on V_n is $2^n - 1$ - the number 2^n of all subsets of V_n minus 1, while the number of all oriented multicuts on V_n is the n-th *ordered Bell number* $Bo(n)$, i.e., the number of ordered partitions of V_n; $Bo(n)$ form the sequence A000670 in [Sloa15]:

$$1, 1, 3, 13, 75, 541, 4683, 47293, \dots.$$

Denote

$$MET_{n;0} = \{(d;(0)) : d \in MET_n\},\ \text{and}\ CUT_{n;0} = \{(d;(0)) : d \in CUT_n\}.$$

So,

$$MET_n \simeq MET_{n;0} \simeq P(MET_{n;0}) \simeq Q(MET_{n;0}),$$

and

$$CUT_n \simeq CUT_{n;0} \simeq P(CUT_{n;0}) \simeq Q(CUT_{n;0}).$$

Denote by $WCUT_n$ the cone $\{(d;w) \in WMET_n \,|\, d \in CUT_n\}$ of all *weighted l_1-semimetrics* on V_n.

Denote
$$e_j = (((0)); w = (w_i = 1_{i=j})) \in WMET_n.$$
So, $2P(e_j)$ is equal to 2 on the position jj, is equal to 1 on the positions ij, ji with $i \neq j$, and is equal to 0, otherwise; $2Q(e_j)$ is equal to -1 on the position ji, is equal to 1 on the position ij (with $i \neq j$ again), and is equal to 0, otherwise.

Denote by $F(C)$ the set of facets of a cone C, and by $R(C)$ the set of its extreme rays.

Theorem 12.1 *The following statements hold:*

(i) the sets $R(WMET_n)$, $R(wPMET_n)$, and $R(wWQMET_n)$ of extreme rays of the cones $WMET_n$, $wPMET_n$, and $wWQMET_n$ are connected with the set $R(MET_{n;0})$ of extreme rays of the cone $MET_{n;0}$ by the formulas
$$R(WMET_n) = \{e_j \,|\, j \in V_n\} \cup R(MET_{n;0}),$$
$$R(wPMET_n) = \{2P(e_j) \,|\, j \in V_n\} \cup P(R(MET_{n;0})),$$
$$R(wWQMET_n) = \{2Q(e_j) \,|\, j \in V_n\} \cup Q(R(MET_{n;0}));$$

(ii) the sets $F(WMET_n)$, $F(wPMET_n)$, and $F(wWQMET_n)$ of facets of the cones $WMET_n$, $wPMET_n$, and $wWQMET_n$ are connected with the set $F(MET_{n;0})$ of facets of the cone $MET_{n;0}$ by the formulas
$$F(WMET_n) = \{w_j \geq 0 \,|\, j \in V_n\} \cup F(MET_{n;0}),$$
$$F(wPMET_n) = \{p_{jj} \geq 0 \,|\, j \in V_n\} \cup F(P(MET_{n;0})),$$
$$F(wWQMET_n) = \{w_j \geq 0 \,|\, j \in V_n\} \cup F(Q(MET_{n;0}));$$

(iii) the incidence number $Inc(2P(e_j))$ of the extreme ray $2P(e_j)$ in the cone $wPMET_n$ is
$$Inc(2P(e_j)) = |F(wPMET_n)| - 1,$$
while the incidence number $Inc(PW_{jj})$ of the facet $PW_{jj} : p_{jj} \geq 0$ in the cone $wPMET_n$ is
$$Inc(PW_{jj}) = |R(wPMET_n)| - 1;$$

(iv) the 1-skeleton graphs and the ridge graphs of the cones $WMET_n$, $wPMET_n$, and $wWQMET_n$ coincide and are connected with the 1-skeleton and the ridge graphs of the cone MET_n by the formulas
$$G_{WMET_n} = G_{wPMET_n} = G_{wWQMET_n} = K_n \times G_{MET_n},$$
$$G^*_{WMET_n} = G^*_{wPMET_n} = G^*_{wWQMET_n} = K_n \times G^*_{MET_n};$$

(v) for the cone $wPMET_n$, its full symmetry group $Is(wPMET_n)$, the diameters $d(G_{wPMET_n})$ and $d(G^*_{wMET_n})$ of its 1-skeleton and ridge graphs, and edge-connectivity are the same, as for the cone MET_n;

(vi) the $\{0,1\}$-valued elements of the cone $wPMET_n$ are $B(n+1)$ $\{0,1\}$-valued elements of $PMET_n$, and $\{0,1\}$-$wPMET_n = CUT_{n;0}$;

(vii) the $\{0,1\}$-valued elements of $WMET_n$ are $2^n \cdot B(n)$ $\{0,1\}$-weighted multicuts on V_n, and $\{0,1\}$-$WMET_n = WCUT_n$; moreover,
$$R(WCUT_n) = \{e_j \,|\, j \in V_n\} \cup R(CUT_{n;0}),$$

and $G_{WCUT_n} = K_{n+S(n,2)}$, *while*

$$F(WCUT_n) = \{w_j \geq 0\, vert\, j \in V_n\} \cup F(CUT_{n;0}),$$

and $G^*_{WCUT_n} = K_n \times G^*_{CUT_n}$; *it has diameter* 2.

Proof. (i) Let $p \in wPMET_n$. We will show that

$$p' = p - \frac{1}{2}\sum_{t=1}^{n} p_{tt} 2P(e_t) \in MET_{n;0}.$$

For example, a well-known weak partial semimetric $i+j$ is the sum of $\sum_t t2P(e_t)$ and the all-zero semimetric $((0))$.

In fact,

$$p'_{ii} = p_{ii} - \frac{1}{2} p_{ii} 2P(e_i)_{ii} = 0.$$

Also, p' satisfies to all triangle inequalities, since for different $i, j, k \in V_n$ we have

$$\begin{aligned} ShT_{ij,k} &= p'_{ij} + p'_{ik} - p'_{jk} \\ &= \left(p_{ij} - \frac{p_{ii}+p_{jj}}{2}\right) + \left(p_{ik} - \frac{p_{ii}+p_{kk}}{2}\right) - \left(p_{jk} - \frac{p_{jj}+p_{kk}}{2}\right) \\ &= p_{ij} + p_{ik} - p_{jk} - p_{ii} \geq 0. \end{aligned}$$

So, $2P(e_i)$, $1 \leq i \leq n$, and the generators of $P(MET_{n;0}) \simeq MET_{n;0}$ (i.e., the zero-extensions of the generators of MET_n) generate $wPMET_n$.

They are, moreover, the generators of $wPMET_n$, since they belong to all n linearly independent facets PW_{ii}; so, their rank in $\mathbb{R}^{\binom{n+1}{2}}$is $(\binom{n}{2} - 1) + n = \binom{n+1}{2} - 1$.

Clearly, any $2P(e_i)$ belongs to all facets of $wPMET_n$, except PW_{ii}, i.e., its incidence is $(n-1) + 3\binom{n}{3}$. So, its rank in $\mathbb{R}^{\binom{n+1}{2}}$ is $\binom{n+1}{2} - 1$.

For $WMET_n$ and $wWQMET_n$, (i) follows similarly, as well as (ii).

(iii), (iv). The ray of $2P(e_i)$ is adjacent to any other extreme ray r, as the set of facets that contain r (with rank $\binom{n+1}{2} - 1$) only loses one element, if we intersect it with the set of facets that contain $2P(e_i)$.

(v). The diameters of $G^*_{wPMET_n}$ and G_{wPMET_n} being 2, their edge-connectivity is equal to their minimal degrees [Ples75]. But this degree is the same as of $G^*_{MET_n}$ (which is regular of degree $\frac{(n-3)(n^2-6)}{2}$, if $n > 3$), and of G_{MET_n}, respectively.

$Is(wPMET_n)$ for $n \geq 5$ is $Sym(n)$, because it contains $Sym(n)$ but cannot be larger than $Is(MET_n) = Sym(n)$.

(vi) If $p \in wPMET_n$ is $\{0,1\}$-valued, then $p_{ij} = 0 < p_{ii} = 1$ is impossible, because $2p_{ij} \geq p_{ii} + p_{jj}$; so, $p \in PMET_n$.

(vii) is implied by (i) and (ii).

□

Any partial semimetric $p \in PMET_n$ induces the partial order on V_n by defining $i \preceq j$ if $p_{ii} = p_{ij}$. This *specialization order* is important in Computer Science applications, where the partial metrics act on certain posets, called *Scott domains*.

In particular, $i_0 \in V_n$ is an *p-maximal* element in V_n, if $p_{ii} = p_{ii_0}$ for all $i \neq i_0$.

It is an *p-minimal* element in V_n, if $p_{i_0 i_0} = p_{i i_0}$ for all $i \neq i_0$.

The *lifting* of $p \in PMET_n$ is the function $p^+ = ((p^+_{ij}))$, $i, j \in V_n \cup \{0\}$, with $p^+_{00} = 0$, $p^+_{0i} = p^+_{i0} = p^+_{ii}$ for $i \in V_n$, and $p^+_{ij} = p_{ij}$ for $i, j \in V_n$.

Clearly, 0 is an p^+-maximal element in the specialization order, induced on $V_n \cup \{0\} = \{0, 1, 2, ..., n\}$ by p^+, since $p^+_{ii} = p_{ii}$, as well as $p^+_{i0} = p_{ii}$ for all $i \in V_n$.

Theorem 12.2 *The following statements hold:*

(i) $sPMET_n = \{p \in PMET_n \,|\, p^+ \in PMET_{n+1}\}$*;*

(ii) $sWMET_n = MET_{n+1} \simeq P(MET_{n+1}) = sPMET_n$*;*

(iii) the $\{0,1\}$*-valued elements of* $sPMET_n$ *are the non-discrete semimetric* $((0))$*, and* $2^n - 1$ *partial* 2*-cuts* $\gamma_{S,\overline{S}}$*,* $S \neq \emptyset$*, generating the cone* $\{0,1\}$*-*$sPMET_n$*; moreover,*

$$\{0,1\}\text{-}sWMET_n = CUT_{n+1} \simeq P(CUT_{n+1}),$$

and $Q(CUT_{n+1}) = OCUT_n$*.*

Proof. We should check for p^+ only facet-defined inequalities PW_{ii}, $ShT_{ij,k}$, and PN_{ij}, involving the new point 0. $2n+1$ of the required inequalities hold as equalities: $p^+_{00} = 0$, and $p^+_{0i} = p^+_{i0} = p^+_{ii} = p_{ii}$ for $i \in V_n$. All $ShT_{0j,i} = p_{ii} + p_{ij} - p_{jj} \geq 0$ hold since PN_{ji} is satisfied. All $ShT_{ij,0} = p_{ii} + p_{jj} - p_{ij} \geq 0$ hold whenever p satisfies PS_{ij}, i.e, for $p \in PMET_n$.

□

Given $p \in sPMET_n$, the semimetric $P^{-1}(p^+) \in MET_{n+2}$ is $P^{-1}(p) \in MET_{n+1}$ with the first point split in two coinciding points. The cone $sPMET_n$ is nothing but the linear image $P(sWMET_n = MET_{n+1})$. So, for $n \geq 4$, $Is(sWMET_n) = Sym(n+1)$ on $\{0, 1, 2, ..., n\}$, acting as $p' = P(\tau(P^{-1}(p)))$ on $sPMET_n$ for any $\tau \in Sym(n+1)$. If τ fixes 0, then $p' = \tau(p)$.

The cone $PMET_n$

Call a partial semimetric $p = ((p_{ij})) \in PMET_n$ *reducible* if $\min_{1 \leq i \leq n} p_{ii} = 0$. For any quasi-semimetric $q = ((q_{ij})) \neq ((0))$ in the cone $WQMET_n$, there exists a weight function $w = (w_i) : V_n \longrightarrow \mathbb{R}_{\geq 0}$ such that $((q_{ij} + w_i))$ is a partial semimetric.

Clearly, $((q_{ij} + w'_i))$ is also a partial semimetric for any (weight) function $w' = (w'_i)$ with

$$w'_i = w_i - \min_{1 \leq j \leq n} w_j + \lambda, \; \lambda \geq 0.$$

In other words, there is a bijection $\mathcal{P}$ between all weightable non-zero quasi-semimetrics $q \in WQMET_n$ and all reducible partial semimetrics $p = \mathcal{P}(q) \in PMET_n$. For $q = ((0))$, set $\mathcal{P}(q) = J$, where $J = ((J_{ij}))$ with $J_{ij} = 1$. So, $\mathcal{P}$ is a bijection between $WQMET_n$ and the set of rays $\{\mathcal{P}(q) + \lambda J\}$, where $\lambda \geq 0$. Here, for $q = ((0))$, $\mathcal{P}(q) = J$ and, otherwise, it is a reducible partial semimetric.

Proposition 12.2 *If* $q = ((q_{ij})) \in WQMET_n$ *with a weight function* $w = (w_i)$ *and* $p = ((q_{ij} + w_i))$*, then* $rank(p) = rank(q) + |\{1 \leq i \leq n : w_i = 0\}|$*.*

Proof. Clearly, p lies on the facet $PN_{ij} : p_{ij} - p_{ii} \geq 0$ with $i \neq j$ of $PMET_n$ if and only if q lies on the facet $NN_{ij} : q_{ij} \geq 0$ of $WQMET_n$.

Also, p lies on the facet $ShT_{ik,j} : p_{ij} + p_{jk} - p_{ik} - p_{jj} \geq 0$ of $PMET_n$ if and only if q lies on the facets $OT_{ik,j} : q_{ij} + q_{jk} - q_{ik} \geq 0$, and $OT_{ki,j} = q_{kj} + q_{ji} - q_{ki} \geq 0$ of $WQMET_n$.

The equality $OT_{ik,j} = OT_{ki,j}$ for weightable quasi-semimetrics is exactly the relaxed symmetry, characterizing weightable quasi-semimetrics. So, all triangle equalities for q stay for p, while equalities $q_{ij} = 0$ for $1 \leq i \neq j \leq n$ transform into $p_{ij} - p_{ii} = 0$. Only the equalities $p_{ii} = 0$, corresponding to the equalities $w_i = 0$, became new; they increase the rank.

□

Above Proposition implies that if $q \in WQMET_n$ represents an extreme ray, then $\mathcal{P}(q) \in PMET_n$ also represents an extreme ray. In fact, the cones have the same dimension. But more extreme rays appear in $PMET_n$. For $q^{(1)}, q^{(2)} \in WQMET_n$, $\mathcal{P}(q^{(1)}) + \mathcal{P}(q^{(2)})$ belongs to the ray $\{\mathcal{P}(q^{(1)} + q^{(2)}) + \lambda J\}$, i.e., $\mathcal{P}(q^{(1)} + q^{(2)}) = \mathcal{P}(q^{(1)}) + \mathcal{P}(q^{(2)}) - \lambda J$ with $\lambda \geq 0$, which is weaker than linearity, corresponding to $\lambda = 0$.

Recall that $MET_n = PMET_n \cap WQMET_n$. Above Proposition implies also that a semimetric represents an extreme ray in $\binom{n+1}{2}$-dimensional cone $PMET_n$ if and only if it represents an extreme ray in $\binom{n}{2}$-dimensional cone MET_n. In fact, exactly n new valid equalities $p_{ii} = 0$ appear for it in $PMET_n$. See the number of such extreme rays for $n \leq 6$ in Table 12.1. For $n \leq 6$, the orbit-representing semimetrics, besides cuts δ_S with $1 \leq |S| \leq \lfloor \frac{n}{2} \rfloor$, are $d(K_{\{1,2\},\{3,4,5\}})$, $d(K_{\{1=6,2\},\{3,4,5\}})$, $d(K_{\{1,2\},\{3,4,5=6\}})$, $d(K_{\{1,2\},\{3,4,5,6\}})$, $d(K_{\{1,2,3\},\{4,5,6\}})$, $d(K_{\{1,2,3\},\{4,5,6\}} - e_{14})$. Here notation $-e_{ij}$ means that the edge ij is deleted.

$\{0,1\}$-valued elements of $PMET_n$ and $dWMET_n$

For a partition $\mathcal{S} = \{S_1, ..., S_t\}$ of V_n and $A \subseteq \{1, ..., t\}$, let us denote $\hat{A} = \bigcup_{h \in A} S_h$, and $w(\hat{A}) = (w_i = 1_{i \in \hat{A}})$. So, weight is constant on each S_h.

As before, for any $S \subset V_n$, denote by $J(S)$ the $\{0,1\}$-valued function with $J(S)_{ij} = 1$ exactly when $i, j \in S$. For any $S_0 \subset V_n$ and partition $\mathcal{S} = \{S_1, \ldots, S_t\}$ of $\overline{S}_0$ define the *partial t-cut* as $\gamma_{S_0;S_1,\ldots,S_t} = J(S_0) + \delta_{S_0,S_1,\ldots,S_t}$. Clearly, $\gamma_{S_0;S_1,\ldots,S_t} \in PMET_n$, and it is $P(d;w)$, where $d = 2\delta_{S_0,S_1,\ldots,S_t} - \delta_{S_0} = \sum_{i=1} \delta_{S_i}$ and $w = (w_i = 1_{i \in S_0})$.

Given integer $n \geq 0$, let $Q(n)$ be the n-th *partition number*, i.e., the number of ways to write i as a sum of positive integers; let $B(n)$ be the n-th *Bell number*, i.e., the number of partitions of $V_n = \{1, 2, ..., n\}$; let $S(n,t)$ be the Stirling number of the second kind, i.e., the number of partitions of $V_n = \{1, 2, ..., n\}$ into t non-empty parts.

Theorem 12.3 *The following statements hold:*
(i) the all $\{0,1\}$-valued elements of $dWMET_n$ are $\sum_{t=1}^{n} 2^t S(n,t)$ elements

$$(\delta_{\mathcal{S}}; w(\hat{A}));$$

the set $R(\{0,1\}\text{-}dWMET_n)$ of extreme rays of the cone $\{0,1\}$-$dWMET_n$ consists of all such elements with $(|A|, t - |A|) = (1,0)$, $(0,2)$, or $(1,1)$, i.e., the set of extreme rays of $\{0,1\}$-$dWMET_n$ contains $(((0)); (1))$ (non-discrete semimeric $((0))$ with weight function $w = (1)$), and 2-cuts $(\delta_S; w)$ with weights $(0), w' = (1_{i \in S})$, or $w'' = (1_{i \notin S})$; there are $1 + 3(2^{n-1} - 1)$ extreme rays of $\{0,1\}$-$dWMET_n$, divided into $\lfloor \frac{3n}{2} \rfloor$ orbits;

(ii) the $\{0,1\}$*-valued elements of* $WQMET_n$ *are*

$$Q(2\delta_S - \delta_{\hat{A}}; w(A)) = \delta_S - \delta^O_{\hat{A}};$$

the set $R(\{0,1\}\text{-}dWMET_n)$ *of extreme rays of the cone* $\{0,1\}$*-*$dWMET_n$ *consists of all such elements with either* $|A| = t - |A| = 1$ *(oriented 2-cuts), or* $2 \le |A|, t - |A| \le n-2$*;*

(iii) the $\{0,1\}$*-valued elements of* $PMET_n$ *are the partial multicuts*

$$P(2\delta_S - \delta_{\hat{A}}; w(A)) = \delta_S + J(\hat{A}) \;\; with \;\; |A| \le 1;$$

there are $B(n+1) = \sum_{i=0}^{n} \binom{n}{i} B(i)$ *of them, divided into* $\sum_{i=0}^{n} Q(i)$ *orbits; the set* $R(\{0,1\}\text{-}PMET_n)$ *of extreme rays of the cone* $\{0,1\}$*-*$PMET_n$ *consists of all such elements except* $B(n) - (2^{n-1} - 1)$ *partial t-cuts with* $|A| = 0, t \neq 2$*, belonging to* $Q(n) - \lfloor \frac{n}{2} \rfloor$ *orbits.*

Proof. (i) the weighted semimetric $(((0)); (1))$ belongs to $R(\{0,1\}\text{-}dWMET_n)$, since its rank is $\binom{n}{2}$ plus $n-1$, the rank of the set of equalities $d_{ij} = w_i - w_j$.

The same holds for $(\delta_S; w)$ with weight $(0), w' = (1_{i \in S})$, or $w'' = (1_{i \notin S})$, since their rank is $\binom{n}{2} - 1$ plus $k \ge 1$ equalities $w_i = 0$, plus, if $k < n$, $n-k$ equalities $d_{i'j} = w_j - w_{i'}$, where $w_{i'} = 0$, and $w_j = 1$.

But the all-ones-weighted 2-cut δ is equal to $\frac{1}{2}((\delta; w') + (\delta; w'') + (((0)); (1))$.

No other $(\delta_{S_1,\dots,S_t}; w)$ belongs to $R(\{0,1\}\text{-}dWMET_n)$, since t should be 2 (otherwise, the rank will be $< \binom{n}{2} - 1 + n$), and the weight should be constant on each S_h, $1 \le h \le t$.

(ii) The $\{0,1\}$-valued elements of $WMET_n$ should be $\{0,1\}$-weighted multicuts δ_S. Now, the inequality $d_{ij} \ge w_i - w_j$, valid on $dWMET_n$, implies (i).

Let $q \in WQMET$ be $\{0,1\}$-valued. Without loss of generality, let

$$\min_{1 \le i \le n} w_i = w_1 = 0.$$

But $q_{1i} + w_1 = q_{i1} + w_i$ for any $i > 1$. So, $w_i = 1$ if and only if $q_{1i} \neq q_{i1}$. The quasi-semimetrics q, restricted on the sets $\{i \,|\, w_i = 0\}$ and $\{i \,|\, w_i = 1\}$, should be $\{0,1\}$-valued semimetrics, i.e., multicuts.

(iii) Let us prove (see for details [DeDe10] and the previous subsection), that all $\{0,1\}$-valued elements of the cone $PMET_n$ are $\sum_{0 \le i \le n} \binom{n}{i} B(n-i)$ (organized into $\sum_{0 \le i \le n} Q(i)$ orbits under $Sym(n)$) elements of the form

$$J(S_0) + \delta_{S_0, S_1, \dots, S_t} = \mathcal{P}\left(\sum_{1 \le i \le t} \delta^O_{S_i} \right),$$

where S_0 is any subset of V_n, and $S_1, \dots, S_t$ is any partition of $\overline{S}_0$.

For it, given an $\{0,1\}$-valued element p of $PMET_n$, let us construct partition $S_0, S_1, \dots, S_t$, such that $p = J(S_0) + \delta_{S_0, S_1, \dots, S_t}$. See, for example, below the partial semimetric p on V_7 (in fact, an $\{0,1\}$-valued extreme ray of $PMET_7$), defined by

$$p = ((p_{ij})) = J(\{6,7\}) + \delta_{\{6,7\},\{1\},\{2,3\},\{4,5\}},$$

and its corresponding weightable quasi-semimetric $q : p = \mathcal{P}(q)$ (in fact, an $\{0,1\}$-valued non-extreme ray of $WQMET_7$), defined by

$$q = ((q_{ij} = p_{ij} - p_{ii})).$$

$$
\begin{array}{ccccccc}
\mathbf{0} & 1 & 1 & 1 & 1 & 1 & 1 \\
1 & \mathbf{0} & 0 & 1 & 1 & 1 & 1 \\
1 & 0 & \mathbf{0} & 1 & 1 & 1 & 1 \\
1 & 1 & 1 & \mathbf{0} & 0 & 1 & 1 \\
1 & 1 & 1 & 0 & \mathbf{0} & 1 & 1 \\
1 & 1 & 1 & 1 & 1 & \mathbf{1} & 1 \\
1 & 1 & 1 & 1 & 1 & 1 & \mathbf{1}
\end{array}
\qquad
\begin{array}{ccccccc}
\mathbf{0} & 1 & 1 & 1 & 1 & 1 & 1 \\
1 & \mathbf{0} & 0 & 1 & 1 & 1 & 1 \\
1 & 0 & \mathbf{0} & 1 & 1 & 1 & 1 \\
1 & 1 & 1 & \mathbf{0} & 0 & 1 & 1 \\
1 & 1 & 1 & 0 & \mathbf{0} & 1 & 1 \\
0 & 0 & 0 & 0 & 0 & \mathbf{1} & 0 \\
0 & 0 & 0 & 0 & 0 & 0 & \mathbf{1}
\end{array}
$$

Set $S_0 = \{1 \leq k \leq n \,|\, p_{kk} = 1\}$; then $p_{kk'} = p_{k'k} = 1$ for any $k \in S_0$, and $1 \leq k' \leq n$ by definition of the facets $PN_{kk'}, PN_{k'k}$.

Let S_1 be a maximal subset of $\overline{S}_0$, such that $p_{kk'} = 0$ for $k, k' \in S_1$, then S_2 be a maximal subset of $\overline{S_0 \cup S_1}$, such that $p_{kk'} = 0$ for $k, k' \in S_2$, and so on.

It remains to show that $p_{kk'} = 1$, if $k \in S_i$, $k' \in S_{i'}$ with different $1 \leq i, i' \leq t$.

Without loss of generality, suppose that $|S_i| \geq 2$ for some $1 \leq i \leq n$, since, otherwise, our statement holds by construction of sets S_i. The inequalities $ShT_{kj,k'} \geq 0$ and $ShT_{k'j,k} \geq 0$, where $k, k' \in S_i$ and $j \in S_{i'}$, imply $p_{kj} = p_{k'j}$, since $p_{kk'} = p_{kk} = p_{k'k'} = 0$. Now, $p_{kj} = p_{k'j} = 0$ is impossible by construction of sets S_i; so, $p_{kj} = p_{k'j} = 1$.

For example, there are $52 = 1 \times 1 + 4 \times 1 + 6 \times 2 + 4 \times 5 + 1 \times 15$ (1+1+2+3+5 orbits) $\{0,1\}$-valued elements of $PMET_4$. Among them, only $\delta_{S_1,\dots,S_t}$ with $t = 1, 3, 4$, i.e., $((0))$, $\delta_{\{1\},\{2\},\{3\},\{4\}}$, and 6 elements of the orbit with $t = 3$ are not representatives of extreme rays.

□

Remark. Easy to see, that the incidence number of the $\{0,1\}$-valued element

$$p = J(S_0) + \delta_{S_0,S_1,\dots,S_t}$$

is the sum of the following values:

- $n - |S_0|$ (to facets $PW_{ii} : p_{ii} \geq 0$);
- $\sum_{1 \leq k \leq t} |S_k|(|S_k| - 1) + (|S_0|(|S_0| - 1) + |S_0|(n - |S_0|)$ (to facets $PN_{ij} : p_{ij} - p_{ii} \geq 0$, $i \neq j$, with $0 - 0 = 0$ and $1 - 1 = 0$, respectively);
- $3 \sum_{1 \leq k \leq t} \binom{|S_k|}{3} + \sum_{1 \leq k \leq t} |S_k|(|S_k| - 1)(n - |S_k|) + |S_0| \sum_{1 \leq k \leq k' \leq t} |S_k||S_{k'}|$ (to facets $ShT_{ij,k} : p_{ik} + p_{kj} - p_{ij} - p_{kk} \geq 0$ with $0 + 0 - 0 - 0 = 0$, $1 + 0 - 1 - 0 = 0$ and $1 + 1 - 1 - 1 = 0$).

It can be checked by direct computation.

On the other hand, the incidence number of the extreme ray, represented by the cut semimetric $\delta_S = \mathcal{P}(\delta_S)$, is

$$3\binom{n}{3} - \frac{n(n - |S|)(|S| - 2)}{2} + |S|^2 = (3\binom{n}{3} + n^2) - \frac{n|S|(n - |S|)}{2} - |S|(n - |S|).$$

The incidence number of the extreme ray, represented by $\mathcal{P}(\delta_S^O) = J(\overline{S}) + \delta_S$, is

$$\left(3\binom{n}{3} + n^2\right) - \frac{n|S|(n - |S|)}{2} - (n - |S|).$$

The case $S = \emptyset$ corresponds to the extreme ray $J = J(V_n) + \delta_\emptyset$ of all-ones.

The orbit size of $\mathcal{P}(\delta_S)$ and $\mathcal{P}(\delta_S^O)$ is $\binom{n}{|S|}$, except the case $|S| = \frac{n}{2}$, when it is $\frac{1}{2}\binom{n}{|S|}$. The incidence number of the extreme ray represented by $\mathcal{P}(\sum_{1\le i\le t}\delta_{\{i\}}^O)$, is

$$t + t(n-t) + (n-t)\binom{n-1}{2}.$$

The size of its orbit is $\binom{n}{t}$. For $t = n-1$, it is a simplicial extreme ray, represented by $P(\beta_{\{n\}}^O)$, where $\beta_{\{n\}}^O$ is an oriented anti-multicut quasi-semimetric (see Chapter 7).

Some other extreme rays of $PMET_n$

Given a *oriented anti-multicut quasi-semimetric* β_S^O (remind, that $(\beta_S^O)_{ij} = 1-(\delta_S^O)_{ij}$, for $1 \le i \ne j \le n$, and $(\beta_S^O)_{ij} = 0$, otherwise), the partial semimetric $((p_{ij})) = \mathcal{P}(\beta_S^O)$ have $p_{ij} = 2$, if $|\{i,j\}\cap S| = 2$, $p_{ii} = 0$, if $i \notin S$, and $p_{ij} = 1$, otherwise.

So, it is the matrix J of all-ones if $S = \emptyset$. If $|S| \ge 2$, then the incidence number of $\mathcal{P}(\beta_S^O)$ is $\frac{n|S|(n-|S|)}{2} + (n-|S|)$. It is the sum of the following values:

- $(n-|S|)$ (to facets PW_{ii}, $i \notin S$);
- $|S|(n-|S|)$ (to facets PN_{ij}, $i \in S, j \notin S$);
- $|S|\binom{n-|S|}{2} + (n-|S|)\binom{|S|}{2}$ (to facets $ShT_{ij,k} : 0 = 1+1-2-0$, or $0 = 1+1-1-1$ for $i,j \in S, k \notin S$, or $i,j \notin S, k \in S$, respectively).

It can be shown, that $\mathcal{P}(\beta_S^O)$ represents an extreme ray of $PMET_n$, and this ray is simplicial if and only if $|S| = 1$.

The partial semimetric $((p_{ij})) = \mathcal{P}(\beta_S)$ is the semimetric $\beta_S = d(K_{S,\overline{S}})$ (remind, that the *anti-cut semimetric* $\beta_S = \beta_S^O + \beta_{\overline{S}}^O$), which represent also an extreme ray in MET_n, if $2 \le |S| \le n-2$. The incidence number of it, as an extreme ray of $PMET_n$, is

$$n + 0 + |S|\binom{n-|S|}{2} + (n-|S|)\binom{|S|}{2} = \frac{n|S|(n-|S|)}{2} + n - |S|(n-|S|).$$

We conjecture, that o-cuts δ_S^O with $1 \le |S| \le n-1$, and o-anticuts β_S^O with $2 \le |S| \le n-2$ are only representatives q of extreme ray in $QMET_n$, such that $\mathcal{P}(q)$ represent an extreme ray in $PMET_n$.

Above formulas for incidence numbers imply that, for any n, the partial semimetrics $P(\beta_{\{i\}}^O) = \sum_{j\in\{\bar{i}\}}\delta_{\{j\}}^O$ form unique orbit of $\{0,1\}$-valued representatives of a *simplicial* (i.e., with incidence number $\binom{n+1}{2} - 1$) extreme ray in $PMET_n$. Besides, $PMET_n$ with $n = 4,5,6$ have, respectively, $0,1,16$ such orbits of size $n!$, $1,3,8$ such orbits of size $\frac{n!}{2}$, and $0,1,1$ such orbits of size $\frac{n!}{3!}$. Also, $PMET_4$ has one such orbit of size 10. Hence, altogether $PMET_n$ with $n = 3,4,5,6$ have $3,16,340,14526$ simplicial extreme rays, organized in $1,2,7,26$ orbits, respectively.

The diameter of the 1-skeleton graph of $PMET_n$ is, perhaps, 2, because the extreme ray J of all ones is incident to all facets, incident to any extreme ray $\{p+\lambda J\}$, except PW_{ii}, whenever $p_{ii} > 0$.

If p is any other $\{0,1\}$-valued partial semimetric, i.e.,

$$p = \mathcal{P}\left(\sum_{1\le k\le t} \delta^O_{S_k}\right),$$

then $n-|S_0|$ such facets are excluded. In particular, for simplicial extreme ray represented by

$$p = \mathcal{P}(\beta^O_{\{i\}}) = \sum_{j\in\{\bar{i}\}} \delta^O_{\{j\}},$$

the common facets are $\binom{n+1}{2} - 1$ facets of p, except $n-1$ facets $p_{jj} = 0$ with $1 \le j \le n, j \ne i$; so, they are not adjacent.

Vertex-splitting and zero-extension operations

The *vertex-splitting* of a function $p = ((p_{ij}))$ on $V_n \times V_n$ is a function $f^{vs} = ((f^{vs}_{ij}))$ on $V_{n+1} \times V_{n+1}$, defined, for $1 \le i, j \le n+1$, by

$$f^{vs}_{n+1\,n+1} = f^{vs}_{n\,n+1} = f^{vs}_{n+1\,n} = 0,\ f^{vs}_{i\,n+1} = p_{in},\ f^{vs}_{n+1\,i} = p_{ni},\ \text{and}\ f^{vs}_{ij} = p_{ij}\,.$$

Vertex-splitting of an o-multicut $\delta^O_{S_1,\dots,S_q}$ is the o-multicut $\delta^O_{S_1,\dots,S_l\cup\{n+1\},\dots,S_q}$, if $n \in S_l$.

Vertex-splitting of a generic $\{0,1\}$-valued element $J(S_0) + \delta_{S_0,S_1,\dots,S_t}$ of $PMET_n$ is $J(S_0) + \delta_{S_0,S_1,\dots,S_l\cup\{n+1\},\dots,S_t} \in PMET_{n+1}$, if $n \in S_l$ with $l \ne 0$, and it is not a partial semimetric, otherwise. So, the only $\{0,1\}$-valued elements, which are not vertex-splittings, are those with $|S_i| = 1$ for all $1 \le i \le t$.

The vertex-splitting of an $\{0,1,2\}$-valued extreme ray, represented by $\mathcal{P}(\beta^O_S)$, is a $\mathcal{P}(\beta^O_S + e_{n\,n+1}) \in PMET_{n+1}$, if $n \notin S$, and it is not a partial semimetric, otherwise.

The orbit O_{18} of extreme rays in $PMET_5$ consists of vertex-splittings of ones from the orbit O_{10} of $\mathcal{P}(\beta^O_{\{1,4\}})$ in $PMET_4$.

The orbits O_{28} and O_{29} of $\{0,1,2,3\}$-valued extreme rays in $PMET_5$ consist of vertex-splittings (two ways) of ones from the orbit O_{11} in $PMET_4$.

Theorem 12.4 *If a partial semimetric p represents an extreme ray of $PMET_n$ and has $p_{nn} = 0$, then its vertex-splitting p^{vs} represents an extreme ray of $PMET_{n+1}$.*

Proof. The condition $p_{nn} = 0$ is needed since, otherwise, p^{vs} violate the inequality

$$p_{n+1\,n+1} - p_{nn} \ge 0,$$

which is valid in $PMET_{n+1}$.

It suffice to present $n+1$ facets which, together with $\binom{n+1}{2} - 1$ linearly independent facets (seen as vectors) containing p, will form $\binom{n+2}{2} - 1$ linearly independent facets, containing p^{vs}.

Such facets are two of types $PW_{ii} : p_{n+1\,n+1} \ge 0$ and $PN_{ij} : p_{n\,n+1} - p_{n+1\,n+1} \ge 0$, and $n-1$ of type $ShT_{in,n+1} : p_{i\,n+1} + p_{n\,n+1} - p_{in} - p_{n+1\,n+1} \ge 0$.

□

Above Theorem gives another proof for the completeness of the list of $\{0,1\}$-valued extreme rays of $PMET_n$.

The cone $\{0,1\}$-$PMET_n$

The subcone $\{0,1\}$-$PMET_n$ of $PMET_n$, generated by all its $\{0,1\}$-valued extreme rays, consists of all partial semimetrics $p=((p_{ij}))$, such that $q=((p_{ij}-p_{ii}))\in OCUT_n$, i.e., the quasi-semimetric q is l_1-embeddable. $\{0,1\}$-$PMET_n$ coincides with $PMET_n$ only for $n=3$.

A *zero-extension of an inequality* $\sum_{1\le i\ne j\le n-1} p_{ij}d_{ij}\ge 0$ is an inequality

$$\sum_{1\le i\ne j\le n} f'_{ij}d_{ij}\ge 0 \text{ with } f'_{ni}=f'_{in}=0, \text{ and } f'_{ij}=p_{ij}, \text{ otherwise.}$$

Easy to see that zero-extension of any facet-defining inequality of $\{0,1\}$-$PMET_{n-1}$ is a valid inequality of $\{0,1\}$-$PMET_n$. We conjecture that, moreover, it is a facet-defining inequality of $\{0,1\}$-$PMET_n$.

Given a sequence $b=(b_1,...,b_n)$ of integers, satisfying the condition $\sum_{i=1}^n b_i\in\{0,1\}$, and an $p=((p_{ij}))\in\{0,1\}$-$PMET_n$, consider the inequalities

$$Hyp_b(p)=-\sum_{1\le i<j\le n} b_ib_jp_{ij}-\frac{1}{2}\sum_{i=1}^n b_i(b_i-1)p_{ii}\ge 0,$$

and

$$Mod_b(p)=-\sum_{1\le i<j\le n} b_ib_jp_{ij}-\frac{1}{2}\sum_{i=1}^n \max\{0,|b_i|(|b_i|+1)-2\}p_{ii}\ge 0.$$

Proposition 12.3 *For the cone* $\{0,1\}$-$PMET_n$ *it holds:*
(i) any inequality $Hyp_b(p)\ge 0$ *is valid on* $\{0,1\}$-$PMET_n$;
(ii) any inequality $Mod_b(p)\ge 0$ *with* $\max_{1\le i\le n}|b_i|\le 2$ *is valid on* $\{0,1\}$-$PMET_n$.

Proof. In fact, it suffices to check its validity for a typical extreme ray of $\{0,1\}$-$PMET_n$ represented by

$$p=J(S_0)+\delta_{S_0,S_1,...,S_t}=\mathcal{P}\left(\sum_{1\le i\le t}\delta^O_{S_i}\right).$$

For any $0\le k\le t$, let $\alpha_k=\sum_{i\in S_k} b_i$; so, $\sum_{i=1}^n b_i=\sum_{k=0}^t \alpha_k$.

(i) Easy to see, that it holds

$$-2\sum_{1\le i<j\le n} b_ib_jp_{ij}=\sum_{i=1}^n b_i^2p_{ii}-\sum_{1\le i,j\le n} b_ib_jp_{ij}$$

$$=\sum_{i\in S_0} b_i^2-\sum_{i\in S_0} b_i\sum_{i=1}^n b_i-\sum_{k=1}^t\left(\sum_{i\in S_k} b_i\right)\left(\sum_{i\notin S_k} b_i\right)$$

$$=\sum_{i\in S_0} b_i^2-\alpha_0\sum_{i=1}^n b_i-\sum_{k=1}^t \alpha_k\left(\left(\sum_{i=1}^n b_i\right)-\alpha_k\right)$$

$$= \sum_{i \in S_0} b_i^2 + \sum_{k=1}^{t} (\alpha_k)^2 - \left(\sum_{i=1}^{n} b_i\right)^2$$

$$= \sum_{i \in S_0} b_i(b_i - 1) + \sum_{k=1}^{t} \alpha_k(\alpha_k - 1) - \left(\sum_{i=1}^{n} b_i\right)\left(\left(\sum_{i=1}^{n} b_i\right) - 1\right).$$

So, $2Hyp_b(p) = \sum_{k=1}^{t} \alpha_k(\alpha_k - 1) \geq 0$, i.e., (i) holds.
(ii) Now,

$$2Mod_b(p) = \left(\sum_{i \in S_0} b_i^2 + \sum_{k=1}^{t} (\alpha_k)^2 - \left(\sum_{i=1}^{n} b_i\right)^2\right) - \left(\sum_{i \in S_0} |b_i|^2 - \sum_{i \in S_0} |b_i| + 2|S_0'|\right)$$

$$= \sum_{k=1}^{t} (\alpha_k)^2 - \left(\sum_{i=1}^{n} b_i\right)^2 + \sum_{i \in S_0} |b_i| - 2|S_0'|,$$

where $S_0' = \{i \in S_0 \mid b_i \neq 0\}$.
If $\sum_{i=1}^{n} b_i = 0$, then $2Mod_b(p) \geq 0$.
If $\sum_{i=1}^{n} b_i = 1$, then either $\sum_{k=1}^{t} (\alpha_k)^2$, or $2|S_0'| - \sum_{i \in S_0} |b_i|$ is at least 1. So, (ii) holds.
□

Remind that the cone $OCUT_n$ of all quasi-semimetrics on V_n, embeddable into $l_{1,or}^m$ for some m, consists of all $n \times n$ matrices $((q_{ij} = p_{ij} - p_{ii}))$, where $((p_{ij})) \in \{0,1\}$-$PMET_n$; so, any such $q = ((q_{ij}))$ is a weightable quasi-semimetric with weights $w_i = p_{ii}$, $1 \leq i \leq n$. Using, that

$$-2 \sum_{1 \leq i < j \leq n} b_i b_j p_{ij} - \sum_{i=1}^{n} b_i^2 p_{ii} = - \sum_{1 \leq i,j \leq n} b_i b_j p_{ij}$$

$$= - \sum_{1 \leq i,j \leq n} b_i b_j q_{ij} - \sum_{i=1}^{n} b_i p_{ii} \sum_{j=1}^{n} b_j$$

$$= - \sum_{1 \leq i,j \leq n} b_i b_j q_{ij} - \left(\sum_{i=1}^{n} b_i\right)\left(\sum_{i=1}^{n} b_i w_i\right),$$

we can reformulate above Proposition as follows.

Proposition 12.4 *Given a sequence $b = (b_1, ..., b_n)$ of integers with $\sum_{i=1}^{n} b_i \in \{0,1\}$, the following two inequalities are valid on the cone $OCUT_n$:*
(i) $-\sum_{1 \leq i,j \leq n} b_i b_j q_{ij} + (1 - \sum_{i=1}^{n} b_i) \sum_{i=1}^{n} b_i w_i \geq 0$;
(ii) $-\sum_{1 \leq i,j \leq n} b_i b_j q_{ij} + \sum_{i=1: b_i \neq 0}^{n} (2 - |b_i| - b_i \sum_{i=1}^{n} b_i) w_i \geq 0$ *with* $\max_{1 \leq i \leq n} |b_i| \leq 2$.

Oriented multicuts and quasi-semimetrics

Proposition 12.5 *The oriented multicuts $\delta^O_{S_1,...,S_t}$ and oriented anti-multicuts $\beta^O_{S_1,...,S_t}$ are $\{0,1\}$-valued quasi-semimetrics, which are weightable if and only if $t \leq 2$. The weight functions of oriented cut δ^O_S and of oriented anti-cut β^O_S are $w_i = 1_{i \notin S}$ and $w_i = 1_{i \in S}$, respectively.*

Proof. In fact, let $i \in S_1$, $j \in S_2$, $k \in S_3$ in the quasi-semimetric $q = \delta^O_{S_1,\dots,S_t}$. If q is weightable, then $q_{ij} = (q_{ji} + w_j) - w_i = w_j - w_i$. It is impossible, since also $q_{ik} = w_k - w_i = 1, q_{jk} = w_k - w_j = 1$. The proof for oriented anti-multicuts is similar.

□

So, for the oriented cut cone $OCUT_n$, generated by all non-zero oriented cuts, it holds $OCUT_n = OMCUT_n \cap WQMET_n$.

Theorem 12.5 *The following statement holds:*

$$OCUT_n = Q(CUT_{n+1} = \{0,1\}\text{-}sWMET_n)$$
$$= Q(\{0,1\}\text{-}dWMET_n) = \{0,1\}\text{-}Q(dWMET_n).$$

Proof. Given a representative $(d;w) = (\delta_S; w'), (\delta_S; w''), (\delta_S; (0)), (\delta_\emptyset; (1))$ of an extreme ray of $\{0,1\}$-$dWMET_n$, we have that $Q(d;w) = (q,w)$ is $(\delta^O_S, w'), (\delta^O_{\overline{S}}, w'')$, $(\delta_S, (0)), (\delta_\emptyset, (1))$, respectively.

But $\delta_S = \delta^O_S + \delta^O_{\overline{S}}$ and $(((0)), t(1))$ are not extreme rays.

□

The above equality $OCUT_n = Q(\{0,1\}\text{-}sWMET_n)$ means that any $q \in OCUT_n$ is $Q(d;w)$, where (d,w) is a semimetric $d' \in CUT_{n+1}$ on $V_n \cup \{0\}$. So,

$$q_{ij} = \frac{1}{2}(d'_{ij} - d'_{0i} + d'_{0j}).$$

But CUT_n is the set of l_1-*semimetrics* on V_n, see [DeLa97]. So, $q \in OCUT_n$ can be seen as l_1-*quasi-semimetrics*; it was realized in [DDP03], [CMM06].

In fact, $OCUT_n$ is the set of quasi-semimetrics q on V_n, for which there exist some $x_1, \dots, x_m \in \mathbb{R}^m$ with all $q_{ij} = ||x_i - x_j||_{1,or}$, where the *oriented* l_1-*norm* is defined by

$$||x||_{1,or} = \sum_{k=1}^{m} \max\{x_k, 0\};$$

the proof is the same as in Proposition 4.2.2 of [DeLa97].

Besides $OCUT_3 = \{0,1\}$-$WQMET_3 = WQMET_3$, and $\{0,1\}$-$WQMET_4 = WQMET_4$, it holds that

$$OCUT_n \subset \{0,1\}\text{-}WQMET_n \subset WQMET_n.$$

We conjecture that

$$G_{OCUT_n} \subset G_{\{0,1\}\text{-}WQMET_n} \subset G_{WQMET_n},$$

and

$$G^*_{\{0,1\}\text{-}WQMET_n} \supset G^*_{WQMET_n} \supset G^*_{MET_n}.$$

The cone $\{0,1\}$-$WQMET_5$ has $OT_{ij,k}$, NN_{ij} and 3 other, all standard, orbits. Those facets give, for permutations of $b = (1,-1,1,-1,1)$, the non-negativity of $-\sum_{1 \le i < j \le 5} b_i b_j q_{ij}$ plus q_{24}, q_{23} or $q_{12} + q_{45}$.

The cone $\{q+q^T : q \in \{0,1\}\text{-}WQMET_n\}$ coincides with MET_n for $n \leq 5$, but for $n = 6$ it has 7 orbits of extreme rays (all those of MET_6 except the one, good representatives of which are not $\{0,1,2\}$-valued as required); its 1-skeleton graph, excluding another orbit of 90 rays, is an induced subgraph of G_{MET_6}. It has 3 orbits of facets including $T_{ij,k}$ (forming $G^*_{MET_6}$ in its ridge graph) and the orbit of

$$\sum_{(ij)\in C_{123456}} d_{ij} + d_{14} + d_{35} - d_{13} - d_{46} - 2d_{25} \geq 0.$$

If $q \in PMET_n$ is $\{0,1\}$-valued with $S = \{i \mid q_{i1} = 1\}$, $S' = \{i \mid q_{1i} = 1\}$, then $q_{ij} = 0$ for $i,j \in \overline{S} \cap \overline{S'}$ (since $q_{i1} + q_{1j} \geq q_{ij}$), and $q_{ij} = q_{ji} = 1$ for $i \in S, j \in \overline{S'}$ (since $q_{ij} + q_{j1} \geq q_{i1}$); so,

$$|\overline{S} \cap \overline{S'}|(|\overline{S} \cap \overline{S'}| - 1) + |S|(|\overline{S}| - 1) + |S'|(|\overline{S'}| - 1) - |S \cap \overline{S'}||\overline{S} \cap S'|$$

elements q_{ij} with $2 \leq i \neq j \leq n$ are defined.

Other results

The proofs of the conjectures below should be tedious but easy.

Conjecture 12.1 *For the 1-skeleton graphs of the cones $OCUT_n$ and $\{0,1\}$-$dWQMET_n$ it holds:*

(i) $G_{OCUT_n} = K_{2S(n,2)}$ and belongs to G_{WQMET_n};

(ii) $\overline{G}_{\{0,1\}\text{-}dWMET_n} = K_{1,S(n,2)} + S(n,2)K_2$; $G_{\{0,1\}\text{-}dWMET_n}$ has diameter 2, all non-adjacencies are of the form $(((0));(1)) \not\sim (\delta^O_S;(0))$, and $(\delta^O_S;w') \not\sim (\delta^O_S;w')$.

Conjecture 12.2 *For the ridge graphs of the cones $PMET_n$ and $WQMET_n$ it holds:*

(i) $G^*_{PMET_n}$ has diameter 2, all non-adjacencies are of the form $PW_{ii} \not\sim PN_{ik}$; $PN_{ij} \not\sim PN_{ji}, PN_{ki}, PN_{jk}, ShT_{ij,k}$; $ShT_{ij,k} \not\sim ShT_{i'j',k'}$ if they conflict;

(ii) $G^*_{WQMET_n}$ has diameter 2; it is $G^*_{PMET_n}$ without vertices PW_{ii}.

Recall that $S(n,2) = 2^{n-1} - 1$ is the Stirling number of the second kind, $G_{CUT_n} = K_{S(n,2)}$, and [DeDe94] $G^*_{MET_n}$, $n \geq 4$, has diameter 2 with $T_{ij,k} \not\sim T_{i'j',k'}$ whenever they are *conflicting*, i.e., have values of different sign on a position (p,q), $p,q \in \{i,j,k\} \cap \{i',j',k'\}$. Clearly, $|\{i,j,k\} \cap \{i',j',k'\}|$ should be 3 or 2, and $T_{ij,k}$ conflicts with 2 and $4(n-3)$ $T_{i'j',k'}$'s, respectively.

It is shown in [DGD12], that $OCUT_n$ and $WQMET_n$ on $\{1,2,...,n\}$ are projections of CUT_{n+1} and MET_{n+1} (defined on $\{0,1,2,...,n\}$) on the subspace, orthogonal to $\delta_{\{0\}}$ (see Chapter 15).

Chapter 13

Cones of hypermetrics

13.1 Preliminaries

For given two hypermetrics d_1 and d_2 on a set X their non-negative linear combination $d = \alpha d_1 + \beta d_2, \alpha, \beta \geq 0$, is a hypermetric on X. Here, as usual, for all $x, y \in X$ it holds

$$(\alpha d_1 + \beta d_2)(x, y) = \alpha d_1(x, y) + \beta d_2(x, y).$$

Then we can speak about the cone of all hypermetrics on n points, in fact, on the set $V_n = \{1, 2, ..., n\}$. We can consider already the similar polytopes and some their generalizations.

In this chapter we consider, for small values of n, the cone of all semimetrics on V_n, the cone of all hypermetrics on V_n, the cone, generated by all cut semimetrics on V_n, the polytope of all semimetrics on V_n, the polytope of all hypermetrics on V_n, the polytope, generated by all cut semimetrics on V_n, and some of their natural generalizations.

Metric, cut and hypermetric cones are among central objects of Discrete Mathematics. While cut semimetrics are important in Analysis and Combinatorics, the hypermetrics have deep connections with Voronoi-Delaunay Theory in Geometry of Numbers; see, for example, [DeTe87], [DeGr93], [DGL95], and Chapters 13-17, 28 in [DeLa97]. So, generalizations of hypermetrics can put those connections in a more general setting.

In this chapter we consider the classical case of hypermetric cone and some its generalizations. We list the facets of the hypermetric cones HYP_n, $n \leq 8$; for $n = 7, 8$ the facets of HYP_n are determined with the help of the connection with Geometry of Numbers and the list of simplices of dimensions 6, 7. We list the extreme rays of HYP_n, $n \leq 8$ (for $n = 7, 8$ by using the list of perfect Delaunay polytopes in dimensions $6, 7$). We compute the vertices and facets of the hypermetric polytope $HYP_n^{\square}$, $n \leq 8$. (See for details [DeDu04], [DeDu13].) Then we shortly consider generalizations of classical case for quasi-hypermetrics, partial hypermetrics and weighted hypermetrics.

The following notations will be used below:

- The *hypermetric cone* HYP_n: the set of all *hypermetrics* on n points, i.e., the functions $d : V_n \times V_n \to \mathbb{R}$, satisfying to all *hypermetric inequalities*

$$Hyp_b : Hyp_b(d) = \sum_{1 \leq i < j \leq n} b_i b_j d(i, j) \leq 0,$$

where $b = (b_1, ..., b_n) \in \mathbb{Z}^n$, and $\sum_{i=1}^{n} b_i = 1$.

- The *metric cone* MET_n: the set of all *semimetrics* on n points, defined by all *triangle inequalities* $d_{ik} \leq d_{ij} + d_{jk}$ on V_n, i.e., the hypermetric inequalities Hyp_b with b being a permutation of $(1, 1, -1, 0, ..., 0)$;

- the *cut cone* CUT_n: the conic hull of all $2^{n-1} - 1$ non-zero *cut semimetrics* on n points;

- the *hypermetric polytope* $HYP_n^{\square}$: the set of all functions $d : V_n \times V_n \to \mathbb{R}$, defined by all, in general, *non-homogeneous, hypermetric inequalities*

$$\sum_{1 \leq i < j \leq n} b_i b_j d(i, j) \leq s(s+1),$$

  where $b = (b_1, ..., b_n) \in \mathbb{Z}^n$, $s \in \mathbb{Z}$, and $\sum_{i=1}^{n} b_i = 2s + 1$;

- the *metric polytope* $MET_n^{\square}$: the set of all $d \in MET_n$, satisfying, in addition, all *perimeter inequalities* $d_{ik} + d_{ij} + d_{jk} \leq 2$ on V_n (i.e., one obtains $MET_n^{\square}$ using in above inequalities for $HYP_n^{\square}$ only b of the form $(1, 1, -1, 0^{n-3})$, and $(1, 1, 1, 0^{n-3})$);

- the *cut polytope* $CUT_n^{\square}$: the convex hull of all 2^{n-1} *cut semimetrics* on n points;

- the *quasi-semimetric cone* $QMET_n$: the set of all *quasi-semimetrics* on n points, defined by all *oriented triangle inequalities* $q_{ik} \leq q_{ij} + q_{jk}$, and all *non-negativity inequalities* $q_{ij} \geq 0$ on V_n;

- The *quasi-hypermetric cone* $QHYP_n$: the set of all *quasi-hypermetrics* on n points, i.e., all $q \in QMET_n$, satisfying, in addition, all *oriented hypermetric inequalities*

$$Hyp_b^O : Hyp_b^O(q) = \sum_{1 \leq i \neq j \leq n} b_i b_j q(i, j) \leq 0,$$

  where $b = (b_1, ..., b_n) \in \mathbb{Z}^n$, and $\sum_{i=1}^{n} b_1 = 1$ (see also the *weightable hypermetric cone* $OWHYP_n$, and the *weighted quasi-hypermetric cone* $WQHYP_n$);

  is the set of all *weightable quasi-hypermetrics* on n points, i.e., all weightable quasi-semimetrics $q \in WQMET_n$, satisfying, in addition, all inequalities $Hyp_{b,O}(q) \geq 0$, and all inequalities $Neg_{b,O}(q) \geq 0$.

- the *oriented cut cone* $OCUT_n$: the conic hull of all non-zero *oriented cut quasi-semimetrics* on n points;

- the *weighted semimetric cone* $WMET_n$: the set of all *weighted semimetrics* $(d; w)$ on n points, defined by all *weight non-negativity conditions* $w_i \geq 0$, and all *triangle inequalities* $d_{ik} + d_{kj} - d_{ij} \geq 0$ on V_n;

- the *weighted hypermetric cone* $WHYP_n$: the set of all *weighted hypermetrics* on n points, i.e., all weighted semimetrics $(d; w) \in WMET_n$, satisfying, in addition, all inequalities $Hyp_b^w(d; w) \geq 0$, and $Hyp_b^{w'}(d; w) \geq 0$, where $b = (b_1, ..., b_n) \in \mathbb{Z}^n$, $\sum_{i=1}^{n} b_i \in \{0, 1\}$,

$$Hyp_b^w(d; w) = -\frac{1}{2} \sum_{1 \leq i, j \leq n} b_i b_j q_{ij} + \left(1 - \sum_{i=1}^{n} b_i\right) \sum_{i=1}^{n} b_i w_i,$$

$$Hyp_b^{w'}(d;w) = -\frac{1}{2}\sum_{1\le i,j\le n} b_ib_jq_{ij} + \left(1+\sum_{i=1}^n b_i\right)\sum_{i=1}^n b_iw_i;$$

- the *weak partial semimetric cone* $wPMET_n$: the set of all *weak partial semimetrics* on n points, generated by all *non-negativity inequalities* $p_{ii} \ge 0$, and all *sharp triangle inequalities* $p_{ik}+p_{kj}-p_{ij}-p_{ii}\ge 0$ on V_n;
- the *partial hypermetric cone* $PHYP_n$: the set of all *partial hypermetrics* on n points, i.e., all *weak partial semimetrics* $p \in wPMET_n$, satisfying, in addition, all inequalities $Hyp_b^w(P^{-1}(p)) \ge 0$, where $b=(b_1,...,b_n)\in\mathbb{Z}^n$, $\sum_{i=1}^n b_i \in \{0,1\}$, $(d;w)=P(p)$ is a weak partition semimetric with $d_{ij}=2p_{ij}-p_{ii}-p_{jj}$, $w_i=p_{ii}$, and, as before,

$$Hyp_b^{w}(d;w) = -\frac{1}{2}\sum_{1\le i,j\le n} b_ib_jq_{ij} + \left(1-\sum_{i=1}^n b_i\right)\sum_{i=1}^n b_iw_i.$$

13.2 Non-oriented case

General definitions

Let $V_n=\{1,2,...,n\}$; let $E_n=\{(i,j)\,|\,i,j\in V_n, i\ne j\}$, where (i,j) denotes the unordered pair of the integers i,j, i.e., $|E_n|=\binom{n}{2}=\frac{n(n-1)}{2}$.

Let d be a *hypermetric* on the set V_n, i.e., a function $d: V_n\times V_n\to\mathbb{R}_{\ge 0}$, for any $i,j\in V_n$ satisfying $d(i,i)=0$, $d(i,j)=d(j,i)$, and all *hypermetric inequalities*

$$Hyp_b: Hyp_b(d)=\sum_{1\le i<j\le n} b_ib_jd(i,j)\le 0,\ \text{where } b\in\mathbb{Z}^n,\ \text{and } \sum_{i=1}^n b_i=1.$$

The hypermetric inequality $Hyp_b(d)=\sum_{1\le i<j\le n} b_ib_jd(i,j)\le 0$ is said to be *pure*, if $|b_i|=0,1$ for all $i\in V_n$. The inequality $Hyp_b(d)=\sum_{1\le i<j\le n} b_ib_jd(i,j)\le 0$ is called an *k-gonal inequality*, if $\sum_{i=1}^n |b_i|=k$ holds. (Note, that k and $\sum_{i=1}^n b_i$ have the same parity.)

So, if b is $\{0,\pm1\}$-valued and has $k+1$ ones, we obtain the *pure* $(2k+1)$*-gonal inequality*. In fact, the case of general b can be seen as some such $\{0,\pm1\}$-valued b on a multiset of $\sum_{i=1}^n |b_i|$ points, in which different points occur $|b_1|,...,|b_n|$ times.

One obtains *triangle inequalities*

$$d(i,j)\le d(i,k)+d(k,j)$$

using only b of the form $(1,1,-1,0^{n-3})$; moreover, the *non-negativity inequality*

$$d(i,j)\ge 0$$

follows from the conditions $d(i,i)=0$ and $d(i,j)\le d(i,k)+d(k,j)$.

So, it holds that any hypermetric is a semimetric, and we can view a hypermetric d as a vector $(d_{ij})_{1\le i<j\le n}\in\mathbb{R}^{E_n}_{\ge 0}$, or as a symmetric $n\times n$ matrix $((d_{ij}))$ with non-negatives entries and zeros on the main diagonal: $d_{ij}=d(i,j)$.

Hence, a hypermetric on V_n can be viewed alternatively as a function on $V_n\times V_n$, as a symmetric $n\times n$ matrix $((d_{ij}))$, or as a vector in $\mathbb{R}^{E_n}$. We will use all three these representations for a semimetric on V_n. Moreover, we will use both symbols $d(i,j)$ and d_{ij} for the values of the hypermetric d between points i and j.

The *hypermetric cone* HYP_n is the set of all hypermetrics on n points.

One can obtain semimetric cone MET_n using only hypermetric inequalities with vectors b of the form $(1, 1, -1, 0^{n-3})$. Hence, the hypermetric cone HYP_n can be defined as the set of all semimetrics on n points, satisfying, in addition, all hypermetric inequalities.

There is an infinity of inequalities, defining the hypermetric cone HYP_n; so, it is not obvious that this cone is polyhedral. The proof of this fact ([DGL93], [DGL95], [DeLa97]) was achieved through the connection with Geometry of Numbers that we are going to explain below.

It is easy to see, that every cut is a hypermetric, and so we have the evident inclusions

$$CUT_n \subseteq HYP_n \subseteq MET_n \subseteq \mathbb{R}^{E_n}_{\geq 0}.$$

The cone HYP_n was introduced in [Deza60]. It holds, that $CUT_n = MET_n$ only for $3 \leq n \leq 4$; also, $CUT_n = HYP_n$ only for $3 \leq n \leq 6$ ([DeLa97]). So, the first proper HYP_n is HYP_7; it was described in [DeDu04]. The next example HYP_8 was computed in [DeDu13a]. See [DeLa97] for a detailed study of those cones and their numerous applications in Combinatorial Optimization, Analysis, and other areas of mathematics. In particular, the hypermetric cone have direct applications in the Geometry of Quadratic Forms.

Hypermetric polytope $HYP_n^{\square}$ on n points is defined as the set of all functions $d : V_n \times V_n \to \mathbb{R}$, satisfying all, in general, *non-homogeneous, hypermetric inequalities*

$$\sum_{1 \leq i < j \leq n} b_i b_j d(i, j) \leq s(s+1),$$

where $b = (b_1, ..., b_n) \in \mathbb{Z}^n$, $s \in \mathbb{Z}$, and $\sum_{i=1}^{n} b_i = 2s + 1$.

One obtains *metric polytope* $MET_n^{\square}$ using only hypermetric inequalities with vectors b of the form $(1, 1, -1, 0^{n-3})$, and $(1, 1, 1, 0^{n-3})$.

In fact, the hypermetric polytope $HYP_n^{\square}$ is defined (see [DeDu13a]) by analogy with the metric and cut polytopes, as a polytope, invariant under the *switching* operations U_S, $S \subset V_n$.

It turns out, that it is polyhedral with the link being established with the centrally symmetric Delaunay polytopes of dimension n; see [DeDu13a] for details.

One can check easily whether a distance matrix d belongs to HYP_n or $HYP_n^{\square}$; this has been used in [DeDu13a] to determine the facets and vertices of $HYP_7^{\square}$ and $HYP_8^{\square}$.

In Table 13.1 we give the information on the structure for the cones CUT_n, HYP_n, MET_n with $n = 3, 4, 5, 6, 7, 8$. The symbol e is used for the number of extreme ray of the given cones; the symbol f is used for the number of their facets. The numbers of orbits under the symmetric group $Sym(n)$ are given in parentheses. Also, the number of vertices (v) and facets (f) in the corresponding polytopes is given for $3 \leq n \leq 8$, with number of orbits under $Sym(n)$ and 2^{n-1} switchings.

Small cones and polytopes of hypermetrics

The case $n = 3$

It holds that $HYP_3 = CUT_3 = MET_3$, and $HYP_3^{\square} = CUT_3^{\square} = MET_3^{\square}$.

C/P	$n=3$	$n=4$	$n=5$	$n=6$	$n=7$	$n=8$
CUT_n, e	3(1)	7(2)	15(2)	31(3)	63(3)	127(4)
CUT_n, f	3(1)	12(1)	40(2)	210(4)	38,780(36)	49,604,520(2,169)
HYP_n, e	3(1)	7(2)	15(2)	31(3)	37,170(29)	242,695,427(9,003)
HYP_n, f	3(1)	12(1)	40(2)	210(4)	3,773(14)	298,592(86)
MET_n, e	3(1)	7(2)	25(3)	296(7)	55,226(46)	119,269,588(3,918)
MET_n, f	3(1)	12(1)	30(1)	60(1)	105(1)	168(1)
$CUT_n^{\square}, v$	4(1)	8(1)	16(1)	32(1)	64(1)	128(1)
$CUT_n^{\square}, f$	4(1)	16(1)	56(2)	368(3)	116,764(11)	217,093,472(147)
$HYP_n^{\square}, v$	4(1)	8(1)	16(1)	32(1)	13,152(6)	1,388,383,872(581)
$HYP_n^{\square}, f$	4(1)	16(1)	56(2)	68(3)	10,396(7)	1,374,560(22)
$MET_n^{\square}, v$	4(1)	8(1)	32(2)	554(3)	275,840(13)	1,550,825,600(533)
$MET_n^{\square}, f$	4(1)	16(1)	40(1)	80(1)	140(1)	224(1)

Table 13.1: The number of extreme rays (vertices) and facets of cones CUT_n, HYP_n, MET_n and polytopes $P = CUT_n^{\square}$, $HYP_n^{\square}$, $MET_n^{\square}$ for $3 \leq n \leq 8$

So, the only facet-defining inequalities for HYP_3 and $HYP_3^{\square}$ are the triangle inequalities, 3 inequalities (from one orbit, obtained by permutations) for HYP_3, and 4 inequalities (from one orbit, obtained by permutations and switchings) for $HYP_3^{\square}$. All the extreme rays of the cone HYP_3 correspond to non-zero cut vectors: there are 3 cut vectors (from one orbit, obtained by permutations). All the vertices of $HYP_3^{\square}$ are 4 cuts (from one orbit, obtained by permutations and switchings).

The case $n = 4$

In the case $n = 4$ the situation is similar: $HYP_4 = CUT_4 = MET_4$, and $HYP_4^{\square} = CUT_4^{\square} = MET_4^{\square}$.

The only facet-defining inequalities for HYP_4 and $HYP_4^{\square}$ are the triangle inequalities, 12 inequalities (from one orbit, obtained by permutations) for HYP_4, and 16 inequalities (from one orbit, obtained by permutations and switchings) for $HYP_4^{\square}$. All the extreme rays of the cone HYP_4 correspond to non-zero cut vectors: there are 7 cut vectors (from two orbits, represented by cuts $\delta_{\{1\}}$ and $\delta_{\{1,2\}}$ and obtained by permutations). All the vertices of $HYP_4^{\square}$ are 8 cuts (from one orbit, obtained by permutations and switchings).

The case $n = 5$

For the case of five points, $HYP_5 \neq MET_5$, i.e., $HYP_5 \subset MET_5$, and $HYP_5^{\square} \subset MET_5^{\square}$ strictly (in general, $HYP_n \subset MET_n$ and $HYP_n^{\square} \subset MET_n^{\square}$ strictly for any $n \geq 5$). But, however, $HYP_5 = CUT_5$, and $HYP_5^{\square} = CUT_5^{\square}$.

In fact, all the facets of the hypermetric cone on 5 points have the form

$$Q(b)^T x = \sum_{ij \in E_n} b_i b_j x_{ij} \leq 0.$$

The cone HYP_5 has 40 facets from 2 orbits (under permutations): 30 triangle inequalities

$$Hyp_{(1,1,-1,0,0)} : Q_3(1,1,-1,0,0)^T d \leq 0,$$

and 10 pentagonal inequalities

$$Hyp_{(1,1,1,-1,-1)} : Q_5(1,1,1,-1,-1)^T d \leq 0.$$

The polytope $HYP_5^{\square}$ has 56 facets from 2 orbits (under permutations and switchings): 40 facets, induced by triangle inequality

$$Hyp_{(1,1,-1,0,0)} : Q_3(1,1,-1,0,0)^T d \leq 0,$$

and 16 facets, induced by pentagonal inequality

$$Hyp_{(1,1,1,-1,-1)} : Q_5(1,1,1,-1,-1)^T d \leq 0.$$

The extreme rays of HYP_5 correspond to 15 non-zero cuts (from two orbits, represented by cuts $\delta_{\{1\}}$ and $\delta_{\{1,2\}}$).

All the vertices of $HYP_5^{\square}$ are 16 cuts (from one orbit, obtained by permutations and switchings).

The case $n = 6$

For the case of six points, $HYP_6 \subset MET_6$ and $HYP_6^{\square} \subset MET_6^{\square}$ strictly, but still $HYP_5 = CUT_5$.

The hypermetric cone HYP_6 has $60 + 60 + 90 = 210$ facets from 4 orbits; the hypermetric polytope $HYP_6^{\square}$ has $80 + 96 + 192 = 368$ facets from 3 orbits. All the facets, up to permutations and switchings, are induced by one of the following inequalities:

- $Hyp_{(1,1,-1,0,0,0)} : Q_6(1,1,-1,0,0,0)^T d \leq 0$ (*triangle inequalities*);
- $Hyp_{(1,1,1,-1,-1,0)} : Q_6(1,1,1,-1,-1,0)^T x \leq 0$ (*pentagonal inequalities*);
- $Hyp_{(2,1,1,-1,-1,-1)} : Q_6(2,1,1,-1,-1,-1)^T x \leq 0$ (*7-gonal inequalities*).

The extreme rays of HYP_6 correspond to 31 non-zero cuts (from three orbits, represented by cuts $\delta_{\{1\}}$, $\delta_{\{1,2\}}$, $\delta_{\{1,2,3\}}$). All the vertices of $HYP_6^{\square}$ are 32 cuts (from one orbit, obtained by permutations and switchings).

The case $n = 7$

In the case $n = 7$, $CUT_7 \subset HYP_7$ strictly, so, it is the first case of the proper hypermetric polyhedra.

Baranovskii, using his method presented in [Bara71], found in [Bara99] the list of all facets of HYP_7: 3,773 facets, divided into 14 orbits.

All 14 orbits F_m, $1 \leq m \leq 14$, of facets of HYP_7, found by Baranovskii, are represented below by the corresponding vector b^m:

- $F_1 : b^1 = (1,1,-1,0,0,0,0)$;
- $F_2 : b^2 = (1,1,1,-1,-1,0,0)$;
- $F_3 : b^3 = (1,1,1,1,-1,-2,0)$;
- $F_4 : b^4 = (2,1,1,-1,-1,-1,0)$;
- $F_5 : b^5 = (1,1,1,1,-1,-1,-1)$;
- $F_6 : b^6 = (2,2,1,-1,-1,-1,-1)$;

- F_7 : $b^7 = (1, 1, 1, 1, 1, -2, -2)$;
- F_8 : $b^8 = (2, 1, 1, 1, -1, -1, -2)$;
- F_9 : $b^9 = (3, 1, 1, -1, -1, -1, -1)$;
- $F_{1)}$: $b^{10} = (1, 1, 1, 1, 1, -1, -3)$;
- F_{11} : $b^{11} = (2, 2, 1, 1, -1, -1, -3)$;
- F_{12} : $b^{12} = (3, 1, 1, 1, -1, -2, -2)$;
- F_{13} : $b^{13} = (3, 2, 1, -1, -1, -1, -2)$;
- F_{14} : $b^{14} = (2, 1, 1, 1, 1, -2, -3)$.

It gives a total of 3,773 inequalities. The first ten orbits are the orbits of hypermetric facets of the cut cone CUT_7; the first four of them come as a zero-extension of facets of the cone HYP_6 (see [DeLa97], Chapter 7). The orbits F_{11}, F_{14} consist of some 19-dimensional simplex faces of CUT_7, becoming simplex facets in HYP_7.

The proof (see [Bara71]) was in terms of volume of simplexes; this result implies that for any facet of HYP_7 the bound $|b_i| \leq 3$ holds.

On the other hand, in [DGL93] 29 orbits of 37,170 extreme rays of HYP_7 were found by classifying the basic simplexes of the Schläfli polytope of the root lattice E_6. In [DeDu04] it was shown that these 37,170 extreme rays contain in the 29 orbits are, in fact, the complete list. So, the hypermetric cone HYP_7 has 37,170 extreme rays, divided into 3 orbits, corresponding to non-zero cuts, and 26 orbits, corresponding to hypermetrics on 7-vertex affine bases of the Schläfli polytope.

As a simple corollary of this result, we get, that *the only perfect Delaunay polytopes of dimension at most six are the 1-simplex and the Schläfli polytope.*

In [DeGr93] were considered extreme rays of HYP_n, which correspond, moreover, to the path-metric of a graph; the Delaunay polytope, generated by such hypermetrics, belongs to an integer lattice and, moreover, to a root lattice. They found, among the 26 non-cut orbits of extreme rays of HYP_7, exactly 12 that are graphic: $O_4, O_5, O_8, O_9, O_{10}, O_{15}, O_{16}, O_{17}, O_{22}, O_{23}, O_{24}, O_{29}$. For example, O_{10}, O_{23} and O_{29} correspond to the graphic hypermetrics on $K_7 - C_5$, $K_7 - P_4$, and $K_7 - P_3$, respectively. Three of the above 12 extreme hypermetrics correspond to polytopal graphs: the 3-polytopal graphs, corresponding to O_4, and 4-polytopal graphs $K_7 - C_5$, $K_7 - P_4$, corresponding to O_{10} and O_{23}.

One has the following properties of adjacency of extreme rays of HYP_7.

Proposition 13.1 *For he cone HYP_7 it holds:*

(i) the restriction of the 1-skeleton graph of HYP_7 on the union of cut orbits $O_1 \cup O_2 \cup O_3$ is the complete graph;

(ii) every non-cut extreme ray of HYP_7 has adjacency 20 (namely, it is adjacent to 20 cuts lying on corresponding non-hypermetric facets of CUT_7);

(iii) any two simplex extreme rays are non-adjacent; any simplex extreme ray (i.e., non-cut ray) has a local graph (i.e., the restriction of the 1-skeleton graph on the set of its neighbors) K_{20};

(iv) the diameter of the 1-skeleton graph graph of HYP_7 is 3.

One also has the following properties of adjacency of facets of HYP_7.

Proposition 13.2 *For the cone HYP_7 it holds:*

(i) any two simplex facets are non-adjacent; any simplex facet (i.e., one amongst $F_9 - F_{14}$) has a local graph K_{20};

(ii) the diameter of the ridge graph of HYP_7 is 3.

The hypermetric polytope $HYP_7^{\square}$ has 113,152 vertices, which are divided in 6 orbits (under a symmetry group $2^6 \cdot 7!$ of $Sym(7)$ and switchings). The hypermetric polytope $HYP_7^{\square}$ has 10396 facets, which form 7 orbits: see Table 13.2, where for each representative of orbit V_i of vertices, the number of incident facets is given.

V/F	F_1	F_2	F_3	F_4	F_5	F_6	F_7
V_1	105	210	35	630	546	147	2100
V_2	8	6	0	4	2	0	1
V_3	11	6	1	2	1	0	0
V_4	12	7	0	2	0	0	0
V_5	15	5	1	0	0	0	0
V_6	14	7	0	0	0	0	0

Table 13.2: Incidence between vertices and facets in $HYP_7^{\square}$.

The case $n = 8$

In the case $n = 8$, $CUT_8 \subset HYP_8$ strictly.

It was proven in [DeDu13a], that the hypermetric cone HYP_8 has $298,592$ facets from 86 orbits. This result is a corollary of the following property: there are exactly 67 types of repartitioning polytopes for $n = 7$; after obtaining the list of 67 repartitioning polytopes, we should look at all the simplices of volume 1 in it, and at the corresponding barycentric coordinates of the remaining vertex.

The 86 orbits of facets of the cone HYP_8 are presented in Tables 13.3, 13.4. The first representatives of each (of 22) switching equivalence classes of facets are boldfaced there. The orbits of simplicial facets are marked by *. About 92% of the total number of facets (60 orbits) are simplicial; they polish the cone. On the other hand, each triangle facet contains about 18.8 millions of extreme rays.

Remark. For facets in HYP_n, greatest common divisor of orbit sizes is 56, if $n = 8$, and is equal to $3, 12, 10, 30, 7$, if $n = 3, 4, 5, 6, 7$, respectively.

The cone HYP_8 has $242,695,427$ extreme rays from $9,003$ orbits. To obtain this result, one should study *perfect Delaunay polytopes.*

The n-dimensional Delaunay polytope D is called *perfect* (*extreme*), if up to scalar multiple there is an unique quadratic form q, having D as Delaunay polytope.

For a given Delaunay polytope D, an *integral affine generating set* S_{aff} is a set $v_1, ..., v_m$ of vertices of D, such that for each vertex v of D there exist such $\lambda_i \in \mathbb{Z}$, that

$$v = \sum_{i=1}^{m} \lambda_i v_i, \text{ and } 1 = \sum_{i=1}^{m} \lambda_i.$$

In the special case $m = n + 1$, the set S_{aff} is called an *affine basis.*

$F_{i,j}$	Representative	$\frac{\lvert F_{i,j}\rvert}{56}$	CUT_8-rank	Inc.([0, 1], 2_{21}, 3_{21}, ER_7)
$F_{1,1}$	$(\mathbf{0,0,0,0,0,-1,1,1})$	3	27	(95, 329734, 737128, 17725428)
$F_{2,1}$	$(\mathbf{0,0,0,-1,-1,1,1,1})$	10	27	(79, 93978, 176058, 3780630)
$F_{3,1}$	$(\mathbf{0,0,-1,-1,-1,1,1,2})$	30	27	(59, 10460, 13052, 209644)
$F_{3,2}$	$(0,0,-1,1,1,1,1,-2)$	15	27	(59, 10460, 13052, 209644)
$F_{4,1}$	$(\mathbf{0,-1,-1,-1,1,1,1,1})$	5	27	(69, 36816, 60480, 1207584)
$F_{5,1}$	$(\mathbf{0,-1,-1,-1,-1,1,1,3})$	15	27	(41, 400, 240, 620)
$F_{5,2}$	$(0,-1,1,1,1,1,1,-3)$	6	27	(41, 400, 240, 620)
$F_{6,1}$	$(\mathbf{0,-1,-1,1,1,1,-2,2})$	60	27	(51, 3567, 3288, 46176)
$F_{6,2}$	$(0,-1,-1,-1,-1,1,2,2)$	15	27	(51, 3650, 4680, 64400)
$F_{6,3}$	$(0,1,1,1,1,-2,-2)$	3	27	(51, 3650, 4680, 64400)
$F_{7,1}$	$(\mathbf{0,-1,-1,-1,1,-2,2,3})$	120	26	(39, 311, 220, 1479)
$F_{7,2}$	$(0,-1,-1,1,1,2,2,-3)$	90	26	(39, 325, 172, 1461)
$F_{7,3}$	$(0,-1,1,1,1,-2,-2,3)$	60	26	(39, 325, 172, 1461)
$F_{7,4}$	$(0,1,1,1,1,-2,2,-3)$	30	26	(39, 311, 220, 1479)
$F_{8,1}$	$(\mathbf{-1,-1,-1,-1,1,1,1,2})$	5	27	(55, 6840, 8526, 141642)
$F_{8,2}$	$(-1,-1,1,1,1,1,1,-2)$	3	27	(55, 6840, 8526, 141642)
$F_{9,1}$	$(\mathbf{-1,-1,-1,-1,-1,1,1,4})$	3	27	$(27,0,0,0)^*$
$F_{9,2}$	$(-1,1,1,1,1,1,1,-4)$	1	27	$(27,0,0,0)^*$
$F_{10,1}$	$(\mathbf{-1,-1,-1,1,1,1,-2,3})$	20	27	(41, 645, 282, 3021)
$F_{10,2}$	$(-1,-1,1,1,1,1,2,-3)$	15	27	(41, 645, 282, 3021)
$F_{10,3}$	$(-1,-1,-1,-1,-1,1,2,3)$	6	27	(41, 495, 828, 5094)
$F_{10,4}$	$(1,1,1,1,1,1,-2,-3)$	1	27	(41, 495, 828, 5094)
$F_{11,1}$	$(\mathbf{-1,-1,-1,1,1,-2,2,2})$	30	27	(45, 1464, 1390, 18310)
$F_{11,2}$	$(-1,1,1,1,1,-2,-2,2)$	15	27	(45, 1464, 1390, 18310)
$F_{11,3}$	$(-1,-1,-1,-1,-1,2,2,2)$	1	27	(45, 2070, 1458, 34956)
$F_{12,1}$	$(\mathbf{-1,-1,1,1,-2,-2,2,3})$	90	27	(37, 293, 166, 1638)
$F_{12,2}$	$(-1,1,1,1,-2,2,2,-3)$	60	27	(37, 293, 166, 1638)
$F_{12,3}$	$(-1,-1,-1,1,2,2,2,-3)$	20	27	(37, 306, 279, 2616)
$F_{12,4}$	$(-1,-1,-1,-1,-2,2,2,3)$	15	27	(37, 300, 285, 2500)
$F_{12,5}$	$(1,1,1,1,-2,-2,-2,3)$	5	27	(37, 306, 279, 2616)
$F_{13,1}$	$(\mathbf{-1,-1,-1,-1,1,-2,2,4})$	30	26	(31, 35, 31, 31)
$F_{13,2}$	$(-1,-1,1,1,1,-2,-2,4)$	30	26	(31, 45, 3, 24)
$F_{13,3}$	$(-1,-1,1,1,1,2,2,-4)$	30	26	(31, 45, 3, 24)
$F_{13,4}$	$(1,1,1,1,1,-2,2,-4)$	6	26	(31, 35, 31, 31)
$F_{14,1}$	$(\mathbf{-1,-1,-1,1,1,2,-3,3})$	60	26	(35, 142, 46, 268)
$F_{14,2}$	$(-1,1,1,1,1,-2,-3,3)$	30	26	(35, 142, 46, 268)
$F_{14,3}$	$(-1,-1,-1,-1,1,-2,3,3)$	15	26	(35, 110, 142, 404)
$F_{14,4}$	$(1,1,1,1,1,2,-3,-3)$	3	26	(35, 110, 142, 404)
$F_{15,1}$	$(\mathbf{-1,-1,-1,-1,1,-3,3,4})$	30	26	$(26,0,0,1)^*$
$F_{15,2}$	$(-1,-1,-1,1,1,3,3,-4)$	30	26	$(26,0,0,1)^*$
$F_{15,3}$	$(-1,1,1,1,1,-3,-3,4)$	15	26	$(26,0,0,1)^*$
$F_{15,4}$	$(1,1,1,1,1,-3,3,-4)$	6	26	$(26,0,0,1)^*$

Table 13.3: The orbits of facets of the cone HYP_8: part 1

$F_{i,j}$	Representative	$\frac{\lvert F_{i,j}\rvert}{56}$	CUT_8-rank	Inc.$([0,1], 2_{21}, 3_{21}, ER_7)$
$F_{16,1}$	$\mathbf{(-1,-1,1,-2,2,2,-3,3)}$	180	26	$(32,63,36,177)$
$F_{16,2}$	$(1,1,1,-2,-2,2,-3,3)$	60	26	$(32,63,36,177)$
$F_{16,3}$	$(-1,1,1,2,2,2,-3,-3)$	30	26	$(32,82,38,272)$
$F_{16,4}$	$(-1,1,1,-2,-2,-2,3,3)$	30	26	$(32,82,38,272)$
$F_{16,5}$	$(-1,-1,-1,-2,-2,2,3,3)$	30	26	$(32,94,26,266)$
$F_{17,1}$	$\mathbf{(-1,-1,1,1,-2,2,-3,4)}$	180	25	$(30,41,6,51)$
$F_{17,2}$	$(-1,1,1,1,-2,2,3,-4)$	120	25	$(30,41,6,51)$
$F_{17,3}$	$(-1,-1,-1,1,2,2,3,-4)$	60	25	$(30,40,22,74)$
$F_{17,4}$	$(-1,-1,-1,1,-2,-2,3,4)$	60	25	$(30,30,32,74)$
$F_{17,5}$	$(1,1,1,1,-2,-2,-3,4)$	15	25	$(30,40,22,74)$
$F_{17,6}$	$(1,1,1,1,2,2,-3,-4)$	15	25	$(30,30,32,74)$
$F_{17,7}$	$(-1,-1,-1,-1,2,2,-3,4)$	15	25	$(30,40,22,72)$
$F_{18,1}$	$\mathbf{(-1,-1,1,-2,2,3,3,-4)}$	180	25	$(27,13,9,20)$
$F_{18,2}$	$(-1,1,1,2,2,-3,3,-4)$	180	25	$(27,15,3,15)$
$F_{18,3}$	$(-1,1,1,-2,-2,-3,3,4)$	180	25	$(27,15,3,15)$
$F_{18,4}$	$(-1,-1,-1,-2,2,-3,3,4)$	120	25	$(27,13,9,18)$
$F_{18,5}$	$(-1,-1,1,2,2,-3,-3,4)$	90	25	$(27,21,1,22)$
$F_{18,6}$	$(1,1,1,-2,2,-3,-3,4)$	60	25	$(27,13,9,20)$
$F_{18,7}$	$(1,1,1,-2,-2,3,3,-4)$	30	25	$(27,21,1,22)$
$F_{19,1}$	$\mathbf{(-1,-1,-1,1,-2,-2,2,5)}$	60	24	$(24,2,1,0)^*$
$F_{19,2}$	$(-1,-1,1,1,2,2,2,-5)$	30	24	$(24,3,0,0)^*$
$F_{19,3}$	$(-1,1,1,1,-2,-2,-2,5)$	20	24	$(24,3,0,0)^*$
$F_{19,4}$	$(1,1,1,1,-2,2,2,-5)$	15	24	$(24,2,1,0)^*$
$F_{20,1}$	$\mathbf{(-1,-1,-1,1,-2,-3,3,5)}$	120	24	$(24,1,1,1)^*$
$F_{20,2}$	$(-1,-1,1,1,2,-3,-3,5)$	90	24	$(24,2,0,1)^*$
$F_{20,3}$	$(-1,-1,-1,1,2,3,3,-5)$	60	24	$(24,2,0,1)^*$
$F_{20,4}$	$(-1,1,1,1,-2,3,3,-5)$	60	24	$(24,2,0,1)^*$
$F_{20,5}$	$(1,1,1,1,2,-3,3,-5)$	30	24	$(24,1,1,1)^*$
$F_{20,6}$	$(1,1,1,1,-2,-3,-3,5)$	15	24	$(24,2,0,1)^*$
$F_{21,1}$	$\mathbf{(-1,1,-2,2,2,-3,-3,5)}$	180	23	$(23,2,1,1)^*$
$F_{21,2}$	$(-1,-1,-2,-2,2,-3,3,5)$	180	23	$(23,3,0,1)^*$
$F_{21,3}$	$(-1,1,2,2,2,-3,3,-5)$	120	23	$(23,2,1,1)^*$
$F_{21,4}$	$(1,1,-2,-2,2,3,3,-5)$	90	23	$(23,2,1,1)^*$
$F_{21,5}$	$(-1,-1,-2,2,2,3,3,-5)$	90	23	$(23,3,0,1)^*$
$F_{21,6}$	$(1,1,-2,-2,-2,-3,3,5)$	60	23	$(23,2,1,1)^*$
$F_{22,1}$	$\mathbf{(-1,-1,1,-2,2,3,4,-5)}$	360	23	$(23,2,1,1)^*$
$F_{22,2}$	$(-1,-1,1,-2,-2,-3,4,5)$	180	23	$(23,2,1,1)^*$
$F_{22,3}$	$(-1,1,1,-2,-2,3,-4,5)$	180	23	$(23,3,0,1)^*$
$F_{22,4}$	$(-1,1,1,2,2,-3,4,-5)$	180	23	$(23,3,0,1)^*$
$F_{22,5}$	$(-1,-1,1,2,2,-3,-4,5)$	180	23	$(23,3,0,1)^*$
$F_{22,6}$	$(1,1,1,-2,2,-3,-4,5)$	120	23	$(23,2,1,1)^*$
$F_{22,7}$	$(-1,-1,-1,-2,2,3,-4,5)$	120	23	$(23,2,1,1)^*$
$F_{22,8}$	$(1,1,1,2,2,3,-4,-5)$	60	23	$(23,2,1,1)^*$
$F_{22,9}$	$(1,1,1,-2,-2,3,4,-5)$	60	23	$(23,3,0,1)^*$

Table 13.4: The orbits of facets of the cone HYP_8: part 2

Given an perfect Delaunay polytope D of associated quadratic form q, and an integral affine generating set $S_{aff} = \{v_1, ..., v_m\}$, the distance function

$$d(i, j) = q(v_i - v_j)$$

defines an extreme ray of HYP_m. It is proved in [DeLa97], that all extreme rays of HYP_n are obtained in this way.

Thus, in order to classify the extreme rays of HYP_n we need the list of perfect Delaunay polytopes of dimension at most $n-1$, and their integral affine generating sets.

The perfect Delaunay polytopes of dimension at most 6 are determined in [DeDu04]. The ones of dimension 7 are considered in [DDD15] (see also [DER07], [Duto07]).

In summary, we obtain the following list.

1. The 1-dimensional polytope $[0, 1]$. It has 4 orbits of 8-point integral affine generating sets. This gives 127 extreme rays of HYP_8.

2. The 6-dimensional *Schläfli polytope* 2_{21}. It has 195 orbits of 8-point integral affine generating sets. This gives a total of $231, 596$ extreme rays in HYP_8.

3. The 7-dimensional *Gosset polytope* 3_{21}. It has 374 orbits of affine basis and gives a total of $7, 126, 560$ extreme rays of HYP_8.

4. The 7-dimensional *Erdahl-Rybnikov polytope* ER_7. It has $8, 430$ orbits of affine basis and gives a total of $235, 337, 144$ extreme rays of HYP_8.

For $P = 3_{21}$ or ER_7, it suffices to enumerate the orbits of 8 vertices in P. We keep the ones that determine a simplex of volume 1, and, thus, are integral affinely generating. We found 374 and $8, 430$ orbits.

For $P = 2_{21}$, the integral affine generating sets can have 7 points with one repeated or 8 points. In the 7-point case we enumerate the affine basis of 2_{21} and consider all ways to duplicate points. In the 8-point case we enumerate all orbits of 8-points and check the ones that integrally affine generates 2_{21}. This gives 195 orbits.

For the interval $[0, 1]$, we simply have to look at the number $2^{n-1} - 1$ of non-zero cut semimetrics on n vertices.

So, altogether HYP_8 has $242, 695, 427$ vertices in $9, 003$ orbits.

The polytope $HYP_8^{\square}$ has 1,374,560 facets, forming 22 orbits under symmetry group of the order $2^7 \cdot 8!$, generated by $Sum(8)$ and switchings. See in Table 13.5 the representatives of all 22 orbits of facets of $HYP_8^{\square}$; the simplicial ones are marked by *.

The hypermetric polytope $HYP_8^{\square}$ has 581 orbits of vertices, which are in details:

- 1 orbit of $2^7 = 128$ cuts, corresponding to the Delaunay polytope $[0, 1]$ (the stabilizer of a cut is isomorphic to $Sym(8)$, the number of classes is 4, the orbit forms a clique, the cone of facets incident to a cut is exactly the hypermetric cone HYP_8, cuts are the only vertices of $HYP_n^{\square}$, having all coordinates integral);

- 24 orbits, corresponding to the Delaunay polytopes 2_{21} and 3_{21} (the denominator of the coordinates is 3 for all vertices);

F_i	Representative	$\frac{\lvert F_i \rvert}{32}$	# of classes	Inc.([0, 1], $\{2_{21}, 3_{21}\}$, ER_7)
F_1	(0, 0, 0, 0, 0, 1, 1, 1)	7	2	(96, 1598784, 80836608)
F_2	(0, 0, 0, 1, 1, 1, 1, 1)	28	3	(80, 383040, 14300640)
F_3	(0, 1, 1, 1, 1, 1, 1, 1)	16	4	(70, 131712, 3975552)
F_4	(0, 0, 1, 1, 1, 1, 1, 2)	168	6	(60, 32160, 590960)
F_5	(0, 1, 1, 1, 1, 1, 2, 2)	336	9	(52, 9600, 122160)
F_6	(1, 1, 1, 1, 1, 1, 1, 2)	32	8	(56, 19656, 370272)
F_7	(0, 1, 1, 1, 1, 1, 1, 3)	112	7	(42, 840, 1120)
F_8	(1, 1, 1, 1, 1, 2, 2, 2)	224	12	(46, 3528, 39906)
F_9	(0, 1, 1, 1, 1, 2, 2, 3)	1, 680	15	(40, 656, 2686)
F_{10}	(1, 1, 1, 1, 1, 1, 2, 3)	224	14	(42, 1323, 6489)
F_{11}	(1, 1, 1, 1, 1, 1, 1, 4)	32	8	(28, 0, 0)*
F_{12}	(1, 1, 1, 1, 1, 2, 3, 3)	672	18	(36, 252, 464)
F_{13}	(1, 1, 1, 1, 2, 2, 2, 3)	1, 120	20	(38, 585, 3210)
F_{14}	(1, 1, 1, 1, 1, 2, 2, 4)	672	18	(32, 66, 36)
F_{15}	(1, 1, 1, 2, 2, 2, 3, 3)	2, 240	24	(33, 120, 302)
F_{16}	(1, 1, 1, 1, 2, 2, 3, 4)	3, 360	30	(31, 62, 82)
F_{17}	(1, 1, 1, 1, 2, 2, 2, 5)	1, 120	20	(25, 3, 0)*
F_{18}	(1, 1, 1, 1, 1, 3, 3, 4)	672	18	(27, 0, 1)*
F_{19}	(1, 1, 1, 2, 2, 3, 3, 4)	6, 720	36	(28, 22, 22)
F_{20}	(1, 1, 1, 1, 2, 3, 3, 5)	3, 360	30	(25, 2, 1)*
F_{21}	(1, 1, 2, 2, 2, 3, 3, 5)	6, 720	36	(24, 3, 1)*
F_{22}	(1, 1, 1, 2, 2, 3, 4, 5)	13, 440	48	(24, 3, 1)*

Table 13.5: Orbits of facets of $HYP_8^{\square}$

- 556 orbits, corresponding to the Delaunay polytope ER_7 (the denominator of the coordinates is 12 for all vertices, the number of incident inequalities and adjacent vertices is 28 for each of them).

Orbits of vertices of the hypermetric polytope $HYP_8^{\square}$, originating from perfect Delaunay polytopes 2_{21} and 3_{21}, are given in Table 13.6. Column 2 is the order $|Stab|$ of the stabilizer; column 4 is the number of orbits under $Sym(8)$. Column 5 is number of orbits of type 2_{21} and 3_{21}, that merged into single orbit V_i in $HYP_8^{\square}$; columns 6, 7 are the number of facets, containing the orbit representative, and the number of vertices of $HYP_8^{\square}$, adjacent to it.

V_i	$\lvert Stab \rvert$	$\frac{\lvert V_i \rvert}{10,752}$	# of orbits	Merging $2 3_{21}$	Incidence	Adjacency
V_1	24	20	36	$2_{21}(8), 3_{21}(12)$	112	848
V_2	48	10	30	$2_{21}(7), 3_{21}(10)$	104	799
V_3	96	5	23	$2_{21}(5), 3_{21}(8)$	94	701
V_4	12	40	48	$2_{21}(11), 3_{21}(16)$	94	758
V_5	8	60	46	$2_{21}(10), 3_{21}(16)$	94	804
V_6	12	40	40	$2_{21}(8), 3_{21}(16)$	92	979
V_7	240	2	18	$2_{21}(2), 3_{21}(5)$	86	926
V_8	12	40	48	$2_{21}(11), 3_{21}(16)$	86	709
V_9	8	60	54	$2_{21}(12), 3_{21}(18)$	86	728
V_{10}	4	120	72	$2_{21}(16), 3_{21}(24)$	86	774
V_{11}	16	30	33	$2_{21}(6), 3_{21}(13)$	84	1,070
V_{12}	4	120	60	$2_{21}(12), 3_{21}(24)$	84	963
V_{13}	48	10	26	$2_{21}(4), 3_{21}(8)$	82	1,023
V_{14}	12	40	48	$2_{21}(7), 3_{21}(18)$	81	1,080
V_{15}	4	120	60	$2_{21}(9), 3_{21}(30)$	79	935
V_{16}	20	24	24	$2_{21}(5), 3_{21}(8)$	78	734
V_{17}	16	30	33	$2_{21}(7), 3_{21}(12)$	78	679
V_{18}	8	60	46	$2_{21}(10), 3_{21}(16)$	78	690
V_{19}	4	120	56	$2_{21}(12), 3_{21}(20)$	78	716
V_{20}	4	120	56	$2_{21}(9), 3_{21}(25)$	78	1,050
V_{21}	60	8	16	$2_{21}(2), 3_{21}(6)$	76	1,070
V_{22}	4	120	48	$2_{21}(9), 3_{21}(20)$	76	941
V_{23}	16	30	29	$2_{21}(5), 3_{21}(11)$	74	1,032
V_{24}	4	120	48	$2_{21}(8), 3_{21}(22)$	74	920

Table 13.6: Orbits of vertices of $HYP_8^{\square}$, originating from perfect Delaunay polytopes 2_{21} and 3_{21}

The 1-skeleton graphs of $HYP_8^\square$ and HYP_8 contain a clique, consisting of all cuts and all non-zero cuts, respectively. Any vertex is adjacent to a cut vertex; so, each of above 1-skeleton graphs have diameter 3.

Theorems and conjectures for general case

Given a quadratic form q, one can define the induced Delaunay tessellation with point set $\mathbb{Z}^n$ ([Voro08], [Schu09], [DSV08], [DSV09]).

It is well known, that there are only a finite number of such tessellations, up to the action of the group $\mathrm{GL}_n(\mathbb{Z})$.

For a generic quadratic form, the tessellation is formed by simplices only; but, importantly, when it is not, this induces linear conditions on the coefficients of the quadratic form. There are a finite number of simplices, up to $\mathrm{GL}_n(\mathbb{Z})$ action, and they have been classified for $n = 7$, extending previous classification for $n \leq 6$ ([RyBa78], [Bara99], [RyBa98]).

Table 13.7 gives key information on all types of Delaunay simplices S in $\mathbb{R}^7$: their volume, the order of their automorphism group, and the number of facets of the *Baranovskii cone*. There are 11 types of such simplices and, in contrast to the lower dimensional cases, two simplices can have the same volume and yet be inequivalent.

S_i	Vol(S_i)	$\lvert Is(S_i)\rvert$	# of facets (orbits)
S_1	1	40,320	298,592(86)
S_2	2	40,320	5,768(9)
S_3	2	1,440	6,590(62)
S_4	3	540	966(9)
S_5	3	1,152	728(9)
S_6	3	240	640(39)
S_7	4	1,440	28(3)
S_8	4	240	153(11)
S_9	4	144	131(10)
S_{10}	5	72	28(6)
S_{11}	5	48	28(8)

Table 13.7: All types of Delaunay simplices S in $\mathbb{R}^7$

Given a simplex S, denote by Bar_S the set of quadratic forms, for which S is contained in a Delaunay polytope of the Delaunay tessellation. It is a polyhedral cone, called a *Baranovskii cone* in [Schu09].

For a quadratic form inside Bar_S, the Delaunay tessellation contains S as a simplex. For a quadratic form on a facet of Bar_S, the simplex S is a part of a *repartitioning polytope*, i.e., a Delaunay polytope with only $n+2$ vertices.

If the simplex S has volume 1, then it is equivalent to the simplex, formed by the vertices $v_1 = 0$, $v_2 = e_1$, ... , $v_{n+1} = e_n$. The quadratic form q is described uniquely by the distance function $d(i,j) = q(v_i - v_j)$ on the vertices v_i.

For a given positive definite quadratic form q, denote by $c(q)$ the center of the sphere, circumscribing S, and by $r(q)$ the radius of this sphere. Since S is of volume 1, a given point $v \in \mathbb{Z}^n$ can be uniquely expressed in barycentric coordinates as $v = \sum_i b_i v_i$ with $1 = \sum_i b_i$, and $b_i \in \mathbb{Z}$.

In [DeLa97](Chapter 14, p. 194–196) the following formula is proved:

$$Q(b)^T d = \sum_{1 \le i < j \le n} b_i b_j d(i,j) = (r(q))^2 - (v - c(q))^2.$$

So, the hypermetric inequalities $Hyp_b : Q(b)^T d = \sum_{1 \le i < j \le n} b_i b_j d(i,j) \le 0$ are connected with the inequalities $(r(q))^2 - (v - c(q))^2 \le 0$, and distance function d corresponds to a quadratic form q, for which S is a part of the Delaunay tessellation, if and only if it belongs to the hypermetric cone.

With this connection, it is possible to prove the polyhedrality of the cone HYP_n. (See the proof below.)

The 1-skeleton graphs of $HYP_n^{\square}$ and HYP_n contain a clique, consisting of all cuts and all non-zero cuts, respectively. We expect, that any vertex is adjacent to a cut vertex (it holds for $n \le 8$); if true, it will imply that each of above 1-skeleton graphs have diameter 3.

The ridge graphs of $MET_n^{\square}$ and MET_n with $n \ge 4$ have diameter 2 ([DeDe94]). We expect that any facet of $HYP_n^{\square}$ and HYP_n is adjacent to a triangle/perimeter facet (it holds for $n \le 7$); if true, it will imply that the ridge graphs of $HYP_n^{\square}$ and HYP_n have diameter 4.

Polyhedrality of the hypermetric Cone

The hypermetric cone HYP_{n+1} is defined by infinitely many inequalities. Hence, a natural question is whether a finite subset of them suffices for describing HYP_{n+1} or, in other words, whether the cone HYP_{n+1} is polyhedral. The answer is *yes*, as stated in the next Theorem, proved by Deza, Grishukhin and Laurent [DGL93] (see aslo [DeLa97], Chapter 14).

Theorem 13.1 *For any $n \ge 2$, the hypermetric cone HYP_{n+1} is polyhedral.*

In [DeLa97] three different proofs for this result are given. In each of them, the image $\xi(HYP_n)$ of the hypermetric cone HYP_{n+1} under the *covariance mapping* ξ is considered instead of the cone HYP_{n+1} itself.

Here, $\xi(HYP_{n+1})$ is the cone, defined by the inequalities

$$\sum_{1 \le i,j \le n} b_i b_j p_{ij} - \sum_{1 \le i \le n} b_i p_{ii} \ge 0 \quad \text{for all} \quad b \in \mathbb{Z}^n.$$

The cone $\xi(HYP_{n+1})$ is shown to be polyhedral in the following three ways: either by showing that it has a finite number of faces, or by showing that is can be decomposed as a finite union of polyhedral cones, or by showing that it coincides with a larger cone, defined by a finite subset of its inequalities.

The first proof was given by Deza, Grishukhin and Laurent (1993) [DGL93]. It is based on the connection, existing between faces of the hypermetric cone and types of Delaunay polytopes, and it uses as an essential tool the fact that the number of types of Delaunay polytopes in any given dimension is finite.

The second proof relies on the fact that the cone $\xi(HYP_{n+1})$ can be decomposed as a finite union of L-type domains. It uses two results of Voronoi; the first one concerns the finiteness of the number of types of lattices in any given dimension and the second one concerns properties of the partition of the *positive semidefinite cone* PSD_n in $\mathbb{R}^{E_n'}$ - the set of vectors $(a_{ij})_{1\le i\le j\le n}$ for which the symmetric $n\times n$ matrix $((a_{ij}))$ is positive semidefinite - into L-type domains (in particular, the fact that each L-type domain is a polyhedral cone).

The third proof is due to Lovász (1994) [Lova94]. It consists of proving directly that, among all the inequalities, defining $\xi(HYP_{n+1})$, only a finite subset of them is necessary; namely, one shows that these inequalities with b bounded in terms of n are sufficient for the description of $\xi(HYP_{n+1})$. For each $p\in\xi(HYP_{n+1})$, the set

$$E_p=\left\{x\in\mathbb{R}^n\,\middle|\,\sum_{1\le i,j\le n}x_ix_jp_{ij}-\sum_{1\le i\le n}x_ip_{ii}=0\right\}$$

is an ellipsoid (possibly degenerate with infinite directions), whose interior is free of integral points. The key argument consists of showing that if p lies on the boundary of the cone $\xi(HYP_{n+1})$, then E_p contains an integral point b which is distinct from 0 and the unit vectors and is "short", which means that all its components are bounded by a constant depending only on n, e.g., $\max_{1\le i\le n}|b_i|\le n!$.

We are going to consider the first proof of above Theorem.

Proof.

Let $d\in HYP_{n+1}$, and let $p=\xi(d)$. The $Ann(p)$ of p is defined as

$$Ann(p)=\left\{b\in\mathbb{Z}^n\,\middle|\,b\ne 0,e_1,...,e_n,\text{ and }\sum_{1\le i,j\le n}b_ib_jp_{ij}-\sum_{1\le i\le n}b_ip_{ii}=0\right\},$$

where $e_1,\dots,e_n$ denote the unit vectors in $\mathbb{R}^n$. Let $F(p)$ denote the smallest face of $\xi(HYP_{n+1})$, containing p, i.e.,

$$F(p)=\xi(HYP_{n+1})\cap\bigcap_{b\in Ann(p)}H_b,$$

where H_b denotes the hyperplane in $\mathbb{R}^{\binom{n+1}{2}}$, defined by the equation

$$\sum_{1\le i,j\le n}b_ib_jp_{ij}-\sum_{1\le i\le n}b_ip_{ii}=0.$$

Clearly, showing that $\xi(HYP_{n+1})$ is polyhedral amounts to showing that the number of its distinct faces is finite or, equivalently, that the number of distinct annulators $Ann(p)$ (for $p\in\xi(HYP_{n+1})$) is finite.

Let P_d denote the Delaunay polytope, associated with d, let L_d be the associated lattice, and let $i\in X\mapsto v_i\in V(P_d)$ be the representation of (X,d) on the sphere S_d with center c, circumscribing P_d. We can assume that $v_0=0$; then, $L_d=\mathbb{Z}(v_1,\dots,v_n)$. For $v\in L_d$, set

$$Z(v)=\left\{b\in\mathbb{Z}^n\,\middle|\,v=\sum_{1\le i\le n}b_iv_i\right\}.$$

Then (see [DeLa97])

$$Ann(p)\cup\{0,e_1,\dots,e_n\}=\bigcup_{v\in V(P_d)}Z(v).$$

Suppose that the polytope P_d has type γ. Let B be a representative basis of the type γ and let Y_γ be the integer matrix, characterizing the type γ, as defined in [DeLa97]. Then,

$$Q_{P_d} = Y_\gamma M_B,$$

where Q_{P_d} denotes the matrix, whose rows are the vectors $v \in V(P_d)$, and M_B denotes the matrix, whose rows are the vectors of B. Let Q denote the $n \times k$ matrix, whose rows are the vectors v_i for $1 \le i \le n$. So, Q may have repeated rows and every row of Q is a row of Q_{P_d}. Then,

$$Q = YM_B$$

for some integer matrix Y. Let y_v $(v \in V(P_d))$ denote the rows of Y_γ. Then the rows of Y are the vectors y_{v_i} for $1 \le i \le n$. Note, that

$$v = \sum_{1\le i\le n} b_i v_i \iff y_v = \sum_{1\le i\le n} b_i y_{v_i}.$$

Therefore, for each $v \in V(P_d)$, it holds

$$Z(v) = \left\{ b \in \mathbb{Z}^n \middle| y_v = \sum_{1\le i\le n} b_i y_{v_i} \right\}.$$

Hence, for $v \in V(P_d)$, $Z(v)$ depends only on $(y_v, y_{v_1}, \dots, y_{v_n})$. Using the fact, that $Ann(p) \cup \{0, e_1, \dots, e_n\}$ is equal to $\bigcup_{v \in V(P_d)} Z(v)$, we deduce that $Ann(p)$ is entirely determined by the matrix Y_γ and the subsystem $(y_{v_1}, \dots, y_{v_n})$ of its rows. In other words, for each $d \in HYP_{n+1}$, the annulator $Ann(\xi(d))$ is completely determined by a pair (γ, θ), where γ is a type of Delaunay polytopes in $\mathbb{R}^k$ with $k \le n$, and θ is a mapping from $\{1, 2, ..., n\}$ to the set of rows of Y_γ.

As Y_γ has $|V(P_d)| \le 2^k$ rows ([DeLa97]), the number of such mappings θ is finite. Moreover, the number of types of Delaunay polytopes in given dimension is finite ([DeLa97]). Therefore, we deduce that the number of distinct annulators $Ann(\xi(d))$ (for $d \in HYP_{n+1}$) is finite. This shows that $\xi(HYP_{n+1})$ is a polyhedral cone.

13.3 Partial and weighted hypermetric cones

Let $V_n = \{1, 2, ..., n\}$; let $E_n^{'} = |\{(i,j) \mid i,j \in V_n\}|$, where (i,j) denotes the unordered pair of the integers i, j, not necessary distinct, i.e., $E_n^{'} = E_n + n = \binom{n+1}{2} = \frac{n(n+1)}{2}$.

For a given sequence $b = (b_1, ..., b_n)$ of integers and an $n \times n$ matrix $A = ((a_{ij}))$, denote by $H_b(a)$ the sum $-\sum_{1\le i,j\le n} b_i b_j a_{ij}$ of the entries of the matrix $-b \cdot A \cdot b^T$.

In this notation, the *hypermetric* on n points can be defined as a semimetric d on V_n with $H_b(d) \ge 0$, whenever $\sum_{i=1}^n b_i = 1$.

Consider a *weighted semimetric* on V_n, i.e., a symmetric function $d : V_n \times V_n \to \mathbb{R}_{\ge 0}$, such that there exists a *weight function* $w = (w_i) : X \longrightarrow \mathbb{R}_{\ge 0}$ on V_n, and for all $i, j, k \in V_n$ it holds $d_{ii} = 0$, and $d_{ij} \le d_{ik} + d_{kj}$.

Denote by $WMET_n$ the set of all weighted semimetrics $(d; w)$ on V_n; it is a full-dimensional convex cone in $\mathbb{R}^{E_n^{'}}$ with n *weight non-negativity facets* $W_i : w_i \ge 0$, and $3\binom{n}{3}$ *triangle facets* $T_{ij,k} : d_{ik} + d_{kj} - d_{ij} \ge 0$.

For a weighted semimetric $(d; w)$ on n points, we will use the following notations:

- $Hyp_b^w(d;w) = -\frac{1}{2}\sum_{1\leq i,j\leq n} b_ib_jd_{ij} + (1-\sum_{i=1}^n b_i)\sum_{i=1}^n b_iw_i =$
$= \frac{1}{2}H_b(d) + (1-\Sigma_b)\langle b,w\rangle;$

- $Hyp_b^{w'}(d;w) = -\frac{1}{2}\sum_{1\leq i,j\leq n} b_ib_jd_{ij} + (1+\sum_{i=1}^n b_i)\sum_{i=1}^n b_iw_i =$
$= \frac{1}{2}H_b(d) + (1+\Sigma_b)\langle b,w\rangle.$

Here $\langle b,w\rangle = \sum_{i=1}^n b_iw_i$ is the *dot product* of vectors $b \in \mathbb{Z}^n$ and $w \in \mathbb{R}^n_{\geq 0}$, and $\Sigma_b = \sum_{i=1}^n b_i \in \{0,1\}$.

A weighted semimetric d on V_n is called *weighted hypermetric*, if it satisfies the inequalities $Hyp_b^w(d;w) \geq 0$ and $Hyp_b^{w'}(d;w) \geq 0$ for all $b \in \mathbb{Z}^n$ with $\sum_{i=1}^n b_i = 1$ or $\sum_{i=1}^n b_i = 0$.

Define the *weighted hypermetric cone* $WHYP_n$ as the set of all *weighted hypermetrics* on n points, i.e., all weighted semimetrics $(d;w) \in WMET_n$, satisfying, in addition, all inequalities $Hyp_b^w : Hyp_b^w(d;w) \geq 0$ and $Hyp_b^{w'} : Hyp_b^{w'}(d;w) \geq 0$ with $\sum_{i=1}^n b_i \in \{0,1\}$.

Consider now a *weak partial semimetric* on the set V_n, i.e., a symmetric function $p : V_n \times V_n \to \mathbb{R}_{\geq 0}$, such that for any $i,j,k \in V_n$ it holds $p_{ik} + p_{kj} - p_{ij} - p_{kk} \geq 0$.

The *weak partial semimetric cone* $wPMET_n$ is the set of all weak partial semimetrics on n points, generated by n *non-negativity inequalities* $PW_{ij} : p_{ii} \geq 0$, and by $3\binom{n}{3}$ *sharp triangle inequalities* $ShT_{ij,k} : p_{ik} + p_{kj} - p_{ij} - p_{kk} \geq 0$.

It is known, that weighted semimetrics and weak partial semimetrics are connected by the map $P : (d,w) \to p$, defined by

$$p_{ij} = \frac{d_{ij} + w_i + w_j}{2}.$$

The map P is an automorphism of the vector space $\mathbb{R}^{\binom{n+1}{2}}$, and $(d;w) = P^{-1}(p)$, where the inverse map P^{-1} is defined by

$$d_{ij} = 2p_{ij} - p_{ii} - p_{jj},\ w_i = p_{ii}.$$

Using the above bijection, we can define a *partial hypermetric* on n points as a weak partial semimetric p on V_n, satisfying all inequalities $Hyp_b^w(P^{-1}(p)) \geq 0$ with $\sum_{i=1}^n b_i = 1$ or $\sum_{i=1}^n b_i = 0$.

It is easy to see, that for $p = P(d;w)$ we have

$$Hyp_b^w(d;w) = H_b(p) + \sum_{i=1}^n b_ip_{ii}.$$

So, a *partial hypermetric* on n points can be defined as a weak partial semimetric p on V_n, satisfying all inequalities

$$Hyp_b^p : H_b(p) + \sum_{i=1}^n b_ip_{ii} \geq 0$$

with $\sum_{i=1}^n b_i = 1$ or $\sum_{i=1}^n b_i = 0$.

Define the *partial hypermetric cone* $PHYP_n$ as the set of all *partial hypermetrics* on n points, i.e., of all weak partial semimetrics $p \in wPMET_n$, satisfying, in addition, all inequalities $Hyp_b^w(P^{-1}(p)) \geq 0$ with $\sum_{i=1}^n b_i \in \{0,1\}$.

By definition, $WHYP_n \subset WMET_n$, and $PHYP_n \subset wPMET_n$. However, it is easy to prove, that, moreover, it holds

$$WHYP_n \subset dWMET_n \subset WMET_n, \text{ and } PHYP_n \subset PMET_n \subset wPMET_n,$$

since the needed inequalities $d_{ij} \geq w_i - w_j$, and $p_{ij} \geq p_{ii}$ (and, in addition, $w_i \geq 0$) are provided by permutations of $Hyp^{w'}_{(1,0,...,0)}(d;w) \geq 0$, and $Hyp^{w}_{(1,-1,0,...,0)}(d;w) \geq 0$. So, any weighted hypermetric is a down-weighted semimetric, and any partial hypermetric is a partial semimetric.

Proposition 13.3 *Besides the cases $PMET_3 = PHYP_3 = \{0,1\}$-$PMET_3$, and $\{0,1\}$-$dWMET_n = WHYP_n$ for $n = 3, 4$, it holds*

$$\{0,1\}\text{-}dWMET_n \subset WHYP_n \subset dWMET_n \simeq PMET_n \supset PHYP_n \supset \{0,1\}\text{-}PMET_n.$$

Proof.
For a given partition $S_0, S_1, ..., S_t$ of V_n and a given $b \in \mathbb{Z}^n$, $\Sigma_b = \sum_{i=1}^{n} b_i \in \{0,1\}$, let r_h be the dot product $\langle b, (1_{i\in S_h})\rangle$. Then we have, that $r_0 = \Sigma_b - \sum_{h=1}^{t} r_h$, and

$$H_b(\delta_{S_0,S_1,...,S_t}) = \frac{1}{2}\sum_{h=0}^{t} H_b(\delta_{S_h,\overline{S}_h}) = \sum_{h=0}^{t} r_h(r_h - \Sigma_b).$$

Let $(d = 2\delta(S_0, S_1, ..., S_t) - \delta(S_0); w = (1_{i\in S_0}))$ be a generic $P^{-1}(p)$, where p is $\{0,1\}$-valued element of $PMET$, belonging to its extreme ray. Then

$$\frac{1}{2}H_b(d) = \frac{1}{2}\left(2\sum_{h=0}^{t} r_h(r_h - \Sigma_b) - 2r_0(r_0 - \Sigma_b)\right) = \sum_{h=1}^{t} r_h(r_h - \Sigma_b)$$

implies for $\Sigma_b \in \{0,1\}$ the inequality

$$Hyp^w_b(d;w) = \sum_{h=1}^{t} r_h(r_h - 1) - \Sigma_b(\Sigma_b - 1) \geq 0.$$

All $\{0,1\}$-valued elements $(d;w)$ of $dWMET_n$, belonging to its extreme rays, are $(((0));(1))$, $(\delta_S;(0))$, $(\delta_S; w' = (1_{i\in S}))$, and $(\delta_S; w'' = (1_{i\notin S}))$. For them, the value $Hyp^{w'}_b(d;w)$ is equal to $(\Sigma_b + 1)\Sigma_b$, $r_S(r_S - \Sigma)$, $r_S(r_S + 1)$, and $(\Sigma_b - r_S)(\Sigma_b - r_S + 1)$, respectively, and so, for $\Sigma_b \in \{0,1\}$ we get the inequality $Hyp^{w'}_b(d;w) \geq 0$.

Assuming polyhedrality of $WHYP_n$, we checked the case $n = 3, 4$ directly; see Proposition below.

□

Proposition 13.4 *The following statements hold:*

(i) all facets of $WHYP_n$, $n \leq 4$, up to $Sym(n)$ and zero-extensions, are Hyp^w_b with $b = (1,-1), (1,1,-1), (1,1,-1,-1)$, and $Hyp^{w'}_b$ with $b = (1)$, $(1,1,-1)$, $(1,1,1,-2)$, $(2,1,-1,-1)$;

(ii) besides $w_i \geq 0$, among the facets of $P^{-1}(PHYP_n)$, $n \leq 5$, up to $Sym(n)$ and zero-extensions, there are facets Hyp^w_b with $b = (1,-1)$, $(1,1,-1)$, $(1,1,-1,-1)$, $(1,1,1,-1,-1)$, $(1,1,1,-1,-2)$, $(2,1,1,-1,-1)$.

Proof. It was obtained by direct computation.

The equality $WHYP_n = \{0,1\}$-$WMET_n$ for $n = 3, 4$ holds, because only inequalities which are requested in $WHYP_n$, appeared among those of $\{0,1\}$-$WMET_n$.

The facets of $PHYP_4$ were deduced by computation using the tightness of the inclusions $\{0,1\}$-$PMET_4 \subset PHYP_4 \subset PMET_4$. In fact, the cone $\{0,1\}$-$PMET_4$ contains exactly one facet (orbit F_5), different from Hyp_b^p and $p_{ii} \geq 0$ (orbit F_1), and the cone $PMET_4$ contained exactly two (orbits O_{10} and O_{11}) non-$\{0,1\}$-valued extreme ray representatives. The 6 rays from O_{10} are removed by 6 representatives of Hyp_b^p with $b = (1, 1, -1, -1)$ (orbit F_4), while the 12 rays from O_{11} are removed by 12 representatives of F_5 (see Tables 12.4, and 12.5 in Chapter 12).

□

In Table 13.8 we present the information on small cones of weighted hypermetrics and partial hypermetrics for $n = 3, 4$.

Cone	Dimension	# of ext. rays (orbits)	# of facets (orbits)	Diameters
$PHYP_3 = PMET_3 = \{0,1\}$-$PMET_3$	6	13(5)	12(3)	3; 2
$PHYP_4$	10	56 (10)	34 (4)	3;2
$WHYP_3 = \{0,1\}$-$dWMET_3$	6	10 (4)	15 (4)	2; 2
$WHYP_4 = \{0,1\}$-$dWMET_4$	10	22 (6)	62 (7)	2; 3

Table 13.8: Main parameters of partial and weighted hypermetrics for $n \leq 4$

13.4 Quasi-hypermetric cones

Weightable quasi-hypermetric cone $OWHYP_n$

At first, consider a *weightable quasi-semimetric* q on the set V_n, i.e., a function $q : V_n \times V_n \to \mathbb{R}$ with a *weight function* $w = (w_x) : V_n \longrightarrow \mathbb{R}_{\geq 0}$, such that for all $i, j, k \in V_n$ it holds $q_{ij} \geq 0$, $q_{ii} = 0$, $q_{ij} \leq q_{ik} + q_{kj}$, and $q_{ij} + w_i = q_{ji} + w_j$ for $i \neq j$.

It is known, that down-weighted semimetrics and weightable quasi-semimetrics are connected by the map $Q : (d, w) \to q$, defined by

$$q_{ij} = \frac{d_{ij} - w_i + w_j}{2}.$$

The map Q is an automorphism of the vector space $\mathbb{R}^{\binom{n+1}{2}}$, and $(d; w) = Q^{-1}(q)$, where the inverse map Q^{-1} is defined by

$$d_{ij} = q_{ij} + q_{ji}.$$

So, we can define a *weightable quasi-hypermetric* q on V_n as a *weightable quasi-semimetric* q on V_n, satisfying, in addition, all inequalities $Hyp_b^w(Q^{-1}(q)) \geq 0$ with $\sum_{i=1}^n b_i = 1$ or $\sum_{i=1}^n b_i = 0$.

It is easy to see, that for $(q, w) = Q(d; w)$ we have

$$Hyp_b^w(d; w) = H_b(q) + \left(1 - \sum_{i=1}^n b_i\right) \sum_{i=1}^n b_i w_i.$$

In other words, a *weightable quasi-hypermetric* q on V_n is a *weightable quasi-semimetric* q on V_n, satisfying, in addition, all inequalities

$$Hyp_b^o : H_b(q) + \left(1 - \sum_{i=1}^{n} b_i\right) \sum_{i=1}^{n} b_i w_i \geq 0$$

with $\sum_{i=1}^{n} b_i = 1$, or $\sum_{i=1}^{n} b_i = 0$.

However, for weightable quasi-semimetrics there exists an other approach to the construction of *weightable quasi-hypermetrics.*

In order to introduce it, let C be any cone, closed under *reversal*, i.e., satisfying the condition

$$q \in C \Rightarrow q^T \in C.$$

For such C, if the linear inequality $\sum_{1 \leq i,j \leq n} f_{ij} q_{ij} = \langle f, q \rangle \geq 0$ is valid on C, then f also defines a face of $\{q + q^T \mid q \in C\}$.

Given a valid inequality

$$G : \sum_{1 \leq i < j \leq n} g_{ij} d_{ij} \geq 0$$

of $\{q + q^T : q \in C\}$, and a *tournament* K_n^O, i.e., an oriented version of the complete graph K_n with an unique arc between any distinct $i, j \in V_n$, let $g^O = ((g_{ij}^O))$ be a matrix with $g_{ij}^O = g_{ij}$, if the arc $\langle i, j \rangle$ belongs to K_n^O, and with $g_{ij}^O = 0$, otherwise.

Consider the inequality

$$G^O : \sum_{1 \leq i < j \leq n} g_{ij}^O d_{ij} \geq 0.$$

Call G^O *standard*, if there exists a permutation $\tau \in Sym(n)$, such that

$$\langle i, j \rangle \in K_n^O \Leftrightarrow \tau(i) < \tau(j).$$

Call G^O *reversal-stable* (*rs* for short), if

$$\langle g^O, q \rangle = \langle g^O, q^T \rangle.$$

Note, that, in general, G^O is not valid on C and does not preserve the rank of G.

For example, the standard $T_{12,3} : q_{13} + q_{23} - q_{12} \geq 0$ is not valid on $OCUT_n$, and the standard $NN_{ij} : q_{ij} \geq 0$ defines a facet in $OCUT_n$, while $G : d_{ij} \geq 0$ only defines a face in MET_n.

If G^O is reversal-stable, then $\langle G^O, q \rangle = \frac{1}{2} \langle G, q + q^T \rangle$, i.e., G^O is valid on C, if G is valid on $\{q + q^T : q \in C\}$.

The facets $OT_{ij,k}$ and NN_{ij} (only 1-st is reversal stable) of $WQMET_n$ are standard and of the form Hyp_b^o, where b is a permutation of $(1, 1, -1, 0, ..., 0)$, or $(1, -1, 0, ..., 0)$, respectively. $OCUT_4$ has one more orbit: six standard, non reversal stable facets of the form Hyp_b^o, where b is a permutation of $(1, 1, -1, -1)$, or $q_{13} + q_{14} + q_{23} + q_{24} - (q_{12} + q_{34}) \geq 0$.

$OCUT_5$ has, up to $Sym(n)$, 3 *new* (i.e., in addition to zero-extensions of the facets of $OCUT_4$) such orbits: one standard reversal stable orbit for $b = (1, 1, 1, -1, -1)$, and two non-standard, non-reversal stable orbits.

$OCUT_6$ has, among its 56 new orbits, two non-standard reversal stable orbits for $b = (2, 1, 1, -1, -1, -1)$, and $b = (1, 1, 1, 1, -1, -2)$.

The adjacencies of cuts in CUT_n are defined only by the facets $Tr_{ij,k}$, and adjacencies of those facets are defined only by cuts. It gives at once $\binom{n}{2} - 1$ linearly independent facets $OTr_{ij,k}$ containing any given pair $(\delta^O_{S_1}, \delta^O_{S_2})$, using that $OTr_{ij,k}$ are reversal stable facets. So, only n more facets are needed to get the adjacencies of o-cuts. It is a way to prove Conjecture 12.1 in Chapter 12.

Call a *tournament K^O_n admissible*, if its arcs can be partitioned into arc-disjoint directed cycles. It does not exists for even n, because then the number of arcs involving each vertex is odd, while each cycle provides 0 or 2 such arcs. But for odd n, there are at least $2^{\frac{n-3}{2}}$ admissible tournaments: take the decomposition of K_n into $\frac{n-1}{2}$ disjoint Hamiltonian cycles and, fixing the order on one them, consider all possible orders on remaining cycles.

For odd n, denote by K^{Oc}_n the *canonic admissible tournament*, consisting of all $\langle i, i+k\rangle$ with $1 \leq i \leq n-1, 1 \leq k \leq \lceil \frac{n}{2} \rceil + 1 - i$, and $\langle i+k, i\rangle$ with $1 \leq i \leq \lfloor \frac{n}{2} \rfloor, \lceil \frac{n}{2} \rceil \leq k \leq n-i$, i.e.,

$$0 = C_{1,2,3,4,5,6,7,\dots} + C_{1,3,5,7,\dots} + C_{1,4,7,\dots} + \dots .$$

The *Kelly conjecture* states that the arcs of every *regular* (i.e., the vertices have the same outdegree) tournament can be partitioned into arc-disjoint directed Hamiltonian cycles.

Zero-extensions of $q_{ij} \geq 0$ and $q_{13} + q_{14} + q_{23} + q_{24} - (q_{12} + q_{34}) \geq 0$ can be seen as the first instances (for $b = (1, -1, 0, \dots, 0)$ and $b = (1, 1, -1, -1, 0, \dots, 0)$, respectively) of the *o-negative type inequality*

$$-\sum_{1 \leq i < j \leq n} b_i b_j q_{a(ij)} \geq 0,$$

where $b = (b_1, \dots, b_n) \in \mathbb{Z}^n$ with $\Sigma_b = 0$, and the arcs $a(ij) \in \{\langle i,j\rangle, \langle j,i\rangle\}$ on the edges (i,j) are by some rule depended of a given tournament K^O_n.

Denote by $OWHYP_n$ the set of all $q \in WQMET_n$, satisfying the two above orbits and all *o-hypermetric inequalities*

$$-\sum_{1 \leq i < j \leq n} b_i b_j q_{a(ij)} \geq 0,$$

where $b = (b_1, \dots, b_n) \in \mathbb{Z}^n$, $\sum_{i=1}^n b_i = 1$, $O = K^O_n$ is an admissible tournament, and the arc $a(ij)$ on the edge (i,j) is the same as in K^O_n, if $b_i b_j \geq 0$, or the opposite one, otherwise.

Formally, define a *weightable quasi-hypermetric* on n points as a weightable quasi-semimetric q on V_n, satisfying all *o-hypermetric inequalities*

$$Hyp_{b,O} : Hyp_{b,O}(q) = -\sum_{1 \leq i < j \leq n} b_i b_j q_{a(ij)} \geq 0$$

with $b = (b_1, \dots, b_n) \in \mathbb{Z}^n$, $\sum_{i=1}^n b_i = 1$, and all *o-negative type inequalities*

$$Neg_{b,O} : Neg_{b,O}(q) = -\sum_{1 \leq i < j \leq n} b_i b_j q_{a(ij)} \geq 0$$

with $b = (b_1, \dots, b_n) \in \mathbb{Z}^n$, $\sum_{i=1}^n b_i = 0$, where $O = K^O_n$ is an admissible tournament, and the arc $a(ij) \in \{\langle i,j\rangle, \langle j,i\rangle\}$ on the edge (i,j) is the same as in K^O_n, if $b_i b_j \geq 0$, or the opposite one, otherwise.

The *weightable quasi-hypermetric cone* $OWHYP_n$ is the set of all *weightable quasi-hypermetrics* on n points, i.e., all weightable quasi-semimetrics $q \in WQMET_n$, satisfying, in addition, all inequalities $Hyp_{b,O}(q) \geq 0$, and all inequalities $Neg_{b,O}(q) \geq 0$.

By definition, it holds, that $OWHYP_n = OCUT_n$ for $n = 3, 4$.

Proposition 13.5 *For $n \geq 5$, it holds*

$$OCUT_n \subset OWHYP_n \subset WQMET_n.$$

Proof. Without loss of generality, let $b_i = 1$ for $1 \leq i \leq \lfloor \frac{n}{2} \rfloor$, and $b_i = -1$, otherwise. The general case means only that we have sets of $|b_i|$ coinciding points.

The inequality $Hyp_{b,O}$ is reversal stable, because Proposition 4 in [DeDe10] implies, that any inequality on a weightable quasi-semimetric $q \in WQMET_n$ is preserved by the reversal of q. So,

$$Hyp_{b,O}(q) = \frac{1}{2} Hyp_b^w(q + q^T).$$

On an o-cut δ_S^O it gives, putting $r = \langle b, (1_{i \in S}) \rangle$, that

$$\frac{1}{2} Hyp_b(\delta_S^O + \delta_{\overline{S}}^O) = Hyp_b(\delta_S) = r(r - \Sigma_b) \geq 0.$$

□

The cone $OWHYP_5$ has, besides o-cuts, 40 extreme rays in two orbits: F_{ab}, F'_{ab}, having 2 on the position ab, 1 on the position ba, 0 on three other positions ka for $k \neq b$ in F_{ab}, or on three other positions bk for $k \neq a$ in F'_{ab}, and ones on other non-diagonal positions.

Also, diameter of G_{OWHYP_5} is equal to 2, as well as diameter of $G^*_{OWHYP_5}$ (see Table 13.9).

Cone	Dimension	# of ext. rays (orbits)	# of facets (orbits)	Diameters
$OWHYP_3 = OCUT_3$	5	6 (2)	9 (2)	1;2
$OWHYP_4 = OCUT_4$	9	14 (3)	30 (3)	1;2
$OWHYP_5$	14	70 (6)	90 (4)	2;2

Table 13.9: Main parameters of $OWHYP_n$, $n \leq 5$

Quasi-hypermetric cone $QHYP_n$

Let $V_n = \{1, 2, ..., n\}$; let $I_n = |\{\langle i, j \rangle \mid i, j \in V_n, i \neq j\}|$, where $\langle i, j \rangle$ denotes the ordered pair of the integers i, j, i.e., $I_n = 2\binom{n}{2} = n(n-1)$.

Let q be a *quasi-hypermetric* on the set V_n, i.e., quasi-semimetric q on V_n, satisfying, in addition, all *oriented hypermetric inequalities*

$$Hyp_b^O : Hyp_b^O(q) = \sum_{1 \leq i \neq j \leq n} b_i b_j q(i, j) \leq 0 \text{ for all } b \in \mathbb{Z}^n, \sum_{i=1}^{n} b_i = 1.$$

As any quasi-hypermetric on V_n is a quasi-semimetric on V_n, it can be viewed alternatively as a function on $V_n \times V_n$, as a vector in $\mathbb{R}^{I_n}$, or as an (in general, non-symmetric) $n \times n$ matrix $((q_{ij}))$ with zeros on the main diagonal. We will use all these representations, as well as both symbols $q(i, j)$ and q_{ij} for the values of the quasi-hypermetric between points i and j.

Denote by $QHYP_n$ the set of all quasi-hypermetrics on V_n. So, $QHYP_n$ forms the cone of quasi-semimetrics q, satisfying $Hyp_b^O(q) \leq 0$ for any sequence $b = (b_1, ..., b_n) \in \mathbb{Z}^n$ with $\sum_{i=1}^{n} b_i = 1$. It means, that $q_{ij} + q_{ji} \in HYP_n$.

The cone $QHYP_n = \{q \in QMET_n : ((q_{ij}+q_{ji})) \in HYP_n\}$ was introduced in [DDP03]. Clearly, it is polyhedral (see Deza, Grishukhin and Laurent (1993) [DGL93]), and the triangle inequality is redundant. It coincides with $QMET_n$ for $n = 3, 4$.

The smallest case when $QHYP_n$ is a proper subcone of $QMET_n$ is $n = 5$; $QHYP_5$ has 90 facets ($20 + 60$ from $QMET_5$ and 10 new facets with $b = (1, 1, 1, -1, -1)$), and 78810 extreme rays; see the list of representatives of all 386 orbits of $QHYP_5$ in http://www.liga.ens.fr/~dutour/. The diameters of G_{QHYP_5} and $G^*_{QHYP_5}$ are 4 and 2 (see Table 13.10).

Cone	Dimension	# of ext. rays (orbits)	# of facets (orbits)	Diameters
$QHYP_3 = QMET_3$	6	12(2)	12(2)	2; 2
$QHYP_4 = QMET_4$	12	164(10)	36(2)	3; 2
$QHYP_5$	20	78810(386)	90(3)	4; 2

Table 13.10: Description of $QHYP_n$, $n \leq 5$

Remind, that the set

$$E_n = \{e + e^T \mid e \text{ is an extreme ray of } QMET_n\}$$

consists, for $n = 3, 4, 5$, of $1, 7, 79$ orbits (amongst of $2, 10, 229$), including $0, 3, 10$ orbits of path-metrics of graphs.

More exactly, a path-metric $d(G)$ belongs to E_4 for the graphs $G = K_4, P_2, C_4, P_4$. Now, $d(G) \in E_5$ for $G = \{K_{2,3}, K_5 - K_3, K_5 - P_2 - P_3, K_5, C_5, \overline{P_2}, \overline{P_3}, \overline{P_4}, \overline{P_5}, \overline{2P_2}\}$, where $d(K_{2,3})$ is an extreme ray of MET_5; $d(K_{2,3})$, $d(K_5 - K_3)$ and $d(K_5 - P_2 - P_3)$ do not belong to CUT_5, and the remaining seven graphs belong to CUT_5. In fact, those seven path metrics $d(G)$ are all of form $e + e^T$, were e is an extreme ray of the cone $QHYP_5$.

Weighted quasi-hypermetric cone $WQHYP_n$ as a projection of HYP_{n+1}

It is shown in [DGD12], that $OCUT_n$ and $WQMET_n$ on $\{1, 2, ..., n\}$ are projections of CUT_{n+1} and MET_{n+1} (defined on $\{0, 1, 2, ..., n\}$) on the subspace, orthogonal to $\delta_{\{0\}}$. Moreover, the *weighted quasi-hypermetric cone* $WQHYP_n$ is defined as such projection of HYP_{n+1}, and it holds

$$OCUT_n \subseteq WQHYP_n \subseteq WQMET_n$$

with equalities $WQHYP_n = WQMET_n$ only for $n = 3$, and $OCUT_n = WQHYP_n$ only for $3 \leq n \leq 5$. See data on $WQHYP_5$ in Table 13.11.

Cone	Dimension	# of ext. rays (orbits)	# of facets (orbits)	Diameters
$WQHYP_3 = OCUT_3 = WQMET_3$	5	6(2)	9(2)	1; 2
$WQHYP_4 = OCUT_4$	9	14(3)	30(3)	1; 2
$WQHYP_5 = OCUT_5$	14	30(4)	130(6)	1; 3

Table 13.11: Small cones $WQHYP_n$

Chapter 14

Cuts over general graphs

14.1 Preliminaries

Polyhedra, generated by cuts and by finite metrics, are central objects of Discrete Mathematics; see, say, [DeLa97]. In particular, they are tightly connected with the well-known NP-hard Optimization Problems, such as the Max-cut Problem and the Unconstrained Quadratic $\{0,1\}$-programming Problem. To find their (mostly unknown) facets is the main approach to these problems.

The classical case of the *metric cone* MET_n, *cut cone* CUT_n, *metric polytope* $MET_n^\square$ and *cut polytope* $CUT_n^\square$ can be considered as an particular case of polyhedra over graphs: $MET_n(K_n)$, $CUT(K_n)$, $MET_n^\square(K_n)$ and $CUT^\square(K_n)$ over the complete graph on n points K_n with the edge set $\{(i,j) \mid i,j \in V_n, i \neq j\}$, where (i,j) denotes the unordered pair of the integers i,j. It is the most interesting and complicated case of the general case of polyhedra $MET_n(G)$, $CUT(G)$, $MET_n^\square(G)$ and $CUT^\square(G)$ over general graph $G = \langle E, V \rangle$.

In this chapter we consider semimetric and cut polytopes over several very symmetric graphs with $15-30$ edges, including skeletons of Platonic solids, $K_{3,3,3}$ $K_{1,4,4}$, $K_{5,5}$, some other $K_{l,m}$, $K_{1,l,m}$, $Prism_n, APrism_n$, Möbius ladder M_{2n}, Heawood graph and Petersen graph.

Using computer search, we list facets for the cut polytopes over these graphs, study the structure and determine the sizes of the orbits of their facets. We study two graphs, the 1-skeleton graph and the ridge graph of these polyhedra: the number of their nodes and edges, their diameters, conditions of adjacency.

The lists of facets for $K_{1,l,m}$ with $(l,m) = (4,4),(3,5),(3,4)$ solve some problems in Quantum Information Theory. (see, for example, [PiSv01], [Fine82], and [DeLa97], Section 5.2).

We define and consider also an analogue of hypermetric inequality for the cut polytope over a graph G. This directly generalizes the corresponding definitions of the metric polytope over a graph and allows to find new facet inequalities in some cases.

The following polyhedra will be studied below:

- The *metric cone* $MET_n(G)$ over a graph $G = \langle E, V \rangle$ of order $|V| = n$ and of size $|E| = m$: the projection of the cone MET_n (of all semimetrics on n points, generated by all triangle inequalities $d_{ik} \leq d_{ij} + d_{jk}$ on V_n) on the subspace $\mathbb{R}^m$, indexed by the edge set E of G;

- The *cut cone* $CUT_n(G)$ over a graph $G = \langle E, V \rangle$ of order $|V| = n$ and of size $|E| = m$: the projection of the cut cone CUT_n (generated by all non-zero cut semimetrics on V_n) on the subspace $\mathbb{R}^m$, indexed by the edge set E of G, or, equivalently, a cone on n points, generated by all non-zero cut vectors of a given graph $G = \langle V, E \rangle$;
- The *metric polytope* $MET_n^{\square}(G)$ over a graph $G = \langle E, V \rangle$ of order $|V| = n$ and of size $|E| = m$: the projection of the polytope $MET_n^{\square}$ (of all semimetrics on n points, generated by all triangle inequalities $d_{ik} \leq d_{ij} + d_{jk}$ and $d_{ij} + d_{ik} + d_{jk} \leq 2$ on V_n) on the subspace $\mathbb{R}^m$, indexed by the edge set E of G;
- The *cut polytope* $CUT_n^{\square}(G)$ over a graph $G = \langle E, V \rangle$ of order $|V| = n$ and of size $|E| = m$: the projection of the cut polytope CUT_n (generated by all cut semimetrics on V_n) on the subspace $\mathbb{R}^m$, indexed by the edge set E of G, or, equivalently, the convex hull of all cut vectors of a given graph $G = \langle V, E \rangle$.

14.2 Metric and cut polyhedra over graphs

Basic results on metric and cuts polyhedra over graphs

Given a graph $G = \langle V, E \rangle$ of order $|V| = n$ and of size $|E| = m$, we can consider its vertex set V as the set $V_n = \{1, 2, ..., n\}$, and its edge set E as a subset of the set $\{(i, j) \,|\, i, j \in V_n, i \neq j\}$, where (i, j) denotes the unordered pair of the integers $i, j \in V_n$; i.e., $|E| \leq E_n$.

For a graph $G = \langle V, E \rangle$, the notion of cut of G is well defined: for a vertex subset $S \subseteq V = \{1, 2, ..., n\}$, the *cut semimetric* $\delta_S(G)$ is a vector (actually, a symmetric $\{0, 1\}$-matrix), defined as

$$(\delta_S(G))_{ij} = \begin{cases} 1, & \text{if } \; ij \in E, \quad \text{and} \quad |S \cap \{i, j\}| = 1, \\ 0, & \text{otherwise.} \end{cases}$$

So, $\delta_S(G)$ can be seen also as the adjacency matrix of a *cut* (into S and $V_n \backslash S = \overline{S}$) *subgraph* of G. Clearly, $\delta_{\overline{S}}(G) = \delta_S(G)$.

The *cut cone* $CUT(G)$ over the graph $G = \langle V, E \rangle$ of order $|V| = n$ and of size $|E| = m$ is defined as the conic hull of all $2^{n-1} - 1$ non-zero semimetric $\delta_S(G)$, $S \subseteq V_n$. Equivalently, $CUT_n(G)$ is the projection of $CUT_{|V|} = CUT_n$ on the subspace $\mathbb{R}^{|E|} = \mathbb{R}^m$, indexed by the edge set E.

The *cut polytope* $CUT^{\square}(G)$ over the graph $G = \langle V, E \rangle$ of order $|V| = n$ and of size $|E| = m$ is defined as the convex hull of all 2^{n-1} cut semimetric $\delta_S(G)$, $S \subseteq V_n$, including zero-cut semimetric $\delta_{\emptyset}(G)$. Equivalently, $CUT_n^{\square}(G)$ is the projection of $CUT_{|V|}^{\square} = CUT_n^{\square}$ on the subspace $\mathbb{R}^{|E|} = \mathbb{R}^m$, indexed by the edge set E.

Metric cone $MET(G)$ over a graph $G = \langle V, E \rangle$ of order $|V| = n$ and of size $|E| = m$ is defined as the projection of $MET_{|V|} = MET_n$ on $\mathbb{R}^{|E|} = \mathbb{R}^m$, indexed by the edges of G.

Metric polytope $MET^{\square}(G)$ over a graph $G = \langle V, E \rangle$ of order $|V| = n$ and of size $|E| = m$ is the projection of $MET_{|V|}^{\square} = MET_n^{\square}$ on $\mathbb{R}^{|E|} = \mathbb{R}^m$, indexed by the edges of G.

The classical case of the cut cone CUT_n and the cut polytope $CUT_n^{\square}$, as well as the metric cone MET_n ad the metric polytope $MET_n^{\square}$, can be considered as $CUT(K_n)$ and $CUT^{\square}(K_n)$ ($MET(K_n)$ and $MET^{\square}(K_n)$, respectively) over complete graph K_n on

n vertices. It is the most interesting and complicated case of $CUT(G)$ and $CUT^{\square}(G)$ ($MET(G)$ and $MET^{\square}(G)$, respectively).

The dimension of $MET_n(G)$, $MET_n^{\square}(G)$, $CUT_n(G)$ and $CUT^{\square}(G)$ is equal to m, i.e., to the *size* $|E|$ of G.

Clearly, it holds

$$CUT(G) \subseteq MET(G), \text{ and } CUT(G) \subseteq MET(G).$$

It is proven in [Seym81] for cones and in [BaMa86] for polytopes (see also [DeLa97], Section 27.3), that $CUT(G) = MET(G)$ *or, equivalently,* $CUT^{\square}(G) = MET^{\square}(G)$ *if and only if* G *does not have any* K_5*-minor.*

Remark. By *Wagner's Theorem* [Wagn37], a finite graph is planar if and only if it has no minors K_5 and $K_{3,3}$.

So, for a connected graph $G = \langle V, E \rangle$, the polytope $CUT^{\square}(G)$ has $2^{|V|-1} = 2^{n-1}$ vertices and dimension $|E| = m$. Obviously, the symmetry group $Aut(G)$ of a graph $G = \langle V, E \rangle$ induces a symmetry of $CUT^{\square}(G)$. For any $U \subset \{1, 2, ..., n\}$, the *switching map* $\delta_S \mapsto \delta_{U \Delta S}$ also defines a symmetry of $CUT^{\square}(G)$, i.e., the switchings still act on $CUT^{\square}(G)$. Together those form the *restricted symmetry group* $Is_{Res}(CUT^{\square}(G))$ of order $2^{n-1}|Aut(G)|$.

The full symmetry group $Is(CUT^{\square}(G))$ might be larger. In [DDD15] the dual description of $CUT^{\square}(G)$ was computed for several graphs; see data on some of them in Tables 14.1 and 14.2. In 3-rd column of these Tables, $A(G)$ denotes $2^{1-n}|Is(CUT^{\square}(G))|$; it is $|Aut(G)|$ for all, but K_4, graphs there.

In Tables 14.1, 14.2 we list information on the cut polytopes of many selected graphs. The number of facets of $CUT^{\square}(G)$ for some graphs G without K_5-minor (two non-planar graphs and the skeletons of Platonic and some semiregular polyhedra) is given in Table 14.1. The similar information about $CUT^{\square}(G)$ for some graphs G with K_5-minor is given in Table 14.2.

For all polytopes computed here, the number of vertices of $CUT^{\square}(G)$ is 2^{n-1}, where n is the *order* $|V|$ of G. The data file of the groups and orbits of facets of considered polytopes is available from [DeDu13].

$G = \langle V, E \rangle$	$\lvert V\rvert, \lvert E\rvert$	$A(G)$	# of facets (orbits)	Cycles
Wagner graph M_8	$8, 12$	16	184(4)	$2, 2, 4, 5$
$K_{3,3} = M_6$	$6, 9$	72	90(2)	$2, 4$
Dodecahedron	$20, 30$	120	$23, 804(5)$	$2, 5, 9, 10, 10$
Icosahedron	$12, 30$	120	$1, 552(4)$	$3, 5, 6, 6$
Cube K_2^2	$8, 12$	48	200(3)	$2, 4, 6$
Octahedron $K_{2,2,2}$	$6, 12$	48	56(2)	$3, 4$
Tetrahedron K_4	$4, 6$	144 *	12(1)	3
Prism_7	$14, 21$	28	$7, 394(6)$	$2, 2, 4, 7, 9, 9$
APrism_6	$12, 24$	24	$2, 032(5)$	$3, 6, 7, 7, 8$
Cuboctahedron	$12, 24$	48	$1, 360(5)$	$3, 4, 6, 6, 8$
Tr. tetrahedron	$12, 18$	24	540(4)	$2, 3, 6, 8$

Table 14.1: Main paramerers of $CUT^{\square}(G) = MET^{\square}(G)$ for some K_5-minor-free graphs

$G = \langle V, E\rangle$	$\|V\|, \|E\|$	$A(G)$	# of facets (orbits)	Cycles
Heawood graph	14, 21	336	5, 361, 194(9)	2, 6, 8
Petersen graph	10, 15	120	3, 614(4)	2, 5, 6
Pyr(APrism$_4$)	9, 24	16	389, 104(17)	3, 3, 3, 4, 5
Möbius ladder M_{14}	14, 21	28	369, 506(9)	2, 2, 4, 8, 10
Tr. octahedron on $\mathbb{P}^2$	12, 18	48	62, 140(7)	2, 2, 4, 6, 6
$K_{5,5}$	10, 25	$2(5!)^2$	16, 482, 678, 610(1, 282)	2, 4
$K_{4,7}$	11, 28	4!7!	271, 596, 584(15)	2, 4
$K_{4,6}$	10, 24	4!6!	23, 179, 008(12)	2, 4
$K_{4,5}$	9, 20	4!5!	983, 560(8)	2, 4
$K_{3,3,3}$	9, 27	$(3!)^4$	624, 406, 788(2, 015)	3, 4
$K_{1,4,4}$	9, 24	$2(4!)^2$	36, 391, 264(175)	3, 4
$K_{1,3,5}$	9, 23	3!5!	71, 340(7)	3, 4
$K_{1,3,4}$	8, 19	3!4!	12, 480(6)	3, 4
$K_{1,1,3,3}$	8, 21	$4(3!)^2$	432, 552(50)	3, 3, 4
$K_{1,1,2,m}, m > 2$	m+4, 4m+5	4m!	$8 + 20m + 8\binom{m}{2}(16m - 5(7)$	3, 3, 3, 4
$K_{m+4} - K_m$, m>1	m+4, 4m+6	4!m!	$8(8m^2 - 3m + 2)(4)$	3, 3
$K_{1,1,1,1,1,3} = K_8 - K_3$	8, 25	360	2, 685, 152(82)	3, 3
$K_7 - K_2$	7, 20	240	31, 400(17)	3, 3

Table 14.2: Main paramerers of $CUT^{\square}(G)$ for some graphs G with K_5-minor

Edge faces and s-cycle faces of metric and cut polyhedra over graphs

If $e = (v_i, v_j) = v_iv_j$ is an edge of a graph $G = \langle V, E\rangle$, and x is a vector in $\mathbb{R}^{|E|}_{\geq 0}$, define x_e as x_{v_i,v_j}. So, if, as usually, $V = V_n = \{1, 2, ..., n\}$, then for a given edge $e = ij$ we have $x_e = x_{ij}$.

Definition 14.1 *Let $G = \langle V, E\rangle$ be a graph.*

(i) Given an edge $e \in E$, the edge inequality (or 2-cycle inequality) is

$$0 \leq x_e \leq 1.$$

(ii) Given a cycle C in G and an odd-sized set F of edges in C, the cycle inequality is

$$f_{C,F} : f_{C,F}(x) = \sum_{e' \in C \setminus F} x_{e'} - \sum_{e \in F} x_e \leq |F| - 1.$$

(iii) Given an s-cycle C, $s \geq 3$, of G (i.e., any cycle in G with s edges), and an edge $e \in C$, the s-cycle inequality is

$$f_{C,e} : f_{C,e}(x) = \sum_{e' \in C \setminus e} x_{e'} - x_e \geq 0.$$

So, s-cycle inequalities are just cycle inequalities with $|F| = 1$. The 3-cycle inequality is just usual triangle inequality.

The edge inequalities and the cycle inequalities are valid on $CUT^{\square}(G)$, since they are, clearly, valid on each cut of G: a cut intersects a cycle in the set of even cardinality. So, they define faces, but not necessarily facets.

In fact, all facets of $MET^{\square}(G)$ are defined ([BaMa86]) by all cycle inequalities $f_{C,F}$ for hordless circuits in G, and all inequalities $0 \leq x_e \leq 1$ for edges e, not belonging to a triangle in G. More exactly, it holds the following Proposition.

Proposition 14.1 *Given a graph $G = \langle V, E\rangle$, it holds:*
(i) $MET(G)$ is defined by all s-cycle inequalities:

$$MET(G) = \{x \in \mathbb{R}_{\geq 0}^{|E|} \mid \sum_{e' \in C \setminus e} x_{e'} - x_e \geq 0\},$$

where C is a cycle in G, and $e \in C$;
(ii) $MET^{\square}(G)$ is defined by all edge and all cycle inequalities:

$$MET^{\square}(G) = \{x \in \mathbb{R}_{\geq 0}^{|E|} \mid x_e \leq 1, \sum_{e' \in C \setminus F} x_{e'} - \sum_{e \in F} x_e \leq |F| - 1\},$$

where C is a cycle in G, F is a subset of the set of the edges of C, $|F|$ is odd;
(iii) the above inequality $\sum_{e' \in C \setminus e} x_{e'} - x_e \geq 0$ defines a facet of $MET(G)$ if and only if C is a chordless circuit; the above inequality $\sum_{e' \in C \setminus F} x_{e'} - \sum_{e \in F} x_e \leq |F| - 1$ defines a facet of $MET^{\square}(G)$ if and only if C is a chordless circuit;
(iv) the inequality $x_e \geq 0$ defines a facet of $MET(G)$ if and only if e does not belong to any triangle of G; the inequality $x_e \leq 1$ defines a facet of $MET^{\square}(G)$ if and only if e does not belong to any triangle of G.

(iii) and (iv) were proved in [BaMa86], (i) and (ii) were proved in [Bara93]. (See also Section 27.3 in [DeLa97].)

As $CUT(G) \subseteq MET(G)$, and $CUT^{\square}(G) \subseteq MET^{\square}(G)$, we obtain the following result for $CUT(G)$ and $CUT^{\square}(G)$.

Proposition 14.2 *Given a graph $G = \langle V, E\rangle$, it holds:*
(i) the inequality $x_e \geq 0$ is facet defining in $CUT(G)$ if and only if e is not contained into an 3-cycle of G; the inequality $x(e) \leq 1$ is facet defining in $CUT^{\square}(G)$ if and only if e is not contained into an 3-cycle of G;
(ii) an s-cycle inequality is facet defining in $CUT(G)$ if and only if the corresponding s-cycle is a chordless circuit; an cycle inequality is facet defining in $CUT^{\square}(G)$ if and only if the corresponding cycle is chordless circuit.

Besides above facets of $MET^{\square}(G)$, some other valid inequalities show up on $CUT^{\square}(G)$. Below we give a way to get general statement for building such inequalities of hypermetric type.

However, it holds the following important result of Seymour [Seym81] and Barahona [Bara93].

Proposition 14.3 *$CUT(G) = MET(G)$ or, equivalently, $CUT^{\square}(G) = MET^{\square}(G)$ if and only if G does not have any K_5-minor.*

As a corollary of this result, we have that the facets of $CUT^{\square}(G)$ (also, of $CUT(G)$) are completely determined by edge inequalities and s-cycle inequalities (by edge inequalities and cycle inequalities, respectively) if and only if G does not have any K_5-minor.

In particular, it holds for any planar graph, because by *Wagner's Theorem* [Wagn37], a finite graph is planar if and only if it has no minors K_5 and $K_{3,3}$. (Closely related *Kuratowski's theorem* [Kura30] states that a finite graph is planar if and only if it does not contain a subgraph that is a *subdivision* of K_5 or of $K_{3,3}$.)

Now it is obviously, that the 3-cycle inequality - the usual triangle inequality - is unique, among edge and all cycle inequalities, to define a facet in the polytope $CUT^{\square}(K_n) = CUT_n^{\square}$.

In general, for a given graph G, a *chordless cycle* is any cycle, which is induced subgraph; so, any triangle, any shortest cycle (a cycle in G with the smallest length), and any cycle, bounding a face in some embedding of G, are chordless.

Let c'_s denote the number of all s-cycles in G; let c_s denote the number of all chordless s-cycles in G.

In these notations, it is easy to check, that for $CUT_n^{\square}(G)$ it holds:

- there are $2|E|$ edge faces, which decompose into orbits, one for each orbit of edges of G under $Aut(G)$;
- there are $2^{s-1}c'_s$ s-cycle faces, which decompose into orbits, one for each orbit of s-cycles of G under $Aut(G)$;
- the incidence of edge faces is $2^{|V|-2}$, and the size of each orbit is twice the size of corresponding orbit of edges;
- he incidence of s-cycle faces is $2^{|V|-s} \cdot s$, and the size of each orbit is 2^{s-1} times the size of corresponding orbit of s-cycles in G.

Automorphism group of metric and cut polyhedra over graphs

Obviously, the symmetry group $Aut(G)$ of a graph $G = \langle V, E\rangle$ induces symmetry of $CUT^{\square}(G)$. Moreover, for any $U \subset \{1, 2, ..., n\}$, the map $\delta_S \mapsto \delta_{U\Delta S}$ also defines a symmetry of $CUT^{\square}(G)$. Together those form the *restricted symmetry group* $Is_{Res}(CUT^{\square}(G))$ of order $2^{|V|-1}|Aut(G)|$. The full symmetry group $Is(CUT^{\square}(G))$ may be larger.

For example, $|Is(CUT^{\square}(K_n))|$ is $2^{n-1} \cdot n!$, if $n \neq 4$, and $6 \cdot 2^3 \cdot 4!$, if $n = 4$. It means, that for all $n \neq 4$ the full symmetry group $Is(CUT^{\square}(K_n))$ coincides with the restricted symmetry group $Is_{Res}(CUT^{\square}(G))$ of order $2^{|V|-1}|Aut(K_n)| = 2^{n-1} \cdot n!$, but for $n = 4$ the full symmetry group $Is(CUT^{\square}(K_4))$ of the cut polytope over skeleton of tetrahedron is larger, that $Is_{Res}(CUT^{\square}(K_4))$:

$$|Is(CUT^{\square}(K_4))| = 6 \times Is_{Res}(CUT^{\square}(K_4)).$$

Denote $2^{1-|V|}|Is(CUT^{\square}(G))|$ by $A(G)$. In this notation, $Is(CUT^{\square}(G)) \neq Is_{Res}(CUT^{\square}(G))$, if $A(G) > |Aut(G)|$. One can get by simple checking the following result.

Proposition 14.4 *For a given graph $G = \langle V, E\rangle$, it holds:*

(i)if $G = \langle V, E\rangle$ is $Prism_m$ ($m \neq 4$), $APrism_m$ ($m > 3$), Möbius ladder M_{2m}, and $Ryr^2(C_m)$ ($m > 3$), then

$$|Aut(G)| = 4m;$$

(ii) if G is a complete multipartite graph with t_1 parts of size a_1, ..., t_r parts of size a_r, with $a_1 < a_2 < ... < a_r$, and all $t_i \geq 1$, then

$$|Aut(G)| = \prod_{i=1}^{r} t_i!(a_i!)^{t_i};$$

(iii) among the cases, considered before, all occurrences of $A(G) > |Aut(G)|$ are:

- $A(G) = m!2^{m-1}|Aut(G)|$ *for* $G = K_{2,m}, m > 2$;
- $A(G) = m!2^{m-1}|Aut(G)|$ *for* $K_{1,1,m}$, $m > 1$;
- $A(G) = 6|Aut(G)| = 6m!$ *for* $G = K_{1,1,1,1}$;
- $A(G) = 6|Aut(G)| = 2m! = 48$ *for* $G = K_{2,2}$;

(iv) if $G = C_m$ $(m > 3)$, *then* $|Aut(G)| = 2m!$, *while* $A(G) = 2m!$ *for* $m = 4$, *and* $A(G) = m! = |V|!$ *for* $m \geq 5$;

(v) if $G = P_m$ $(m > 3)$, *then* $|Aut(G)| = 2$, *while* $A(G) = (m-1)! = (|V|-1)!$.

14.3 Cut polytopes over some graphs

The *girth* and *circumference* of a graph, having cycles, are the length of its shortest and longest cycle, respectively. In a graph G, a *chordless cycle* is any cycle, which is induced subgraph; so, any triangle, any shortest cycle and any cycle, bounding a face in some embedding of G, are chordless.

Let, as before, c'_s and c_s denote the number of all or of chordless s-cycles in G, respectively.

Let G be embedded in some oriented surface; so, we get a map $\langle V, E, F\rangle$, where F is the set of faces of G in this embedding.

Let $\vec{p} = (..., p_i, ...)$ denote the *p-vector* of the map $\langle V, E, F\rangle$, enumerating the number $p_i > 0$ of faces of all sizes i, existing in G.

Call *face-bounding* any s-cycle of G, bounding a face in the map $\langle V, E, F\rangle$. Call an s-cycle of G *i-face-containing*, *edge-containing*, or *point-containing*, if all its interior points form just i-gonal face, edge, or point, respectively.

Call *equator* any cycle C, the interior of which (plus C) is isomorphic to the exterior (plus C).

Platonic and semiregular polyhedra

The chordless $4, 6, 5, 9$-cycles in the skeletons of octahedron, cube, icosahedron and dodecahedron, respectively, are exactly their vertex-containing $4, 6, 5, 9$-cycles.

For octahedron and cube, they are exactly all 3 and 4 equators (i.e., cycles, the interior of which is isomorphic to the exterior), respectively, which are, apropos, the *central circuits* and *zigzags* (see [DDS15]), respectively.

All c_6 chordless 6-cycles of icosahedron are exactly their 30 edge-containing ones and 10 face-containing ones, which are exactly all 10 equators and the *weak zigzags* (see [DDS15]).

All c_{10} chordless 10-cycles of dodecahedron are 30 edge-containing ones and 6 face-containing ones, which are exactly all 6 equators (and zigzags).

Proposition 14.5 *If G is the skeleton of a Platonic solid, then all possible facets of $CUT^{\square}(G)$ are edge facets, and s-cycle facets, coming from all face-bounding cycles and from all (if they exist and not listed before) vertex-, edge-, face-containing cycles. In particular, we get the following results.*

(i) If $G = K_4$ (skeleton of tetrahedron), then $CUT^{\square}(G)$ has unique orbit of $2^2 \cdot p_3 = 16$ (simplicial) 3-cycle facets (from all $|F| = p_3 = 4$ face-bounding cycles of G).

(ii) If $G = K_{2,2,2}$ (skeleton of octahedron), then $CUT^{\square}(G)$ has 56 facets, namely:

- *orbit of* $2^2 \cdot p_3$ *3-cycle facets (from all* $|F| = p_3 = 8$ *face-bounding cycles);*
- *orbit of* $2^3 \cdot c_4$ *4-cycle facets (from all* $c_4 = \frac{|V|}{2} = 3$ *vertex-containing 4-cycles).*

(iii) If $G = K_2^3$ *(skeleton of cube), then* $CUT^{\square}(G)$ *has 200 facets, namely:*

- *orbit of* $2|E| = 24$ *edge facets;*
- *orbit of* $2^3 \cdot p_4$ *4-cycle facets (from all* $|F| = p_4 = 6$ *face-bounding cycles);*
- *orbit of* $2^5 \cdot c_6 = 128$ *6-cycle facets (from all* $c_6 = 4$ *vertex-containing 6-cycles).*

(iv) If G is the skeleton of icosahedron, then $CUT^{\square}(G)$ *has 1,552 facets, namely:*

- *orbit of* $2^2 \cdot p_3 = 80$ *3-cycle facets (from all* $|F| = p_3 = 20$ *face-bounding cycles);*
- *orbit of* $2^4 c_5 = 192$ *5-cycle facets (from all* $c_5 = 12$ *vertex-containing 5-cycles);*
- *orbit of* $2^5|E| = 960$ *6-cycle facets (from* $|E| = 30$ *edge-containing 6-cycles);*
- *orbit of 320 6-cycle facets (from* $\frac{|F|}{2} = 10$ *face-containing 6-cycles).*

(v) If G is the skeleton of dodecahedron, then $CUT^{\square}(G)$ *has 23,804 facets in 5 orbits, namely:*

- *orbit of* $2|E| = 60$ *edge facets;*
- *orbit of* $2^4 \cdot p_5 = 192$ *5-cycle facets (from all* $|F| = p_5 = 12$ *face-bounding cycles);*
- *orbit of* $2^8 \cdot c_9 = 5,120$ *9-cycle facets (from all* $c_9 = 20$ *vertex-containing 9-cycles);*
- *orbit of* $2^9|E| = 15,360$ *10-cycle facets (from 30 edge-containing 10-cycles);*
- *orbit of* $2^9 \cdot 6 = 3,072$ *10-cycle facets (from* $\frac{|F|}{2} = 6$ *face-containing 10-cycles).*

In a truncated tetrahedron, call *ring-edges* those bounding a triangle, and *rung-edges* all 6 other ones.

Proposition 14.6 *For the skeletons of truncated tetrahedron and of cuboctahedron, the following results hold.*

(i) If G is the skeleton of truncated tetrahedron, then $CUT^{\square}(G)$ *has 540 facets, namely:*

- *orbit of* $2 \cdot 6$ *edge facets (from all 6 rung-edges);*
- *orbit of* $2^3 \cdot p_3$ *3-cycle facets (from all* $p_3 = 4$ *3-face-bounding cycles);*
- *orbit of* $2^5 \cdot p_6$ *6-cycle facets (from all* $p_6 = 4$ *6-face-bounding cycles);*
- *orbit of* $2^7 \cdot 3 = 384$ *8-cycle facets (from* $\frac{1}{2}\binom{4}{2}$ *rung-edge-containing 8-cycles, which are also the equators).*

(ii) If G is the skeletion of cuboctahedron, then $CUT^{\square}(G)$ *has 1,360 facets, namely:*

- *orbit of* $2^2 \cdot p_3$ *3-cycle facets (from all* $p_3 = 8$ *3-face-bounding cycles);*

- *orbit of* $2^3 \cdot p_4$ *4-cycle facets (from all* $p_4 = 6$ *4-face-bounding cycles);*
- *orbit of* $2^5 \cdot |V|$ *6-cycle facets (from all* 12 *vertex-containing* 6*-cycles);*
- *orbit of* $2^5 \cdot 4 = 128$ *6-cycle facets (from all* $\frac{p_3}{2}$ *3-face-containing* 6*-cycles, which are also equators and central circuits);*
- *orbit of* $2^7 \cdot p_4 = 768$ *8-cycle facets (from all* 6 *4-face-containing* 8*-cycles, which are also zigzags).*

Given a $Prism_m$ ($m \neq 4$), or an $APrism_m$ ($m \neq 3$), we call *rung-edges* the edges, connecting two m-gons, and we call *ring-edges* other $2m$ edges.

Let P be an ordered partition $X_1 \cup ... \cup X_{2t} = \{1, 2, ..., m\}$ of V_m into ordered sets X_i of $|X_i| \geq 3$ consecutive integers. Call *P-cycle of $Prism_m$* the chordless $(m+2t)$-cycle, obtained by taking the path X_1 on, say, the first m-gon, then rung edge (in the same direction), then path X_2 on the second m-gon, etc., till returning to the path X_1. Any vertex of $Prism_m$ can be taken as the first element of X_1, in order to fix a P-cycle.

So, an P-cycle defines an orbit of $2^{m+2t-1} \cdot 2m$ $(m+2t)$-cycle facets of $CUT^{\square}(Prism_m)$, except the case $\langle |X_1|, ..., |X_{2t}| \rangle = \langle |X_2|, ..., |X_{2t}|, |X_1| \rangle$, when the orbit is twice smaller.

An *P-cycle of $APrism_m$* is defined similarly, but we ask only $|X_i| \geq 2$, and rung edges, needed to change m-gon, should be selected, in the cases $|X_i| = 2, 3$, so that they not lead to a ring edge, i.e., a chord on P.

Clearly, P-cycles are all possible chordless t-cycles with $t \neq 4, m$ for $Prism_m$, and with $t \neq 2, m$ for $APrism_m$.

Proposition 14.7 *For the graphs* $Prism_m$*,* $m \geq 5$*, and* $APrism_m$*,* $m \geq 4$*, the following results hold.*

(i) If G *is* $Prism_m$*,* $m \geq 5$*, then all facets of* $CUT^{\square}(G))$ *are:*

- *orbit of* $2m$ *edge facets (from all* m *rung-edges);*
- *orbit of* $4m$ *edge facets (from all* $2m$ *ring-edges);*
- *orbit of* $2^3 \cdot p_4 = 8m$ *4-cycle facets (from all* m *4-face-bounding* 4*-cycles);*
- *orbit of* $2^{m-1} \cdot p_m$ *of* m*-cycle facets (from both* m*-face-bounding* m*-cycles);*
- *orbits of cycle facets for all possible* P*-cycles.*

(ii) If G *is* $APrism_m$*,* $m \geq 4$*, then all facets of* $CUT^{\square}(G))$ *are:*

- *orbit of* $2^2 \cdot p_3 = 8m$ *3-cycle facets (from all* $2m$ *3-face-bounding* 3*-cycles);*
- *orbit of* $2^{m-1} \cdot p_m$ *of* m*-cycle facets (from both* m*-face-bounding* m*-cycles);*
- *orbits of cycle facets for all possible* P*-cycles.*

Möbius ladders

Any Möbius ladder M_{2m} is *toroidal*, i.e., its vertices can be placed on a torus such that no edges cross. Moreover, the Möbius ladder $M_6 = K_{3,3}$ is 1-*planar*, i.e., can be drawn in the Euclidean plane in such a way that each edge has at most one crossing point, where it crosses a single additional edge.

Given the Möbius ladder M_{2m}, call *ring-edges* $2m$ those belonging to the $(2m)$-cycle $C_{1,...,2m}$, and *rung-edges* all other ones, i.e., $(i, i+m)$ for $i = 1, ..., m$.

For any odd t dividing m, denote by $C(m,t)$ the $(m+t)$-cycle of M_{2m}, having, up to a cyclic shift, the form

$$1, ..., 1+\frac{m}{t}, 1+\frac{m}{t}+m, ..., 1+\frac{2m}{t}+m, 1+\frac{2m}{t}+2m, ..., 1+\frac{3m}{t}+2m, ...,$$

i.e., t consecutive sequences of $\frac{2m}{t}-1$ ring-edges, followed by a rung-edge. Such $C(m,1)$ exists for any $m \geq 3$; for $t > 1$, their existence requires divisibility of m by t. Clearly, the number of $(m+t)$-cycles $C(m,t)$ is $\frac{2m}{t}$.

Conjecture 14.1 *If $G = M_{2m}$, $m \geq 4$, then among facets of $CUT^{\square}(G)$ there are:*

- *two orbits of $4m$ and $2m$ edge facets (from all $2m$ ring- and m rung-edges);*
- *orbit of $2^3 \cdot c_4 = 8m$ 4-cycles facets (from all m 4-cycles);*
- *orbit of $2^m \cdot 2m$ $(m+1)$-cycle facets (from all $2m$ $(m+1)$-cycles $C(m,1)$);*
- *for any odd divisor $t > 1$ of m, orbit of $2^{m+t} \cdot \frac{m}{t}$ $(m+t)$-cycle facets (from all $(m+t)$-cycles $C(m,t)$).*

There are no other orbits for $m = 3, 4$; for $m = 3$ first two orbits unite into one of 18 edge facets, while all other orbits unite into one of $2^3 \cdot c_4 = 72$ 4-cycle facets.

$CUT^{\square}(M_{10})$ has only one more orbit: the orbit of 2^{10} facets of incidence 15 (i.e., simplicial facets), defined by a cyclic shift of

$$\sum_{i=1}^{10} \frac{1}{2}\left(3-(-1)^i\right) x_{i,i+1} + \sum_{i=0}^{m} x_{i,i+m} - 2(x_{5,10} + 2x_{1,2} + x_{3,8}).$$

$CUT^{\square}(M_{12})$ also has only one more orbit: $2^{12} \cdot 6$ similar facets of incidence 20.

Petersen graph

Petersen graph is both, toroidal and 1-planar.

It has three circuit double covers: by six 5-gons (actually, zigzags), by five cycles of lengths $9, 6, 5, 5, 5$, and by 5 cycles of lengths $8, 6, 6, 5, 5$. It can be embedded in projective plane, in torus and in Klein bottle with corresponding sets of six, five and five faces.

Petersen graph have only $5, 6, 8$ and 9-cycles; it has $c_5 = 12$, and $c_6 = 10$.

Proposition 14.8 *If G is Petersen graph, then $CUT^{\square}(G)$ has four orbits of facets:*

- $2|E| = 30$ *edge facets;*

- $2^4 \cdot c_5 = 192$ *5-cycle facets;*
- $2^5 \cdot c_6 = 320$ *5-cycle facets;*
- *orbit of* $2^{10} \cdot 3$ *simplices, represented by*

$$(C_{12345} - 2x_{15}) - (C_{1'4'2'5'3'} - x_{1'4'} - x_{2'5'}) + 2 \sum_{1 \le i \le 5} x_{ii'},$$

where Petersen graph is seen as $C_{12345}(x) + C_{1'4'2'5'3'}(x) + \sum_{1 \le i \le 5} x_{ii'}$, *and* $C_{1...m}(x) = x_{12} \dots + x_{m-1,m} + x_{m1}$.

Heawood graph

Heawood graph, having 14 vertices and 21 edges, can be defined as $(3,6)$-cage, i.e., a graph with the fewest possible number of vertices, in which each vertex has exactly 3 neighbors, and in which the shortest cycle has length exactly 6. It is also toroidal and 1-planar; it has the girth 6, and $c_6 = 28$.

Proposition 14.9 *If* G *is Heawood graph, then three of nine orbits of* $CUT^{\square}(G)$ *are:*

- $2|E| = 42$ *edge facets;*
- $2^5 \cdot 28$ *6-cycle facets;*
- $2^7|E|$ *8-cycle facets.*

Complete-like graphs

It is known, that complete graph K_n is planar only for for $n \le 4$, is 1-planar only for $n = 5, 6$, and is toroidal only for $n = 5, 6, 7$.

Among complete multipartite graphs G, the planar ones are $K_{2,m}$, $K_{1,1,m}$, $K_{1,2,2}$, $K_{1,1,1,1} = K_4$, and their subgraphs. The 1-planar complete multipartite graphs G are, besides above, K_6, $K_{1,1,1,6}$, $K_{1,1,2,3}$, $K_{2,2,2,2}$, $K_{1,1,1,2,2}$, and their subgraphs.

Given sets $A_1, ..., A_t$ with $t \ge 2$, and $1 \le |A_1| \le ... \le |A_t|$, let G be complete multipartite graph $K_{a_1,...,a_t}$ with $a_i = |A_i|$ for $1 \le i \le t$.

All possible chordless cycles in G are $c_3 = \sum_{1 \le i < j < k \le t} a_i a_j a_k$. triangles and $c_4 = \sum_{1 \le i \le t} \binom{a_i}{2}\binom{a_j}{2}$ quadrangles.

Hence, $c_3 > 0$ if and only if $t > 2$, and $c_4 > 0$ if and only if $(a_1, t) \ne (1, 2)$.

So, among edge and s-cycle facets of $CUT^{\square}(G)$, only three such orbits are possible:

- $2|E|$ edge facets if $t = 2$;
- $4 \cdot c_3$ 3-cycle facets, if $t \ge 3$;
- $8 \cdot c_4$ 4-cycle facets, if $(a_1, t) \ne (1, 2)$.

All cases, when there are no other facets, i.e., when G has no K_5-minor, are given in Table 14.1; note that the facets are simplices for $G = K_{2,2}$, and $G = K_{1,1,1,1}$.

In particular, $G = K_{m+i} - K_m$, $m > 1$, has no K_5-minor only for $i = 1, 2, 3$. The facets of $CUT^{\square}(G)$ form: the orbit of $2m$ edge facets for $i = 1$; the orbit of $2m$ 3-cycle facets for $i = 2$; at last, two orbits (of sizes $12m$ and 4) of 3-cycle facets for $i = 3$.

For $G = K_{m+4} - K_m = K_{1,1,1,1,m}, m > 1$, and $G = K_{1,1,2,m}$, $m > 1$, the number of orbits stays constant for any m: 4 and 7, respectively.

Some of remaining cases are presented in Table 14.2.

Given sequence $b = (b_1, ..., b_n)$ of integers, which sum to 1, let us denote by HYP_b and call (when it is applicable) *hypermetric* the inequality

$$HYP_b(x) = \sum_{1 \le i,j \le n} b_i b_j x_{ij} \le 0.$$

Note, that $HYP_{(1,1,-1,0,...,0)}$ is the usual *triangle inequality* $T_{12,3}$, and, in general, HYP_b with all non-zero b_x being $b_i = b_j = 1 = -b_k$ is the triangle inequality $T_{ij,k}$. Similarly, HYP_b with all non-zero b_x being $b_i = b_j = b_k = 1 = -b_u = -b_v$, is the *pentagonal inequality*; denote it by $Pent_{ijk,uv}$.

If $G = K_{1,1,2,m}$ with $m \ge 3$, then $CUT^{\square}(G)$ has $8 + 20m + 8\binom{m}{2}(16m - 15)$ facets in 7 orbits: 3 orbits of $8, 4m, 16m$ 3-cycle facets; one orbit of $8\binom{m}{2}$ 4-cycle facets; at last, 3 orbits of $64\binom{m}{2}, 64\binom{m}{2}$, $384\binom{m}{3}$ $\{0, \pm 1\}$-valued non-s-cycle facets, having 4 values -1, and $11, 11, 12$ values of 1, respectively. (The partition is $\{1\}, \{2\}, \{3, 4\}, \{5, ..., m + 4\}$.)

$CUT^{\square}(K_{1,1,2,2})$ has 184 facets in 4 orbits: 2 orbits of $8+8, 32$, 3-cycle facets; one orbit of 8 4-cycle facets; at last, one orbit of 2^7 facets, represented by

$$HYP_{(1,1,1,-1,-1,0)}(x) + HYP_{(0,0,1,1,0,-0,-1)}(x) \le 0.$$

The graph $G = K_{m+t} - K_m = K_{1,...,1,m}$ has an K_5-minor only if $t \ge 4$. If $m \ge 3$, then $CUT^{\square}(G)$ has 2 orbits of $4m\binom{t}{2}$ and $4\binom{t}{3}$ 3-cycle facets, and, for $t < 4$ only, no other facets. (The partition is $\{1\}, ..., \{t\}, \{t + 1, ..., t + m\}$.)

If $G = K_{m+4} - K_m$, then $CUT^{\square}(G)$ has $8(8m^2 - 3m + 2)$ facets in 4 orbits: 2 orbits of $24m$, 16 3-cycle facets, and 2 orbits of sizes $16m, 128\binom{m}{2}$, represented, respectively, by

$$Pent_{125,34} : HYP_{(1,1,-1,-1,1,0,...,0)}(x) \le 0, \text{ and}$$

$$HYP_{(1,1,-1,0,1,-1,0,...,0)}(x) + HYP_{(0,0,0,-1,1,1,0,...,0)}(x) \le 0.$$

If $G = K_{m+5} - K_m$ (i.e., $G = K_{1,1,1,1,1,m}$), then among many orbits of facets of $CUT^{\square}(G)$, there are 2 orbits of $40, 40m$ 3-cycle facets, and 3 orbits of $16, 80m, 20m(m-1)$ facets, represented, respectively, by

$$HYP_{(1,1,1,-1,-1,0,...,0)}(x) \le 0,$$

$$HYP_{(1,1,-1,-1,0,1,0,...,0)}(x) \le 0, \quad \text{and}$$

$$HYP_{(1,-1,-1,0,0,1,1,0,...,0)}(x) + HYP_{(0,0,0,1,0,1,-1,0,...,0)}(x) \le 0.$$

Among 12 remaining orbits for $K_7 - K_2$, two (of sizes $2^7 \cdot 30, 2^7 \cdot 60$) are $\{0, \pm 1\}$-valued; they are represented, respectively, by

$$HYP_{(1,1,-1,-1,1,1,-1)}(x) + (x_{34} + x_{47} - x_{27} + x_{12} - x_{13}) \le 0, \text{ and}$$

$$(x_{13} + x_{34} + x_{45} + x_{15}) + (x_{23} + x_{36} + x_{67} + x_{27})$$
$$-(x_{14} + x_{47} - x_{57} + x_{25} + x_{26} + x_{16}).$$

Let $G = Pyr^2(C_m)$: the graph $Wheel_m = Pyr_m$ (i.e., a circuit C_m with a new point, connected to each point of cycle) with another new point, connected to all $m+1$ points of Pyr_m. Clearly, it is K_4, K_5 if $m = 2, 3$, respectively.

For $m \geq 4$, it holds, that $A(G) = 4m$, and all chordless cycles are $3m$ triangles, and unique m-cycle. Any of $3m+1$ edges belongs to a triangle. So, among orbits of facets of $CUT^{\square}(G)$, there are two (of size $8m$ and $4m$) orbits of 3-cycle facets, and one orbit of 2^{m-1} m-cycle facets. All other facets for $m \leq 7$ are $\{0, \pm 1\}$-valued.

For $Pyr^2(C_{1234})$, the unique remaining orbit consists of 2^7 facets, represented by $Pent_{355,12} + T_{12,4}$.

Among remaining orbits for $Pyr^2(C_{1\dots m})$ with $m = 5$ and 7, there is an orbit of 2^{m+1} facets represented, respectively, by

$$C_{12345}(x) - 2((x_{45} + x_{67}) + (x_{16} + x_{17} + x_{36} + x_{37})) \leq 0, \quad \text{and}$$

$$C_{12345}(x) - 2((x_{12} + x_{19}) + (x_{29} + x_{38} + x_{49} + x_{58} + x_{69} + x_{78})) \leq 0,$$

where $C_{1\dots m}(x) = x_{12} + \dots + x_{m-1,m} + x_{m1}$.

For $m = 5$, two remaining orbits (each of size $2^6 \cdot 5$) are represented, respectively, by

$$C_{12345}(x) - 2x_{15} + x_{67} + ((x_{17} - x_{16}) - (x_{37} - x_{36}) + (x_{47} - x_{46})) \leq 0, \quad \text{and}$$

$$C_{12345}(x) - 2x_{15} + x_{67} + ((x_{17} - x_{16}) - (x_{47} - x_{46}) + (x_{57} - x_{56})) \leq 0.$$

For $m = 6$, one of 4 remaining orbits (of size $2^7 \cdot 6$) is represented by

$$C_{123456}(x) - 2x_{12} + x_{78} + ((x_{17} - x_{18}) + (x_{57} - x_{58}) - (x_{67} - x_{68})) \leq 0.$$

Note that $K_7 - C_5 = Pyr^2(C_5)$.

Now, $G = K_7 - C_{1234} = K_{\{7\},\{6\},\{5\},\{1,3\},\{2,4\}}$ has $c_3 = 19$; $CUT^{\square}(G)$ has four orbits of facets: three (of sizes $48, 24, 4$) of 3-cycle facets, and one orbit of size 32, represented by $Pent_{456,27}$. Each of K_5-minors, $K_{\{2,4,5,6,7\}}$ and $K_{\{1,3,5,6,7\}}$, provides 16 of above 32 facets.

$G = K_7 - C_7$ has $c_3 = c_4 = 7$; $CUT^{\square}(G)$ has three orbits of facets: one (of size 28) of 3-cycle facets, one (of size 56) of 4-cycle facets and one of size 64, represented by

$$(K_7(x) - C_{1234567}(x)) - 2(x_{15} + P_{27364}(x)) \leq 0,$$

where $K_n(x) = \sum_{1 \leq i < j \leq n} x_{ij}$, and $P_{1\dots n}(x) = x_{12} + \dots + x_{n-1,n}$.

14.4 Some general results about metric and cut polyhedra over graphs

Hypermetrics on graphs

It was discussed above, that in order to obtain $CUT^{\square}(G)$ over a given a graph $G = \langle V, E \rangle$, it suffices to restrict the cut semimetrics on the edges of G.

The notion of $MET_n^{\square}$ can also be extended to the graph setting but requires more work: for a cycle C and an odd sized set F of edges in C, the *cycle inequality* $f_{C,F}$ has the form

$$\sum_{e' \in C \setminus F} x_{e'} - \sum_{e \in F} x_e \leq |F| - 1,$$

and the *metric polytope* $MET^{\square}(G)$ *over the graph* G can be defined as the polytope, generated by all cycle inequalities $f_{C,F}$, and all the *edge inequalities* $0 \leq x_e \leq 1$. In fact, it is the projection of MET_n on $\mathbb{R}^{|E|}$, indexed by the edges of G.

Besides above inequalities of $MET^{\square}(G)$, some valid inequalities of hypermetric type show up on $CUT^{\square}(G)$. Below we give a way to get such inequalities.

Proposition 14.10 *Let us take a valid inequality on* $CUT_n^{\square}$ *of the form*

$$f(x) = \sum_{1 \leq i < j \leq n} a_{ij} x_{ij} \leq A.$$

Suppose that we have n *vertices* $v_1, ..., v_n$ *of* G *with any two vertices* v_i, v_j *being joined by a such path* P_{ij} *that:*

- *the edge set of all paths* P_{ij} *are disjoint;*
- *if* $a_{ij} > 0$, *then* P_{ij} *is reduced to an edge.*

Then, the following inequality

$$i_{f,G} : i_{f,G}(x) = \sum_{1 \leq i < j \leq n} a_{ij} \left(\sum_{e \in P_{ij}} x_e \right) \leq A$$

is valid on $CUT^{\square}(G)$.

Proof. Let us take a cut of $G = \langle V, E \rangle$, defined by $S \subset V$. If S cuts the paths P_{ij} in at most one edge, then the inequality on $CUT^{\square}(G)$ reduces to the one on $CUT_n^{\square}$ and so, is valid.

In the general case, we will create a new cuts S', which will allow us to prove the required inequality. If S cuts P_{ij} in more than one edge, then $a_{ij} \leq 0$. If both v_i and v_j are in the same part of the partition $(S, V \backslash S)$, then we set all vertices of P_{ij} to be in the same part of the partition $(S', V \backslash S')$. Otherwise, there exists an edge $e = (w_i, w_j) \in P_{ij}$, cut by S. If $w_j \in S'$, then we set the vertices between w_j and v_j to belong to S', and we set the vertices from w_i to v_i not to belong to it. Due to the sign condition on a_{ij}, one obtains

$$i_{f,G}(\delta_S) \leq i_{f,G}(\delta_{S'}).$$

Then, from S' we can obtain very simply a cut S'' of K_n; this gives

$$i_{f,G}(\delta_{S'}) = f(\delta_{S''}) \leq A,$$

which proves the required inequality.

□

When applied to the triangle inequalities of $CUT_n^{\square} = CUT^{\square}(K_n)$ and taking switchings, the above proposition gives us the metric polytope $MET^{\square}(G)$. Therefore, one can try to define the *hypermetric polytope* $HYP^{\square}(G)$ *over a graph* G by the switchings of all inequalities $i_{f,G}$ from above proposition, constructed for f being any hypermetric inequality.

What is not clear: when $CUT^{\square}(G) = HYP^{\square}(G)$, and whether there is a nice characterization of such hypermetrics. (Above discussion is applied also to cones.)

Any K_m-subgraph of G will satisfy the hypothesis and the facets of $CUT_m^{\square}$ will give facets of $CUT^{\square}(G)$. Proposition 14.10 gives valid inequality induced by a class of graphs homeomorphic to K_m. Here, a graph H is *homeomorphic* to a subgraph of G, if H can be mapped to G so that the edges of H are mapped to disjoint paths in G; a graph homeomorphic to K_m is a special case of an K_n-minor.

In [Seym81] it is proved that $CUT^{\square}(G) = MET^{\square}(G)$ if and only if G has no K_5 as a minor. However, the proof appears non-constructive and does not seem to be able to give hypermetric inequalities, or their generalization, in a straightforward way.

Correlation and diversity polyhedra over graphs

Let $V_n = \{1, 2, ..., n\}$; let $E_n^{'} = |\{(i,j) \,|\, i,j \in V_n\}|$, where (i,j) denotes the unordered pair of the integers i,j, not necessary distinct, i.e., $E_n^{'} = E_n + n = \binom{n+1}{2} = \frac{n(n+1)}{2}$. In the following, we often identify V_n with the set of diagonal pairs ii for $i = 1, 2, ..., n$. In other words, a vector $p \in \mathbb{R}^{|V_n| \cup E_n}$ can be supposed to be indexed by the pairs ij for $1 \leq i \leq j \leq n$.

The main objects considered in this subsection are the correlation cone and polytope, that we now introduce. Let S be a subset of V_n. Let us define the vector $\pi(S) = (\pi(S)_{ij})_{1 \leq i \leq j \leq n} \in \mathbb{R}^{E_n^{'}}$ by

$$\pi(S)_{ij} = \begin{cases} 1, & \text{if } i,j \in S, \\ 0, & \text{otherwise,} \end{cases}$$

where $1 \leq i \leq j \leq n$; $\pi(S)$ is called a *correlation vector.*

The cone in $\mathbb{R}^{E_n^{'}}$, generated by all correlation vectors $\pi(S)$ for $S \subseteq V_n$, is called the *correlation cone* and is denoted by COR_n.

The polytope in $\mathbb{R}^{E_n^{'}}$, defined as the convex hull of the correlation vectors $\pi(S)$ for $S \subseteq V_n$, is called the *correlation polytope* and is denoted by $COR_n^{\square}$. Hence,

$$COR_n = \left\{ \sum_{S \subseteq V_n} \lambda_S \pi(S) \,\middle|\, \lambda_S \geq 0 \text{ for all } S \subseteq V_n \right\},$$

$$COR_n^{\square} = \left\{ \sum_{S \subseteq V_n} \lambda_S \pi(S) \,\middle|\, \sum_{S \subseteq V_n} \lambda_S = 1, \text{ and } \lambda_S \geq 0 \text{ for all } S \subseteq V_n \right\}.$$

It is sometimes convenient to consider an arbitrary finite subset X instead of V_n. Then, the correlation cone is denoted by $COR(X)$, and the correlation polytope by $COR^{\square}(X)$. Hence,

$$COR(V_n) = COR_n, \text{ and } COR^{\square}(V_n) = COR_n^{\square}.$$

The correlation polytope has been considered in the literature in connection with many different problems, arising in various fields. We mention some of them below.

The correlation polytope plays, for instance, an important role in Combinatorial Optimization. Indeed, it permits to formulate a well-known NP-hard Optimization Problem, namely, the *Unconstrained Quadratic* $\{0,1\}$*-programming Problem*:

$$\begin{array}{ll} \max & \sum_{1 \leq i \leq j \leq n} c_{ij} x_i x_j, \\ \text{s.t.} & x \in \{0,1\}^n, \end{array}$$

where $c_{ij} \in \mathbb{R}$ for all i, j. Clearly, this problem can be reformulated as:

$$\begin{array}{ll} \max & c^T p \\ \text{s.t.} & p \in COR_n^\square. \end{array}$$

The members of $COR_n^\square$ can be interpreted as joint correlations of events in some probability space. This fact explains the name "correlation polytope", which was introduced by Pitowsky [Pito86]. For $n = 3$, the correlation polytope $COR_3^\square$ is called there the *Bell-Wigner polytope.* In this context, the correlation polytope occurs in connection with the *Boole problem.*

The correlation polytope also arises in the field of Quantum Mechanics, in connection with the so-called Representability Problem for density matrices of order 2.

It turns out that the correlation cone (polytope) is very closely related to the cut cone (polytope). In fact, it is nothing but its image under a linear bijective mapping: the cut cone CUT_{n+1} (respectively, the cut polytope $CUT_{n+1}^\square$) is in one-to-one correspondence with the correlation cone COR_n (respectively, the correlation polytope $COR_n^\square$) via the *covariance mapping*

$$\xi : \mathbb{R}^{E_{n+1}} \longrightarrow \mathbb{R}^{E_n'},$$

defined as follows:

$$p = \xi(d)$$

for $d = (d_{ij})_{1 \le i < j \le n+1}$, and $p = (p_{ij})_{1 \le i \le j \le n}$ with

$$\begin{cases} p_{ii} = d_{i,n+1} & \text{for } 1 \le i \le n, \\ p_{ij} = \frac{1}{2}(d_{i,n+1} + d_{j,n+1} - d_{ij}) & \text{for } 1 \le i < j \le n, \end{cases}$$

or, equivalently,

$$\begin{cases} d_{i,n+1} & = p_{ii} & \text{for } 1 \le i \le n, \\ d_{ij} & = p_{ii} + p_{jj} - 2p_{ij} & \text{for } 1 \le i < j \le n. \end{cases}$$

Therefore,

$$\xi(CUT_{n+1}) = COR_n, \text{ and } \xi(CUT_{n+1}^\square) = COR_n^\square.$$

In the same way, given a finite subset X and an element $x_0 \in X$, the cut cone $CUT(X)$ and the correlation cone $COR(X \setminus \{x_0\})$ (respectively, the cut polytope $CUT^\square(X)$ and the correlation polytope $COR^\square(X \setminus \{x_0\})$) are in one-to-one linear correspondence via the covariance mapping ξ, pointed at the position x_0.

In Quantum Physics and Quantum Information Theory, *Bell inequalities*, involving joint probabilities of two probabilistic events, are exactly inequalities valid for the correlation polytope $COR^\square(G)$ (called also *Boolean quadric polytope* $BQP(G)$) *of a graph* G.

In particular, $COR^\square(K_{n,m})$ is seen in Quantum Theory as the set of possible results of a series of Bell experiments with a non-entangled (separable) quantum state shared by two distant parties, where one party has n choices of possible two-valued measurements, and the other party has m choices.

A valid inequality of $COR^\square(K_{n,m})$ is called a *Bell inequality* and, if facet inducing, a *tight Bell inequality.* This polytope is linearly isomorphic (via the covariance map) to the cut polytope $CUT^{\square}(K_{1,n,m})$ ([DeLa97], Section 5.2). Similarly, $CUT^{\square}(K_{1,n,m,l})$ represents three-party Bell inequalities.

The symmetry group of $COR^{\square}(K_{n,m})$ and $CUT^{\square}(K_{1,n,m})$ has order $2^{1+n+m}n!m!$.

Table 14.2 gives the number of facets of $COR^{\square}(K_{n,m})$ with $(n,m) = (4,4), (3,5), (3,4)$. The cases $(n,m) = (2,2), (3,3)$ were settled [Fine82] and [PiSv01], respectively. For $n = 4$, only partial lists of facets were known; our 175 orbits of 36, 391, 264 facets of $COR^{\square}(K_{4,4})$ finalize this case.

In contrast to the Bell inequalities, which probe entanglement between spatially-separated systems, the *Leggett-Garg inequalities* test the correlations of a single system measured at different times. The polytope, defined by those inequalities for n observables, is, actually the cut polytope $CUT_n^{\square}$.

The *diversity cone* DIV_n is the set of all *diversities on* n *points*, i.e., the functions $f : \{A \mid A \subseteq \{1,2,...,n\}\} \Rightarrow \mathbb{R}$, satisfying all $f(A) \geq 0$ with equality if $|A| \leq 1$, and all

$$f(A \cup B) + f(B \cup C) \geq f(A \cup C), \text{ if } B \neq \emptyset.$$

The *induced diversity metric* d_{ij} is $f(\{i,j\})$.

The *cut diversity cone* $CDIV_n$ is the conic hull of all *cut diversities* $\delta_S(A)$, where $A \subseteq \{1,2,...,n\}$, which are defined, for any $S \subseteq \{1,2,...,n\}$, as

$$\delta_S(A) = \begin{cases} 1, & \text{if } A \cap S \neq \emptyset, \text{ and } A \setminus S \neq \emptyset, \\ 0, & \text{otherwise.} \end{cases}$$

In fact, $CDIV_n$ is the set of all diversities from DIV_n, which are isometrically embeddable into an *l_1-diversity*, i.e., one defined on $\mathbb{R}^m$ with $m \leq \binom{n}{\lfloor \frac{n}{2} \rfloor}$ by

$$f_m(A) = \sum_{i=1}^{m} \max_{a,b \in A}\{|a_i - b_i|\}.$$

These two cones are natural "hypergraph" extensions of $MET(G)$ and $CUT(G)$.

Similar *Vitanyi multiset-metric* represents an generalization of the notion of diversity. It was proposed by Vitanyi (2011) and is defined as follows.

Given two multisets m and m', let $n = mm'$ be the multiset, consisting of the elements of the multisets m and m', that is, if x occurs once in m and once in m', then it occurs twice in n. The *Vitanyi multiset-metric* is a function $d : M \Rightarrow \mathbb{R}$ on the set M of all non-empty finite multisets, satisfying the following conditions:

1. $d(m) = 0$ if all elements of m are equal, and $d(m) > 0$, otherwise (*positive definiteness*);

2. $d(m)$ is invariant under all permutations of m (*totally symmetry*);

3. $d(mm') \leq d(mm'') + d(m''m')$ (*multiset triangle inequality*).

The usual metric between two elements results, if the multiset m has two elements in 1. and 2., and the multisets m, m', m'' have one element each in 3.

An example is the set of all non-empty finite multisets m of integers with $d(m) = \max\{x \mid x \in m\} - \min\{x \mid x \in m\}$. Cohen–Vitanyi, 2012, defined another multiset-metric, generalizing *normalized web distance* ([DeDe14], Chapter 22).

Any diversity is a *Vitanyi multiset-metric*, restricted to subsets. But much of Bryant–Tupper's theory of diversities does not extend on multisets.

Chapter 15

Connections between generalized metrics polyhedra

15.1 Preliminaries

In this chapter we consider important connections between classical symmetric case of semimetric polyhedra (in particular, of cut semimetric polyhedra) and asymmetric case of quasi-semimetric polyhedra (in particular, of oriented cut semimetric polyhedra).

We show that the cone $WQMET_n$ of weighted n-point quasi-semimetrics and the cone $OCUT_n$, generated by all non-zero oriented cuts on n points, are projections along an extreme ray of the metric cone MET_{n+1} and of the cut cone CUT_{n+1}, respectively. This projection is such that if one knows all faces of an original cone, then one knows all faces of the projected cone.

Oriented (or directed) distances are encountered very often, for example, these are one-way transport routes, rivers with quick flow and so on.

The notions of directed distances, quasi-metrics and oriented cuts are generalizations of the notions of distances, metrics and cuts, respectively (see, for example, [DeLa97]), which are central objects in Graph Theory and Combinatorial Optimization.

Quasi-metrics are used in Semantics of Computations (see, for example, [Seda97]) and in Computational Geometry (see, for example, [AACLMP9797]). Oriented distances have been used already by Hausdorff in 1914, see [Haus14].

The cones of quasi-semimetrics, weightable quasi-semimetrics, oriented cuts, oriented multicuts and other related generalizations of metrics were studied in [DePa99], [DDP03], [DeDe10], [DDV11], [DDD15]. The polytope of oriented cuts was considered in [AvMe11].

In [CMM06], authors give an example of directed metric, derived from a metric as follows. Let d be a metric on a set $V_n \cup \{0\}$, where 0 is a distinguished point. Then a quasi-metric q on the set V_n is given as

$$q_{ij} = d_{ij} + d_{i0} - d_{j0}.$$

This quasi-metric belongs to a special important subclass of quasi-metrics, namely, to a class of *weighted quasi-metrics*. We show (see also [DDV11]), that any weighted quasi-metric is obtained by a slight generalization of this method.

All semimetrics on a set of cardinality n form a *metric cone* MET_n. There are two important subcones of MET_n, namely, the cone HYP_n of *hypermetrics*, and the cone

CUT_n of ℓ_1*-metrics*. These three cones form the following nested family (see [DeLa97]):

$$CUT_n \subseteq HYP_n \subseteq MET_n.$$

We introduce a space Q_n, called a *space of weighted quasi-metrics*, and define in it a cone $WQMET_n$. Elements of this cone satisfy triangle and non-negativity inequalities. Among extreme rays of the cone $WQMET_n$ there are rays, spanned by *o-cut vectors*, i.e., incidence vectors of *oriented cut semimetrics*.

We define in the space Q_n a cone $OCUT_n$ as the conic hull of all o-cut vectors. Elements of the cone $OCUT_n$ are weighted ℓ_1-quasi-semimetrics.

Let semimetrics in the cone MET_{n+1} be defined on the set $V_n \cup \{0\}$. The *cut cone* CUT_{n+1} (the cone of ℓ_1-semimetrics on V_n) is the conic hull of cut semimetrics δ_S for all $S \subset V_n \cup \{0\}$. The cut semimetrics δ_S are extreme rays of all the three cones MET_{n+1}, HYP_{n+1}, and CUT_{n+1}. In particular, $\delta_{\{0\}} = \delta_{V_n}$ is an extreme ray of these three cones.

In this chapter, it is shown that the cones $WQMET_n$ and $OCUT_n$ are projections of the corresponding cones MET_{n+1} and CUT_{n+1} along the extreme ray δ_{V_n}. Moreover, we define a cone $WQHYP_n$ of *weighted quasi-hypermetrics* as projection along δ_{V_n} of the cone HYP_{n+1}. So, we obtain a nested family

$$OCUT_n \subseteq WQHYP_n \subseteq WQMET_n.$$

15.2 Decomposition of real vector spaces

Spaces $\mathbb{R}^{E_n}$ and $\mathbb{R}^{E_n^O}$

Let V_n be the set $\{1, 2, ..., n\}$. Let

$$\mathcal{E}_n = \{(i,j) \mid i,j \in V_n, i \neq j\} \quad \text{and} \quad \mathcal{E}_n^O = \{\langle i,j\rangle \mid i,j \in V_n, i \neq j\}$$

be sets of all unordered (ij) and ordered ij pairs of distinct elements $i, j \in V_n$, respectively. Let $E_n = |\mathcal{E}_n|$, and $E_n^O = |\mathcal{E}_n^O|$, i.e.,

$$E_n = \binom{n}{2} = \frac{n(n-1)}{2}, \quad \text{and} \ , \ E_n^O = I_n = 2\binom{n}{2} = n(n-1).$$

Consider two Euclidean spaces $\mathbb{R}^{E_n}$ and $\mathbb{R}^{E_n^O}$ of vectors $d \in \mathbb{R}^{E_n}$ and $g \in \mathbb{R}^{E_n^O}$ with coordinates $d_{(ij)}$ and g_{ij}, where $(ij) = (i,j) \in \mathcal{E}_n$ and $ij = \langle i,j\rangle \in \mathcal{E}_n^O$, respectively.

Denote by

$$\langle d, t\rangle = \sum_{(ij)\in\mathcal{E}_n} d_{(ij)} t_{(ij)}$$

the *dot product* of vectors $d, t \in \mathbb{R}^{E_n}$. Similarly,

$$\langle f, g\rangle = \sum_{ij\in\mathcal{E}_n^O} f_{ij} g_{ij}$$

is the dot product of vectors $f, g \in \mathbb{R}^{E_n^O}$.

Let $\{e_{(ij)} \mid (ij) \in \mathcal{E}_n\}$ and $\{e_{ij} \mid ij \in \mathcal{E}_n^O\}$ be orthonormal bases of $\mathbb{R}^{E_n}$ and $\mathbb{R}^{E^O_n}$, respectively. Then, for $f \in \mathbb{R}^{E_n}$ and $q \in \mathbb{R}^{E_n^O}$, we have

$$\langle e_{(ij)}, f\rangle = f_{(ij)}, \text{ and } \langle e_{ij}, q\rangle = q_{ij}.$$

For $f \in \mathbb{R}^{E_n^O}$, define $f^r \in \mathbb{R}^{E_n^O}$ as follows:

$$f^r_{ij} = f_{ji} \text{ for all } ij \in \mathcal{E}_n^O.$$

Each vector $g \in \mathbb{R}^{E_n^O}$ can be decompose into *symmetric* g^s and *antisymmetric* g^a *parts* as follows:

$$g^s = \frac{1}{2}(g + g^r),\ g^a = \frac{1}{2}(g - g^r),\ g = g^s + g^a.$$

Call a vector g *symmetric*, if $g^r = g$, and *antisymmetric*, if $g^r = -g$. Let $\mathbb{R}_s^{E_n^O}$ and $\mathbb{R}_a^{E_n^O}$ be subspaces of the corresponding vectors.

Note, that the spaces $\mathbb{R}_s^{E_n^O}$ and $\mathbb{R}_a^{E_n^O}$ are mutually orthogonal. In fact, for $p \in \mathbb{R}_s^{E_n^O}$ and $f \in \mathbb{R}_a^{E^O}$, we have

$$\langle p, f\rangle = \sum_{ij\in\mathcal{E}_n^O} p_{ij}f_{ij} = \sum_{(ij)\in\mathcal{E}_n} (p_{ij}f_{ij} + p_{ji}f_{ji}) = \sum_{(ij)\in\mathcal{E}_n} (p_{ij}f_{ij} - p_{ij}f_{ij}) = 0.$$

Hence,

$$\mathbb{R}^{E_n^O} = \mathbb{R}_s^{E_n^O} \oplus \mathbb{R}_a^{E_n^O},$$

where $\oplus$ is the direct sum.

Obviously, there is an isomorphism φ between the spaces $\mathbb{R}^{E_n}$ and $\mathbb{R}_s^{E_n^O}$. Let $d \in \mathbb{R}^{E_n}$ have coordinates $d_{(ij)}$. Then

$$d^O = \varphi(d) \in \mathbb{R}_s^{E_n^O}, \text{ such that } d^O_{ij} = d^O_{ji} = d_{(ij)}.$$

In particular,

$$\varphi(e_{(ij)}) = e_{ij} + e_{ji}.$$

The map φ is invertible. In fact, for $q \in \mathbb{R}_s^{E_n^O}$, we have $\varphi^{-1}(q) = d \in \mathbb{R}^{E_n}$, such that $d_{(ij)} = q_{ij} = q_{ji}$. The isomorphism φ will be useful in what follows.

Space of weights Q_n^w

One can consider the sets $\mathcal{E}_n$ and $\mathcal{E}_n^O$ as sets of edges (ij) and arcs ij of an unordered and ordered complete graphs K_n and K_n^O on the vertex set V_n, respectively. The graph K_n^O has two arcs ij and ji between each pair of vertices $i, j \in V_n$.

It is convenient to consider vectors $g \in \mathbb{R}^{E_n^O}$ as functions on the set of arcs $\mathcal{E}_n^O$ of the graph K_n^O. So, the decomposition $\mathbb{R}^{E^O} = \mathbb{R}_s^{E_n^O} \oplus \mathbb{R}_a^{E_n^O}$ is a decomposition of the space of all functions on arcs in $\mathcal{E}_n^O$ onto the spaces of symmetric and antisymmetric functions.

Besides, there is an important direct decomposition of the space $\mathbb{R}_a^{E_n^O}$ of antisymmetric functions into two subspaces. In the Theory of Electric Networks, these spaces are called spaces of *tensions* and *flows* (see also [Aign79]).

The *tension space* relates to *potentials* (*weights*) w_i, given on vertices $i \in V_n$ of the graph K_n^O. The corresponding antisymmetric function g^w is determined as

$$g^w_{ij} = w_i - w_j.$$

It is called *tension* on the arc ij. Obviously,

$$g^w_{ji} = w_j - w_i = -g^w_{ij}.$$

Denote by Q_n^w the subspace of $\mathbb{R}^{E_n^O}$, generated by all tensions on arcs $ij \in \mathcal{E}_n^O$. We call Q_n^w a *space of weights.*

Each tension function g^w is represented as weighted sum of elementary *potential* functions q_k, for $k \in V_n$, as follows:

$$g^w = \sum_{k \in V_n} w_k q_k,$$

where

$$q_k = \sum_{j \in V_n \setminus \{k\}} (e_{kj} - e_{jk}), \text{ for all } k \in V_n, \tag{15.1}$$

are basic functions, that generate the space of weights Q_n^w. Hence, the values of the basic functions q_k on arcs are as follows:

$$(q_k)_{ij} = \begin{cases} 1, & \text{if } i = k, \\ -1, & \text{if } j = k, \\ 0, & \text{otherwise.} \end{cases} \tag{15.2}$$

We obtain

$$g_{ij}^w = \sum_{k \in V_n} w_k (q_k)_{ij} = w_i - w_j.$$

It is easy to verify, that

$$q_k^2 = \langle q_k, q_k \rangle = 2(n-1) \text{ for all } k \in V_n, \ \langle q_k, q_l \rangle = -2 \text{ for all } k, l \in V_n, k \neq l,$$

$$\text{and } \sum_{k \in V_n} q_k = 0.$$

Hence, there are only $n-1$ independent functions q_k, that generate the space Q_n^w.

The weighted quasi-metrics lie in the space $\mathbb{R}_s^{E^O} \oplus Q_n^w$, that we denote as Q_n. Direct complements of Q_n^w in $\mathbb{R}_a^{E_n^O}$ and Q_n in $\mathbb{R}^{E_n^O}$ is a *space* Q_n^c *of circuits (flows)*.

Space of circuits Q_n^c

The *space of circuits* (*space of flows*) is generated by characteristic vectors of oriented circuits in the graph K_n^O. Arcs of K_n^O are ordered pairs ij of vertices $i, j \in V_n$. The arc ij is oriented from the vertex i to the vertex j. Recall, that K_n^O has both the arcs ij and ji for each pair of vertices $i, j \in V_n$.

Let $G_s \subset K_n$ be a subgraph of K_n with a set of edges $E(G_s) \subset \mathcal{E}_n$. We relate to the graph G_s a directed graph $G \subset K_n^O$ with the arc set $E_n^O(G) \subset \mathcal{E}_n^O$ as follows. An arc ij belongs to G, i.e., $ij \in E_n^O(G)$, if and only if $(ij) = (ji) \in E(G)$. This definition implies that in this case, the arc ji belongs to G also, i.e., $ji \in E_n^O(G)$.

Let C_s be a circuit in the complete graph K_n. The circuit C_s is determined by a sequence of distinct vertices $i_k \in V_n$, where $1 \leq k \leq p$, and p is the length of C_s. The edges of C_s are unordered pairs (i_k, i_{k+1}), where indices are taken modulo p. By above definition, an *oriented bicircuit* C of the graph K_n^O relates to the circuit C_s. Arcs of C are ordered pairs $i_k i_{k+1}$ and $i_{k+1} i_k$, where indices are taken modulo p.

Take an orientation of C. Denote by $-C$ the *opposite* circuit with opposite orientation. Denote an arc of C *direct* or *opposite* if its direction coincides with or is opposite to the given orientation of C, respectively. Let C^+ and C^- be subcircuits of C, consisting of direct and opposite arcs, respectively.

The following vector f^C is the *characteristic vector of the bicircuit* C:

$$f^C_{ij} = \begin{cases} 1, & \text{if } ij \in C^+, \\ -1, & \text{if } ij \in C^-, \\ 0, & \text{otherwise.} \end{cases}$$

Note, that $f^{-C} = (f^C)^r = -f^C$, and $f^C \in \mathbb{R}^{E^O}_a$.

Denote by Q^c_n the space, linearly generated by circuit vectors f^C for all bicircuits C of the graph K^O_n. It is well known, that characteristic vectors of *fundamental circuits* form a basis of Q^c_n. Fundamental circuits are defined as follows.

Let T be a *spanning tree* of the graph K_n. Since T is spanning, its vertex set $V(T)$ is the set of all vertices of K_n, i.e., $V(T) = V_n$. Let $E(T) \subset \mathcal{E}_n$ be the set of edges of T. Then any edge $e = (ij) \notin E(T)$ closes an unique path in T between vertices i and j into a circuit C^e_s. This circuit C^e_s is called *fundamental.* Call corresponding oriented bicircuit C^e also *fundamental.*

There are $E_n - |E(T)| = \frac{n(n-1)}{2} - (n-1)$ fundamental circuits. Hence,

$$\dim Q^c_n = \frac{n(n-1)}{2} - (n-1), \text{ and } \dim Q_n + \dim Q^c_n = n(n-1) = \dim \mathbb{R}^{E^O_n}.$$

This implies, that Q^c_n is an orthogonal complement of Q^w_n in $\mathbb{R}^O_a$ and Q_n in $\mathbb{R}^{E^O_n}$, i.e.

$$\mathbb{R}^{E^O_n}_a = Q^w_n \oplus Q^c_n, \text{ and } \mathbb{R}^{E^O_n} = Q_n \oplus Q^c_n = \mathbb{R}^{E^O_s} \oplus Q^w_n \oplus Q^c_n.$$

Cut and o-cut vector set-functions

The space Q_n is generated also by incidence vectors of oriented cuts, which we define in this subsection.

Each subset $S \subset V_n$ determines cuts of the graphs K_n and K^O_n, that are subsets of edges and arcs of these graphs.

A *cut*$(S) \subset \mathcal{E}_n$ is a subset of edges (ij) of K_n, such that $(ij) \in cut(S)$ if and only if $|\{i,j\} \cap S| = 1$.

A *cut* $cut^O(S) \subset \mathcal{E}^O_n$ is a subset of arcs ij of K^O_n, such that $ij \in cut^O(S)$ if and only if $|\{i,j\} \cap S| = 1$. So, if $ij \in cut^O(S)$, then $ji \in cut^O(S)$ also.

An *oriented cut o-cut*$(S) \subset E^O_n$ is a subset of arcs ij of K^O_n, such that $ij \in o\text{-}cut(S)$ if and only if $i \in S$, and $j \notin S$.

We relate to these three types of cuts their characteristic vectors $\delta_S \in \mathbb{R}^{E_n}$, $\delta^O_S \in \mathbb{R}^{E^O_n}_s$, $q_S \in \mathbb{R}^{E^O_n}_a$, and $\delta^O_S \in \mathbb{R}^{\mathcal{E}^O_n}$ as follows.

For *cut*(S), we set

$$\delta_S = \sum_{i \in S, j \in \overline{S}} e_{(ij)}, \text{ such that } (\delta_S)_{(ij)} = \begin{cases} 1, & \text{if } |\{i,j\} \cap S| = 1, \\ 0, & \text{otherwise,} \end{cases}$$

where $\overline{S} = V_n \backslash S$. For $cut^O(S)$, we set

$$\delta^O_S = \varphi(\delta_S) = \sum_{i \in S, j \in \overline{S}} (e_{ij} + e_{ji}), \text{ and } q_S = \sum_{i \in S, j \in \overline{S}} (e_{ij} - e_{ji}).$$

Hence,

$$(\delta_S^O)_{ij} = \begin{cases} 1, & \text{if } |\{i,j\} \cap S| = 1, \\ 0, & \text{otherwise,} \end{cases} \quad \text{and } (q_S)_{ij} = \begin{cases} 1, & \text{if } i \in S, j \notin S, \\ -1, & \text{if } j \in S, i \notin S, \\ 0, & \text{otherwise.} \end{cases}$$

Note, that, for one-element sets $S = \{k\}$, the function $q_{\{k\}}$ is q_k, defined before. It is easy to see, that

$$\langle \delta_S^O, q_T \rangle = 0 \text{ for any } S, T \subset V_n.$$

For the oriented cut *o-cut*(S), we set

$$\delta_S^O = \sum_{i \in S, j \in \overline{S}} e_{ij}.$$

Hence,

$$(\delta_S^O)_{ij} = \begin{cases} 1, & \text{if } i \in S, j \notin S, \\ 0, & \text{otherwise.} \end{cases}$$

Obviously, it holds $\delta_\emptyset^O = \delta_{V_n}^O = \mathbf{0}$, where $\mathbf{0} = (0, ..., 0) \in \mathbb{R}^{E_n^O}$ is the *zero-vector*, whose all $\frac{n(n-1)}{2}$ coordinates are equal zero. We have

$$(\delta_S^O)^r = \delta_{\overline{S}}^O, \; \delta_S^O + \delta_{\overline{S}}^O = \delta_S^O, \tag{15.3}$$

$$\delta_S^O - \delta_{\overline{S}}^O = q_S, \; \delta_S^O = \frac{1}{2}(\delta_S^O + q_S). \tag{15.4}$$

Besides, we have

$$(\delta_S^O)^s = \frac{1}{2}\delta_S^O, \; (\delta_S^O)^a = \frac{1}{2}q_S.$$

Recall, that a set-function $f = f(S)$, defined on all $S \subset V_n$, is called *submodular* if, for any $S, T \subset V_n$, the following *submodular inequality* holds:

$$f(S) + f(T) - (f(S \cap T) + f(S \cup T)) \geq 0.$$

It is well known, that the cut vector set-function $\delta \in \mathbb{R}^{E_n}$ (here $\delta(S) = \delta_S$) is submodular (see, for example, [Aign79]). The above isomorphism φ of the spaces $\mathbb{R}^{E_n}$ and $\mathbb{R}_s^{E_n^O}$ implies that the vector set-function $\delta^O = \varphi(\delta) \in \mathbb{R}_s^{E_n^O}$ is submodular also.

A set-function $f = f(S)$, defined on all $S \subset V_n$, is called *modular* if, for any $S, T \subset V_n$, the above submodular inequality holds as equality. This equality is called *modular equality*. It is well known (and can be easily verified), that antisymmetric vector set-function $f^a(S)$ is modular for any oriented graph G. Hence, our antisymmetric vector set-function $q_S \in \mathbb{R}_a^{E^O}$ for the oriented complete graph K_n^O is modular also.

Note, that the set of all submodular set-functions on a set V_n forms a cone in the space $\mathbb{R}^{2^n}$. Therefore, the last equality in (15.4) implies, that the vector set-function $\delta_S^O \in \mathbb{R}^{E_n^O}$ is submodular.

The modularity of the antisymmetric vector set-function q_S is important for what follows. It is well-known (see, for example, [Birk67]) (and it can be easily verified using modular equality), that a modular set-function $m(S)$ is completely determined by its values on the empty set and on all one-element sets. Hence, a modular set-function $m(S)$ has the following form:

$$m(S) = m_0 + \sum_{i \in S} m_i,$$

where $m_0 = m(\emptyset)$, and $m_i = m(\{i\}) - m(\emptyset)$. For brevity, we set $f(\{i\}) = f(i)$ for any set-function $f(S)$. Since $q_\emptyset = q_{V_n} = 0$, we have

$$q_S = \sum_{k \in S} q(k), \ S \subset V_n, \text{ and } q_{V_n} = \sum_{k \in V_n} q_k = 0. \tag{15.5}$$

Using equations (15.4) and (15.5), we obtain

$$\delta_S^O = \frac{1}{2}(\delta_S^O + \sum_{k \in S} q_k). \tag{15.6}$$

Now we show, that o-cut vectors δ_S^O, taken for all $S \subset V_n$, linearly generate the space $Q_n \subseteq \mathbb{R}^{E_n^O}$. In fact, the space, generated by δ_S^O, consists of the following vectors:

$$c = \sum_{S \subset V_n} \lambda_S \delta_S^O, \text{ where } \lambda_S \in \mathbb{R}.$$

Recall, that $\delta_S^O = \frac{1}{2}(\delta_S^O + q_S)$. Hence, we have

$$c = \frac{1}{2} \sum_{S \subset V_n} \lambda_S(\delta_S^O + q_S) = \frac{1}{2} \sum_{S \subset V_n} \lambda_S \delta_S^O + \frac{1}{2} \sum_{S \subset V_n} \lambda_S q_S = \frac{1}{2}(d^O + q),$$

where $d^O = \varphi(d)$ for $d = \sum_{S \subset V_n} \lambda_S \delta_S$. For a vector q, we have

$$q = \sum_{S \subset V_n} \lambda_S q_S = \sum_{S \subset V_n} \lambda_S \sum_{k \in S} q_k = \sum_{k \in V_n} w_k q_k, \text{ where } w_k = \sum_{k \in S \subset V_n} \lambda_S.$$

Since $q_{ij} = \sum_{k \in V_n} w_k (q_k)_{ij} = w_i - w_j$, we have

$$c_{ij} = \frac{1}{2}(d_{ij}^O + w_i - w_j). \tag{15.7}$$

It is well-known (see, for example, [DeLa97]) that the cut vectors $\delta_S \in \mathbb{R}^{E_n}$, taken for all $S \subset V_n$, linearly generate the full space $\mathbb{R}^{E_n}$. Hence, the vectors $\delta_S^O \in \mathbb{R}_s^{E_n^O}$, taken for all $S \subset V_n$, linearly generate the full space $\mathbb{R}_s^{E_n^O}$.

According to Proposition 15.9, antisymmetric parts of o-cut vectors δ_S^O generate the space Q_n^w. This implies, that the space $Q_n = \mathbb{R}_s^{E_n^O} \oplus Q_n^w$ is generated by δ_S^O, taken for all $S \subset V_n$.

Properties of the space Q_n

Let $x \in Q_n$, and let f^C be the characteristic vector of a bicircuit C. Since f^C is orthogonal to Q_n, we have

$$\langle x, f^C \rangle = \sum_{ij \in C} f_{ij}^C x_{ij} = 0.$$

This equality implies, that each point $x \in Q_n$ satisfies, for any bicircuit C, the following equalities:

$$\sum_{ij \in C^+} x_{ij} = \sum_{ij \in C^-} x_{ij}.$$

Let $K_{1,n-1} \subset K_n$ be a *spanning star* of K_n, consisting of all $n-1$ edges, incident to a vertex of K_n. Let this vertex be 1. Each edge of $K_n - K_{1,n-1}$ has the form (ij), where $i \neq 1 \neq j$. The edge (ij) closes a *fundamental triangle* with edges $(1i), (1j), (ij)$. The corresponding bitriangle $T(1ij)$ generates the equality

$$x_{1i} + x_{ij} + x_{j1} = x_{i1} + x_{1j} + x_{ji}.$$

These equalities are the case $k = 3$ of *k-cyclic symmetry*, considered in [DeDe10]. They were derived by another way in [AvMe11]. They correspond to fundamental bitriangles $T(1ij)$, $i, j \in V_n \backslash \{1\}$, and are all $\frac{n(n-1)}{2} - (n-1)$ independent equalities, determining the space, where the Q_n lies.

Above coordinates x_{ij} of a vector $x \in Q_n$ are given in the orthonormal basis

$$\{e_{ij} \,|\, ij \in \mathcal{E}_n^{O}\}.$$

But, for what follows, it is more convenient to consider vectors $q \in Q_n$ in another basis.

Recall, that $\mathbb{R}_s^{E^O} = \varphi(\mathbb{R}^{E_n})$. Let, for $(ij) \in \mathcal{E}_n$, $\varphi(e_{(ij)}) = e_{ij} + e_{ji} \in \mathbb{R}_s^{E_n^O}$ be basic vectors of the subspace $\mathbb{R}_s^{E_n^O} \subset Q_n$. Let $q_i \in Q_n^w$, $i \in V_n$, be basic vectors of the space $Q_n^w \subset Q_n$. Then, for $q \in Q_n$, we set

$$q = q^s + q^a, \text{ where } q^s = \sum_{(ij) \in \mathcal{E}_n} q_{(ij)} \varphi(e_{(ij)}), \; q^a = \sum_{i \in V_n} w_i q_i.$$

Now, we obtain an important expression for the dot product $\langle g, q \rangle$ of vectors $g, q \in Q_n$. Recall, that

$$\langle \varphi(e_{(ij)}), q_k \rangle = \langle (e_{ij} + e_{ji}), q_k \rangle = 0$$

for all $(ij) \in \mathcal{E}_n$, and all $k \in V_n$. Hence,

$$\langle g^s, q^a \rangle = \langle g^a, q^s \rangle = 0,$$

and we have

$$\langle g, q \rangle = \langle g^s, q^s \rangle + \langle g^a, q^a \rangle.$$

Besides, we have

$$\langle (e_{ij} + e_{ji}), (e_{kl} + e_{lk}) \rangle = 0, \text{ if } (ij) \neq (kl), \; (e_{ij} + e_{ji})^2 = \langle (e_{ij} + e_{ji}), (e_{ij} + e_{ji}) \rangle = 2,$$

and

$$(q_i, q_j) = -2 \text{ if } i \neq j, \; (q_i)^2 = \langle q_i, q_i \rangle = 2(n-1).$$

Let v_i, $i \in V_n$, be weights of the vector g. Then we have

$$\langle g, q \rangle = 2 \sum_{(ij) \in \mathcal{E}_n} g_{(ij)} q_{(ij)} + 2(n-1) \sum_{i \in V_n} v_i w_i - 2 \sum_{i \neq j \in V_n} v_i w_j.$$

For the last sum, we have

$$\sum_{i\neq j\in V_n} v_iw_j = \left(\sum_{i\in V_n} v_i\right)\left(\sum_{i\in V_n} w_i\right) - \sum_{i\in V_n} v_iw_i.$$

Since weights are defined up to an additive scalar, we can choose weights v_i such that $\sum_{i\in V_n} v_i = 0$. Then the last sum in the product $\langle g,q\rangle$ is equal to $-\sum_{i\in V_n} v_iw_i$. Finally, we obtain, that the sum of antisymmetric parts is equal to $2n\sum_{i\in V_n} v_iw_i$. So, for the product of two vectors $g,q \in Q_n$ we have the following expression:

$$\langle g,q\rangle = \langle g^s,q^s\rangle + \langle g^a,q^a\rangle = 2\left(\sum_{(ij)\in\mathcal{E}_n} g_{(ij)}q_{(ij)} + n\sum_{i\in V_n} v_iw_i\right), \text{if } \sum_{i\in V_n} v_i = 0, \text{ or } \sum_{i\in V_n} w_i = 0.$$

In what follows, we consider inequalities $\langle g,q\rangle \geq 0$. In this case we can delete the multiple 2, and rewrite such inequality as follows:

$$\sum_{(ij)\in\mathcal{E}_n} g_{(ij)}q_{(ij)} + n\sum_{i\in V_n} v_iw_i \geq 0, \text{ where } \sum_{i\in V_n} v_i = 0. \tag{15.8}$$

Below we consider some cones in the space Q_n. Since the space Q_n is orthogonal to the space of circuits Q_n^c, each facet vector of a cone in Q_n is defined up to a vector of the space Q_n^c. Of course, each vector $g' \in \mathbb{R}^{E_n^O}$ can be decomposed as $g' = g + g^c$, where $g \in Q_n$, and $g^c \in Q_n^c$. Call the vector $g \in Q_n$ *canonical representative* of the vector g'. Usually, we will use canonical facet vectors. But sometimes not canonical representatives of a facet vector are useful.

Cones CON, that will be considered, are invariant under the operation $q \to q^r$. In other words, $CON^r = CON$. This operation changes signs of weights:

$$q_{ij} = q_{(ij)} + w_i - w_j \to q_{(ij)} + w_j - w_i = q_{(ij)} - w_i + w_j.$$

Let $\langle g,q\rangle \geq 0$ be an inequality, determining a facet F of a cone $CON \subset Q_n$. Since $CON = CON^r$, the cone CON has, together with the facet F, also a facet F^r. The facet F^r is determined by the inequality $\langle g^r,q\rangle \geq 0$.

15.3 Construction of projections of cones on $n+1$ points

Projections of cones CON_{n+1}

Recall, that $Q_n = \mathbb{R}_s^{E_n^O} \oplus Q_n^w$, $\mathbb{R}_s^{E_n^O} = \varphi(\mathbb{R}^{E_n})$, and dim $Q_n = \frac{n(n+1)}{2} - 1$. Let $0 \notin V_n$ be an additional point. Then the set of unordered pairs (ij) for $i,j \in V_n \cup \{0\}$ is $\mathcal{E}_n \cup \mathcal{E}_0$, where $\mathcal{E}_0 = \{(0i) : i \in V_n\}$. Obviously, $E_0 = |\mathcal{E}_0| = n$, $\mathbb{R}^{E_n+E_0} = \mathbb{R}^{E_n} \oplus \mathbb{R}^{E_0}$, and dim $\mathbb{R}^{E_n+E_0} = \frac{n(n+1)}{2}$.

The space $\mathbb{R}^{E_n+E_0}$ contains the following three important cones: the cone MET_{n+1} of semimetrics, the cone HYP_{n+1} of hypermetrics, and the cone CUT_{n+1} of ℓ_1-semimetrics, all on the set $V_n \cup \{0\}$. Denote by CON_{n+1} any of these cones.

Recall, that a semimetric $d = ((d_{(ij)})) \in MET_{n+1}$ is called *metric* if $d_{(ij)} \neq 0$ for all $(ij) \in \mathcal{E}_n \cup \mathcal{E}_0$. For brevity sake, in what follows, we call elements of the cones CON_{n+1} simply metrics (hypermetrics, ℓ_1-metrics), assuming that they can be semimetrics.

Note, that if $d \in CON_{n+1}$ is a metric on the set $V_n \cup \{0\}$, then a restriction d^{V_n} of d on the set V_n is a point of the cone $CON_n = CON_{n+1} \cap \mathbb{R}^{E_n}$ of metrics on the set V_n. In other words, we can suppose that $CON_n \subset CON_{n+1}$.

The cones MET_{n+1}, HYP_{n+1} and CUT_{n+1} contain the cut vectors δ_S, that span extreme rays for all $S \subset V_n \cup \{0\}$. Denote by l_0 the extreme ray, spanned by the cut vector $\delta_{V_n} = \delta_{\{0\}}$. Consider a projection $\pi(\mathbb{R}^{E_n+E_0})$ of the space $\mathbb{R}^{E_n+E_0}$ along the ray l_0 onto a subspace of $\mathbb{R}^{E_n+E_0}$, that is orthogonal to δ_{V_n}. This projection is such that $\pi(\mathbb{R}^{E_n}) = \mathbb{R}^{E_n}$, and $\pi(\mathbb{R}^{E_n+E_0}) = \mathbb{R}^{E_n} \oplus \pi(\mathbb{R}^{E_0})$.

Note, that $\delta_{V_n} \in \mathbb{R}^{E_0}$, since $\delta_{V_n} = \sum_{i \in V_n} e_{(0i)}$. For simplicity sake, set

$$e_0 = \delta_{\{0\}} = \delta_{V_n} = \sum_{i \in V_n} e_{(0i)}.$$

Recall, that the vector e_0 spans the extreme ray l_0. Obviously, the space $\mathbb{R}^{E_n}$ is orthogonal to l_0, and, therefore, $\pi(\mathbb{R}^{E_n}) = \mathbb{R}^{E_n}$.

Let $x \in \mathbb{R}^{E_n}$. We decompose this point as follows:

$$x = x^{V_n} + x^0,$$

where $x^{V_n} = \sum_{(ij) \in \mathcal{E}_n} x_{(ij)} e_{(ij)} \in \mathbb{R}^{E_n}$, and $x^0 = \sum_{i \in V_n} x_{(0i)} e_{(0i)} \in \mathbb{R}^{E_0}$. We define a map π as follows:

$$\pi(e_{(ij)}) = e_{(ij)} \text{ for } (ij) \in \mathcal{E}_n, \text{ and } \pi(e_{(0i)}) = e_{(0i)} - \frac{1}{n} e_0 \text{ for } i \in V_n.$$

So, we have

$$\pi(x) = \pi(x^{V_n}) + \pi(x^0) = \sum_{(ij) \in \mathcal{E}_n} x_{(ij)} e_{(ij)} + \sum_{i \in V_n} x_{(0i)} (e_{(0i)} - \frac{1}{n} e_0). \tag{15.9}$$

Note, that the projection π transforms the positive orthant of the space $\mathbb{R}^{E_0}$ onto the whole space $\pi(\mathbb{R}^{E_0})$.

Now we describe, how faces of a cone in the space $\mathbb{R}^{E_n+E_0}$ are projected along one of its extreme rays.

Let l be an extreme ray, and F be a face of a cone in $\mathbb{R}^{E_n+E_0}$. Let π be the projection along l. Let $\dim F$ be dimension of the face F. Then the following equality holds:

$$\dim \pi(F) = \dim F - \dim (F \cap l). \tag{15.10}$$

Let $g \in \mathbb{R}^{E_n+E_0}$ be a facet vector of a facet G, and e be a vector, spanning the line l. Then $\dim (G \cap l) = 1$, if $\langle g, e \rangle = 0$, and $\dim (G \cap l) = 0$, if $\langle g, e \rangle \neq 0$.

Theorem 15.1 *Let G be a face of the cone $\pi(CON_{n+1})$. Then $G = \pi(F)$, where F is a face of CON_{n+1}, such that there is a facet of CON_{n+1}, containing F, and the extreme ray l_0, spanned by $e_0 = \delta_{V_n}$.*

In particular, G is a facet of $\pi(CON_{n+1})$ if and only if $G = \pi(F)$, where F is a facet of CON_{n+1}, containing the extreme ray l_0. Similarly, l' is an extreme ray of $\pi(CON_{n+1})$ if and only if $l' = \pi(l)$, where l is an extreme ray of CON_{n+1}, lying in a facet of CON_{n+1}, that contains l_0.

Proof. Let $\mathcal{F} = F(CON_{n+1})$ be a set of all facets of the cone CON_{n+1}. Then $\cup_{F\in\mathcal{F}}\pi(F)$ is a covering of the projection $\pi(CON_{n+1})$. By (15.10), if $l_0 \subset F \in \mathcal{F}$, then $\pi(F)$ is a facet of $\pi(CON_{n+1})$. If $l_0 \not\subset F$, then there is an one-to-one correspondence between points of F and $\pi(F)$. Hence, dim $\pi(F) = n$, and $\pi(F)$ cannot be a facet of $\pi(CON_{n+1})$, since $\pi(F)$ fills an n-dimensional part of the cone $\pi(CON_{n+1})$.

If F' is a face of CON_{n+1}, then $\pi(F')$ is a face of the above covering. If F' belongs only to facets $F \in \mathcal{F}$, such that $l_0 \not\subset F$, then $\pi(F')$ lies inside of $\pi(CON_{n+1})$. In this case, it is not a face of $\pi(CON_{n+1})$. This implies that $\pi(F')$ is a face of $\pi(CON_{n+1})$ if and only if $F' \subset F$, where F is a facet of CON_{n+1}, such that $l_0 \subset F$. Suppose that dimension of F' is $n-1$, and $l_0 \not\subset F'$. Then dim $\pi(F') = n-1$. If F' is contained in a facet F of CON_{n+1}, such that $l_0 \subset F$, then $\pi(F') = \pi(F)$. Hence, $\pi(F')$ is a facet of the cone $\pi(CON_{n+1})$, that coincides with the facet $\pi(F)$.

Now, the assertions of Theorem about facets and extreme rays of $\pi(CON_{n+1})$ follows. □

Theorem 15.1 describes all faces of the cone $\pi(CON_{n+1})$, if one knows all faces of the cone CON_{n+1}.

Recall, that we consider $CON_n = CON_{n+1} \cap \mathbb{R}^{E_n}$ as a subcone of CON_{n+1}, and, therefore, $\pi(CON_n) \subset \pi(CON_{n+1})$. Since $\pi(\mathbb{R}^{E_n}) = \mathbb{R}^{E_n}$, we have $\pi(CON_n) = CON_n$.

Let $\langle f, x\rangle \geq 0$ be a facet-defining inequality of a facet F of the cone CON_{n+1}. Since $CON_{n+1} \subset \mathbb{R}^{E_n} \oplus \mathbb{R}^{E_0}$, we represent vectors $f, x \in \mathbb{R}^{E_n+E_0}$ as $f = f^{V_n} + f^0, x = x^{V_n} + x^0$, where $f^{V_n}, x^{V_n} \in \mathbb{R}^{E_n}$, and $f^0, x^0 \in \mathbb{R}^{E_0}$. Hence, the above facet-defining inequality can be rewritten as

$$\langle f, x\rangle = \langle f^{V_n}, x^{V_n}\rangle + \langle f^0, x^0\rangle \geq 0.$$

It turns out, that CON_{n+1} has always a facet F with its facet vector $f = f^{V_n} + f^0$, such that $f^0 = 0$. Since f^{V_n} is orthogonal to $\mathbb{R}^{E_0}$, the hyperplane $\langle f^{V_n}, x\rangle = \langle f^{V_n}, x^{V_n}\rangle = 0$, supporting the facet F, contains the whole space $\mathbb{R}^{E_0}$. The equality $\langle f^{V_n}, x^{V_n}\rangle = 0$ defines a facet $F^{V_n} = F \cap \mathbb{R}^{E_n}$ of the cone CON_n.

A facet F of the cone CON_{n+1} with a facet vector $f = f^{V_n} + f^0$ is called *zero-lifting* of a facet F^{V_n} of CON_n, if $f^0 = 0$, and $F \cap \mathbb{R}^{E_n} = F^{V_n}$.

Similarly, a facet $\pi(F)$ of the cone $\pi(CON_{n+1})$ with a facet vector f is called *zero-lifting* of F^{V_n}, if $f = f^{V_n}$, and $\pi(F) \cap \mathbb{R}^{E_n} = F^{V_n}$.

It is well-known (see, for example, [DeLa97]), that each facet F^{V_n} with facet vector f^{V_n} of the cone CON_n can be zero-lifted up to a facet F of CON_{n+1} with the same facet vector f^{V_n}.

Proposition 15.1 *Let a facet F of CON_{n+1} be a zero-lifting of a facet F^{V_n} of CON_n. Then $\pi(F)$ is a facet of $\pi(CON_{n+1})$, that is also zero-lifting of F^{V_n}.*

Proof. Recall, that the hyperplane $\{x \in \mathbb{R}^{E_n \cup E_0} : \langle f^{V_n}, x\rangle = 0\}$, supporting the facet F, contains the whole space $\mathbb{R}^{E_0}$. Hence, the facet F contains the extreme ray l_0, spanned by the vector $e_0 \in \mathbb{R}^{E_0}$. By Theorem 15.1, $\pi(F)$ is a facet of $\pi(CON_{n+1})$. The facet vector of $\pi(F)$ can be written as $f = f^{V_n} + f'$, where $f^{V_n} \in \mathbb{R}^{E_n}$, and $f' \in \pi(\mathbb{R}^{E_0})$. Since the hyperplane, supporting the facet $\pi(F)$, is given by the equality $\langle f^{V_n}, x\rangle = 0$ for $x \in \pi(\mathbb{R}^{E_n+E_0})$, we have $f' = 0$. Besides, obviously, $\pi(F) \cap \mathbb{R}^{E_n} = F^{V_n}$. Hence, $\pi(F)$ is zero-lifting of F^{V_n}. □

Cones $\psi(CON_{n+1})$

Note, that basic vectors of the space $\mathbb{R}^{E_n+E_0}$ are $e_{(ij)}$ for $(ij) \in \mathcal{E}_n$, and $e_{(0i)}$ for $(0i) \in \mathcal{E}_0$. Since $\pi(e_0) = \sum_{i \in V_n} \pi(e_{(0i)}) = 0$, we have dim $\pi(\mathbb{R}^{E_0}) = n - 1 =$ dim Q_n^w. Note, that $\pi(\mathbb{R}^{E_n}) = \mathbb{R}^{E_n}$. Hence, there is a bijection χ between the spaces $\pi(\mathbb{R}^{E_n+E_0})$ and Q_n.

We define this bijection $\chi : \pi(\mathbb{R}^{E_n+E_0}) \to Q_n$ as follows:

$$\chi(\mathbb{R}^{E_n}) = \varphi(\mathbb{R}^{E_n}) = \mathbb{R}_s^{E_n^O}, \text{ and } \chi(\pi(\mathbb{R}^{E_0})) = Q_n^w,$$

where

$$\chi(e_{(ij)}) = \varphi(e_{(ij)}) = e_{ij} + e_{ji}, \text{ and } \chi(\pi(e_{(0i)})) = \chi(e_{(0i)} - \frac{1}{n}e_0) = q_i,$$

where q_i is defined in (15.1).

Note, that $(e_{ij} + e_{ji})^2 = 2 = 2e_{(ij)}^2$, and

$$\langle q_i, q_j \rangle = -2 = 2n\langle (e_{(0i)} - \frac{1}{n}e_0), (e_{(0j)} - \frac{1}{n}e_0)\rangle, \; q_i^2 = 2(n-1) = 2n(e_{(0i)} - \frac{1}{n}e_0)^2.$$

Roughly speaking, the map χ is a homothety that extends vectors $e_{(0i)} - \frac{1}{n}e_0$ up to vectors q_i by the multiple $\sqrt{2n}$.

Setting $\psi = \chi \circ \pi$, we obtain a map $\psi : \mathbb{R}^{E_n+E_0} \to Q_n$, such that

$$\psi(e_{(ij)}) = e_{ij} + e_{ji} \text{ for } (ij) \in \mathcal{E}_n, \; \psi(e_{(0i)}) = q_i \text{ for } i \in V_n. \tag{15.11}$$

Now we show how a point $x = x^{V_n} + x^0 \in \mathbb{R}^{\mathcal{E}_n+\mathcal{E}_0}$ is transformed into a point $q = \psi(x) = \chi(\pi(x)) \in Q_n$. We have

$$\pi(x) = x^{V_n} + \pi(x^0),$$

where, according to (15.9),

$$x^{V_n} = \sum_{(ij)\in\mathcal{E}_n} x_{(ij)}e_{(ij)} \in \pi(\mathbb{R}^{E_n}) = \mathbb{R}^{E_n}, \quad \text{and}$$

$$\pi(x^0) = \sum_{i\in V_n} x_{(0i)}(e_{(0i)} - \frac{1}{n}e_0) \in \pi(\mathbb{R}^{E_0}).$$

Obviously, $\chi(x^{V_n} + \pi(x^0)) = \chi(x^{V_n}) + \chi(\pi(x^0))$, and

$$\psi(x^{V_n}) = \chi(x^{V_n}) = \sum_{(ij)\in\mathcal{E}_n} x_{(ij)}(e_{ij} + e_{ji}) = \varphi(x^{V_n}) = q^s, \; \chi(\pi(x^0)) = \sum_{i\in V_n} x_{(0i)}q_i = q^a.$$

Recall, that $q^s = \sum_{(ij)\in\mathcal{E}_n} q_{(ij)}(e_{ij} + e_{ji})$, and $q^a = \sum_{i\in V_n} w_i q_i$. Hence,

$$q_{(ij)} = x_{(ij)}, \; (ij) \in \mathcal{E}_n, \text{ and } w_i = x_{(0i)}. \tag{15.12}$$

Let $f \in \mathbb{R}^{E_n+E_0}$ be a facet vector of a facet F of the cone CON_{n+1},

$$f = f^{V_n} + f^0 = \sum_{(ij)\in\mathcal{E}_n} f_{(ij)}e_{(ij)} + \sum_{i\in V_n} f_{(0i)}e_{(0i)}.$$

Let $\langle f, x\rangle \geq 0$ be the inequality, determining the facet F. The inequality $\langle f, x\rangle \geq 0$ takes on the set $V_n \cup \{0\}$ the following form:

$$\langle f, x\rangle = \sum_{(ij)\in\mathcal{E}_n} f_{(ij)}x_{(ij)} + \sum_{i\in V_n} f_{(0i)}x_{(0i)} \geq 0.$$

Since $x_{(ij)} = q_{(ij)}$, $x_{(0i)} = w_i$, we can rewrite this inequality as follows:

$$\langle f, q\rangle = \langle f^{V_n}, q^s\rangle + \langle f^0, q^a\rangle \equiv \sum_{(ij)\in\mathcal{E}_n} f_{(ij)}q_{(ij)} + \sum_{i\in V_n} f_{(0i)}w_i \geq 0. \tag{15.13}$$

Comparing the inequality (15.13) with (15.8), we see, that a canonical form of the facet vector f is $f = f^s + f^a$, where

$$f^s_{(ij)} = f_{(ij)} \text{ for } (ij) \in \mathcal{E}_n,\ f^a_{ij} = v_i - v_j, \text{ for } ij \in \mathcal{E}^{\mathcal{O}}_n, v_i = \frac{1}{n}f_{(0i)}, \text{ for } i \in V_n. \tag{15.14}$$

Theorem 15.2 *Let F be a facet of the cone CON_{n+1}. Then $\psi(F)$ is a facet of the cone $\psi(CON_{n+1})$ if and only if the facet F contains the extreme ray l_0, spanned by the vector e_0.*

Let $l \neq l_0$ be an extreme ray of CON_{n+1}. Then $\psi(l)$ is an extreme ray of $\psi(CON_{n+1})$ if and only if the ray l belongs to a facet, containing the extreme ray l_0.

Proof. By Theorem 15.1, the projection π transforms the facet F of CON_{n+1} into a facet of $\pi(CON_{n+1})$ if and only if $l_0 \subset F$. By the same Theorem, the projection $\pi(l)$ is an extreme ray of $\pi(CON_{n+1})$ if and only if l belongs to a facet, containing the extreme ray l_0.

Recall, that the map χ is a bijection between the spaces $\mathbb{R}^{E_n+E_0}$ and Q_n. This implies the assertion of this Theorem for the map $\psi = \chi \circ \pi$.

□

By Theorem 15.2, the map ψ transforms the facet F in a facet of the cone $\psi(CON_{n+1})$ only if F contains the extreme ray l_0, i.e., only if the equality $\langle f, e_0\rangle = 0$ holds. Hence, the facet vector f should satisfy the equality $\sum_{i\in V_n} f_{(0i)} = 0$.

The inequalities (15.13) give all facet-defining inequalities of the cone $\psi(CON_{n+1})$ from known facet-defining inequalities of the cone CON_{n+1}.

A proof of Proposition 15.2 below will be given later for each of the cones MET_{n+1}, HYP_{n+1} and CUT_{n+1} separately.

Proposition 15.2 *Let F be a facet of CON_{n+1} with facet vector $f = f^{V_n} + f^0$, such that $\langle f^0, e_0\rangle = 0$. Then CON_{n+1} has also a facet F^r with facet vector $f^r = f^{V_n} - f^0$.*

Proposition 15.2 implies the following important fact.

Proposition 15.3 *For $q = q^s + q^a \in \psi(CON_{n+1})$, the map $q = q^s + q^a \to q^r = q^s - q^a$ preserves the cone $\psi(CON_{n+1})$, i.e.*

$$(\psi(CON_{n+1}))^r = \psi(CON_{n+1}).$$

Proof. Let F be a facet of CON_{n+1} with facet vector f. By Proposition 15.2, if $\psi(F)$ is a facet of $\psi(CON_{n+1})$, then F^r is a facet of CON_{n+1} with facet vector f^r. Let $q \in \psi(CON_{n+1})$. Then q satisfies as the inequality $\langle f, q\rangle = \langle f^{V_n}, q^s\rangle + \langle f^0, q^a\rangle \geq 0$ (see (15.13)), so, the inequality $\langle f^r, q\rangle = \langle f^{V_n}, q^s\rangle - \langle f^0, q^a\rangle \geq 0$. But it is easy to see that $\langle f, q\rangle = \langle f^r, q^r\rangle$, and $\langle f^r, q\rangle = \langle f, q^r\rangle$. This implies that $q^r \in \psi(CON_{n+1})$.

□

The assertion of the following Proposition 15.4 is implied by the equality $(\psi(CON_{n+1}))^r = \psi(CON_{n+1})$.

Call a facet G of the cone $\psi(CON_{n+1})$ *symmetric*, if $q \in F$ implies $q^r \in F$, and call a facet G of the cone $\psi(CON_{n+1})$ *asymmetric*, if it is not symmetric.

Proposition 15.4 *Let $g \in Q_n$ be a facet vector of an asymmetric facet G of the cone $\psi(CON_{n+1})$; let $G^r = \{q^r \mid q \in G\}$. Then G^r is a facet of $\psi(CON_{n+1})$, and g^r is its facet vector.*

Recall, that CON_{n+1} has facets, that are zero-lifting of facets of CON_n. Call a facet G of the cone $\psi(CON_{n+1})$ *zero-lifting* of a facet F^{V_n} of CON_n, if $G = \psi(F)$, where F is a facet of CON_{n+1}, which is zero-lifting of F^{V_n}.

Proposition 15.5 *Let $g \in Q_n$ be a facet vector of a facet G of the cone $\psi(CON_{n+1})$. Then the following assertions are equivalent:*

(i) $g = g^r$;

(ii) the facet G is symmetric;

(iii) $G = \psi(F)$, where F is a facet of CON_{n+1}, which is zero-lifting of a facet F^{V_n} of CON_n;

(iv) G is a zero-lifting of a facet F^{V_n} of CON_n.

Proof. (i)⇒(ii). If $g = g^r$, then $g = g^s$. Hence, $q \in G$ implies

$$\langle g, q\rangle = \langle g^s, q\rangle = \langle g^s, q^s\rangle = \langle g, q^r\rangle = 0.$$

This means that $q^r \in G$, i.e., G is symmetric.

(ii)⇒(i). By Proposition 15.3, the map $q \to q^r$ is an automorphism of $\psi(CON_{n+1})$. This map transforms a facet G with facet vector g into a facet G^r with facet vector g^r. If G is symmetric, then $G^r = G$, and, therefore, $g^r = g$.

(iii)⇒(i). Let $f = f^{V_n} + f^0$ be a facet vector of a facet F of CON_{n+1}, such that $f^0 = 0$. Then the facet F is zero-lifting of the facet $F^{V_n} = F \cap \mathbb{R}^{E_n}$ of the cone CON_n. In this case, f^{V_n} is also a facet vector of the facet $G = \psi(F)$ of $\psi(CON_{n+1})$. Obviously, $(f^{V_n})^r = f^{V_n}$.

(iii)⇒(iv). This implication is implied by definition of zero-lifting of a facet of the cone $\psi(CON_{n+1})$.

(iv)⇒(i). The map χ induces a bijection between $\pi(F)$ and $\psi(F)$. Since $\pi(F)$ is zero-lifting of F^{V_n}, the facet vector of $\pi(F)$ belongs to $\mathbb{R}^{E_n}$. This implies that the facet vector g of $\psi(F)$ belongs to $\mathbb{R}^{E_n}$, i.e., $g^r = g$.

□

The symmetry group of CON_{n+1} is the symmetric group $Sym(n+1)$ of permutations of indices (see [DeLa97]). The group $Sym(n)$ is a subgroup of the symmetry group of the cone $\psi(CON_{n+1})$. The full symmetry group of $\psi(CON_{n+1})$ is $Z_2 \times Sym(n)$, where Z_2 corresponds to the map $q \to q^r$ for $q \in \psi(CON_{n+1})$. By Proposition 15.4, the set of facets of $\psi(CON_{n+1})$ is partitioned into pairs G, G^r. But it turns out that there are pairs, such that $G^r = \sigma(G)$, where $\sigma \in Sym(n)$.

Projections of hypermetric facets

The metric cone MET_{n+1}, the hypermetric cone HYP_{n+1} and the cut cone CUT_{n+1}, lying in the space $\mathbb{R}^{E_n+E_0}$, have an important class of *hypermetric* facets, that contains the class of *triangular* facets.

Let b_i, $i \in V_n$, be integers, such that $\sum_{i\in V_n} b_i = \mu$, where $\mu = 0$, or $\mu = 1$. Usually these integers are denoted as a sequence $b = (b_1, b_2, ..., b_n)$, where $b_i \geq b_{i+1}$. If, for some i, we have $b_i = b_{i+1} = ... = b_{i+m-1}$, then the sequence is shortened as $(b_1, ..., b_i^m, b_{i+m}, ..., b_n)$.

One relates to this sequence the following inequality of type $(b_1, ..., b_n)$:

$$\langle f(b), x\rangle = -\sum_{i,j\in V_n} b_i b_j x_{(ij)} \geq 0,$$

where $x = (x_{(ij)}) \in \mathbb{R}^{E_n}$, and the vector $f(b) \in \mathbb{R}^{E_n}$ has coordinates $f(b)_{(ij)} = -b_i b_j$. This inequality is called of *negative* or *hypermetric* type if in the sum $\sum_{i\in V_n} b_i = \mu$ we have $\mu = 0$, or $\mu = 1$, respectively.

The set of hypermetric inequalities on the set $V_n \cup \{0\}$ determines the *hypermetric cone* HYP_{n+1}. There are infinitely many hypermetric inequalities for metrics on $V_n \cup \{0\}$. But it is proved in [DeLa97], that only finite number of these inequalities determines facets of HYP_{n+1}. Since triangle inequalities are inequalities $\langle f(b), x\rangle \geq 0$ of type $b = (1^2, 0^{n-3}, -1)$, the hypermetric cone HYP_{n+1} is contained in MET_{n+1}, i.e.,

$$HYP_{n+1} \subseteq MET_{n+1}$$

with equality only for $n = 2$.

The hypermetric inequality $\langle f(b), x\rangle \geq 0$ takes the following form on the set $V_n \cup \{0\}$:

$$-\sum_{i,j\in V_n\cup\{0\}} b_i b_j x_{(ij)} = -\sum_{(ij)\in\mathcal{E}_n} b_i b_j x_{(ij)} - \sum_{i\in V_n} b_0 b_i x_{(0i)} \geq 0. \quad (15.15)$$

If we decompose the vector $f(b)$ as $f(b) = f^{V_n}(b) + f^0(b)$, then $f^{V_n}(b)_{(ij)} = -b_i b_j$, $(ij) \in \mathcal{E}_n$, and $f^0(b)_{(0i)} = -b_0 b_i$, $i \in V_n$.

Let, for $S \subset V_n$, the equality $\sum_{i\in S} b_i = 0$ hold. Denote by b^S a sequence, such that $b_i^S = -b_i$, if $i \in S$, and $b_i^S = b_i$, if $i \notin S$. The sequence b^S is called *switching* of b by the set S.

The hypermetric cone HYP_{n+1} has the following property (see [DeLa97]): if an inequality $\langle f(b), x\rangle \geq 0$ defines a facet and $\sum_{i\in S} b_i = 0$ for some $S \subset V_n \cup \{0\}$, then the inequality $\langle f(b^S), x\rangle \geq 0$ defines a facet, too.

Proof of Proposition 15.2 for HYP_{n+1}.

Consider the inequality (15.15), where $\langle f^0(b), e_0\rangle = -\sum_{i\in V_n} b_0 b_i = 0$. Then $\sum_{i\in V_n} b_i = 0$. Hence, the cone HYP_{n+1} has similar inequality for b^{V_n}, where $b_i^{V_n} = -1$ for all $i \in V_n$. Hence, if one of these inequalities defines a facet, so does another. Obviously, $f^0(b^{V_n}) = -f^0(b)$. Hence, these facets satisfy the assertion of Proposition 15.2.

□

Theorem 15.3 *Let $\langle f(b), x\rangle \geq 0$ define a hypermetric facet of a cone in the space $\mathbb{R}^{E_n+E_0}$. Then the map ψ transforms it either in a hypermetric facet, if $b_0 = 0$, or in a distortion of a facet of negative type, if $b_0 = 1$. Otherwise, the projection is not a facet.*

Proof. By Section 15.3, the map ψ transforms the hypermetric inequality (15.15) for $x \in \mathbb{R}^{E_n+E_0}$ into the inequality

$$-\sum_{(ij)\in\mathcal{E}_n} b_i b_j q_{(ij)} - b_0 \sum_{i\in V_n} b_i w_i \geq 0$$

for $q = \sum_{(ij)\in\mathcal{E}_n} q_{(ij)}\varphi(e_{(ij)}) + \sum_{i\in V_n} w_i q_i \in Q_n$.

Since $f(b)$ determines a hypermetric inequality, we have $b_0 = 1 - \sum_{i\in V_n} b_i = 1 - \mu$. So, the above inequality takes the form

$$\sum_{(ij)\in\mathcal{E}_n} b_i b_j q_{(ij)} \leq (\mu - 1)\sum_{i\in V_n} b_i w_i.$$

By Theorem 15.1, this facet is projected by the map ψ into a facet if and only if $\langle f(b), e_0\rangle = 0$, where $e_0 = \sum_{i\in V_n} e_{(0i)}$. Hence, we have

$$\langle f(b), e_0\rangle = \sum_{i\in V_n} f(b)_{(0i)} = -\sum_{i\in V_n} b_0 b_i = -b_0\mu = (\mu - 1)\mu.$$

This implies, that the hypermetric facet-defining inequality $\langle f(b), x\rangle \geq 0$ is transformed into a facet-defining inequality if and only if either $\mu = 0$, and then $b_0 = 1$, or $\mu = 1$, and then $b_0 = 0$. So, we get:

- if $\mu = 1$ and $b_0 = 0$, then the above inequality is an usual hypermetric inequality in the space $\psi(\mathbb{R}^{E_n}) = \varphi(\mathbb{R}^{E_n}) = \mathbb{R}_s^{E_n^O}$;
- if $\mu = 0$ and $b_0 = 1$, then the above inequality is the following distortion of an inequality of negative type:

$$-\sum_{(ij)\in\mathcal{E}_n} b_i b_j q_{(ij)} - \sum_{i\in V_n} b_i w_i \geq 0, \text{ where } \sum_{i\in V_n} b_i = 0. \tag{15.16}$$

□

Comparing (15.8) with the inequality (15.16), we see, that a canonical facet vector $g(b)$ of a facet of $\psi(HYP_{n+1})$ has the form $g(b) = g^s(b) + g^a(b)$, where $g_{ij}(b) = g_{(ij)}(b) + v_i - v_j$, and

$$g_{(ij)}(b) = -b_i b_j, \; v_i = -\frac{1}{n} b_i \text{ for all } i \in V_n.$$

Define the *cone of weighted quasi-hypermetrics* as $WQHYP_n = \psi(HYP_{n+1})$, we can apply Proposition 15.3, in order to obtain the following assertion.

Proposition 15.6 *The map $q \to q^r$ preserves the cone $WQHYP_n$, i.e.*

$$(WQHYP_n)^r = WQHYP_n.$$

In other words, if $q \in WQHYP_n$ has weights $w_i, i \in V_n$, then the cone $WQHYP_n$ has a point q^r with weights $-w_i, i \in V_n$.

15.4 Projections of MET_{n+1} and CUT_{n+1}

Generalizations of metrics

The metric cone MET_{n+1} is defined in the space $\mathbb{R}^{E_n+E_0}$. It has an extreme ray, which is spanned by the vector $e_0 = \sum_{i \in V_n} e_{(0i)} \in \mathbb{R}^{E_0}$. Facets of MET_{n+1} are defined by the following set of triangle inequalities, where $d \in MET_{n+1}$:

- triangle inequalities of the subcone MET_n, defining facets of MET_{n+1}, that are zero-lifting and contain e_0:

$$d_{(ik)} + d_{(kj)} - d_{(ij)} \geq 0, \ i, j, k \in V_n; \tag{15.17}$$

- triangle inequalities, defining facets, that are not zero-lifting and contain the extreme ray l_0, spanned by the vector e_0:

$$d_{(ij)} + d_{(j0)} - d_{(i0)} \geq 0, \text{ and } d_{(ij)} + d_{(i0)} - d_{(j0)} \geq 0, \ i, j \in V_n; \tag{15.18}$$

- triangle inequalities, defining facets, that do not contain the extreme ray l_0, and do not define facets of MET_n:

$$d_{(i0)} + d_{(j0)} - d_{(ij)} \geq 0, \ i, j \in V_n. \tag{15.19}$$

One can say, that the cone $MET_n \in \mathbb{R}^{E_n}$ is lifted into the space $\mathbb{R}^{E_n+E_0}$, using restrictions (15.18) and (15.19). Note, that the inequalities (15.18) and (15.19) imply the following inequalities of non-negativity:

$$d_{(i0)} \geq 0, \ i \in V_n. \tag{15.20}$$

A cone, defined by inequalities (15.17) and (15.20) is denoted by $WMET_n$ and is called the *cone of weighted metrics* (d, w), where $d \in MET_n$, and $w_i = d_{(0i)}$, $i \in V_n$, are *weights.*

If weights $w_i = d_{(0i)}$ satisfy additionally to the inequalities (15.20) also the inequalities (15.18), then the weighted metrics (d, w) form a cone $dWMET_n$ of *down-weighted metrics.* If metrics have weights that satisfy the inequalities (15.20) and (15.19), then these metrics are called *up-weighted* metrics. See details in Chapter 12 and in [DeDe10], [DDV11].

Above defined generalizations of metrics are functions on unordered pairs $(ij) \in \mathcal{E}_n \cup \mathcal{E}_0$. Generalizations of metrics as functions on ordered pairs $ij \in \mathcal{E}_n^{\mathcal{O}}$ are called *quasi-metrics.*

The cone $QMET_n$ of quasi-metrics is defined in the space $\mathbb{R}^{E_n^{\mathcal{O}}}$ by non-negativity inequalities $q_{ij} \geq 0$ for all $ij \in E_n^{\mathcal{O}}$, and by triangle inequalities $q_{ij} + q_{jk} - q_{ik} \geq 0$ for all ordered triples ijk for each $q \in QMET_n$. Below we consider in $QMET_n$ a subcone $WQMET_n$ of weighted quasi-metrics.

Cone of weighted quasi-metrics

We call a quasi-metric q *weighted* if it belongs to the subspace $Q_n \subset \mathbb{R}^{E_n^O}$. So, we define

$$WQMET_n = QMET_n \cap Q_n.$$

A *quasi-metric* q is called *weightable*, if there are weights $w_i \geq 0$ for all $i \in V_n$, such that the following equalities hold for all $i, j \in V_n$, $i \neq j$:

$$q_{ij} + w_i = q_{ji} + w_j.$$

Since $q_{ij} = q^s_{ij} + q^a_{ij}$, we have

$$q_{ij} + w_i = q^s_{ij} + q^a_{ij} + w_i = q^s_{ji} + q^a_{ji} + w_j,$$

i.e., $q^a_{ij} - q^a_{ji} = 2q^a_{ij} = w_j - w_i$, what means that, up to multiple $\frac{1}{2}$ and sign, the antisymmetric part of q_{ij} is $w_i - w_j$. So, weightable quasi-metrics are weighted.

Note, that weights of a weighted quasi-metric are defined up to an additive constant. So, if we take weights non-positive, we obtain a weightable quasi-metric. Hence, sets of weightable and weighted quasi-metrics coincide.

By definition of the cone $WQMET_n$ and by symmetry of this cone, the triangle inequality $q_{ij} + q_{jk} - q_{ik} \geq 0$ and non-negativity inequality $q_{ij} \geq 0$ determine facets of the cone $WQMET_n$. Facet vectors of these facets are

$$t_{ijk} = e_{ij} + e_{jk} - e_{ik} \text{ and } e_{ij},$$

respectively. It is not difficult to verify, that $t_{ijk}, e_{ij} \notin Q_n$. Hence, these facet vectors are not canonical. Below, we give canonical representatives of these facet vectors.

Let $T(ijk) \subset K_n^O$ be a triangle of K_n^O with direct arcs ij, jk, ki and opposite arcs ji, kj, ik. Hence,

$$f^{T(ijk)} = (e_{ij} + e_{jk} + e_{ki}) - (e_{ji} + e_{kj} + e_{ik}).$$

Proposition 15.7 *Canonical representatives of facet vectors t_{ijk} and e_{ij} are*

$$t_{ijk} + t^r_{ijk} = t_{ijk} + t_{kji}, \text{ and } g(ij) = (e_{ij} + e_{ji}) + \frac{1}{n}(q_i - q_j),$$

respectively.

Proof. We have $t_{ijk} - f^{T(ijk)} = e_{ji} + e_{kj} - e_{ki} = t_{kji} = t^r_{ijk}$. This implies, that the facet vectors t_{ijk} and t_{kji} determine the same facet, and the vector $t_{ijk} + t_{kji} \in \mathbb{R}_s^{E_n^O}$ is a canonical representative of facet vectors of this facet. We obtain the first assertion of Proposition.

Consider now the facet vector e_{ij}. It is more convenient to take the doubled vector $2e_{ij}$. We show that the vector

$$g(ij) = 2e_{ij} - \frac{1}{n} \sum_{k \in V_n \setminus \{i,j\}} f^{T(ijk)}$$

is a canonical representative of the facet vector $2e_{ij}$. It is sufficient to show that $g(ij) \in Q_n$, i.e., $g_{kl}(ij) = g^s_{kl}(ij) + w_k - w_l$. In fact, we have

$$g_{ij}(ij) = 2 - \frac{n-2}{n} = 1 + \frac{2}{n}, \quad g_{ji}(ij) = \frac{n-2}{n} = 1 - \frac{2}{n},$$

$$g_{ik}(ij) = -g_{ki}(ij) = \frac{1}{n}, \; g_{jk}(ij) = -g_{kj}(ij) = -\frac{1}{n}, \; g_{kk'}(ij) = 0.$$

Hence, we have

$$g^s(ij) = e_{ij} + e_{ji}, \; w_i = -w_j = \frac{1}{n}, \text{ and } w_k = 0 \text{ for all } k \in V_n \backslash \{i, j\}.$$

These equalities imply the second assertion of Proposition. □

Let τ_{ijk} be a facet vector of a facet of MET_n, determined by the inequality $d_{(ij)} + d_{(jk)} - d_{(ik)} \geq 0$. Then $t_{ijk} + t_{kji} = \varphi(\tau_{ijk})$, where the map $\varphi : \mathbb{R}^{E_n} \to \mathbb{R}_s^{E_n^O}$ is defined above. Obviously, a triangular facet is symmetric.

Recall, that $q_{ij} = q_{(ij)} + w_i - w_j$, if $q \in WQMET_n$. Let $i, j, k \in V_n$. It is not difficult to verify, that the following equalities hold:

$$q_{ij}^s + q_{jk}^s - q_{ik}^s = q_{ij} + q_{jk} - q_{ij} \geq 0. \tag{15.21}$$

Since $q_{ij}^s = q_{ji}^s = q_{(ij)}$, these inequalities show that the symmetric part q^s of the vector $q \in WQMET_n$ is a semimetric. Hence, if $w_i = w$ for all $i \in V_n$, then the quasi-semimetric $q = q^s$ itself is a semimetric. This implies, that the cone $WQMET_n$ contains the cone of semimetric MET_n. Moreover,

$$MET_n = WQMET_n \cap \mathbb{R}_s^{E_n^O}.$$

Now we show explicitly how the map ψ transforms the cones MET_{n+1} and $dWMET_n$ into the cone $WQMET_n$; see also Lemma 1 (ii) in [DDV11].

Theorem 15.4 *The following equalities hold:*

$$\psi(MET_{n+1}) = \psi(dWMET_n) = WQMET_n, \; \textit{and} \; WQMET_n^r = WQMET_n.$$

Proof. All facets of the metric cone MET_{n+1} on the set $V_n \cup \{0\}$ are given by triangular inequalities $d_{(ij)} + d_{(ik)} - d_{(kj)} \geq 0$. They are hypermetric inequalities $\langle f(b), d\rangle \geq 0$, where b has only three non-zero values $b_j = b_k = 1$, and $b_i = -1$, for some $i, j, k \in V_n \cup \{0\}$. By Theorem 15.3, the map ψ transforms this facet into a hypermetric facet, i.e., into a triangular facets of the cone $\psi(MET_{n+1})$, if and only if $b_0 = 0$, i.e., if $0 \notin \{i, j, k\}$. If $0 \in \{i, j, k\}$, then, by the same theorem, the equality $b_0 = 1$ should be satisfied. This implies $0 \in \{j, k\}$. In this case, the facet-defining inequality has the form (15.16), that in the case $k = 0$ is

$$q_{(ij)} + w_i - w_j \geq 0.$$

This inequality is just the non-negativity inequality $q_{ij} \geq 0$.

If $b_i = 1, b_j = -1$, and $k = 0$, the inequality $d_{(ij)} + d_{(j0)} - d_{(0i)} \geq 0$ is transformed into the inequality

$$q_{(ij)} + w_j - w_i \geq 0, \text{ i.e., } q_{ij}^r \geq 0.$$

This inequality and inequalities (15.21) imply the last equality of this Theorem.

The inequalities (15.19) define facets F of MET_{n+1} and $dWMET_n$, that do not contain the extreme ray l_0. Hence, by Theorem 15.3, $\psi(F)$ are not facets of $WQMET_n$. But recall, that the cone $dWMET_n$ contains all facets of MET_{n+1}, excluding facets, defined by the inequalities (15.19). Instead of these facets, the cone $dWMET_n$ has facets G_i, defined by the non-negativity equalities (15.20) with facet vectors $e_{(0i)}$ for all $i \in V_n$. Obviously, all these facets do not contain the extreme ray l_0. Hence, by Theorem 15.2, $\psi(G_i)$ is not a facet of $\psi(dWMET_n)$. Hence, we have also the equality $WQMET_n = \psi(dWMET_n)$.

□

Remark. Facet vectors of facets of MET_{n+1}, that contain the extreme ray l_0, spanned by the vector e_0, are $\tau_{ijk} = \tau^{V_n}_{ijk}$, $\tau_{ij0} = \tau^{V_n} + \tau^0$, and $\tau_{ji0} = \tau^{V_n} - \tau^0$, where

$$\tau^{V_n} = e_{(ij)}, \quad \text{and} \quad \tau^0 = e_{(j0)} - e_{(i0)}.$$

Hence, Proposition 15.2 is true for MET_{n+1}, and we can apply Proposition 15.3 in order to obtain the equality $WQMET^r_n = WQMET_n$ of Theorem 15.4.

The cone CUT_{n+1}

The cut vectors $\delta_S \in \mathbb{R}^{E_n+E_0}$, constructed for all $S \subset V_n \cup \{0\}$, span all extreme rays of the cut cone $CUT_{n+1} \subset \mathbb{R}^{E_n+E_0}$. In other words, CUT_{n+1} is the conic hull of all cut vectors. Since the cone CUT_{n+1} is full-dimensional, its dimension is dimension of the space $\mathbb{R}^{E_n+E_0}$, that is $\frac{n(n+1)}{2}$.

Recall, that $\delta_S = \delta_{V_n \cup \{0\} \setminus S}$. Hence, we can consider only S, such that $S \subset V_n$, i.e., $0 \notin S$. Moreover, by Section 15.2, it holds

$$\delta_S = \sum_{i \in S, j \notin S} e_{(ij)} = \sum_{i \in S, j \in V_n \setminus S} e_{(ij)} + \sum_{i \in S} e_{(0i)} = \delta^{V_n}_S + \sum_{i \in S} e_{(0i)}, \tag{15.22}$$

where $\delta^{V_n}_S$ is the restriction of δ_S on the space $\mathbb{R}^{E_n} = \psi(\mathbb{R}^{E_n})$. Note, that

$$\delta_{V_n} = \delta_{\{0\}} = \sum_{i \in V_n} e_{(0i)} = e_0.$$

Consider a facet F of CUT_{n+1}. Let f be facet vector of F. Set

$$R^*(F) = \{S \subset V_n : \langle f, \delta_S \rangle = 0\}.$$

For $S \in R^*(F)$, the vector δ_S is called *root* of the facet F. By (15.22), for $S \in R^*(F)$, we have

$$\langle f, \delta_S \rangle = \langle f, \delta^{V_n}_S \rangle + \sum_{i \in S} f_{(0i)} = 0. \tag{15.23}$$

We represent each facet vector of CUT_{n+1} as $f = f^{V_n} + f^0$, where $f^{V_n} \in \mathbb{R}^{E_n}$, and $f^0 \in \mathbb{R}^{E_0}$.

The set of facets of the cone CUT_{n+1} is partitioned onto equivalence classes by *switchings* (see [DeLa97]). For each $S, T \subset V_n \cup \{0\}$, the switching by the set T transforms the cut vector δ_S into the vector $\delta_{S \Delta T}$, where Δ is symmetric difference, i.e., $S \Delta T = S \cup T - S \cap T$. It is proved in [DeLa97] that if $T \in R^*(F)$, then $\{\delta_{S \Delta T} \mid S \in R^*(F)\}$ is the set of roots of the switched facet F^{δ_T} of CUT_{n+1}. Hence, $R^*(F^{\delta_T}) = \{S \Delta T \mid S \in R^*(F)\}$.

Let F be a facet of CUT_{n+1}. Then F contains the vector $e_0 = \delta_{V_n}$ if and only if $V_n \in R^*(F)$. Hence, Proposition 15.8 below is an extended reformulation of Proposition 15.2.

Proposition 15.8 *Let F be a facet of CUT_{n+1}, such that $V_n \in R^*(F)$. Let $f = f^{V_n} + f^0$ be facet vector of F. Then the vector $f^r = f^{V_n} - f^0$ is the facet vector of switching $F^{\delta_{V_n}}$ of the facet F, and $V_n \in R^*(F^{\delta_{V_n}})$.*

Proof. Since $V_n \in R^*(F)$, $F^{\delta_{V_n}}$ is a facet of CUT_{n+1}. Since $S\Delta V_n = V_n \backslash S = \overline{S}$ for $S \subset V_n$, we have

$$R^*(F^{\delta_{V_n}}) = \{\overline{S} \,|\, S \in R^*(F)\}.$$

Since $\emptyset \in R^*(F)$, the set $\emptyset\Delta V_n = V_n \in R^*(F^{\delta_{V_n}})$. Now, using (15.23) for $S \in R^*(F^{\delta_{V_n}})$, we have

$$\langle f^r, \delta_S\rangle = \langle (f^{V_n} - f^0), \delta_S\rangle = \langle f^{V_n}, \delta_S^{V_n}\rangle - \sum_{i\in S} f_{(0i)}.$$

Note, that $\delta_{\overline{S}}^{V_n} = \delta_S^{V_n}$, and, since $V_n \in R^*(F)$, $\delta_{V_n} = \delta_{\{0\}}$, we have

$$\langle f, \delta_{V_n}\rangle = \sum_{i\in V_n} f_{(0i)} = 0.$$

Hence, $\sum_{i\in\overline{S}} f_{(0i)} = -\sum_{i\in S} f_{(0i)}$.

It is easy to see, that $(f^r, \delta_S) = (f, \delta_{\overline{S}})$. Since $S \in R^*(F^{\delta_{V_n}})$ if and only if $\overline{S} \in R^*(F)$, we see, that f^r is a facet vector of $F^{\delta_{V_n}}$.

□

The set of facets of CUT_{n+1} is partitioned into orbits under action of the permutation group $Sym(n+1)$. But some permutation non-equivalent facets are equivalent under switchings. We say that two facets F, F' of CUT_{n+1} belong to the same *type* if there are a permutation $\sigma \in Sym(n+1)$, and $T \subset V_n$, such that $\sigma(F') = F^{\delta_T}$.

Cone $OCUT_n$

Denote by $OCUT_n \subset \mathbb{R}^{E_n^O}$ the cone, whose extreme rays are spanned by o-cut vectors δ_S^O for all $S \subset V_n$, $S \neq \emptyset, V_n$. In other words, let

$$OCUT_n = \left\{ c \in Q_n \,\middle|\, c = \sum_{S\subset V_n} \lambda_S \delta_S^O,\ \lambda_S \geq 0 \right\}.$$

Coordinates c_{ij} of a vector $c \in OCUT_n$ are given in (15.7), where $w_i \geq 0$ for all $i \in V_n$. Hence, $OCUT_n \subset Q_n$. Recall, that

$$\delta_S^O = \frac{1}{2}\left(\delta_S^O + \sum_{i\in S} q_i\right), \tag{15.24}$$

where $\delta_S^O = \varphi(\delta_S^{V_n})$. Note, that $\delta_{\overline{S}}^O = \delta_S^O$, and $q_{\overline{S}} = -q_S$, where $\overline{S} = V_n \backslash S$.

Denote by $CUT_n^O = \varphi(CUT_n)$ the cone, generated by δ_S^O for all $S \subset V_n$. The vectors δ_S^O for all $S \subset V_n$, $S \neq \emptyset, V_n$, are all extreme rays of the cone CUT_n^O, that we identify with CUT_n, embedded into the space $\mathbb{R}^{E_n^O}$.

Proposition 15.9 *For $S \subset V_n$, the following equality holds:*

$$\psi(\delta_S) = 2\delta_S^O.$$

Proof. According to Section 15.3, $\psi(\delta_S^{V_n}) = \varphi(\delta_S^{V_n}) = \delta_S^O$. Besides, $\psi(e_{(0i)}) = q_i$ for all $i \in V_n$. Hence, using (15.22), we obtain

$$\psi(\delta_S) = \psi(\delta_S^{V_n}) + \sum_{i \in S} \psi(e_{(0i)}) = \varphi(\delta_S^{V_n}) + \sum_{i \in S} q_i = \delta_S^O + q_S.$$

Recall, that $\psi(\delta_{V_n}) = \psi(e_0) = \mathbf{0}$, and $\delta_{V_n}^O = 0$. Hence, according to (15.24), we obtain

$$\psi(\delta_S) = 2\delta_S^O \text{ for all } S \subset V_n.$$

Proposition is proved. □

Theorem 15.5 *The following equalities hold:*

$$\psi(CUT_{n+1}) = OCUT_n, \text{ and } OCUT_n^r = OCUT_n.$$

Proof. Recall, that the conic hull of vectors δ_S for all $S \subset V_n \cup \{0\}$ is the cone CUT_{n+1}. The conic hull of vectors δ_S^O for all $S \subset V_n$ is the cone $OCUT_n$. Since $\psi(\delta_{V_n}) = \delta_{V_n}^O = \mathbf{0}$, the first result follows.

The equality $OCUT_n^r = OCUT_n$ is implied by the equalities $(\delta_S^O)^r = \delta_{\overline{S}}^O$ for all $S \subset V_n$.

By Proposition 15.8, the equality $OCUT_n^r = OCUT_n$ is the special case $CON_{n+1} = CUT_{n+1}$ of Proposition 15.3. □

Facets of $OCUT_n$

Proposition 15.10 *Let F be a facet of CUT_{n+1}. Then $\psi(F)$ is a facet of $OCUT_n$ if and only if $V_n \in R^*(F)$.*

Proof. By Theorem 15.2, $\psi(F)$ is a facet of $OCUT_n$ if and only if $e_0 = \delta_{V_n} \subset F$, i.e., if and only if $V_n \in R^*(F)$. □

For a facet G of $OCUT_n$ with facet vector g, we set

$$R^*(G) = \{S \subset V_n, | \langle g, \delta_S^O \rangle = 0\}$$

and call the vector δ_S^O for $S \in R^*(G)$ by *root* of the facet G.

Note, that $\delta_\emptyset = \mathbf{0}$, and $\delta_\emptyset^O = \delta_{V_n}^O = \mathbf{0}$. Hence, $\emptyset \in R^*(F)$ and $\emptyset \in R^*(G)$ for all facets F of CUT_{n+1} and all facets G of $OCUT_n$.

The roots $\delta_\emptyset = \mathbf{0}$ and $\delta_\emptyset^O = \delta_{V_n}^O = \mathbf{0}$ are called *trivial roots.*

Proposition 15.11 *For a facet F of CUT_{n+1}, let $G = \psi(F)$ be a facet of $OCUT_n$. Then the following equality holds:*

$$R^*(G) = R^*(F).$$

Remark. We give two proofs of this equality. Both are useful.

First proof. According to Section 15.3, the map ψ transforms an inequality $\langle f, x\rangle \geq 0$, defining a facet of CUT_{n+1}, into the inequality (15.13), defining the facet $G = \psi(F)$ of $OCUT_n$. Recall, the the inequality (15.13) relates to the representation of vectors $q \in Q_n$ in the basis $\{\varphi(e_{ij}), q_i\}$, i.e.,

$$q = \sum_{(ij)\in\mathcal{E}_n} q_{(ij)}\varphi(e_{(ij)}) + \sum_{i\in V_n} w_i q_i.$$

Let $q = \delta^O_S$ for $S \in R^*(G)$. Then, according to (15.24), we have $q_{(ij)} = \frac{1}{2}(\delta^{V_n}_S)_{(ij)}$, $w_i = \frac{1}{2}$ for $i \in S$, and $w_i = 0$ for $i \in \overline{S}$. Hence, omitting the multiple $\frac{1}{2}$, the inequality in (15.13) gives the equality

$$\sum_{(ij)\in\mathcal{E}_n} f_{(ij)}(\delta^{V_n}_S)_{(ij)} + \sum_{i\in S} f_{(0i)} = 0,$$

which coincides with (15.23). This implies the assertion of this Proposition.

Second proof. By Theorem 15.2, $\psi(l)$ is an extreme ray of $\psi(F)$ if and only if l is an extreme ray of F, and $l \neq l_0$. Since l is spanned by δ_S for some $S \in R^*(F)$, and $\psi(l)$ is spanned by $\psi(\delta_S) = \delta^O_S$, we have $R^*(G) = \{S \subset V_n \,|\, S \in R^*(F)\}$. Since $\delta^O_{V_n} = \mathbf{0}$, we can suppose, that $V_n \in R^*(G)$, and then $R^*(G) = R^*(F)$.

□

Remark. Note, that $\delta_{V_n} = \delta_{\{0\}} = e_0 \neq \mathbf{0}$ is a non-trivial root of F, i.e., $V_n \in R^*(F)$. But $\delta^O_{V_n} = \psi(\delta_{V_n}) = \mathbf{0}$ is a trivial root of $R^*(G)$.

Recall, that, for a subset $T \subset V_n$, we set $\overline{T} = V_n \backslash T$. Note, that $\overline{T} = V_n \Delta T$, and $\overline{T} \neq V_n \cup \{0\} \backslash T$.

Proposition 15.12 *Let F be a facet of CUT_{n+1}, and $T \in R^*(F)$. Then the image $\psi(F^{\delta_T})$ of the switched facet F^{δ_T} is a facet of $OCUT_n$ if and only if $\overline{T} \in R^*(F)$.*

Proof. By Lemma 15.10, $\psi(F^{\delta_T})$ is a facet of $OCUT_n$ if and only if $V_n \in R^*(F^{\delta_T})$, i.e., if and only if $V_n \Delta T = \overline{T} \in R^*(F)$.

□

For a facet G of $OCUT_n$, define G^{δ_T} as the conic hull of $\delta^O_{S\Delta T}$ for all $S \in R^*(G)$. Since each facet G of $OCUT_n$ is $\psi(F)$ for some facet F of CUT_{n+1}, Proposition 15.12 and Proposition 15.11 imply the following assertion.

Theorem 15.6 *Let G be a facet of $OCUT_n$. Then G^{δ_T} is a facet of $OCUT_n$ if and only if $T, \overline{T} \in R^*(G)$, and then $R^*(G^{\delta_T}) = \{S\Delta T \,|\, S \in R^*(G)\}$.*

Theorem 15.6 asserts that the set of facets of the cone $OCUT_n$ is partitioned onto equivalence classes by switchings $G \to G^{\delta_T}$, where $T, \overline{T} \in R^*(G)$.

The case $T = V_n$ in Theorem 15.6 plays a special role. Recall, that $V_n \in R^*(F)$ if F is a facet of CUT_{n+1}, such that $\psi(F)$ is a facet of $OCUT_n$. Hence, Proposition 15.8 and Proposition 15.3 imply the following fact.

Proposition 15.13 *Let F be a facet of CUT_{n+1}, such that $\psi(F)$ is a facet of $OCUT_n$. Let $g = g^s + g^a$ be a facet vector of the facet $\psi(F)$. Then the vector $g^r = g^s - g^a$ is a facet vector of the facet $\psi(F^{\delta_{V_n}}) = (\psi(F))^r = (\psi(F))^{\delta_{V_n}}$, such that*

$$R((\psi(F))^r) = \{\overline{S} \,|\, S \in R^*(F)\}.$$

Recall, that roughly speaking $OCUT_n$ is projection of CUT_{n+1} along the vector $\delta_{V_n} = \delta_{\{0\}}$.

Let $\sigma \in Sym(n)$ be a permutation of the set V_n. For a vector $q \in \mathbb{R}^{E_n^O}$, we have $\sigma(q)_{ij} = q_{\sigma(i)\sigma(j)}$. Obviously if g is a facet vector of a facet G of $OCUT_n$, then $\sigma(g)$ is the facet vector of the facet $\sigma(G) = \{\sigma(q) \,|\, q \in G\}$.

Note, that, by Proposition 15.13, the switching by V_n is equivalent to the operation $q \to q^r$. Hence, the symmetry group of $OCUT_n$ contains the group $Z_2 \times Sym(n)$, where Z_2 relates to the map $q \to q^r$ for $q \in OCUT_n$.

Theorem 15.7 *The group $Z_2 \times Sym(n)$ is the symmetry group of the cone $OCUT_n$.*

Proof. Let γ be a symmetry of $OCUT_n$. Then γ is a symmetry of the set $F(e_0)$ of facets F of the cone CUT_{n+1}, containing the vector e_0. The symmetry group $\Gamma(e_0)$ of the set $F(e_0)$ is a subgroup of the symmetry group of the cut polytope $CUT^{\square}_{n+1}$. In fact, $\Gamma(e_0)$ is stabilizer of the edge e_0 of the polytope $CUT^{\square}_{n+1}$. But it is well-known, that $\Gamma(e_0)$ consists of the switchings by V_n and permutations $\sigma \in Sym(n+1)$, leaving the edge e_0 non-changed. The map ψ transforms these symmetries of $F(e_0)$ into symmetries $\sigma \in Sym(n)$ and $q \to q^r$ of the cone $OCUT_n$. □

The set of all facets of $OCUT_n$ is partitioned onto orbits of facets that are equivalent by the symmetry group $Z_2 \times Sym(n)$. It turns out that, for some facets G, subsets $S \in R^*(G)$ and permutations $\sigma \in Sym(n)$, we have $G^{\delta_S} = \sigma(G)$.

By Proposition 15.5, if a facet of CUT_{n+1} is zero-lifting of a facet F^{V_n} of CUT_n, then the facet $G = \psi(F)$ of $OCUT_n$ is symmetric, and $G = G^r = G^{\delta_{V_n}}$ is zero-lifting of F^{V_n}.

So, there are two important classes of orbits of facets of $OCUT_n$. Namely, the orbits of symmetric facets, that are zero-lifting of facets of CUT_n, and orbits of asymmetric facets, that are ψ-images of facets of CUT_{n+1}, and are not zero-lifting.

15.5 Cases $3 \leq n \leq 6$

Compare results of this section with Table 12.1 in Chapter 12 (see also Table 2 of [DDV11]).

Most of described below facets are hypermetric or negative type. We give here the corresponding vectors b.

The case n=3

It is known, that $CUT_4 = HYP_4 = MET_4$. Hence,

$$OCUT_3 = WQHYP_3 = WQMET_3.$$

All these cones have two orbits of facets: one orbit of non-negativity facets with $b = (1, 0, -1)$, and another orbit of triangular facets with $b = (1^2, -1)$.

The case $n = 4$

We have $CUT_5 = HYP_5 \subset MET_5$. Hence,

$$OCUT_4 = WQHYP_4 \subset WQMET_4.$$

The cones $HYP_5 = CUT_5$ have two orbits of facets: triangular and pentagonal facets. Recall, that a triangular facet with facet vector τ_{ijk} is zero-lifting if $0 \not\in \{i, j, k\}$. Hence, the cones $WQHYP_4 = OCUT_4$ have three orbits of facets: non-negativity facets with $b = (1, 0^2, -1)$, triangular facets with $b = (1^2, 0, -1)$, and weighted version of negative type inequalities with $b = (1^2, -1^2)$.

The case n=5

We have again $CUT_6 = HYP_6 \subset MET_6$. Hence,

$$OCUT_5 = WQHYP_5 \subset WQMET_5.$$

The cones $HYP_6 = CUT_6$ have four orbits of facets, all are hypermetric: triangular facets with $b = (1^2, 0^3, -1)$, pentagonal facets with $b = (1^3, 0, -1^2)$, and two more types, one with $b = (2, 1^2, -1^3)$, and its switching with $b = (1^4, -1, -2)$. These four types provide 6 orbits of facets of the cones $WQHYP_5 = OCUT_5$: non-negativity facets with $b = (1, 0^3, -1)$, triangular facets with $b = (1^2, 0^2, -1)$, facets of negative type with $b = (1^2, 0, -1^2)$, pentagonal facets with $b = (1^3, -1^2)$, and two facets of negative type with $b = (2, 1, -1^3)$, and $b = (1^3, -1, -2)$.

The last two types belong to the same orbit of the full symmetry group $Z_2 \times Sym(5)$. Hence, the cone $OCUT_5$ has 5 orbits of facets under action of its symmetry group.

The case n=6

.

Now, we have $CUT_7 \subset HYP_7 \subset MET_7$. Hence,

$$OCUT_6 \subset WQHYP_6 \subset WQMET_6.$$

The cone CUT_7 has 36 orbits of facets that are equivalent under action of the permutation group $Sym(7)$. Switchings contract these orbits into 11 types F_k, $1 \le k \le 11$, of facets that are switching equivalent (see [DeLa97], Section 30.6). Vidali compute orbits of facets of $OCUT_6$ under action of the group $Sym(6)$. Using these computations, we give in Table below numbers of orbits of facets of cones CUT_7 and $OCUT_6$ (see also Figure 30.6.1 of [DeLa97]).

The first row of Table 15.1 gives types of facets of CUT_7. In the second row of Table, for each type F_k, the numbers of orbits of facets of CUT_7 of type F_k under action of the group $Sym(7)$ are given. The third row of Table gives, for each type F_k, the numbers of orbits of facets of $OCUT_6$, that are obtained from facets of type F_k under action of the group $Sym(6)$. The fourth row gives, for each type F_k, the numbers of orbits of facets of $OCUT_6$, that are obtained from facets of type F_k under action of the group $Z_2 \times Sym(6)$.

The last column of Table gives the total numbers of orbits of facets of the cones CUT_7 and $OCUT_6$.

types	F_1	F_2	F_3	F_4	F_5	F_6	F_7	F_8	F_9	F_{10}	F_{11}	|Size|
$Sym(7)$	1	1	2	1	3	2	4	7	5	3	7	36
$Sym(6)$	2	2	4	1	3	2	7	13	6	6	15	61
$Z_2 \times Sym(6)$	2	2	3	1	2	1	4	7	3	4	8	37

Table 15.1: The description of facets of CUT_7 and $OCUT_6$

The first three types F_1, F_2, F_3 relate to 4 orbits of hypermetric facets of CUT_7, that are zero-lifting, where $b = (1^2, 0^4, -1)$, $b = (1^3, 0^2, -1^2)$, and $b = (2, 1^2, 0, -1^3)$, $b = (1^4, 0, -1, -2)$. Each of these four orbits of facets of CUT_7 under action of $Sum(7)$ gives two orbits of facets of $OCUT_6$ under action of the group $Sym(6)$.

The second three types F_4, F_5, F_6 relate to 6 orbits of hypermetric facets of CUT_7, that are not zero-lifting. Each of these 6 orbits gives one orbit of facets of $OCUT_6$ under action of the group $Sym(6)$.

The third three types F_7, F_8, F_9 relate to 16 orbits of facets of clique-web types $CW_1^7(b)$. These 16 orbits give 26 orbits of facets of $OCUT_6$ under action of $Sym(6)$.

The last two types $F_{10} = Par_7$ and Gr_7 are special (see [DeLa97]). They relate to 10 orbits of CUT_7, that give 21 orbits of facets of $OCUT_6$ under action of $Sym(6)$.

The subgroup Z_2 of the full symmetry group $Z_2 \times Sym(6)$ contracts some pairs of orbits of the group $Sym(6)$ into one orbit of the full group. The result is given in the forth row of Table.

Note, that the symmetry groups of CUT_7 and $OCUT_6$ have 36 and 37 orbits of facets, respectively.

Appendixes

APPENDIX 1: EXTREME RAYS OF THE CONE $QMET_5$

By direct computation 43590 extreme rays of $QMET_5$ were found. Under group action we got 229 orbits.

See below the full list of the representatives of orbits, presented as 5×5 matrices. The numbers before each matrix are, respectively, the orbit number, the adjacency of a member of the orbit, the incidence of a member of the orbit, and orbit size. Remark, that adjacency is greater or equal to incidence number with equality only for the last 23 orbits.

1, 8313, 64, 10	**2, 6903, 56, 20**	**3, 2534, 48, 30**	**4, 2394, 48, 60**	**5, 1947, 52, 40**	**6, 1932, 52, 20**
0 0 0 0 1	0 0 0 1 1	0 0 1 1 1	0 0 1 1 1	0 0 0 1 1	0 1 1 1 1
0 0 0 0 1	0 0 0 1 1	0 0 1 1 1	0 0 1 1 1	0 0 0 1 1	0 0 0 0 1
0 0 0 0 1	0 0 0 1 1	0 0 0 1 1	0 0 0 0 1	0 0 0 1 1	0 0 0 0 1
0 0 0 0 1	0 0 0 0 0	0 0 0 0 0	0 0 0 0 1	0 0 0 0 1	0 0 0 0 1
0 0 0 0 0	0 0 0 0 0	0 0 0 0 0	0 0 0 0 0	0 0 0 0 0	0 0 0 0 0

7, 1109, 44, 120	**8, 1107, 44, 120**	**9, 701, 40, 120**	**10, 461, 38, 120**	**11, 444, 36, 120**	**12, 413, 40, 60**
0 0 1 1 1	0 1 1 1 1	0 1 1 1 1	0 1 1 1 1	0 1 1 1 1	0 1 1 1 1
0 0 1 1 1	0 0 0 1 1	0 0 1 1 1	0 0 1 1 1	0 0 1 1 1	0 0 0 1 1
0 0 0 1 1	0 0 0 1 1	0 0 0 1 1	0 0 0 0 1	0 0 0 1 1	0 0 0 1 1
0 0 0 0 1	0 0 0 0 1	0 0 0 0 1	0 0 0 0 1	0 0 0 0 1	0 0 0 0 1
0 0 0 0 0	0 0 0 0 0	0 0 0 0 0	0 0 1 1 0	0 0 0 1 0	0 0 0 1 0

13, 404, 41, 120	**14, 401, 40, 60**	**15, 398, 40, 120**	**16, 376, 32, 30**	**17, 355, 36, 120**	**18, 345, 37, 120**
0 1 1 1 2	0 0 1 1 1	0 0 1 1 2	0 1 1 1 1	0 1 1 1 1	0 1 1 1 2
0 0 0 1 1	0 0 1 1 1	0 0 1 1 2	0 0 1 1 0	0 0 1 1 1	0 0 1 1 1
0 0 0 1 1	0 0 0 1 1	0 0 0 1 1	0 0 0 1 0	0 0 0 1 1	0 0 0 1 1
0 0 0 0 1	0 0 0 0 1	0 0 0 0 1	0 0 1 0 0	0 0 0 0 0	0 0 0 0 1
0 0 0 0 0	0 0 0 1 0	0 0 0 0 0	1 1 1 1 0	0 0 1 1 0	0 0 0 0 0

19, 342, 40, 60	**20, 316, 38, 60**	**21, 292, 34, 60**	**22, 287, 33, 120**	**23, 286, 32, 60**	**24, 248, 29, 120**
0 0 1 1 1	0 1 1 1 1	0 0 1 1 1	0 1 1 2 1	0 1 1 1 1	0 1 1 1 1
0 0 1 1 1	0 0 0 1 1	0 0 1 1 1	0 0 1 1 1	0 0 1 1 1	0 0 1 1 1
0 0 0 1 1	0 0 0 1 1	0 0 0 1 0	0 0 0 1 1	0 0 0 1 0	0 0 0 1 0
0 0 0 0 0	0 0 0 0 0	0 0 1 0 0	0 0 0 0 0	0 0 1 0 0	0 0 1 0 0
0 0 1 1 0	0 1 1 1 0	1 1 1 1 0	0 0 1 1 0	0 1 1 1 0	1 1 1 1 0

25, 186, 33, 240	**26, 175, 34, 240**	**27, 175, 30, 120**	**28, 174, 31, 120**	**29, 172, 34, 120**	**30, 172, 34, 120**
0 1 1 1 1	0 1 1 1 2	0 1 1 2 1	0 1 1 2 1	0 0 1 2 1	0 0 1 1 1
0 0 1 1 1	0 0 1 1 1	0 0 1 1 1	0 0 1 2 1	0 0 1 2 1	0 0 1 1 1
0 0 0 1 1	0 0 0 1 1	0 0 0 1 0	0 0 0 1 1	0 0 0 1 0	0 0 0 1 0
0 0 0 0 1	0 0 0 0 1	0 0 1 0 0	0 0 0 0 0	0 0 1 0 0	0 0 1 0 0
0 0 1 1 0	0 0 0 1 0	0 1 1 1 0	0 0 1 1 0	1 1 1 1 0	1 1 1 2 0

31, 160, 30, 120	**32, 149, 35, 240**	**33, 143, 28, 240**	**34, 138, 29, 240**	**35, 138, 24, 20**	**36, 133, 26, 60**
0 1 1 2 1	0 1 1 1 2	0 1 1 2 1	0 1 1 1 1	0 1 1 1 1	0 1 2 2 1
0 0 1 1 1	0 0 1 1 2	0 0 1 1 1	0 0 1 1 1	0 0 1 1 0	0 0 1 1 1
0 0 0 1 1	0 0 0 1 1	0 0 0 1 0	0 0 0 1 0	0 1 0 1 0	0 0 0 1 0
0 0 0 0 0	0 0 0 0 1	0 0 1 0 0	0 0 1 0 0	0 1 1 0 0	0 0 1 0 0
0 1 1 1 0	0 0 0 0 0	1 1 1 1 0	1 1 1 2 0	1 1 1 1 0	0 1 1 1 0

37, 131, 25, 20	**38, 128, 33, 120**	**39, 122, 26, 120**	**40, 116, 27, 120**	**41, 113, 25, 120**	**42, 113, 24, 240**
0 1 2 1 1	0 1 1 1 1	0 1 2 1 1	0 1 1 1 1	0 1 2 2 1	0 1 2 2 1
0 0 1 1 1	0 0 1 1 1	0 0 1 1 1	0 0 1 1 1	0 0 1 1 1	0 0 1 1 1
0 0 0 0 0	0 0 0 1 1	0 0 0 1 0	0 0 0 1 0	0 0 0 1 0	0 0 0 1 0
0 1 1 0 1	0 0 0 0 0	0 1 1 0 0	0 1 1 0 0	0 0 1 0 0	0 1 1 0 0
0 1 1 1 0	0 1 1 1 0	1 1 1 1 0	1 1 1 1 0	1 1 1 1 0	1 1 1 1 0

43, 109, 27, 240	44, 107, 31, 120	45, 102, 25, 240	46, 98, 29, 120	47, 95, 31, 240	48, 91, 30, 240
0 1 1 2 1	0 1 1 1 1	0 1 2 2 2	0 1 2 2 1	0 1 1 2 1	0 1 1 2 1
0 0 1 1 1	0 0 1 1 0	0 0 2 1 1	0 0 1 1 0	0 0 1 2 1	0 0 1 2 1
0 0 0 1 0	0 0 0 1 0	0 0 0 1 0	0 1 0 2 1	0 0 0 2 1	0 0 0 1 1
0 1 1 0 0	0 0 1 0 0	0 1 1 0 0	0 1 0 0 0	0 0 0 0 1	0 0 0 0 1
1 1 1 1 0	1 1 1 2 0	1 1 1 1 0	1 1 1 1 0	0 0 1 1 0	0 0 1 1 0

49, 91, 30, 120	50, 91, 24, 120	51, 90, 26, 120	52, 88, 28, 40	53, 87, 31, 240	54, 84, 21, 40
0 1 1 2 2	0 1 1 1 1	0 1 2 1 1	0 1 1 1 1	0 1 1 2 1	0 2 2 2 2
0 0 1 1 1	0 0 1 1 0	0 0 2 1 0	0 0 1 1 1	0 0 1 1 1	0 0 2 1 0
0 0 0 1 1	0 1 0 1 0	0 0 0 1 0	0 0 0 1 1	0 0 0 1 1	0 1 0 2 0
0 0 0 0 1	0 1 1 0 0	0 1 1 0 0	0 0 1 0 1	0 0 0 0 1	0 2 1 0 0
0 0 0 1 0	1 1 1 2 0	1 1 2 1 0	0 0 1 1 0	0 0 1 1 0	2 2 2 2 0

55, 83, 33, 240	56, 83, 32, 240	57, 81, 31, 240	58, 75, 30, 240	59, 75, 27, 120	60, 72, 27, 120
0 1 1 2 2	0 1 1 2 2	0 1 1 2 1	0 1 1 2 1	0 1 1 2 1	0 1 2 2 2
0 0 1 1 2	0 0 1 1 2	0 0 1 2 1	0 0 1 1 1	0 0 1 1 1	0 0 1 2 1
0 0 0 1 2	0 0 0 1 2	0 0 0 1 1	0 0 0 1 0	0 0 0 1 0	0 0 0 1 1
0 0 0 0 2	0 0 0 0 1	0 0 0 0 0	0 0 1 0 0	0 1 1 0 0	0 0 0 0 0
0 0 0 0 0	0 0 0 0 0	0 0 1 2 0	0 1 1 2 0	1 1 2 1 0	0 0 1 1 0

61, 71, 25, 240	62, 70, 25, 240	63, 70, 22, 240	64, 69, 26, 240	65, 68, 28, 240	66, 68, 27, 240
0 1 2 2 1	0 1 2 2 1	0 1 2 2 2	0 1 2 2 2	0 1 1 1 1	0 1 2 1 1
0 0 2 1 1	0 0 1 1 1	0 0 2 1 1	0 0 1 2 1	0 0 1 1 1	0 0 2 1 1
0 0 0 1 0	0 0 0 1 0	0 0 0 1 0	0 0 0 1 1	0 0 0 1 0	0 0 0 1 0
0 1 1 0 0	0 0 1 0 1	0 1 1 0 0	0 0 0 0 0	0 0 1 0 1	0 1 1 0 0
1 1 1 1 0	0 1 1 1 0	1 1 2 1 0	0 1 1 1 0	1 1 1 1 0	1 1 1 1 0

67, 65, 30, 240	68, 64, 30, 240	69, 64, 29, 240	70, 64, 29, 240	71, 64, 29, 240	72, 64, 26, 240
0 1 1 2 1	0 1 1 2 2	0 1 1 1 1	0 1 1 2 1	0 1 1 2 1	0 1 2 2 2
0 0 1 2 1	0 0 1 1 1	0 0 1 1 1	0 0 1 1 1	0 0 1 1 1	0 0 1 2 1
0 0 0 1 0	0 0 0 1 1	0 0 0 1 1	0 0 0 1 0	0 0 0 1 1	0 0 0 1 0
0 0 1 0 1	0 0 0 0 0	0 0 1 0 0	0 0 1 0 1	0 0 0 0 1	0 0 1 0 0
0 1 1 1 0	0 1 1 1 0	1 1 1 2 0	0 1 1 1 0	0 1 1 1 0	0 1 1 1 0

73, 64, 26, 240	74, 63, 28, 120	75, 63, 28, 120	76, 63, 24, 120	77, 63, 24, 120	78, 62, 28, 240
0 1 2 2 1	0 1 1 1 1	0 1 2 2 1	0 1 1 2 1	0 1 2 1 1	0 1 1 2 1
0 0 1 1 1	0 0 1 1 1	0 0 1 2 1	0 0 1 1 0	0 0 1 1 1	0 0 1 2 1
0 0 0 1 0	0 0 0 1 0	0 0 0 1 1	0 1 0 1 0	0 0 0 0 1	0 0 0 1 0
0 0 1 0 0	0 1 1 0 0	0 0 0 0 0	0 1 1 0 1	0 1 1 0 1	0 0 1 0 0
0 1 1 2 0	1 1 1 2 0	0 1 1 1 0	1 1 1 1 0	0 1 1 1 0	1 1 1 1 0

79, 62, 25, 240	80, 61, 28, 240	81, 60, 32, 240	82, 58, 30, 240	83, 58, 23, 240	84, 57, 33, 120
0 1 2 2 1	0 1 1 2 1	0 1 1 2 2	0 1 1 1 1	0 1 1 2 1	0 1 1 2 2
0 0 1 2 1	0 0 1 1 1	0 0 1 1 1	0 0 1 1 1	0 0 1 1 0	0 0 1 1 2
0 0 0 1 1	0 0 0 1 0	0 0 0 1 1	0 0 0 1 0	0 1 0 1 1	0 0 0 1 1
0 0 1 0 0	0 0 1 0 0	0 0 0 0 0	0 0 1 0 1	0 1 1 0 0	0 0 0 0 1
0 1 1 1 0	1 1 2 1 0	0 0 1 1 0	0 1 1 1 0	1 1 1 1 0	0 0 0 0 0

85, 54, 28, 240	86, 54, 25, 120	87, 54, 25, 120	88, 53, 27, 240	89, 53, 24, 240	90, 51, 28, 20
0 1 2 2 2	0 1 2 2 1	0 1 2 1 2	0 1 2 1 1	0 1 2 2 1	0 1 1 1 1
0 0 1 2 1	0 0 1 1 1	0 0 1 1 1	0 0 2 1 0	0 0 1 2 1	0 0 1 1 1
0 0 0 1 1	0 0 0 0 0	0 0 0 0 0	0 0 0 1 0	0 0 0 1 1	0 0 0 0 0
0 0 0 0 0	0 1 1 0 1	0 1 1 0 1	0 1 1 0 0	0 0 1 0 0	0 1 1 0 1
0 0 1 2 0	0 1 2 1 0	0 1 1 1 0	1 2 2 1 0	0 1 1 2 0	0 1 1 1 0

91, 51, 27, 240	92, 51, 25, 240	93, 51, 25, 240	94, 48, 25, 80	95, 47, 31, 120	96, 47, 29, 120
0 1 2 2 2	0 1 2 1 1	0 1 2 2 2	0 2 2 2 2	0 1 1 2 2	0 1 2 2 3
0 0 1 2 2	0 0 2 1 0	0 0 1 2 1	0 0 2 2 2	0 0 1 1 1	0 0 1 2 2
0 0 0 2 1	0 0 0 1 0	0 0 0 1 0	0 0 0 2 1	0 0 0 1 1	0 0 0 1 2
0 0 0 0 0	0 1 2 0 0	0 0 1 0 0	0 0 1 0 2	0 0 0 0 0	0 0 0 0 1
0 1 1 1 0	1 2 2 1 0	0 1 1 2 0	0 0 2 1 0	0 1 1 2 0	0 0 0 0 0

97, 47, 27, 240	98, 47, 26, 240	99, 47, 26, 240	100, 47, 26, 240	101, 47, 26, 120	102, 47, 23, 120
0 1 2 3 2	0 1 1 2 1	0 1 2 2 2	0 1 2 2 1	0 1 2 2 2	0 1 1 1 1
0 0 2 2 1	0 0 1 1 1	0 0 1 2 1	0 0 1 1 1	0 0 1 1 1	0 0 1 1 0
0 0 0 1 1	0 0 0 1 0	0 0 0 1 0	0 0 0 1 1	0 0 0 1 0	0 1 0 1 0
0 0 0 0 0	0 0 1 0 0	0 1 1 0 0	0 0 1 0 0	0 0 1 0 0	0 1 1 0 1
0 1 1 1 0	1 1 1 2 0	0 1 2 1 0	0 1 1 2 0	0 1 1 1 0	1 1 1 1 0

103, 46, 25, 240	104, 45, 22, 240	105, 45, 22, 120	106, 44, 26, 240	107, 44, 20, 240	108, 44, 20, 120
0 2 2 2 2	0 1 1 2 1	0 1 2 2 1	0 1 1 2 1	0 2 2 2 2	0 2 3 2 1
0 0 1 1 1	0 0 1 1 0	0 0 1 1 1	0 0 1 1 1	0 0 2 1 0	0 0 1 2 0
0 0 0 1 0	0 1 0 1 1	0 0 0 1 0	0 0 0 1 0	0 1 0 2 0	0 1 0 1 1
0 0 1 0 0	0 1 1 0 0	0 1 1 0 0	0 1 1 0 0	0 2 1 0 0	0 2 1 0 0
0 1 1 2 0	1 1 1 2 0	1 1 2 1 0	1 1 1 2 0	2 2 2 3 0	2 2 2 2 0

109, 43, 25, 240	110, 43, 25, 240	111, 42, 26, 240	112, 42, 23, 120	113, 41, 27, 240	114, 40, 26, 240
0 1 2 2 1	0 1 2 2 1	0 2 2 2 1	0 1 2 1 1	0 1 1 2 1	0 1 3 2 2
0 0 1 1 0	0 0 1 1 1	0 0 1 2 1	0 0 1 1 1	0 0 1 2 1	0 0 2 1 1
0 1 0 2 1	0 0 0 1 0	0 0 0 2 0	0 0 0 1 0	0 0 0 1 0	0 0 0 1 0
0 1 0 0 0	0 0 1 0 1	0 0 1 0 0	0 1 1 0 1	0 1 1 0 0	0 1 1 0 0
1 1 1 2 0	1 1 1 1 0	1 1 1 3 0	1 1 1 1 0	1 1 2 1 0	1 1 1 1 0

115, 40, 26, 240
0 1 2 1 1
0 0 2 1 0
0 0 0 1 0
0 1 1 0 0
1 1 3 2 0

116, 40, 26, 240
0 1 2 2 1
0 0 1 2 1
0 0 0 2 1
0 0 1 0 0
0 1 1 1 0

117, 39, 26, 240
0 1 2 2 2
0 0 1 2 1
0 0 0 2 0
0 0 1 0 0
0 1 1 2 0

118, 39, 23, 240
0 1 1 2 1
0 0 1 1 1
0 1 0 2 0
0 1 1 0 0
1 1 1 2 0

119, 39, 23, 240
0 1 3 2 1
0 0 2 1 1
0 0 0 1 0
0 1 1 0 0
1 1 2 1 0

120, 39, 23, 240
0 1 2 2 2
0 0 1 2 1
0 0 0 1 0
0 1 1 0 0
1 1 2 1 0

121, 39, 23, 240
0 1 2 2 2
0 0 1 1 1
0 0 0 1 0
0 1 1 0 0
1 1 2 1 0

122, 38, 27, 240
0 1 2 2 2
0 0 1 1 1
0 0 0 1 0
0 0 1 0 0
0 1 1 2 0

123, 38, 27, 240
0 1 2 3 1
0 0 1 2 1
0 0 0 2 0
0 0 0 0 0
0 1 2 2 0

124, 38, 25, 120
0 1 2 2 2
0 0 1 2 1
0 0 0 1 1
0 0 0 0 0
0 1 1 2 0

125, 38, 24, 240
0 2 2 2 1
0 0 1 2 1
0 0 0 2 0
0 0 1 0 0
1 1 1 2 0

126, 38, 24, 240
0 1 1 1 1
0 0 1 1 0
0 1 0 1 1
0 1 1 0 0
1 1 1 2 0

127, 38, 24, 240
0 1 2 2 1
0 0 1 2 1
0 0 0 1 1
0 1 1 0 0
1 1 1 1 0

128, 37, 26, 240
0 1 2 2 1
0 0 1 2 1
0 0 0 1 1
0 0 0 0 1
0 1 1 2 0

129, 37, 24, 240
0 1 2 1 1
0 0 2 1 0
0 0 0 1 0
0 1 1 0 0
1 1 2 2 0

130, 37, 24, 240
0 1 2 1 2
0 0 2 1 1
0 0 0 1 0
0 1 1 0 1
1 1 1 1 0

131, 36, 25, 240
0 1 1 2 1
0 0 1 1 0
0 1 0 1 0
0 1 1 0 1
1 1 2 1 0

132, 36, 25, 240
0 1 2 1 1
0 0 1 1 1
0 0 0 1 0
0 1 1 0 1
0 1 2 1 0

133, 35, 27, 240
0 1 1 2 1
0 0 1 2 1
0 0 0 1 0
0 1 1 0 0
1 1 1 1 0

134, 35, 26, 240
0 1 2 3 3
0 0 2 2 2
0 0 0 2 1
0 0 0 0 0
0 1 1 1 0

135, 35, 25, 240
0 2 2 2 2
0 0 1 2 2
0 0 0 2 1
0 0 1 0 0
0 1 2 1 0

136, 35, 25, 240
0 1 2 2 2
0 0 1 1 2
0 0 0 0 1
0 1 1 0 2
0 1 1 0 0

137, 34, 27, 240
0 1 2 3 2
0 0 1 2 1
0 0 0 1 0
0 0 1 0 0
0 1 1 1 0

138, 34, 24, 240
0 1 1 2 1
0 0 1 1 0
0 1 0 2 1
0 1 1 0 0
1 1 1 1 0

139, 34, 24, 120
0 1 1 2 1
0 0 1 1 0
0 1 0 1 0
0 1 1 0 0
1 1 2 1 0

140, 32, 26, 240
0 1 3 1 1
0 0 2 1 0
0 0 0 1 0
0 1 2 0 0
1 2 2 1 0

141, 32, 26, 240
0 2 2 3 2
0 0 1 1 1
0 0 0 1 0
0 0 1 0 0
0 1 2 1 0

142, 32, 26, 240
0 1 2 2 2
0 0 1 2 2
0 0 0 1 1
0 0 0 0 0
0 1 1 2 0

143, 32, 26, 120
0 1 2 2 1
0 0 1 1 1
0 0 0 0 0
0 1 1 0 1
0 1 2 2 0

144, 32, 26, 120
0 1 3 3 2
0 0 2 3 1
0 0 0 1 1
0 0 0 0 0
0 1 1 2 0

145, 32, 25, 240
0 1 2 2 1
0 0 1 2 0
0 1 0 2 1
0 1 0 0 0
1 1 1 3 0

146, 32, 25, 240
0 1 2 3 2
0 0 2 2 2
0 0 0 1 0
0 1 1 0 0
0 1 2 1 0

147, 32, 25, 240
0 1 2 3 1
0 0 1 2 1
0 0 0 1 1
0 0 1 0 0
0 1 1 2 0

148, 32, 23, 240
0 1 2 2 1
0 0 1 1 1
0 0 0 1 0
0 1 1 0 1
0 1 2 1 0

149, 31, 26, 120
0 2 2 3 3
0 0 2 2 1
0 0 0 2 1
0 0 0 0 0
0 1 1 3 0

150, 31, 25, 240
0 1 2 2 2
0 0 1 2 1
0 0 0 1 0
0 0 1 0 0
1 1 1 1 0

151, 31, 25, 240
0 1 2 2 1
0 0 1 2 1
0 0 0 1 0
0 0 1 0 0
1 1 1 1 0

152, 31, 23, 240
0 1 3 2 2
0 0 3 2 1
0 0 0 2 0
0 1 1 0 0
2 2 2 2 0

153, 31, 23, 240
0 1 2 1 1
0 0 2 1 1
0 0 0 1 0
0 1 1 0 0
1 1 2 1 0

154, 31, 22, 240
0 1 2 2 1
0 0 1 1 1
0 0 0 1 0
0 1 1 0 1
1 1 1 1 0

155, 31, 22, 120
0 1 1 1 1
0 0 1 1 0
0 1 0 1 0
0 1 1 0 0
1 1 2 2 0

156, 30, 24, 240
0 2 2 2 1
0 0 2 2 1
0 0 0 2 0
0 0 1 0 1
1 1 3 2 0

157, 30, 24, 240
0 1 3 2 1
0 0 2 1 1
0 0 0 1 0
0 1 1 0 0
1 2 2 1 0

158, 30, 24, 120
0 1 2 2 2
0 0 1 2 1
0 0 0 1 0
0 1 2 0 0
1 2 2 1 0

159, 30, 24, 120
0 1 2 2 2
0 0 1 1 1
0 0 0 1 0
0 1 2 0 0
1 1 2 1 0

160, 29, 25, 40
0 2 2 2 2
0 0 2 2 1
0 0 0 0 0
0 1 2 0 2
0 2 2 1 0

161, 29, 24, 240
0 1 2 3 1
0 0 1 2 0
0 1 0 1 1
0 1 1 0 0
1 1 1 2 0

162, 29, 24, 240
0 1 2 2 1
0 0 1 1 0
0 1 0 2 1
0 1 1 0 0
1 1 1 1 0

163, 29, 24, 240
0 1 2 1 1
0 0 1 1 1
0 0 0 1 1
0 1 1 0 0
1 1 2 1 0

164, 29, 24, 120
0 1 2 1 1
0 0 1 1 1
0 0 0 1 0
0 1 1 0 1
1 1 1 2 0

165, 29, 24, 120
0 1 2 3 2
0 0 2 2 1
0 0 0 1 1
0 0 0 0 0
0 1 1 2 0

166, 29, 23, 240
0 1 1 2 1
0 0 1 1 1
0 1 0 1 0
1 1 1 0 0
1 1 2 1 0

167, 29, 23, 240
0 1 2 2 1
0 0 1 2 0
0 1 0 2 1
0 1 0 0 0
1 1 1 2 0

168, 29, 23, 240
0 1 2 1 1
0 0 2 1 0
0 0 0 1 0
0 1 1 0 0
1 2 2 2 0

169, 29, 23, 240
0 1 2 2 1
0 0 1 1 1
0 0 0 1 0
0 0 1 0 0
1 1 1 2 0

170, 29, 23, 240
0 1 2 2 2
0 0 2 2 1
0 0 0 1 0
0 1 1 0 0
0 1 2 1 0

171, 29, 21, 240
0 1 2 2 1
0 0 2 1 1
0 0 0 1 0
0 1 1 0 0
1 1 2 1 0

172, 29, 21, 120
0 2 2 2 1
0 0 2 1 1
0 0 0 1 0
0 1 1 0 0
1 1 2 1 0

173, 28, 24, 240
0 1 2 2 2
0 0 1 1 1
0 0 0 1 0
0 1 1 0 0
1 1 1 1 0

174, 28, 24, 240
0 1 2 1 2
0 0 1 1 2
0 0 0 1 1
0 1 1 0 2
0 1 1 0 0

175, 27, 23, 120
0 1 2 2 2
0 0 1 2 2
0 1 0 2 1
0 1 0 0 0
1 1 1 1 0

176, 27, 23, 120
0 1 2 2 2
0 0 2 2 1
0 0 0 0 0
0 1 1 0 1
0 1 2 1 0

177, 26, 24, 240
0 1 2 3 2
0 0 1 2 1
0 0 0 1 0
0 1 1 0 0
1 1 2 1 0

178, 26, 24, 240
0 1 2 2 1
0 0 1 2 1
0 0 0 1 0
0 1 1 0 1
0 1 2 1 0

179, 26, 24, 120
0 1 1 2 1
0 0 1 1 1
0 0 0 1 0
0 1 1 0 0
1 1 2 2 0

180, 26, 23, 240
0 2 2 2 1
0 0 1 2 1
0 0 0 1 1
0 1 2 0 0
1 1 2 1 0

181, 26, 23, 240
0 1 2 3 2
0 0 2 2 1
0 0 0 1 1
0 1 1 0 0
1 1 1 1 0

182, 26, 23, 240
0 2 2 2 2
0 0 1 2 1
0 0 0 1 1
0 0 1 0 0
0 1 2 1 0

183, 26, 23, 240
0 1 2 3 2
0 0 2 2 1
0 0 0 1 1
0 0 1 0 0
0 1 1 1 0

184, 26, 22, 240
0 1 2 2 1
0 0 1 1 0
0 1 0 1 0
0 1 1 0 1
1 1 1 1 0

185, 26, 22, 60
0 2 3 3 2
0 0 2 1 2
0 0 0 0 0
0 1 3 0 1
0 2 2 1 0

186, 26, 22, 40
0 2 4 2 2
0 0 2 2 1
0 0 0 0 0
0 1 2 0 2
0 2 2 1 0

187, 25, 23, 240
0 1 1 2 1
0 0 1 1 0
0 1 0 1 1
0 1 1 0 0
1 1 2 1 0

188, 25, 23, 240
0 1 3 2 2
0 0 2 1 2
0 0 0 1 0
0 1 1 0 1
1 1 1 1 0

189, 25, 23, 240
0 1 2 1 2
0 0 1 1 1
0 0 0 1 0
0 1 1 0 1
1 1 2 1 0

190, 25, 23, 240
0 1 2 2 1
0 0 1 2 1
0 0 0 1 0
0 1 1 0 1
1 1 1 1 0

191, 25, 22, 240
0 1 2 2 1
0 0 1 2 0
0 1 0 1 1
0 1 1 0 0
1 1 1 2 0

192, 24, 23, 240
0 1 3 2 1
0 0 3 1 1
0 0 0 1 0
0 1 2 0 0
1 2 2 1 0

193, 24, 23, 240
0 1 2 3 2
0 0 2 2 2
0 0 0 1 0
0 1 2 0 0
1 2 2 1 0

194, 24, 23, 240
0 1 2 2 2
0 0 2 1 1
0 0 0 1 0
0 1 2 0 0
1 1 3 1 0

195, 23, 22, 240
0 1 2 2 2
0 0 1 2 1
0 1 0 1 0
1 1 1 0 0
1 1 2 1 0

196, 23, 22, 240
0 1 1 1 2
0 0 1 0 1
0 1 0 1 1
1 1 1 0 2
1 1 1 0 0

197, 23, 22, 240
0 1 4 3 2
0 0 3 2 2
0 0 0 2 0
0 1 1 0 0
2 2 2 2 0

198, 23, 22, 240
0 1 2 2 2
0 0 1 2 1
0 0 0 2 0
0 1 1 0 0
2 2 3 2 0

199, 23, 22, 240
0 2 2 3 2
0 0 2 1 1
0 0 0 1 0
0 1 1 0 0
1 1 2 1 0

200, 23, 22, 240
0 1 2 2 1
0 0 2 1 1
0 0 0 1 0
0 1 1 0 0
1 2 2 1 0

201, 23, 22, 240
0 1 3 2 2
0 0 2 1 1
0 0 0 1 0
0 1 1 0 0
1 1 2 1 0

202, 23, 22, 240
0 1 2 2 1
0 0 1 2 1
0 0 0 1 0
0 1 1 0 0
1 1 2 1 0

203, 23, 22, 120
0 1 2 2 2
0 0 1 2 1
0 0 0 1 0
0 1 2 0 0
1 2 2 3 0

204, 22, 22, 240
0 2 3 2 1
0 0 1 1 1
0 1 0 2 0
1 1 1 0 0
1 1 2 2 0

205, 22, 21, 240
0 2 2 2 2
0 0 2 1 0
0 1 0 2 0
0 2 1 0 0
2 2 4 3 0

206, 22, 21, 240
0 1 2 2 1
0 0 1 1 0
0 1 0 1 1
0 1 1 0 0
1 1 1 2 0

207, 22, 21, 240
0 1 2 2 2
0 0 2 1 1
0 0 0 1 0
0 1 1 0 1
1 1 1 1 0

208, 21, 21, 240
0 1 3 2 1
0 0 2 1 1
0 0 0 1 0
0 1 2 0 0
1 2 2 1 0

209, 21, 21, 240
0 1 2 2 2
0 0 2 2 1
0 0 0 1 0
0 1 2 0 0
1 2 2 1 0

210, 21, 21, 240
0 1 2 2 1
0 0 2 1 1
0 0 0 1 0
0 1 2 0 0
1 2 2 1 0

211, 21, 21, 240
0 1 2 2 2
0 0 2 1 1
0 0 0 1 0
0 1 2 0 0
1 1 2 1 0

212, 21, 21, 120
0 2 3 2 1
0 0 2 1 1
0 0 0 1 0
0 1 1 0 0
1 1 2 1 0

213, 20, 20, 240
0 1 2 2 1
0 0 1 1 1
0 1 0 1 0
1 1 1 0 0
1 1 2 1 0

214, 20, 20, 240
0 2 2 2 2
0 0 2 1 1
0 0 0 1 0
0 1 1 0 0
1 1 2 1 0

215, 20, 20, 240
0 1 2 2 1
0 0 1 2 1
0 1 0 1 1
0 1 1 0 0
1 1 1 1 0

216, 19, 19, 240
0 2 2 3 1
0 0 2 1 0
0 1 0 1 1
1 1 1 0 0
1 3 2 2 0

217, 19, 19, 240
0 2 3 3 1
0 0 1 2 0
0 2 0 1 1
0 2 2 0 0
2 2 2 2 0

218, 19, 19, 240
0 2 3 2 1
0 0 2 2 0
0 1 0 2 1
0 2 1 0 0
2 2 2 2 0

219, 19, 19, 240
0 2 3 2 2
0 0 2 1 0
0 1 0 2 1
0 2 1 0 0
2 2 2 3 0

220, 19, 19, 240
0 2 3 3 1
0 0 1 2 0
0 1 0 1 1
0 2 2 0 0
2 2 2 2 0

221, 19, 19, 240
0 2 3 2 1
0 0 1 2 0
0 1 0 1 1
0 2 2 0 0
2 2 2 2 0

222, 19, 19, 240
0 2 3 3 1
0 0 1 2 0
0 1 0 1 1
0 2 1 0 0
2 2 2 2 0

223, 19, 19, 240
0 2 3 2 1
0 0 1 2 0
0 1 0 2 1
0 2 1 0 0
2 2 2 2 0

224, 19, 19, 240
0 2 2 3 2
0 0 2 1 0
0 1 0 2 0
0 2 1 0 0
2 2 3 2 0

225, 19, 19, 240
0 2 2 3 2
0 0 2 1 0
0 1 0 2 0
0 2 1 0 1
2 2 2 2 0

226, 19, 19, 240
0 2 2 2 2
0 0 2 1 0
0 1 0 2 1
0 2 1 0 0
2 2 2 3 0

227, 19, 19, 240
0 2 2 2 2
0 0 2 1 0
0 1 0 2 0
0 2 1 0 1
2 2 2 2 0

228, 19, 19, 240
0 2 2 2 2
0 0 2 1 0
0 1 0 2 0
0 2 1 0 0
1 2 3 3 0

229, 19, 19, 240
0 2 2 2 2
0 0 2 1 0
0 1 0 2 0
0 2 1 0 0
1 2 2 3 0

APPENDIX 2: FACETS OF $OMCUT_5$

See below the list of representatives of all 194 orbits of $OMCUT_5$, presented as 5×5 matrices. The numbers before each matrix are, respectively, the orbit number, the adjacency of a member of the orbit, the incidence of a member of the orbit, and the orbit size.

1, 10695, 307, 20	**2, 6451, 307, 60**	**3, 1590, 198, 60**	**4, 940, 111, 120**	**5, 444, 70, 60**	**6, 352, 111, 120**
0 0 0 0 0	0 0 0 0 0	0 −1 0 1 0	0 −1 0 1 1	0 −1 0 1 1	0 −1 0 0 1
0 0 0 0 0	−1 0 0 0 1	−1 0 0 1 0	−1 0 0 1 1	−1 0 0 1 1	−1 0 0 0 1
0 0 0 0 0	0 0 0 0 0	0 0 0 0 0	0 0 0 0 0	−1 −1 0 1 1	−1 −1 0 1 1
0 0 0 0 1	0 0 0 0 0	0 0 0 0 1	0 0 0 0 0	1 1 0 0 −1	1 1 0 0 −1
0 0 0 0 0	1 0 0 0 0	1 1 0 −1 0	1 1 0 −1 0	1 1 0 −1 0	1 1 0 0 0

7, 269, 111, 60	**8, 166, 56, 30**	**9, 162, 125, 120**	**10, 160, 112, 240**	**11, 154, 39, 40**	**12, 129, 61, 240**
0 0 0 0 0	0 −1 1 1 0	0 −1 0 1 0	0 −1 1 0 0	0 −1 −1 1 2	0 0 0 1 0
−1 0 0 1 1	−1 0 1 1 0	−1 0 1 0 0	−1 0 0 1 0	−1 0 −1 1 2	−1 0 0 1 1
−1 −1 0 1 1	0 0 0 −1 1	0 0 0 1 0	−1 0 0 1 1	−1 −1 0 1 2	−1 0 0 1 0
1 1 0 0 −1	0 0 −1 0 1	0 1 −1 0 1	1 0 0 0 0	1 1 1 0 −2	1 0 0 0 0
1 1 0 −1 0	1 1 0 0 0	1 0 1 −1 0	1 1 0 −1 0	1 1 1 −1 0	1 0 1 −1 0

13, 110, 63, 240	**14, 107, 39, 20**	**15, 99, 81, 120**	**16, 98, 88, 60**	**17, 94, 55, 240**	**18, 92, 61, 240**
0 −1 0 1 1	0 −1 −1 2 2	0 −2 0 1 1	0 −1 1 0 0	0 −1 0 1 0	0 0 −1 1 1
−1 0 0 1 0	−1 0 −1 2 2	−1 0 1 0 0	−1 0 1 0 0	−1 0 1 1 0	−1 0 1 0 1
−1 0 0 1 1	−1 −1 0 2 2	−1 −2 0 2 2	−1 −1 0 2 2	0 0 0 0 0	0 0 0 1 0
1 0 0 0 0	1 1 1 0 −2	1 2 0 0 −1	1 1 0 0 −1	0 1 −1 0 1	0 0 0 0 0
1 1 0 −1 0	1 1 1 −2 0	1 2 0 −1 0	1 1 0 −1 0	1 0 1 0 0	1 0 1 −1 0

19, 92, 59, 240	**20, 90, 43, 120**	**21, 89, 85, 240**	**22, 88, 73, 120**	**23, 86, 43, 80**	**24, 83, 74, 240**
0 −1 1 1 0	0 −2 0 2 2	0 −1 0 1 0	0 1 −2 0 1	0 0 −1 1 1	0 0 1 0 0
−1 0 1 0 1	−1 0 0 1 1	−1 0 1 2 −1	−1 0 0 1 1	−1 0 0 1 1	−1 0 1 1 1
0 0 0 0 0	−1 0 0 1 1	−1 0 0 0 1	−2 1 0 −1 2	0 −1 0 1 1	−1 −1 0 1 2
0 1 −1 0 1	1 1 0 0 −1	1 1 −1 0 1	1 0 1 0 −1	0 0 0 0 0	1 1 0 0 −2
1 0 0 0 0	1 1 0 −1 0	1 1 0 −1 0	2 −1 1 1 0	1 1 1 −1 0	2 2 −1 −2 0

25, 81, 41, 240	**26, 80, 51, 120**	**27, 79, 56, 120**	**28, 78, 51, 240**	**29, 77, 51, 240**	**30, 69, 42, 240**
0 0 0 1 1	0 −1 1 0 1	0 1 −1 0 0	0 0 0 1 0	0 0 1 1 0	0 0 1 0 0
−1 0 0 1 2	−1 0 1 0 1	−1 0 0 1 1	−1 0 1 0 0	−1 0 0 1 1	−1 0 0 0 1
−1 −1 0 1 2	−1 −1 0 2 1	−2 0 0 1 1	0 0 0 1 1	−1 0 0 1 1	−1 −1 0 2 1
1 1 0 0 −2	1 1 0 0 −1	2 0 1 0 −1	0 0 0 0 0	1 0 0 0 −1	1 1 0 0 −1
2 1 0 −2 0	1 1 0 −1 0	2 0 1 −1 0	1 1 0 −1 0	2 0 1 −2 0	1 1 0 0 0

31, 67, 55, 240	**32, 67, 38, 120**	**33, 66, 54, 10**	**34, 65, 44, 120**	**35, 65, 39, 240**	**36, 64, 51, 240**
0 −1 0 1 1	0 −1 0 0 1	0 −1 −1 1 1	0 −1 1 1 1	0 0 0 0 0	0 −2 0 2 1
−1 0 1 0 0	−1 0 0 0 1	−1 0 −1 1 1	−1 0 1 1 1	−1 0 1 1 0	−1 0 1 1 0
0 0 0 1 1	−1 −1 0 1 2	−1 −1 0 1 1	−1 −1 0 1 2	0 −1 0 1 2	0 0 0 0 0
0 0 0 0 0	1 1 1 0 −2	1 1 1 0 −1	1 1 0 0 −2	1 1 1 0 −2	0 1 −1 0 1
1 1 0 −1 0	2 2 1 −1 0	1 1 1 −1 0	2 2 −1 −2 0	1 2 1 −2 0	1 1 1 −1 0

37, 62, 39, 120	**38, 60, 51, 120**	**39, 60, 40, 120**	**40, 59, 73, 120**	**41, 59, 67, 120**	**42, 58, 47, 240**
0 −1 0 2 2	0 0 −1 1 0	0 −1 0 1 0	0 −1 2 1 0	0 −1 0 0 1	0 −1 0 1 1
−1 0 0 2 2	−1 0 1 1 0	−1 0 0 1 0	−1 0 2 1 0	−1 0 1 −1 1	−1 0 1 0 1
−1 −1 0 1 2	0 0 0 0 0	−1 −1 0 1 2	−1 −1 0 1 3	−1 0 0 1 1	−1 −1 0 1 1
1 1 0 0 −2	0 0 0 0 1	1 1 1 0 −1	1 1 0 0 −2	1 0 1 0 −1	0 1 0 0 0
2 2 0 −3 0	1 0 1 0 0	2 2 1 −2 0	3 3 −2 −3 0	1 1 −1 1 0	2 1 0 −1 0

43, 57, 67, 120	**44, 57, 53, 120**	**45, 57, 43, 240**	**46, 56, 42, 240**	**47, 56, 37, 240**	**48, 53, 63, 40**
0 −1 1 0 1	0 −1 1 1 0	0 −1 −1 2 1	0 −1 0 0 1	0 0 1 −1 0	0 −1 −1 1 1
−1 0 1 0 1	−1 0 1 1 0	−1 0 0 1 1	−1 0 1 −1 2	−1 0 0 1 1	−1 0 −1 1 1
−1 −1 0 1 2	0 0 0 0 0	−1 −1 0 2 1	−1 0 0 1 0	0 −1 0 1 1	−1 −1 0 1 1
1 1 0 0 −2	0 0 0 0 1	0 1 0 0 0	1 0 1 0 0	0 1 1 0 −1	0 0 0 0 1
2 2 −1 −1 0	1 1 −1 0 0	2 1 1 −2 0	1 1 −1 1 0	1 0 0 1 0	2 2 2 −2 0

49, 53, 45, 240	**50, 52, 40, 240**	**51, 50, 44, 120**	**52, 50, 34, 240**	**53, 50, 27, 120**	**54, 49, 63, 40**
0 0 1 0 0	0 −2 0 2 2	0 0 1 0 0	0 −1 0 1 1	0 −2 0 1 1	0 0 −1 1 0
−1 0 0 0 1	−1 0 1 1 2	−1 0 1 0 1	−1 0 −1 2 1	−1 0 1 0 0	−1 0 0 1 0
−1 −1 0 2 1	−1 0 0 1 0	0 −1 0 1 1	−1 0 0 2 1	−1 0 0 1 1	0 −1 0 1 0
1 1 0 0 −1	1 1 0 0 −1	0 1 0 0 −1	1 0 1 0 −1	2 1 1 0 −1	0 0 0 0 2
1 1 1 −1 0	1 2 −1 −1 0	1 2 −1 0 0	1 1 1 −2 0	2 1 1 −1 0	1 1 1 −1 0

55, 49, 51, 120	**56, 49, 47, 120**	**57, 48, 47, 240**	**58, 48, 43, 120**	**59, 47, 32, 240**	**60, 46, 61, 60**
0 −1 0 1 1	0 −1 1 1 1	0 −2 1 2 0	0 −1 1 0 1	0 −2 1 1 1	0 −1 1 −1 1
−1 0 1 0 0	−1 0 1 0 1	−1 0 1 1 0	−1 0 1 0 1	−1 0 1 0 0	−1 0 1 −1 1
−1 −1 0 2 2	−1 0 0 1 1	0 1 0 0 0	−1 −1 0 1 1	−1 −1 0 1 2	−1 −1 0 2 1
1 1 0 0 −1	1 0 0 0 −1	0 1 −1 0 1	1 1 −1 0 0	1 1 0 0 −1	1 1 −1 0 0
1 1 0 −1 0	2 1 −1 −1 0	1 0 1 −1 0	1 1 0 0 0	1 2 0 0 0	1 1 0 1 0

61, 46, 30, 240	**62, 45, 49, 240**	**63, 44, 38, 240**	**64, 43, 43, 120**	**65, 42, 44, 240**	**66, 42, 38, 240**
0 2 −2 0 1	0 −1 1 0 0	0 −2 1 1 1	0 −2 0 1 1	0 −2 0 1 2	0 0 1 0 0
−1 0 0 1 1	−1 0 0 0 1	−1 0 1 0 0	−1 0 0 1 1	−1 0 1 0 1	−1 0 0 0 1
−2 1 0 0 1	−1 1 0 1 1	0 0 0 1 0	−1 −2 0 2 2	−2 −2 0 2 2	0 −1 0 1 1
1 0 1 0 0	1 1 0 0 −1	0 1 0 0 1	1 2 0 0 −1	1 2 0 0 −1	0 1 1 0 −1
2 −1 1 0 0	2 0 1 −1 0	1 1 0 −1 0	1 2 0 −1 0	2 2 0 −1 0	1 1 −1 1 0

67, 42, 36, 240	**68, 42, 36, 240**	**69, 41, 41, 240**	**70, 41, 40, 240**	**71, 41, 40, 240**	**72, 41, 39, 240**
0 −1 0 1 1	0 −1 1 0 0	0 −2 1 1 1	0 −1 0 1 1	0 0 1 2 0	0 0 0 0 0
−1 0 1 0 1	−1 0 0 0 1	−1 0 0 1 1	−1 0 1 2 0	−1 0 0 2 2	−1 0 1 1 1
−1 0 0 1 1	−1 0 0 1 2	0 0 0 0 1	−1 0 0 0 1	−1 1 0 1 0	−1 −1 0 1 2
1 0 1 0 −1	1 1 1 0 −2	0 1 1 0 −1	1 1 −1 0 0	1 0 0 0 −1	1 1 1 0 −2
1 1 −1 0 0	2 1 1 −1 0	1 1 −1 0 0	1 1 0 −1 0	2 0 2 −3 0	2 2 0 −2 0

73, 41, 37, 240	**74, 41, 35, 240**	**75, 40, 44, 240**	**76, 40, 41, 120**	**77, 40, 29, 240**	**78, 39, 38, 240**
0 0 1 −1 1	0 −2 0 2 2	0 −2 2 0 1	0 −1 2 1 0	0 1 2 0 0	0 −2 0 2 2
−1 0 0 1 1	−1 0 −1 2 2	−1 0 1 0 0	−1 0 2 1 0	−1 0 1 0 1	−1 0 0 1 0
−1 −1 0 1 1	−1 −1 0 2 1	0 0 0 1 0	−1 −1 0 1 3	0 −1 0 2 1	−1 1 0 1 2
1 0 0 0 0	1 1 0 0 −1	0 1 0 0 1	1 1 0 0 −2	1 1 0 0 −2	1 0 1 0 −1
1 1 0 0 0	1 2 1 −2 0	1 1 −1 0 0	2 2 −1 −2 0	1 2 −1 0 0	1 1 0 −1 0

79, 39, 37, 120	**80, 39, 36, 240**	**81, 39, 36, 240**	**82, 39, 32, 240**	**83, 39, 31, 240**	**84, 38, 38, 120**
0 −1 1 0 0	0 0 1 1 0	0 0 1 0 0	0 0 1 0 0	0 −1 −1 2 2	0 −2 0 2 2
−1 0 1 0 0	−1 0 1 1 1	−1 0 1 0 0	−1 0 1 1 1	−1 0 0 2 2	−1 0 1 0 0
−1 −1 0 2 3	−1 −1 0 1 2	0 −1 0 1 2	−1 −1 0 1 2	−1 −1 0 1 2	−1 0 0 2 2
1 1 1 0 −2	1 1 0 0 −2	0 1 0 0 0	1 1 0 0 −2	1 1 1 0 −2	1 1 0 0 −1
2 2 1 −2 0	2 1 0 −2 0	1 1 0 −1 0	2 1 0 −1 0	1 2 0 −2 0	1 1 0 −1 0

85, 38, 30, 240	**86, 37, 45, 40**	**87, 35, 39, 240**	**88, 35, 39, 120**	**89, 35, 36, 240**	**90, 35, 33, 240**
0 0 2 −1 1	0 −1 −1 2 1	0 −1 0 1 1	0 −1 0 1 2	0 −1 0 1 1	0 1 −2 1 1
−1 0 1 1 1	−1 0 −1 2 1	−1 0 1 1 0	−1 0 0 1 2	−1 0 1 1 1	−1 0 0 1 1
0 1 0 1 −2	−1 −1 0 2 1	0 0 0 0 1	−2 −2 0 2 3	−2 −2 0 2 2	−2 1 0 0 2
1 −1 2 0 1	0 0 0 0 1	0 0 −1 0 1	2 2 0 0 −3	1 2 0 0 −1	1 0 1 0 −1
1 1 −3 1 0	2 2 2 −3 0	1 2 0 −1 0	2 2 0 −2 0	2 2 0 −2 0	2 −1 1 0 0

91, 34, 36, 240	**92, 34, 33, 240**	**93, 34, 32, 240**	**94, 33, 34, 240**	**95, 32, 36, 240**	**96, 32, 33, 120**
0 −1 1 1 0	0 0 1 1 0	0 −1 0 2 2	0 −2 1 2 1	0 0 1 1 0	0 0 0 1 0
−1 0 1 1 0	−1 0 0 1 2	−1 0 1 2 1	−1 0 1 1 1	−1 0 −2 1 3	−1 0 −1 1 1
−1 0 0 1 2	−1 −1 0 1 2	−1 −1 0 1 2	0 1 0 0 −1	−1 −2 0 1 2	−1 −1 0 1 1
2 2 0 0 −2	1 1 0 0 −2	1 1 0 0 −2	0 1 −1 0 1	1 2 1 0 −2	0 1 1 0 0
3 2 0 −3 0	2 1 0 −2 0	2 2 0 −3 0	1 1 0 −1 0	2 0 0 0 0	2 2 2 −2 0

97, 32, 31, 240	**98, 32, 27, 240**	**99, 31, 41, 240**	**100, 31, 41, 240**	**101, 31, 37, 240**	**102, 31, 36, 120**
0 −1 1 2 2	0 0 1 1 1	0 −1 −1 2 1	0 0 1 1 0	0 0 0 0 1	0 −1 −1 1 1
−1 0 1 2 1	−1 0 1 2 1	−1 0 1 0 1	−1 0 1 1 0	−1 0 1 0 1	−1 0 −1 1 1
−1 0 0 1 1	−1 −1 0 2 2	−1 −1 0 2 1	−1 0 0 1 2	−1 0 0 0 1	−1 −1 0 1 1
2 1 −1 0 −2	2 1 0 0 −3	0 1 0 0 0	1 0 0 0 −1	1 −1 1 0 0	0 1 1 0 0
2 1 0 −3 0	2 2 −1 −3 0	2 1 1 −2 0	2 2 −1 −2 0	1 1 −1 1 0	2 2 2 −2 0

103, 31, 35, 240	**104, 31, 31, 240**	**105, 31, 31, 120**	**106, 31, 30, 240**	**107, 31, 30, 240**	**108, 31, 27, 240**
0 −1 1 0 1	0 −2 1 0 1	0 −1 −2 2 3	0 −1 −2 3 2	0 −2 0 1 1	0 1 1 0 −1
−1 0 0 1 1	−1 0 1 0 1	−1 0 −2 2 3	−1 0 −2 3 2	−1 0 2 −1 2	−1 0 −2 1 2
−1 0 0 1 1	−1 0 0 1 1	−1 −1 0 3 3	−1 −2 0 3 3	−1 0 0 1 0	−1 −2 0 2 1
1 0 1 0 −1	1 1 0 0 −1	1 1 2 0 −3	1 1 2 0 −2	1 0 2 0 1	1 1 1 0 0
1 1 0 −1 0	2 1 −1 1 0	1 1 3 −3 0	1 2 2 −3 0	1 2 −2 1 0	1 1 2 −1 0

109, 30, 41, 120	**110, 30, 37, 120**	**111, 30, 36, 240**	**112, 30, 33, 120**	**113, 30, 30, 120**	**114, 30, 29, 120**
0 −1 0 0 1	0 1 1 −1 0	0 0 2 0 0	0 0 1 −1 1	0 −1 1 1 0	0 −1 −1 2 1
−1 0 2 −2 2	−1 0 −1 2 1	−1 0 1 1 0	−1 0 0 1 1	−1 0 1 1 0	−1 0 −1 2 2
−2 0 0 2 1	−1 −1 0 2 1	−1 −1 0 1 3	0 −1 0 1 0	−1 −1 0 1 3	−1 −1 0 2 2
2 0 1 0 −1	1 1 1 0 −1	1 1 1 0 −2	0 1 1 0 0	1 1 1 0 −2	0 1 1 0 −1
2 1 −1 1 0	1 1 1 −2 0	2 2 −1 −2 0	1 1 −1 0 0	3 3 0 −3 0	2 1 1 −2 0

115, 30, 28, 240	**116, 29, 32, 240**	**117, 29, 31, 40**	**118, 29, 30, 240**	**119, 29, 28, 240**	**120, 28, 41, 240**
0 0 −2 2 0	0 −2 2 1 1	0 −1 −1 1 2	0 1 1 1 0	0 1 −2 0 1	0 −2 1 2 0
−1 0 0 1 1	−1 0 1 0 0	−1 0 −1 1 2	−1 0 1 1 1	−1 0 0 1 1	−1 0 1 0 1
−2 1 0 1 0	−1 0 0 1 1	−1 −1 0 1 2	−1 0 0 2 1	−2 1 0 0 1	0 1 0 1 −1
2 1 1 0 0	1 1 0 0 0	0 0 0 0 0	2 1 −1 0 −2	2 0 1 0 −1	0 1 −1 0 2
3 1 1 −2 0	2 1 −1 −1 0	2 2 2 −2 0	2 1 0 −3 0	2 0 1 0 0	1 0 1 −1 0

121, 28, 36, 240	**122, 28, 33, 240**	**123, 28, 33, 240**	**124, 28, 30, 240**	**125, 28, 28, 240**	**126, 28, 27, 240**
0 −1 1 0 1	0 −1 0 0 1	0 −2 0 2 1	0 0 1 2 0	0 −1 2 2 1	0 −2 0 1 1
−1 0 −1 2 1	−1 0 1 0 1	−1 0 1 1 0	−1 0 0 1 2	−1 0 2 2 1	−1 0 1 0 1
−1 0 0 2 1	−1 −1 0 1 2	−1 −2 0 2 2	−1 1 0 1 0	−1 0 0 1 1	−1 −1 0 1 1
1 0 1 0 −1	1 1 1 0 −2	1 2 0 0 −1	1 −1 0 0 0	2 2 −2 0 −2	1 1 1 0 0
1 1 1 −2 0	2 2 0 −1 0	1 2 1 −2 0	2 0 2 −2 0	3 2 −2 −3 0	1 2 −1 0 0

127, 27, 37, 120	**128, 27, 34, 240**	**129, 27, 33, 120**	**130, 27, 32, 120**	**131, 27, 31, 240**	**132, 27, 29, 240**
0 0 0 1 0	0 −1 −2 2 1	0 −2 −2 3 2	0 −1 0 1 1	0 1 1 −1 1	0 −1 0 2 1
−1 0 −1 1 1	−1 0 −2 2 2	−1 0 −1 2 1	−1 0 0 1 1	−1 0 1 2 1	−1 0 1 2 1
−1 −1 0 1 1	−1 0 0 1 0	−1 −1 0 2 1	−1 −1 0 1 1	0 2 0 0 −2	−1 0 0 1 0
0 0 0 0 1	1 1 2 0 −1	0 1 1 0 0	0 0 0 0 1	1 −2 3 0 1	1 1 −1 0 0
2 2 2 −2 0	2 2 3 −3 0	2 2 2 −3 0	2 2 1 −2 0	1 1 −3 1 0	1 1 1 −2 0

133, 27, 29, 240
$$\begin{matrix} 0 & -2 & 2 & 0 & 1 \\ -1 & 0 & 1 & 0 & 0 \\ -1 & 0 & 0 & 2 & 1 \\ 1 & 1 & 0 & 0 & 0 \\ 1 & 1 & 0 & -1 & 0 \end{matrix}$$

134, 27, 29, 240
$$\begin{matrix} 0 & -2 & 1 & 0 & 1 \\ -1 & 0 & 2 & -1 & 2 \\ -1 & 1 & 0 & 1 & 0 \\ 1 & -1 & 2 & 0 & 1 \\ 1 & 2 & -2 & 1 & 0 \end{matrix}$$

135, 26, 38, 240
$$\begin{matrix} 0 & 1 & -1 & 1 & 0 \\ -1 & 0 & 0 & 1 & 1 \\ -2 & 0 & 0 & 2 & 1 \\ 1 & 0 & 0 & 0 & 0 \\ 2 & 1 & 1 & -2 & 0 \end{matrix}$$

136, 26, 36, 120
$$\begin{matrix} 0 & -1 & 2 & 2 & 0 \\ -1 & 0 & 2 & 2 & 0 \\ -1 & -1 & 0 & 1 & 3 \\ 1 & 1 & 0 & 0 & -2 \\ 2 & 2 & 0 & -3 & 0 \end{matrix}$$

137, 26, 36, 120
$$\begin{matrix} 0 & 2 & 2 & 0 & -2 \\ -1 & 0 & -1 & 1 & 3 \\ -1 & -1 & 0 & 1 & 3 \\ 1 & 1 & 1 & 0 & -3 \\ 2 & 1 & 1 & -2 & 0 \end{matrix}$$

138, 26, 35, 240
$$\begin{matrix} 0 & -1 & 0 & 1 & 0 \\ -1 & 0 & 1 & 1 & 0 \\ -1 & -1 & 0 & 1 & 2 \\ 1 & 1 & 1 & 0 & -1 \\ 2 & 2 & 0 & -2 & 0 \end{matrix}$$

139, 26, 33, 120
$$\begin{matrix} 0 & -1 & 1 & 1 & 1 \\ -1 & 0 & 1 & 1 & 1 \\ -2 & -2 & 0 & 2 & 3 \\ 2 & 2 & 0 & 0 & -3 \\ 2 & 2 & 0 & -2 & 0 \end{matrix}$$

140, 26, 32, 240
$$\begin{matrix} 0 & 0 & 2 & 1 & -1 \\ -1 & 0 & 1 & 2 & 0 \\ 0 & 0 & 0 & -2 & 2 \\ 1 & 0 & -2 & 0 & 1 \\ 1 & 1 & 0 & 0 & 0 \end{matrix}$$

141, 26, 32, 240
$$\begin{matrix} 0 & 0 & -1 & 1 & 0 \\ -1 & 0 & 0 & 1 & 1 \\ 0 & -1 & 0 & 1 & 1 \\ 0 & 0 & 0 & 0 & 1 \\ 1 & 1 & 1 & -1 & 0 \end{matrix}$$

142, 26, 31, 240
$$\begin{matrix} 0 & 0 & 2 & 0 & 0 \\ -1 & 0 & 1 & 1 & 0 \\ -1 & -1 & 0 & 1 & 3 \\ 1 & 1 & 0 & 0 & -1 \\ 2 & 2 & -1 & -2 & 0 \end{matrix}$$

143, 26, 30, 240
$$\begin{matrix} 0 & -1 & -2 & 1 & 2 \\ -1 & 0 & -2 & 2 & 2 \\ -1 & 0 & 0 & 0 & 1 \\ 1 & 1 & 2 & 0 & -2 \\ 2 & 2 & 3 & -2 & 0 \end{matrix}$$

144, 26, 29, 240
$$\begin{matrix} 0 & 0 & 2 & -2 & 2 \\ -1 & 0 & 0 & 1 & 1 \\ 0 & -1 & 0 & 1 & 0 \\ 0 & 0 & 1 & 0 & 1 \\ 1 & 1 & -1 & 1 & 0 \end{matrix}$$

145, 26, 29, 240
$$\begin{matrix} 0 & 0 & 1 & 0 & 0 \\ -1 & 0 & 1 & -1 & 1 \\ -1 & -1 & 0 & 2 & 1 \\ 1 & 1 & -1 & 0 & 0 \\ 1 & 1 & 0 & 1 & 0 \end{matrix}$$

146, 26, 29, 240
$$\begin{matrix} 0 & 1 & 0 & -1 & 2 \\ -1 & 0 & 0 & 0 & 1 \\ -2 & 1 & 0 & 1 & 2 \\ 1 & 1 & -1 & 0 & 0 \\ 2 & -1 & 1 & 1 & 0 \end{matrix}$$

147, 26, 29, 120
$$\begin{matrix} 0 & -1 & 1 & 2 & 1 \\ -1 & 0 & 1 & 2 & 1 \\ -1 & -1 & 0 & 1 & 2 \\ 1 & 1 & 0 & 0 & -2 \\ 2 & 2 & 0 & -3 & 0 \end{matrix}$$

148, 26, 29, 120
$$\begin{matrix} 0 & -1 & 0 & 1 & 1 \\ -1 & 0 & 0 & 1 & 1 \\ -2 & -2 & 0 & 2 & 2 \\ 1 & 1 & 0 & 0 & 0 \\ 2 & 2 & 1 & -2 & 0 \end{matrix}$$

149, 26, 29, 120
$$\begin{matrix} 0 & 1 & 2 & -1 & -1 \\ -1 & 0 & -3 & 2 & 2 \\ -1 & -2 & 0 & 2 & 2 \\ 1 & 1 & 2 & 0 & -1 \\ 1 & 1 & 2 & -1 & 0 \end{matrix}$$

150, 26, 28, 240
$$\begin{matrix} 0 & -3 & 1 & 2 & 0 \\ -1 & 0 & 1 & 1 & -1 \\ -1 & 2 & 0 & 1 & 1 \\ 0 & 1 & -1 & 0 & 2 \\ 2 & 2 & 0 & 0 & 0 \end{matrix}$$

151, 26, 27, 120
$$\begin{matrix} 0 & -1 & 2 & 1 & 1 \\ -1 & 0 & 2 & 1 & 1 \\ -2 & -2 & 0 & 2 & 4 \\ 2 & 2 & 0 & 0 & -4 \\ 3 & 3 & -1 & -3 & 0 \end{matrix}$$

152, 25, 39, 120
$$\begin{matrix} 0 & -1 & 1 & 0 & 0 \\ -1 & 0 & 1 & -1 & 2 \\ -1 & 0 & 0 & 1 & 1 \\ 1 & 1 & 1 & 0 & -1 \\ 2 & 0 & -1 & 1 & 0 \end{matrix}$$

153, 25, 34, 240
$$\begin{matrix} 0 & -1 & 1 & 1 & 0 \\ -1 & 0 & 1 & 0 & 0 \\ -1 & -2 & 0 & 3 & 2 \\ 1 & 2 & 0 & 0 & -1 \\ 1 & 2 & 1 & -2 & 0 \end{matrix}$$

154, 25, 33, 240
$$\begin{matrix} 0 & 1 & 2 & 0 & 0 \\ -1 & 0 & 1 & 1 & 0 \\ -1 & -2 & 0 & 2 & 4 \\ 1 & 2 & 2 & 0 & -3 \\ 2 & 2 & 0 & -3 & 0 \end{matrix}$$

155, 25, 33, 240
$$\begin{matrix} 0 & 1 & 1 & 0 & 0 \\ -1 & 0 & 1 & 0 & 1 \\ 0 & -1 & 0 & 2 & 1 \\ 1 & 1 & 0 & 0 & -2 \\ 1 & 1 & 1 & -2 & 0 \end{matrix}$$

156, 25, 31, 120
$$\begin{matrix} 0 & -2 & 0 & 1 & 1 \\ -1 & 0 & 1 & 0 & 0 \\ -1 & 0 & 0 & 1 & 1 \\ 2 & 2 & 0 & 0 & -1 \\ 2 & 2 & 0 & -1 & 0 \end{matrix}$$

157, 25, 29, 240
$$\begin{matrix} 0 & -2 & 0 & 2 & 1 \\ -1 & 0 & 1 & 0 & 0 \\ -1 & -1 & 0 & 2 & 1 \\ 0 & 2 & 0 & 0 & 1 \\ 2 & 1 & 1 & -2 & 0 \end{matrix}$$

158, 25, 28, 240
$$\begin{matrix} 0 & -2 & 1 & 1 & 1 \\ -1 & 0 & 1 & -1 & 1 \\ -1 & 0 & 0 & 1 & 1 \\ 1 & 1 & 0 & 0 & -1 \\ 1 & 1 & 0 & 1 & 0 \end{matrix}$$

159, 25, 28, 120
$$\begin{matrix} 0 & -2 & -2 & 2 & 3 \\ -1 & 0 & -1 & 1 & 2 \\ -1 & -1 & 0 & 1 & 2 \\ 0 & 1 & 1 & 0 & -1 \\ 2 & 2 & 2 & -2 & 0 \end{matrix}$$

160, 24, 35, 120
$$\begin{matrix} 0 & -1 & -1 & 2 & 0 \\ -1 & 0 & 1 & 0 & 1 \\ -1 & -1 & 0 & 1 & 1 \\ 0 & 1 & 1 & 0 & 1 \\ 2 & 1 & 0 & -1 & 0 \end{matrix}$$

161, 24, 34, 240
$$\begin{matrix} 0 & -1 & 1 & 1 & 0 \\ -1 & 0 & 2 & 1 & 0 \\ -1 & -1 & 0 & 1 & 3 \\ 1 & 1 & 1 & 0 & -2 \\ 3 & 3 & -1 & -3 & 0 \end{matrix}$$

162, 24, 31, 240
$$\begin{matrix} 0 & 0 & -1 & 1 & 1 \\ -1 & 0 & 1 & 1 & 2 \\ -2 & -2 & 0 & 2 & 3 \\ 1 & 2 & 0 & 0 & -2 \\ 2 & 2 & 0 & -2 & 0 \end{matrix}$$

163, 24, 29, 240
$$\begin{matrix} 0 & 1 & 2 & 0 & 0 \\ -1 & 0 & 1 & 1 & 0 \\ -1 & -2 & 0 & 2 & 4 \\ 1 & 2 & 0 & 0 & -1 \\ 2 & 2 & 0 & -3 & 0 \end{matrix}$$

164, 24, 29, 240
$$\begin{matrix} 0 & 1 & 1 & 0 & 0 \\ -1 & 0 & 0 & 1 & 1 \\ 0 & 1 & 0 & 1 & -2 \\ 1 & -1 & 1 & 0 & 1 \\ 1 & 1 & -2 & 1 & 0 \end{matrix}$$

165, 24, 29, 120
$$\begin{matrix} 0 & -1 & -1 & 2 & 0 \\ -1 & 0 & 1 & 1 & 0 \\ -1 & -1 & 0 & 1 & 1 \\ 0 & 1 & 0 & 0 & 1 \\ 2 & 1 & 1 & -1 & 0 \end{matrix}$$

166, 24, 28, 240
$$\begin{matrix} 0 & 1 & 1 & 2 & 0 \\ -1 & 0 & 2 & 1 & 0 \\ -1 & 0 & 0 & 3 & 2 \\ 2 & 1 & -1 & 0 & -2 \\ 2 & 2 & 0 & -4 & 0 \end{matrix}$$

167, 24, 28, 120
$$\begin{matrix} 0 & 1 & 3 & 0 & 0 \\ -2 & 0 & 1 & 0 & 2 \\ 0 & -2 & 0 & 3 & 1 \\ 1 & 2 & 0 & 0 & -2 \\ 2 & 3 & -2 & 1 & 0 \end{matrix}$$

168, 24, 25, 240
$$\begin{matrix} 0 & 2 & -3 & 1 & 1 \\ -1 & 0 & 0 & 1 & 1 \\ -3 & 2 & 0 & 0 & 3 \\ 1 & 0 & 2 & 0 & -1 \\ 3 & -2 & 1 & 1 & 0 \end{matrix}$$

169, 23, 32, 240
$$\begin{matrix} 0 & -1 & -3 & 2 & 2 \\ -1 & 0 & -3 & 3 & 3 \\ -1 & 1 & 0 & 0 & 0 \\ 1 & 1 & 3 & 0 & -2 \\ 2 & 2 & 4 & -3 & 0 \end{matrix}$$

170, 23, 29, 240
$$\begin{matrix} 0 & -2 & 0 & 1 & 1 \\ -1 & 0 & 1 & 0 & 0 \\ -1 & -2 & 0 & 2 & 3 \\ 1 & 2 & 1 & 0 & -2 \\ 2 & 3 & 1 & -2 & 0 \end{matrix}$$

171, 23, 29, 120
$$\begin{matrix} 0 & -1 & 1 & 1 & 0 \\ -1 & 0 & 1 & 1 & 0 \\ -2 & -2 & 0 & 3 & 5 \\ 2 & 2 & 2 & 0 & -4 \\ 4 & 4 & 2 & -5 & 0 \end{matrix}$$

172, 23, 27, 240
$$\begin{matrix} 0 & 1 & 1 & -2 & 2 \\ -1 & 0 & 1 & 0 & 1 \\ 0 & 0 & 0 & 2 & 1 \\ 0 & 2 & 0 & 0 & -1 \\ 1 & 0 & -1 & 1 & 0 \end{matrix}$$

173, 23, 26, 240
$$\begin{matrix} 0 & -2 & 1 & 1 & 0 \\ -1 & 0 & 1 & 1 & 0 \\ 0 & 0 & 0 & 1 & 1 \\ 0 & 1 & -1 & 0 & 1 \\ 2 & 1 & 0 & -1 & 0 \end{matrix}$$

174, 23, 26, 120
$$\begin{matrix} 0 & -1 & 1 & 1 & 0 \\ -1 & 0 & 1 & 1 & 0 \\ -1 & -1 & 0 & 2 & 3 \\ 2 & 2 & 1 & 0 & -3 \\ 3 & 3 & 1 & -4 & 0 \end{matrix}$$

175, 23, 26, 120
$$\begin{matrix} 0 & -1 & 2 & 2 & 0 \\ -1 & 0 & 2 & 2 & 0 \\ 0 & 0 & 0 & -2 & 2 \\ 1 & 1 & -3 & 0 & 1 \\ 1 & 1 & 0 & 0 & 0 \end{matrix}$$

176, 23, 25, 240
$$\begin{matrix} 0 & -2 & 0 & 1 & 1 \\ -1 & 0 & 1 & 0 & 0 \\ -1 & 0 & 0 & 1 & 1 \\ 2 & 1 & 1 & 0 & -1 \\ 2 & 2 & 0 & -1 & 0 \end{matrix}$$

177, 22, 30, 120
$$\begin{matrix} 0 & -1 & 1 & 1 & 2 \\ -1 & 0 & 1 & 1 & 2 \\ -2 & -2 & 0 & 2 & 4 \\ 2 & 2 & 0 & 0 & -4 \\ 3 & 3 & -1 & -3 & 0 \end{matrix}$$

178, 22, 29, 120
$$\begin{matrix} 0 & -1 & 2 & 2 & 1 \\ -1 & 0 & 2 & 2 & 1 \\ -2 & -2 & 0 & 3 & 4 \\ 2 & 2 & 0 & 0 & -4 \\ 4 & 4 & -2 & -5 & 0 \end{matrix}$$

179, 22, 28, 120
$$\begin{matrix} 0 & -2 & 3 & 1 & 1 \\ -1 & 0 & 3 & 2 & 2 \\ 0 & 0 & 0 & 1 & 1 \\ 2 & 3 & -3 & 0 & -2 \\ 2 & 3 & -3 & -2 & 0 \end{matrix}$$

180, 22, 28, 120
$$\begin{matrix} 0 & -1 & 3 & 1 & 0 \\ -1 & 0 & 3 & 1 & 0 \\ 0 & 0 & 0 & -2 & 2 \\ 1 & 1 & -3 & 0 & 1 \\ 2 & 2 & -2 & 0 & 0 \end{matrix}$$

181, 22, 26, 240
$$\begin{matrix} 0 & -1 & 1 & 1 & 1 \\ -1 & 0 & 2 & -1 & 2 \\ -1 & -1 & 0 & 1 & 1 \\ 0 & 1 & -1 & 0 & 1 \\ 2 & 1 & 0 & 0 & 0 \end{matrix}$$

182, 22, 25, 240
$$\begin{matrix} 0 & 1 & -2 & 1 & 0 \\ -1 & 0 & -2 & 2 & 3 \\ -2 & 2 & 0 & 1 & 0 \\ 1 & 0 & 2 & 0 & -1 \\ 2 & 0 & 2 & -1 & 0 \end{matrix}$$

183, 22, 25, 240
$$\begin{matrix} 0 & 1 & -2 & 0 & 2 \\ -1 & 0 & 0 & 1 & 0 \\ -2 & 1 & 0 & -1 & 3 \\ 1 & 1 & 1 & 0 & 0 \\ 2 & -2 & 1 & 2 & 0 \end{matrix}$$

184, 22, 25, 80
$$\begin{matrix} 0 & 0 & -1 & 1 & 2 \\ -1 & 0 & 0 & 1 & 2 \\ 0 & -1 & 0 & 1 & 2 \\ 1 & 1 & 1 & 0 & -3 \\ 1 & 1 & 1 & -2 & 0 \end{matrix}$$

185, 22, 25, 40
$$\begin{matrix} 0 & -1 & -1 & 3 & 3 \\ -1 & 0 & -1 & 3 & 3 \\ -1 & -1 & 0 & 3 & 3 \\ 1 & 1 & 1 & 0 & -3 \\ 2 & 2 & 2 & -5 & 0 \end{matrix}$$

186, 22, 23, 120
$$\begin{matrix} 0 & -3 & -1 & 2 & 2 \\ -1 & 0 & 1 & 1 & 0 \\ -3 & -3 & 0 & 3 & 3 \\ 2 & 3 & 0 & 0 & -1 \\ 2 & 3 & 1 & -1 & 0 \end{matrix}$$

187, 21, 25, 120
$$\begin{matrix} 0 & -4 & 0 & 3 & 3 \\ -2 & 0 & 1 & 1 & 1 \\ -4 & -4 & 0 & 4 & 4 \\ 3 & 4 & 0 & 0 & -2 \\ 3 & 4 & 0 & -2 & 0 \end{matrix}$$

188, 21, 24, 240
$$\begin{matrix} 0 & 1 & -2 & 1 & 1 \\ -1 & 0 & 0 & 1 & 0 \\ -2 & 2 & 0 & 0 & 3 \\ 1 & 0 & 2 & 0 & -1 \\ 2 & -2 & 1 & 1 & 0 \end{matrix}$$

189, 21, 23, 120
$$\begin{matrix} 0 & 2 & 2 & -2 & -1 \\ -1 & 0 & -1 & 3 & 2 \\ -1 & -1 & 0 & 3 & 2 \\ 2 & 1 & 1 & 0 & -3 \\ 2 & 2 & 2 & -4 & 0 \end{matrix}$$

190, 21, 23, 120
$$\begin{matrix} 0 & -1 & 1 & 2 & 2 \\ -1 & 0 & 1 & 2 & 2 \\ -1 & -1 & 0 & 2 & 2 \\ 2 & 2 & -1 & 0 & -3 \\ 2 & 2 & 0 & -4 & 0 \end{matrix}$$

191, 21, 22, 240
$$\begin{matrix} 0 & -1 & -2 & 3 & 3 \\ -1 & 0 & 1 & 2 & 2 \\ -2 & 0 & 0 & 1 & 1 \\ 1 & 1 & 0 & 0 & -1 \\ 2 & 1 & 1 & -2 & 0 \end{matrix}$$

192, 20, 21, 240
$$\begin{matrix} 0 & -3 & 1 & 4 & 3 \\ -1 & 0 & 2 & 4 & 2 \\ -1 & 1 & 0 & -2 & 2 \\ 1 & 3 & -3 & 0 & -1 \\ 2 & 2 & 0 & -2 & 0 \end{matrix}$$

193, 20, 21, 120
$$\begin{matrix} 0 & -2 & -2 & 2 & 2 \\ -1 & 0 & 0 & 1 & 1 \\ -3 & 1 & 0 & 1 & 1 \\ 4 & 3 & 1 & 0 & -2 \\ 4 & 3 & 1 & -2 & 0 \end{matrix}$$

194, 19, 19, 40
$$\begin{matrix} 0 & -1 & -1 & 2 & 3 \\ -1 & 0 & -1 & 2 & 3 \\ -1 & -1 & 0 & 2 & 3 \\ 1 & 1 & 1 & 0 & -3 \\ 2 & 2 & 2 & -4 & 0 \end{matrix}$$

Bibliography

[AACLMP9797] Aichholzer O., Aurenhammer F., Chen D.Z., Lee D.T., Mukhopadhyay A. and Papadopoulou E. *Voronoi Diagrams for Direction-sensitive Distances*, Proceedings of 13-th Annual ACM Symposium Computational Geometry, Nice, France, 1997.

[Aign79] Aigner M. *Combinatorial Theory*, Springer-Verlag, Berlin, 1979.

[AAADDS15] Alahmadi A., Alhazmi H., Ali S., Deza M.M., Dutour Sikirić M. and Sole P. *Hypercube Emulation of Interconnection Networks Topologies*, preprint at http://arxiv.org/abs/1507.02147 279, 2015.

[Alev99] Alevras D. *Small Min-cut Polyhedra*, Math. Op. Research, Nr. 24, 1999.

[Asso80] Assouad P. *Plongements Isomètriques dans L_1: Aspect Analytique*, in *Séminaire d'Initiation à l'Analyse* (G. Choquet, H. Fakhoury, J. Saint-Raymond eds.), Nr. 14, Université Paris VI, 1979–1980.

[Asso84] Assouad P. *Sur les Inègalités Valides dans L_1*, European Journal of Combinatorics, Nr. 5, 1984.

[AsDe80] Assouad P. and Deza M.M. *Espaces Métriques Plongeables dans un Hypercube: Aspects Combinatoires*, Annals of Discrete Mathematics, Nr. 8, 1980.

[AsDe82] Assouad P. and Deza M.M. *Metric Subspaces of L_1*, in *Publications Mathèmatiques d'Orsay*, Nr. 82-03, Université de Paris-Sud, Orsay, 1982.

[AvDe91] Avis D. and Deza M.M. *The Cut Cone, L_1-embeddability, Complexity and Multicommodity Flows*, Networks, Nr. 21, 1991.

[AvMe11] Avis D. and Meagher C. *On the Directed Cut Cone and Polytope*, Manuscript, Draft Nr. MAPR-D-11-00057, 2011.

[AvMu89] Avis D. and Mutt *All the Facets of the Six-point Hamming Cone*, European Journal of Combinatorics, Nr. 10(4), 1989.

[Bara93] Barahona F. *On Cuts and Mutchings in Planar Graphs and Polytope*, Math. Programming, Nr. 60, 1993.

[BaMa86] Barahona F. and Mahjoub A.R. *On the Cut Polytope*, Math. Programming, Nr. 36(2), 1986.

[Bara71] Baranovskii E.P. *Simplexes of L-subdivisions of Euclidean Spaces*, Mathematical Notes, Nr. 10, 1971.

[Bara99] Baranovskii E.P. *The Conditions for a Simplex of a* 6*-dimensional Lattice to be an* L*-simplex* (in Russian), Mathematica, Nr. 2, 1999.

[Bara00] Baranovskii E.P. and Kononenko P.G. *A Method of Deducing* L*-polyhedra for* n*- lattices*, Mathematical Notes, Nr. 68(6), 2000.

[Bass95] Bassalygo L.A. *Supports of a Code*, in *Applied Algebra, Algebraic Algorithms and Error-Correcting Codes*, Proceedings of 11-th International Symposium, Paris, 1995 (G. Cohen, M.T. Giusti eds.), Springer-Verlag, 1995.

[BhRo03] Bhutani K.R. and Rosenfeld A. *Geodesics in Fuzzy Graphs*, Electron Notes Discrete Mathematics, 15:51G54, 2003.

[Birk67] Birkhoff G. *Lattice Theory*, AMS, Providence, Rhode Island, 1967.

[Blum53] Blumenthal L.M. *Theory and Applications of Distance Geometry*, Oxford University Press, Oxford, 1953.

[Boxe97] Boxer L. *On Hausdirff-like Metirc for Fuzzy Sets*, Pattern Recognition Letters, Nr. 18, 1997.

[Bras02] Brass P. *On the Noexistence of Hausdorff-like Metrics for Fuzzy Sets*, Pattern Recognition Letters, Nr. 23, 2002.

[BDPRS14] Bremner D., Dutour Sikirić M., Pasechnik D.V., Rehn T. and Schürmann A. *Computing Symmetry Groups of Polyhedra*, LMS Journal Comput. Mathematics, Nr. 17(1), 2014.

[BDS09] Bremner D., Dutour Silirić M. and Schüurmann A. *Polyhedral Representation Conversion up to Symmetries*, in *Polyhedral computation*, Vol. 48 of CRM Proc. Lecture Notes, Amer. Math. Soc., Providence, RI, 2009.

[BDK66] Bretagnolle J., Dacunha-Castelle D. and Krivine J.-L. *Lois Stables et Espaces* L_p, Ann. Inst. H. Poincaré, Nr. 2, 1966.

[Borg86] Borgefors G. *Distance Transformations in Digital Images*, Comp. Vision, Graphic and Image Processing, Vol. 34, 1986.

[Brya85] Bryant V. *Metric Spaces: Iteration and Application*, Cambridge Univ. Press, 1985.

[CMM06] Charikar M., Makarychev M. and Makarychev Y. *Directed Metrics and Directed Graph Partitioning Problems*, Proc. of 17-th ACM-SIAM Symposium on Discrete Algorithms, 2006.

[CJTW93] Chartrand G., Johns G.L., Tian S. and Winters S.J. *Directed Distance in Digraphs: Centers and Medians*, Journal of Graph Theory, Nr. 17, 1993.

[ChFi98] Chepoi V. and Fichet B. *A Note on Circular Decomposable Metrics*, Geom. Dedicata, Vol. 69, 1998.

[CPK96] Cho Y.J., Park S.C. and Khan M.S. *Coincidence Theorems in* 2*-metric Spaces*, SEA Bull. Math., Nr. 20, 1996.

[ChRe96] Christof T. and Reinelt G. *Combinatorial Optimization and Small Polytopes*, Top (Spanish Statistical and Operations Research Society), Nr. 4, 1996.

[ChRe01] Christof T. and Reinelt G. *Decomposition and Parallelization Techniques for Enumerating the Facets of Combinatorial Polytopes*, International Journal Comput. Geom. Appl., Nr. 11(4), 2001.

[DaCh88] Das P.P. and Chatterji B.N. *Knight's Distance in Digital Geometry*, Pattern Recognition Letters, Vol. 7, 1988.

[DaMu90] Das P.P. and Mukherjee J. *Metricity of Super-knight's Distance in Digital Geometry*, Pattern Recognition Letters, Vol. 11, 1990.

[DeDe94] Deza A. and Deza M.M. *The Ridge Graph of the Metric Polytope and Some Relatives*, in *Polytopes: Abstract, Convex and Computational* (T. Bisztriczky, P. McMullen, R. Schneider, A. Ivic Weiss eds.), 1994.

[DeDe95] Deza A. and Deza M.M. *The Combinatorial Structure of Small Cut and Metric Polytopes*, in *Combinatorics and Graph Theory* (T.H. Ku ed.), World Scientific, Singapore, 1995.

[DDF96] Deza A., Deza M.M. and Fukuda K. *On Skeletons, Diameters and Volumes of Metric Polyhedra*, in *Combinatorics and Computer Science*, Lecture Notes in Computer Science, Springer-Verlag, Berlin, Nr. 1120, 1996.

[DFM03] Deza A., Fukuda K., Mizutani T. and Vo C. *On the Face Lattice of the Metric Polytope*, Lecture Notes in Computer Science, Vol. 2866, Springer-Verlag, Berlin, 2003.

[DFPS01] Deza A., Fukuda K., Pasechnik D.V. and Sato M. *On the Skeleton of the Metric Polytope*, Lecture Notes in Computer Science, Springer-Verlag, Berlin, Vol. 2098, 2001.

[DGP06] Deza A., Goldengorin B. and Pasechnik D.V. *The Isometries of the Cut, Metric and Hypermetric Cones*, Journal of Algebraic Combinatorics, Nr. 23, 2006.

[Deza60] Tylkin M.E. (Deza M.M.) *Hamming Geometry of Unitary Cubes*, Doklady Akademii Nauk SSSR, Nr. 134–5, 1960. (English translation in *Cybernetics and Control Theory*, 134–5, 1961.)

[Deza62] Tylkin M.E. (Deza M.M.) *Realiability of Distance Matrices in Unit Cubes*, Problemy Kubernetiki, Nr. 7, 1962.

[Deza73] Deza M.M. *Matrices des Formes Quadratiques non négatives pour des Arguments Binaires*, Comptes Rendus de l'Académie des Sciences de Paris, Nr. 277(A), 1973.

[DeGr93] Deza M.M. and Grishukhin V.P. *Hypermetric Graphs*, Quarterly Journal of Mathematics, Oxford, Nr. 2, 1993.

[DGD12] Deza M.M., Grishukhin V.P. and Deza E. *Cones of Weighted Quasi-metrics, Weighted Quasi-hypermetrics and of Oriented Cuts*, in *Mathematics of Distances and Applications*, ITHEA, Sofia, 2012.

[DGL91] Deza M., Grishukhin V.P. and Laurent M. *The Symmetries of the Cut Polytope and of some Relatives*, in *DIMACS Series in Discrete Mathematics and Theoretical Computer Science* (P. Gritzmann, P. Sturmfels eds.), Vol. 4, 1991.

[DGL92] Deza M.M., Grishukhin V.P. and Laurent M. *Extreme Hypermetrics and L-polytopes*, in *Sets, Graphs and Numbers* (G. Halasz et al. eds.), Budapest, Hungary, 1991.

[DGL93] Deza M.M., Grishukhin V.P. and Laurent M. *The Hypermetric Cone is Polyhedral*, Combinatorica, Nr. 13, 1993.

[DGL95] Deza M.M., Grishukhin V.P. and Laurent M. *Hypermetrics in Geometry of Numbers*, in *Combinatorial Optimization* (W. Cook, L. Lovász, P. Seymour eds.), DIMACS Series in Discrete Mathematics and Theoretical Computer Science, Nr. 20, AMS, 1995.

[DeDe06] Deza M.M., Deza E. *Dictionary of Distances*, Elsevier, 2006.

[DeDe09] Deza M.M., Deza E. *Encyclopedia of Distances*, Springer-Verlag, 2009.

[DeDe12] Deza M.M., Deza E. *Encyclopedia of Distances*, 2-nd ed., Springer-Verlag, 2012.

[DeDe14] Deza M.M., Deza E. *Encyclopedia of distances*, 3-rd ed., Springer-Verlag, 2014.

[DeDe10] Deza M.M. and Deza E. *Cones of Partial Metrics*, Contributions in Discrete Mathematics, Nr. 6, 2010.

[DDV11] Deza M.M., Deza E. and Vidali J. *Cones of Weighted and Partial Metrics*, in *Algebra 2010: Advances in Algebraic Structures*, World Scientific, 2011.

[DeDu03] Deza M.M. and Dutour M. *Cones of Metrics, Hemi-metrics and Supermetrics*, Annales of European Academy of Sciences, 2003.

[DeDu04] Deza M.M. and Dutour M. *The Hypermetric cone on Seven vertices*, Experimental Mathematics, Nr. 12, 2004.

[DeDu13] Deza M.M. and Dutour Sikirić M. *Enumeration of the Facets of cut polytopes over Some Highly Symmetric Graphs*, preprint at arxiv:arXiv:1501.05407, 2013.

[DeDu13a] Deza M.M. and Dutour Sikirić M. *The hypermetric Cone on* 8 *Vertices and Some Generalizations*, preprint at arxiv:arXiv:1503.04554, 2013.

[DDD15] Deza M.M., Dutour Sikirić M. and Deza E. *Computations of Metric/Cut Polyhedra and Their Relatives*, in *Handbook of Geometric Constraint Systems Principles*, in preparation.

[DDM04)] Deza M.M., Dutour M., Maehara H. *On Volume-measures as Hemimetrics*, Ryukyu Mathematical Journal, Nr. 17, 2004.

[DDP03] Deza M.M., Dutour M and Panteleeva E.I. (Deza E.) *Small Cones of Oriented Semimetrics*, American Journal of Mathematics and Management Science, Nr. 22–3,4, 2003.

[DDS15] Deza M.M., Dutour Sikirić M. and Shtogrin M. *Geometric Structure of Chemistry-relevant Graphs: Zigzags and Central circuits*, Springer-Verlag, Berlin, 2015.

[DeLa97] Deza M.M. and Laurent M. *Geometry of Cuts and Metrics*, Springer-Verlag, Berlin, 1997; paperback edition 2014.

[DePa99] Deza M.M. and Panteleeva E.I. (Deza E.) *Quasi-metrics, Directed Multicuts and Related Polyhedra*, European Journal of Combinatorics, Special Issue *Discrete Metric Spaces*, Nr. 21(6), 2000.

[DeRo00] Deza M.M. and Rosenberg I.G. *n-semimetrics*, European Journal of Combinatorics, Special Issue *Discrete Metric Spaces*, Nr. 21(6), 2000.

[DeRo02] Deza M.M. and Rosenberg I.G. *Small Cones of Hemimetrics and Partition Hemimetrics*, Discrete Mathematics, Special Issue in honor of R. Fraisse, Nr. 291, 2005.

[DeTe87] Deza M.M. and Terwilliger P. *The Classification of Finite Connected Hypermetric Spaces*, Graphs and Combinatorics, Nr. 3, 1987.

[Diat96] Diatta J. *Une Extension de la Classification Hierarchique: les Quasi-hierarchies*, Ph. D. thesis, University de Provence, Aix-Marseille I, 1996.

[Dill77] Dillahunty J. *On Ternary Operations and Semigroups*, Journal of Undergraduate Mathematics, Nr. 9, 1997.

[Duto02] Dutour M. *The Symmetries of Metric, Quasi-metric and Hemimetric cones*, 2002.

[Duto07] Dutour Sikirić M. *The Seven Dimensional Perfect Delaunay polytopes and Delaunay Simplices*, Manuscript, 2007.

[Duto08] Dutour Sikirić M. *Cut and Metric Cones*, 2008, http://www.liga.ens.fr/~dutour/Metric/CUT_MET/index.html

[Duto10] Dutour Sikirić M. *Polyhedral*, 2010, http://www.liga.ens.fr/~dutour/polyhedral

[DER07] Dutour Sikirić M., Erdahl R. and Rybnikov K. *Perfect Delaunay Polytopes in Low Dimensions*, Integers: Electronic Journal of Conbinatorical Number Theory, Nr. 7, 2007.

[DSV07] Dutour Sikirić M., Schürmann A. and Vallentin F. *Classification of Eight-dimensional Perfect Forms*, Electron. Res. Announc. Amer. Math. Soc., Nr. 13, 2007.

[DSV08] Dutour Sikirić, Schürmann A. and Vallentin F. *A Generalization of Voronois Reduction Theory and its Application*, Duke Math. Journal, Vol. 142(1), 2008.

[DSV09] Dutour Sikirić M., Schürmann A. and Vallentin F. *Complexity and Algorithms for Computing Voronoi Cells of Lattices*, Math. Comp., Nr. 78(267), 2009.

[DSV10] Dutour Sikirić M., Schürmann A. and Vallentin F. *The Contact Polytope of the Leech lattice*, Discrete Comput. Geom., Nr. 44(4), 2010.

[Erda87] Erdahl R.M. *Representability Conditions*, in *Density Matrices and Density Functionals* (R.M. Erdahl, V.H. Smith eds.), D. Reidel, 1987.

[Ernv85] Ernvall S. *On the Modular Distance*, IEEE Trans. Inf. Theory, Vol. IT-31, Nr. 4, 1985.

[Fan98] Fan J.L. *Note on Hausdorff-like Metrics for Fuzzy Sets*, Pattern Recognition Letters, Nr. 19, 1998.

[Fine82] Fine A. *Hidden Variables, Joint Probability, and the Bell Inequalities*, Phys. Rev. Letters, Vol. 48(5), 1982.

[FrMa90] Frankl P. and Maehara H. *Simplices with Given 2-face Areas*, European Journal of Combinatorics, Nr. 11, 1990.

[Frec06] Fréchet M. *Sur Quelques Points du Calcul Fonctionnel*, Rend. Circolo mat. Palermo, Nr. 22, 1906.

[Frod58] Froda A. *Espaces p-métriques et leur Topologie*, Comptes Rendus de l'Acad. Sci. Paris, Nr. 247, 1958.

[Fuku95] Fukuda K. *Cdd Reference Manual, Version 0.56*, ETH Zentrum, Zürich, Switzerland, 1995.

[Gahl63] Gähler S. *2-metric Spaces and their Topological Structure* (in German), Math Nachr., Nr. 26, 1963.

[Gahl90] Gähler S. *Literature for the Theory of n-metric Spaces*, 1990.

[GaGa65] Gähler S. and Gähler W. *Espaces 2-métriques et Localement 2-métriques*, Ann. Sci. École Norm. Sup., Nr. 82–3, 1965.

[Gap] The GAP group. *GAPGroups, Algorithms, and Permutations*, Version 4.4.6.

[GHKLMS03] Gierz G., Hofmann K.H., Keimel K., Lawson J.D., Mislove M. and Scott D.S. *Continuous Lattices and Domains*, Encyclopedia of Mathematics and its Applications, 93, Cambridge University Press, 2003.

[GoKu95] Good A.C. and Kuntz I.D. *Investigating the Extension of Pairwise Distance Pharmacore Measures to Triplet-based Descriptors*, Journal Computer-Aided Molecular Design, Nr. 9, 1995.

[GKP94] Graham R.L., Knuth D.E., Patashnik O. *Concrete Mathematics*, 2-nd ed., Reading, MA: Addison-Wesley Professional,1994.

[Gris90] Grishukhin V.P. *All Facets of the Cut Cone C_n for $n = 7$ are known*, European Journal of Combinatorics, Nr. 11, 1990.

[Gris92] Grishukhin V.P. *Computing Extreme Rays of the Metric Cone for Seven Points*, European Journal of Combinatorics, Nr. 13, 1992.

[GrYe04] Gross J.L. and Yellen J. *Handbook of Graph Theory*, CRC Press, 2004.

[GuRi05] Güldürek A. and Richmond T. *Every Finite Topology is Generated by a partial pseudometric*, Order, Nr. 22, 2005.

[Haus14] Hausdorff F. *Grundzüge der Mengenlehre*, Leipzig, Veit, 1914.

[Heck99] Heckmann R. *Approximation of Metric Spaces by Partial Metric Spaces*, Applied Categorical Structures, Nr. 7, 1999.

[Hitz01] Hitzler P. *Generalized Metrics and Topology in Logic Programming Semantics*, PhD Thesis, Dept. Mathematics, National University of Ireland, University College Cork, 2001.

[JoCa95] Joly S. and Le Calvé G. *Three-way distances*, Journal of Classification, Nr. 12, 1995.

[Kell70] Kelly J.B. *Metric Inequalities and Symmetric Differences*, in *Inequalities II* (O. Shisha ed.), Academic Press, New Jork, 1970.

[KPR86] Klement E.P., Puri M.L. and Ralescu D.A. *Limit Theorems for Fuzzy Random Variables*, Proc. Roy. Soc. London, Ser. A, Vol. 407, 1986.

[Klei88] Klein R. *Voronoi Diagrams in the Moscow Metric*, Graph Theoretic Concepts in Comp. Sci., Vol. 6, 1988.

[Kura30] . Kuratowski K. *Sur le Probleme des Courbes Gauches en Topologie* (in French), Fund. Math., Nr. 15, 1930.

[LaDa84] Lal S.N. and Das M. *Invariant Semi 2-metrics on Abelian Groups*, Math. Nachr., Nr. 117, 1984.

[Laur96] Laurent M. *Graphic Vertices of the Metric Polytope*, Discrete Mathematics, Nr. 151, 1996.

[LaPo92] Laurent M. and Poljak S. *The Metric Polytope*, in *Integer Programming and Combinatorial Optimization* (E. Balas, G. Cornuejols eds.), Carnegie Mellon University, GSIA, Pittsburgh, 1992.

[LLR94] Linial N., London E. and Rabinovich J. *The Geometry of Graphs and some of its Algorithmic Applications*, Combinatorica, Nr. 15(2), 1995.

[Lova94] Lovász L. *Personal communication*, 1994.

[Matt92] Matthews S.G. *Partial Metric Topology*, Research Report 212, Dept. of Computer Science, University of Warwick, 1992.

[McKa15] McKay B. *The Nauty Program*, http://cs.anu.edu.au/people/bdm/nauty/

[MeRu74] Melter R.A. and Rudeanu S. *Geometry of 3-rings*, Proceedings of Lattice Theory Conf. Szeged, Colloquia Math. Soc., J. Bolyai, Nr. 14, 1974.

[Meng28] Menger K. *Untersuchungen über Allgemeine Metrik*, Math. Ann., Nr. 100, 1928.

[PaHa72] Patrinos A.N. and Hakimi S.L. *Distance Matrix of a Graph and its Tree Realization*, Quarterly of Applied Mathematics, Nr. 30, 1972.

[Pete98] Petersen J. *Sur le Theoreme de Tait*, L'Intermediaire des Mathematiciens, Nr. 5, 1898.

[Pito86] Pitowsky I. *The Range of Quantum Probability*, Journal of Mathematical Physics, Nr. 27, 1986.

[PiSv01] Pitowsky I. and Svozil K. *Optimal Tests of Quantum Nonlocality*, Phys. Rev., Vol. A(3), Nr. 64(1), 2001.

[Ples75] Plesník J. *Critical Graphs of Given Diameter*, Acta Math. Univ. Comenian, Nr. 30, 1975.

[PoTo96] Popescu D. and Tomescu I. *Bonferroni Inequalities and Negative Cycles in Large Complete Signed Graphs*, European Journal of Combinatorics, Nr. 17(5), 1996.

[Rior68] Riordan J. *Combinatorial Identities*, J. Wiley & Sons, New York, 1968.

[RyBa78] Ryskov S.S. and Baranovski E.P. *C-types of n-dimensional Lattices and 5-dimensional Primitive Parallelohedra (with Application to the Theory of Coverings)*, Proc. Steklov Inst. Math., Vol. 4, 1978.

[RyBa98] Ryshkov S.S. and Baranovski E.P. *Repartitioning Complexes in n-dimensional Lattices (with full Description for $n \leq 6$)*, in *Voronoi Impact on Modern Science*, Institute of Mathematics, Kyiv, 1998.

[Scho38] Schoenberg I.J. *Metric Spaces and Positive Definite Functions*, Trans. AMS, 44, 1938.

[Schr86] Schrijver A. *Theory of Linear and Integer Programming*, Wiley, 1986.

[Schu09] Schürmann A. *Computational Geometry of Positive Definite Quadratic Forms*, University Lecture Series, Vol. 48, American Mathematical Society, Providence, RI, 2009.

[Seda97] Seda A.K. *Quasi-metrics and the Semantic of Logic Programs*, Fundamenta Informaticae, Vol. 29, 1997.

[Serf99] Serfati M. *The Lattice Theory of r-ordered Partitions*, Discrete Mathematics, Nr. 194, 1999.

[Seym81] Seymour P.D. *Matroids and Multicommodity Flows*, European Journal of Combinatorics, Nr. 2(3), 1981.

[ShLi95] Shi L. and Li R. *Max 2SAT and Directed Multicut*, in *Combinatorics and Graph Theory* (T.H. Ku ed.), World Scientific, Singapore, 1995.

[Sloa15] Sloane N. *The On-Line Encyclopedia of Integer Sequences*, http://www.research.att.com/~njas/sequences

[Smit89] Smith T.E. *Shortest-path Distances: an Axiomatic Approach*, Geographical Analysis, Nr. 21, 1989.

[Stol69] Stoltenberg R.A. *On Quasi-metric Spaces*, Duke Math. Journal, Nr. 36, 1969.

[Sylv78] Sylvester J.J. *Chemistry and Algebra*, Nature, Nr. 17, 1878.

[Sylv78a] Sylvester J.J. *On an Application of the new Atomic Theory to the Graphical Representation of the Invariants and Covariants of Binary Quantics, - with three Appendices*, American Journal of Mathematics, Pure and Applied, Vol. 1(1), 1878.

[Teta91] Tetali P. *Random Walks and the Effective Resistance of Networks*, Journal of Theoretical Probability, Nr. 4, 1991.

[Vida15] Vidali J. *Cones of Weighted and Partial Metrics*, http://lkrv.fri.uni-lj.si/~janos/cones/

[Vito99] Vitolo P. *The Representation of Weighted Quasi-Metric Spaces*, Rend. Inst. Mat.Univ. Trieste, Nr. 31, 1999.

[Voro08] Voronoi G. *Nouvelles Applications des Paramètres Continus à la Théorie des Formes Quadratiques. Deuxième Mémoire, Recherches sur les Parallélloèdres Primitifs*, Journal Reine Angew. Math, Nr. 134(1), 1908.

[Wagn37] Wagner K. *Über eine Eigenschaft der ebene Komplexe*, Math. Annal., Nr. 114, 1937.

[Wils31] Wilson W.A. *On Quasi-metric Spaces*, American Journal of Math., Nr. 53, 1931.

Printed in the United States
By Bookmasters